# Lecture Notes in Physics

Edited by H. Araki, Kyoto, J. Ehlers, München, K. Hepp, Zürich
R. Kippenhahn, München, H. A. Weidenmüller, Heidelberg
J. Wess, Karlsruhe and J. Zittartz, Köln
Managing Editor: W. Beiglböck

## 305

K. Nomoto (Ed.)

# Atmospheric Diagnostics of Stellar Evolution: Chemical Peculiarity, Mass Loss, and Explosion

Proceedings of the 108th Colloquium of the
International Astronomical Union, Held at
the University of Tokyo, Japan, 1–4 September 1987

# Springer-Verlag

Berlin Heidelberg New York London Paris Tokyo

**Editor**

Ken'ichi Nomoto
Department of Earth Science and Astronomy
University of Tokyo, Tokyo 153, Japan

ISBN 3-540-19478-9 Springer-Verlag Berlin Heidelberg New York
ISBN 0-387-19478-9 Springer-Verlag New York Berlin Heidelberg



Printing: Druckhaus Beltz, Hemsbach/Bergstr.
Binding: J. Schäffer GmbH & Co. KG., Grünstadt
2158/3140-543210

# PREFACE

IAU Colloquium No. 108, *Atmospheric Diagnostics of Stellar Evolution: Chemical Peculiarity, Mass Loss, and Explosion,* was attended by 63 participants from Japan and 65 participants from 20 other countries. In all, 34 invited talks, 32 contributed talks, and 49 poster papers were presented, and the meeting was widely regarded as a great success.

This colloquium was organized with the following original intention: Recent detailed observations and extensive theoretical modeling, both of the stellar atmosphere and the stellar interior, have opened up new possibilities of establishing evolutionary models of the chemical and dynamical behavior of stars that are unified from the central core to the outermost layers. In particular, there is a common interest in the atmospheric phenomena of chemical peculiarities, mass loss, and explosion, all of which are strongly related to the hydrodynamical evolution of the stellar interior. Instead of analyzing these phenomena separately for specific stellar types or masses, we intended to gain a sequential view of the different evolutionary paths of stars of various masses by bringing researchers of different specialities together.

On 23 February 1987, a supernova was observed in the Large Magellanic Cloud. This supernova (SN 1987A) has revealed many new exciting features and provided us with a unique and excellent opportunity to chart the evolution and explosion of massive stars. Our colloquium greatly benefitted from this timely occurrence, and became the first IAU meeting to discuss intensively SN 1987A. In the session on *explosion,* historic underground neutrino observations, ground-based optical observations, and satellite observations of SN 1987A were collected and compared with theoretical models. Among the highlights was the exciting announcement of the detection of x-rays from the supernova, which was much earlier than the theoretical prediction.

The progenitor of SN 1987A was, surprisingly, a B3 supergiant. The mass loss and the change in the surface abundances during the progenitor's evolution were directly related to the discussion of the two other major topics of the colloquium, *chemical peculiarity* and *mass loss.* In these, there were extensive discussions on the interaction between the evolving stellar interior and the atmosphere for various types of stars that exhibit peculiar surface abundances and a loss of mass . Abundance anomalies of He, Li, CNO and s-process elements in main-sequence stars, AGB stars, OB supergiants, Wolf-Rayet stars and interacting binaries were interpreted by the combined effects of diffusion, meridional circulation, convection, and mass loss. Mass loss mechanisms and self-consistent atmospheric models with stellar wind were discussed in the light of new observations over all wave bands for a wide range of stars in the HR diagram.

Every evening, ample time for discussion (and a beer !) was provided, which proved to be very popular among participants. At the banquet, we celebrated the 80th birthday of Yoshio Fujita and his pioneering work on *carbon stars.*

Tokyo                                                                                             Ken'ichi Nomoto

IAU Colloquium No. 108
Atmospheric Diagnostics of Stellar Evolution:
Chemical Peculiarity, Mass Loss, and Explosion

## IAU COLLOQUIUM No. 108

## ATMOSPHERIC DIAGNOSTICS OF STELLAR EVOLUTION:
### Chemical Peculiarity, Mass Loss, and Explosion

**Scientific Organizing Committee:**
> R.P. Kudritzki and K. Nomoto (Joint Chairmen)
> A.A. Boyarchuk, R. Cayrel, C. Chiosi, A.P. Cowley, C. de Jager, C. de Loore,
> B. Gustafsson, I. Iben, Jr., Y. Kondo, D.L. Lambert, A. Maeder, G. Michaud,
> A.J. Willis, S.E. Woosley

**Local Organizing Committee:**
> T. Tsuji (Chairman)
> H. Ando, Y. Eriguchi, R. Hirata, K. Kodaira, T. Kogure, K. Nariai,
> K. Nomoto, T. Onaka, Y. Osaki, H. Shibahashi, K. Suda, D. Sugimoto,
> T. Watanabe, Y. Yamashita

**Meeting Place:** Sanjo Conference Hall, University of Tokyo, Bunkyo-ku, Tokyo 113

**Supported by:** IAU Commissions 35 and 36
> IAU Commissions 27 and 29 (co-sponsors)
> Japan Society for the Promotion of Science
> Tokyo Astronomical Observatory
> Tokyo Geographical Society
> Astronomical Society of Japan

**Donated by:** The Commemrative Association for the Japan World Exposition
> Inoue Foundation of Science
> The Kajima Foundation
> The Mitsubishi Foundation
> Yoshida Foundation for Science and Technology

v

# TABLE OF CONTENTS



---

* *(I)* : Invited Talk

# Part II. MASS-LOSING STARS IN DIFFERENT STAGES OF EVOLUTION

## II-1. HOT STARS

# Part III. CHEMICAL AND DYNAMICAL STRUCTURES OF EXPLODING STARS

## III-1. NOVAE AND DWARF NOVAE

# III-2. SUPERNOVAE

## Part IV. CONCLUDING REMARKS

# I. Chemical Peculiarities as Probe of Stellar Evolution

*Chair: C. de Jager and G. Michaud*

# MAIN SEQUENCE ABUNDANCES:
# OBSERVATIONAL ASPECTS


Jun JUGAKU

*Tokyo Astronomical Observatory*



Abstract: Although once it was thought that main-sequence stars are remarkably homogeneous with respect to their chemical composition, the upper main-sequence stars ($30000 > T_e > 7000$) show a variaety of chemically peculiar stars besides the so-called normal stars. Those include the Am, Ap, $\lambda$ Bootis, He-deficient, and He-rich stars. This review summarizes the current data, which are necessary to construct and test the theoretical models of these stars. In the second half of the review we concentrate on Li. In the lower main-sequecnce stars abundances of Li have been determined in hundreds of stars. Some of the remarkable results are: (1) A uniform upper abundance value irrespective of stellar effective temperature, (2) abundance gap in the F stars of the Hyades, and (3) increasing depletion with smaller stellar mass for the Hyades.

MAIN SEQUENCE ABUNDANCES, MASS LOSS AND MERIDIONAL CIRCULATION

Georges Michaud
Département de Physique, Université de Montréal
Montréal, Québec, CANADA H3C 3J7

ABSTRACT  Constraints that abundance anomalies observed on main sequence stars put on turbulence, meridional circulation and mass loss are reviewed.  The emphasis is on recent observations of Li abundances.

Upper limits to turbulence are obtained from the Be abundance in the Sun and from underabundances of Ca and Sc in FmAm stars.  The Li abundance in G type stars suggests the presence of turbulence below convection zones.

The abundance anomalies, both over and underabundances, observed in FmAm and $\lambda$ Booti stars can be explained by diffusion in the presence of mass loss.  A mass loss rate of $10^{-15}$ $M_0$ yr$^{-1}$ is required to explain the FmAm stars while a mass loss rate of $10^{-13}$ $M_0$ yr$^{-1}$ is required by the $\lambda$ Booti stars.

The position and width of the Li abundance gap observed in Hyades and other open clusters is explained by diffusion.  A detailed reproduction of the Li($T_{eff}$) curve seems to require a mass loss rate of slightly more than $10^{-15}$ $M_0$ yr$^{-1}$, of the same order as the mass loss rate required by the FmAm stars.  In the presence of such a mass loss only small overabundances of heavy elements are expected.  The observed variations in the Li abundance as a function of the age of clusters suggests that the Li abundance observed in old halo stars <u>does not</u> represent the cosmological abundance.

Detailed two dimensional calculations of diffusion in presence of meridional circulation for HgMn and FmAm stars lead to a cut-off of about 100 km s$^{-1}$ for the maximum equatorial rotational velocity at which abundance anomalies are expected in these objects. This agrees with observations.  A similar calculation for the F stars of the Hyades where Li underabundances are observed leads to a contradiction, unless meridional circulation patterns are modified by the presence of convection zones once they become as large as in late F stars.  There remains a possibility that meridional circulation would be responsible for some of the reduction of the Li abundance as observed in the Hyades and UMa. Further observations are suggested to distinguish the effects of settling and nuclear destruction.

## I.  Particle Transport Velocities

It is necessary to compare atomic diffusion and mass loss velocities to other possible transport velocities in order to evaluate the relative importance of each.  Each process is briefly presented in this section.

The results of Tassoul and Tassoul (1982, 1984) will be used to evaluate meridional circulation. The two dimensional meridional circulation velocity patterns are well determined by their calculations and will be compared to the other velocity fields. They apply strictly to stars with outer radiative zones only.

Following Schatzman et al. (1981), the turbulent diffusion velocity can be approximated by:

$$v_T = D_T \left( - \frac{\partial \ln c}{\partial r} \right) = R_* D_{11} \left( - \frac{\partial \ln c}{\partial r} \right) \qquad (1)$$

where $D_T$ is the turbulent diffusion coefficient taken to be $R_*$ times larger than the atomic self diffusion coefficient, $D_{11}$. For a trace element, the quantity c is defined by $c(A) = N(A)/N_P$ where $N_P$ is the particle number density exclusive of electrons. Particle transport by turbulent diffusion is speculative because $D_T$ is only assumed (Michaud 1985).

The mass loss rate, dM/dt, is assumed small enough to have negligible effect on stellar structure. It merely introduces, throughout the static stellar envelope, a global outward velocity of matter, $v_w$, expressing the conservation of the flux of the main constituent. A trace element diffusing in the presence of mass loss must satisfy the conservation equation, in which both $v_w$ and the diffusion velocity appear (Michaud and Charland 1986, Paquette et al. 1986).


## II. The Sun

While the Li solar photospheric abundance is some 100 to 200 times smaller than the interstellar value, the Be abundance is about normal (Boesgaard, 1976). This observation can be coupled with Li abundance observations in field stars and in clusters (Cayrel et al. 1984). Since no element separation is expected in the convection zone itself (Schatzman 1969), those observations show that, in solar type stars, Li has probably been carried below the convection zone by some mild turbulence (Schatzman 1977; Vauclair et al. 1978b; Baglin et al. 1985) to the region where it can burn ($T=2.5\ 10^6$ K). Turbulence must be weak enough not to carry Be to the slightly deeper region where it can burn ($T=3.5\ 10^6$ K). The Li abundance then gives a value for the turbulence below the convection zone ($R_*=50$; see Michaud 1985) and an upper limit to the turbulence ($R_*<90$) a little deeper in (between 2.5 and $3.5\ 10^6$ K). While the value of $R_*=50$ depends on the exact depth of the convection zone and so is model dependent, the upper limit is well established since it depends only on the temperatures at which Li and Be burn.


## III. The F Stars

Boesgaard an Tripicco (1986a) have recently obtained striking observations showing a clear hole in the Li abundance of the Hyades F stars at $T_{eff}=6700$ K. It is perhaps two orders of magnitude deep but only 300 K wide in $T_{eff}$ (see their Fig. 2). The low Li abundance stars at $T_{eff}=6700$ K are clearly separated from the stars showing a progressive decrease of the Li abundance with $T_{eff}$ as observed below 6000 K. The separation of the gap from the cooler stars seems to imply that 2 processes are involved. The nuclear burning of Li transported by turbulence is probably involved below 6000 K (see the preceding section), but a different process seems required around 6700 K.

The gravitational settling of Li explains without arbitrary parameters the presence and width of the hole in the Li abundance at this $T_{eff}$. As can be seen from Fig. 2 of Michaud (1986), the depth of the superficial convection zone increases by more than two orders of magnitude as one goes from 7000 to 6500 K. While Li has one electron left at the T of the bottom of the convection zone in the model for $T_{eff}$ =7000 K, it has none in the model for 6500 K. The radiative acceleration on Li at the bottom of the convection zone decreases from being about 5 times larger than gravity in the model with $T_{eff}$ = 7000 K to being negligible in models with $T_{eff}$ = 6800 K or less (see also Vauclair, this conference). If the $T_{eff}$ is slightly smaller than 7000 K, Li settles gravitationally while it is supported by radiative acceleration at higher $T_{eff}$. At $T_{eff}$ =6800 K, the Li abundance is reduced by gravitational settling by a factor of 20 according to Michaud (1986). In still cooler stars, the effect is smaller as convection zones become progressively deeper and the diffusion time scale increases: the bigger the reservoir, the longer it takes to empty. At the age of the Hyades, diffusion has had little effect in stars with $T_{eff}$ < 6500 K. The diffusion time scale varies approximately as $_\Delta M^{0.5}$ (see Michaud 1977). It varies from 2.8 $10^8$ yr at $T_{eff}$ = 6900 K to 4 $10^9$ yr at 6300 K and $10^{10}$ yr at 6000 K (assuming $\alpha$=1.4).

Observations of other clusters can lead to a better understanding of the evolution of the Li abundance. In the Pleiades, Pilachowski and Hobbs (1987) have observed less than a factor of 1.5 decrease of the Li abundance in the gap (see also Duncan and Jones 1983, Duncan 1981). Since the Pleiades are about ten times younger than the Hyades, a scaling of the exponential dependence of the abundance reduction leads to exp(3.4/10)=1.4; this is reasonable agreement.

Other clusters have now been observed: Coma by Boesgaard and Tripicco (1987) and NGC 752 and M 67 by Hobbs and Pilachowski (1986a, 1986b). In Coma and NGC 752 a dip occurs at the same $T_{eff}$ as in the Hyades, though not as well defined. In the older clusters, NGC 752 and M 67 the Li abundance around $T_{eff}$ =6200 K appears to have been lowered by diffusion and/or burning. Within the error bars, the interstellar matter Li abundance is consistent with the original Li abundance in young clusters (Hobbs 1984).

In the UMa moving group and in a new study of the Hyades, Boesgaard, Budge and Burck (1988) and Boesgaard (1987) obtain upper limits of $10^{-3}$ the original Li abundance in some stars. Since there is a measurement error of $\pm2$ mA (Boesgaard, Budge and Burck, 1988 §II), and since an underabundance by a factor of 30 leads to a line of about 2 mA in the middle of the gap (see Fig. 4 of Boesgaard and Tripicco 1986a), it appears to me that no underabundance by a factor of more than 30 can be determined from these spectra. The presence of a blend can only make the determination of the upper limit more difficult. It cannot be used to reduce the upper limit, as was apparently done here.

Spite and Spite (1982) and Spite et al. (1984) have determined the Li abundance in Halo stars (see also Hobbs and Duncan 1987). It is about 8 times smaller than the current value in young stars before it is affected by diffusion or burning. But if one uses the diffusion time scale at $T_{eff}$ = 6300 K ($\tau$= 4 $10^9$ yr as seen above) at an age of 8 $10^9$ yr, which is a minimum for halo stars, one obtains a factor of exp(-2)=7.4 reduction in the Li abundance at $T_{eff}$ = 6300 K by diffusion alone. This suggests that the original Li abundance in halo stars may well have been about the same as the original Li abundance in young clusters today. In the cooler of the halo stars nuclear burning would have reduced the Li abundance. There would be a large $T_{eff}$ interval over which the Li abundance would be about constant because these two effects would combine to form a plateau.

That the Li abundance observed by Spite and Spite (1982) could not be the cosmological abundance but had been reduced by a factor of at least 4 by either diffusion or burning was first noted by Michaud, Fontaine and Beaudet (1984). These authors also emphasized that this plateau is constant over a surprisingly large $T_{eff}$ interval. This remains a problem requiring further study.

That the plateau should be shifted to lower $T_{eff}$ in halo stars can be understood rather easily as due to a Z dependence of the depth of the convection zone at a given $T_{eff}$ (Michaud, Fontaine and Beaudet 1984).

The interpretation of the Li abundance gap using a diffusion model has been questioned because of the observed absence of abundance anomalies of heavy elements in F stars (Boesgaard and Lavery 1986; Thévenin, Vauclair and Vauclair 1986; Tomkin, Lambert and Balachandran 1985) where Be has been observed to be underabundant. Such anomalies had been predicted on account of the diffusion calculations in the absence of any mass loss (Michaud et al. 1976, Vauclair et al. 1978b). It has recently been shown that even a very small mass loss was sufficient to reduce considerably any expected overabundance in F stars. On Fig. 2c of Michaud and Charland (1986), it is shown that a mass loss rate of $10^{-15}$ $M_o$ yr$^{-1}$ is sufficient to keep the Sr overabundance below a factor of 1.5 while Sr would be expected to be more than 100 times overabundant in the absence of mass loss (Michaud et al. 1976). The presence of even a very small mass loss rate considerably limits any overabundance when the radiative acceleration and gravity are close to each other as is the case for heavy elements in stars cooler than $T_{eff}$ = 7000 K. The same small mass loss rate reduces the Li overabundance in stars of $T_{eff}$ = 7000 K or more where Li is supported. As shown in Fig. 4 of Michaud (1986), the same mass loss rate of $10^{-15}$ $M_o$ yr$^{-1}$ eliminates the Li overabundance of a factor of 10 expected in the absence of mass loss at $T_{eff}$ = 7000 K. It has now been verified that the presence of mass loss cannot increase the Li underabundance that diffusion leads to beyond a total factor of 30 underabundance.

Detailed calculations of radiative accelerations for a few selected elements are currently underway at Montréal to define a test of this model. Nitrogen and oxygen seem specially promising (see Fig. 1): they are in the He configuration when Li is not supported. This is based on calculations carried out as described by Michaud et al. (1976) but needs to be confirmed by more detailed calculations.

The introduction of a mass loss rate may seem arbitrary. However in this case the position and depth of the observed Li abundance gap is explained without arbitrary para-

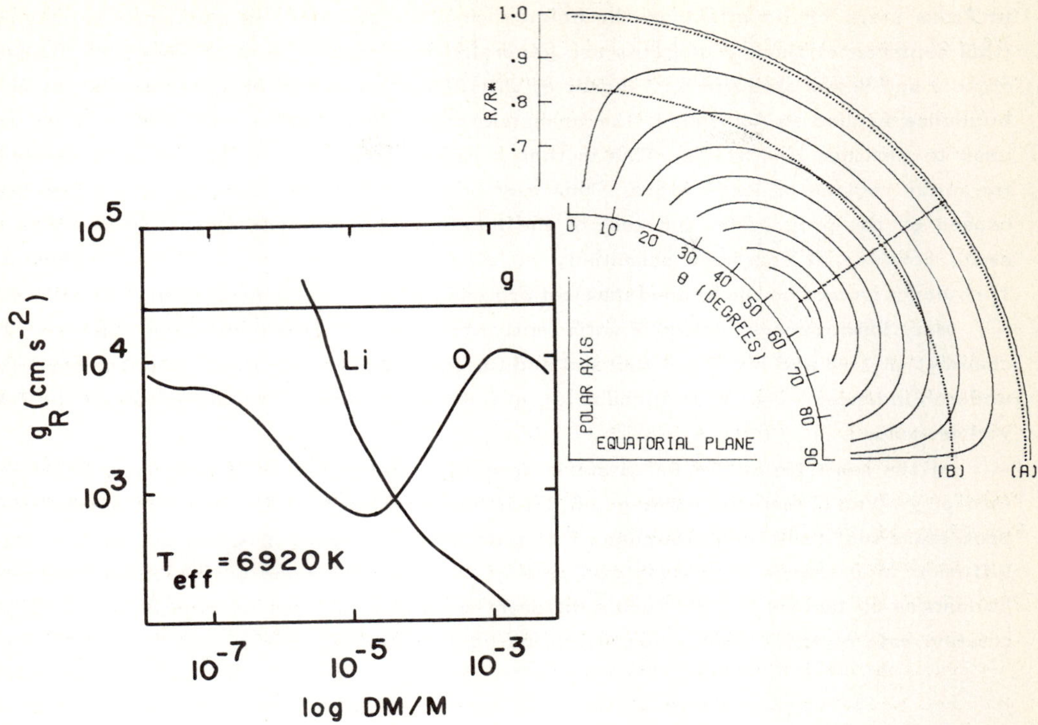

Fig. 1     Radiative accelerations of Li and oxygen as a function of the mass above the point of interest.   When the radiative acceleration on Li is smaller than gravity, so is that on O.   According to these calculations oxygen should sink if Li sinks.

Fig. 2     Meridional circulation stream lines (full lines) as a function of the angle from the polar axis.  The position of convection zones is indicated by dotted lines.  That identified by A is for a 12000 K main sequence star while that indicated by B is for a 6400 K star.

meters, other than the uncertainty on $\alpha$, and the introduction of the mass loss is only necessary to improve the fit of the shape of the Li abundance curve with $T_{eff}$.  Furthermore, as I will now show, such mass loss rates are also required to explain the observed abundance anomalies in FmAm stars which are probably the hot continuation of the F stars of the Hyades with Li underabundances.


## IV. The FmAm and $\lambda$ Booti stars.


If one assumes that the outer region of an Am star is perfectly stable and calculates the abundance anomalies produced, one obtains values that exceed the observed overabundances by more than a factor of 10 (Michaud et al.1976).  Some perturbing process appears to be important.

It is possible to get an upper limit to the turbulence below the H convection zone in FmAm stars by its effect on the abundance of Sc and Ca. As mentioned by Jugaku (this conference) these 2 elements are generally observed to be underabundant in these objects and it was obtained by Vauclair et al. (1978a) that to explain the observed underabundance implied $R_* \lesssim 3$ even if the uncertainty on the radiative acceleration of Sc was used to maximise this value. This is then a rather strict upper limit for turbulence in the outer regions of FmAm stars. This may be related to the presence of a $\mu$ gradient caused by the progressive increase of the He abundance inward. Once turbulence is so small, however, it can have essentially no effect on the abundance of heavy elements. Turbulence cannot be the hydrodynamical process reducing overabundances in FmAm stars.

Mass loss appears to reduce sufficiently the observed overabundances (Fig. 3 of Michaud et al. 1983 and Fig. 2d of Michaud and Charland 1986) while maintaining the observed underabundances (Fig. 4 of Michaud et al. 1983). A mass loss rate of about $10^{-15}$ $M_0$ yr$^{-1}$ is needed.

At the same $T_{eff}$ as the FmAm stars showing overabundances of most heavy elements (see e. g., Van't Veer-Menneret et al. 1985, Burkhart et al. 1987), there are also the $\lambda$ Booti stars that have underabundances of most heavy elements (Baschek and Searle 1969). Diffusion in presence of a mass loss rate of $10^{-13}$ $M_0$ yr$^{-1}$ leads to generalized underabundances by factors of $\sim 3$. Such a different mass loss rate may be caused by the higher rotation rate of the $\lambda$ Booti stars as compared to the FmAm stars.


## V.  Meridional Circulation


Following the derivation by Tassoul and Tassoul (1982) of a physically consistent meridional circulation velocity field, it became possible to do detailed calculations of gravitational settling in presence of circulation. The first comparison (Michaud 1982, Michaud et al. 1983) was encouraging, though based on a very rough approximation of the meridional circulation velocity fields (Figure 2) and justified the effort of a detailed two dimensional calculation of diffusion in presence of meridional circulation. The aim becomes to test a well defined hydrodynamic model as precisely as possible using observed abundance anomalies. We test whether the meridional circulation patterns obtained by Tassoul and Tassoul (1982) for radiative models explain the disappearance of abundance anomalies at about 100 km s$^{-1}$ for HgMn and FmAm stars. We then test for the Li abundance gap observed in young stellar clusters. As the external convection zone becomes deeper, the model of Tassoul and Tassoul assuming a purely radiative outer region should break down at some point.

The calculations turned out to require a grid of 20 (horizontal)x 100 (vertical). Details of the calculations may be found in Charbonneau and Michaud (1987a).

For the HgMn stars, Wolff and Preston (1978) obtain an upper limit of 100 km s$^{-1}$ for the V sin i at which they are observed. The meridional circulation is slow enough to allow the disappearance of the He convection zone by He settling for rotational velocities up to 75 km s$^{-1}$ (Charbonneau and Michaud 1987a). Once the He convection zone has

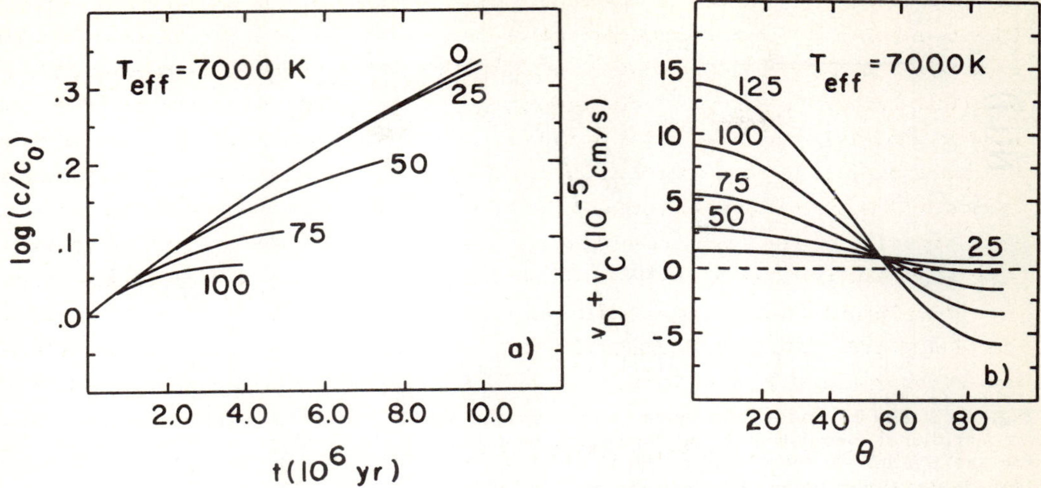


Fig. 3    On part a) is shown the Li abundance as a function of time in a $T_{eff}$ = 7000 K star.  The curves are identified by the equatorial rotational velocity in km s⁻¹.  The effect of meridional circulation on the Li abundance starts to be felt for rotational velocities of 50 km s⁻¹.  This can be understood from part b) of the figure where the total vertical transport velocity is shown as a function of the angle from the rotation axis. For rotational velocities of up to some 40 km s⁻¹, the transport velocity is everywhere positive.

disappeared, the atmospheric region becomes stable and diffusion can cause the abundance anomalies observed on HgMn stars (Michaud 1982).  For FmAm stars, Abt and Levy (1985) obtain an upper limit of 0.9 day for the orbital period of the binary systems in which there are FmAm stars while the limit obtained by Charbonneau and Michaud (1987a) is 0.7 day or $V_e$=100 km s⁻¹ assuming synchronous rotation.   In my opinion this constitutes excellent agreement and suggests that the meridional circulation patterns of Tassoul and Tassoul (1982) constitute the main velocity field opposing chemical separation.  In particular it appears that turbulence does not play a major role.  Note that in obtaining their solution Tassoul and Tassoul had to assume a non negligible turbulent viscosity though the solution did not depend on the value chosen (the dependence was only as $\mu^{1/11}$).

One can similarly use the meridional circulation fields to test its effect on the diffusion of Li in the F stars of clusters (Charbonneau and Michaud (1987).  It turns out however that the upper limit of the equatorial rotation velocity is much smaller.  This can be traced to the increase in the depth of the convection zone.  The diffusion velocity decreases considerably due to the $\rho^{-1}$ dependence of the diffusion coefficient while the meridional circulation velocity is nearly constant as one goes deeper in the star.  While the critical velocity in the middle of the gap is about 15 km s⁻¹, there are stars in the middle of the gap of the Hyades with a V sin i of 50 km s⁻¹ (Boesgaard 1987).  These stars have very low Li abundance and if the low Li abundance in the gap is to be explained by diffusion it is clear that the calculations of Tassoul and Tassoul (1982) do not apply to F stars.

It is however interesting to consider the alternate possibility that in those stars

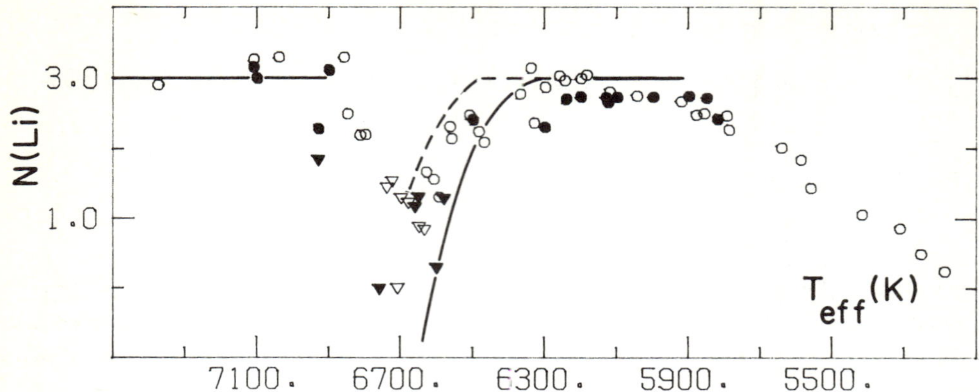

Fig. 4    The Li underabundance caused by the matter brought, to the convection zone, by meridional circulation from the region where Li burns, is shown as a function of $T_{eff}$ for the Hyades (8 10⁸ yr) and UMa (4 10⁸ yr). It is compared to observations for these two clusters (Boesgaard, Budge and Burck 1988).

rotating fast enough to stop diffusion, meridional circulation brings to the surface matter from which Li has been burned. Calculations for this model were carried out using the same circulation velocity fields as previously. The burning of Li was assumed to be complete at $T=2.5$ $10^6$ K so that matter arriving from that depth has no Li. When after a time $t_0$, matter that was originally deeper than $T= 2.5$ $10^6$ K arrives in the convection zone, the Li abundance starts decreasing in the convection zone. It then decreases with a time constant $\Theta$, obtained from the time it takes to replace the mass of the convection zone by new material. The time evolution of the abundance is then:

$$c = c_0 \exp((t-t_0)/\theta) \qquad \text{for } t > t_0.$$

The effect of diffusion was completely neglected in this calculation. Where the radiative acceleration on Li is negligible, the settling would increase the effect of the burning and so taking it into account could only strengthen the argument. When however the radiative acceleration is larger than gravity, the situation is a little more complex. Figure 3 shows the effect of meridional circulation on the superficial Li abundance in a case when Li is supported. The meridional circulation starts reducing the Li overabundance only for equatorial rotational velocities larger than about 50 km s⁻¹. On part b of Figure 3 meridional circulation and diffusion velocities are compared as a function of the angle from the rotation axis. For V> 35 km s⁻¹, the upward diffusion velocity is everywhere larger than the downward circulation velocity, so shielding the surface Li by keeping it in the convection zone. For stars with $T_{eff}$ >6900 K and V<50 km s⁻¹, the surface Li is shielded from the burning zone.

On Fig. 4 is shown the Li abundances to be expected from such a model at the age of the Hyades and of Uma. They are compared to the observations of Boesgaard, Budge and Burck (1988). To do these calculations the equatorial rotation velocity is needed. I used V= 50 km s⁻¹ at $T_{eff}$=6700 K and V=25 km s⁻¹ at $T_{eff}$ =6350 K, from an average of the observed rotational velocities in the appropriate $T_{eff}$ range taking the effect of sin i

Table 1<br>
<u>Stars that are a Problem for Meridional Circulation</u>

| Star | Cluster | $T_{eff}$ (K) | v sin i | $v_e$(crit) | log(N(Li)) |
|---|---|---|---|---|---|
| TR111R | Coma | 6400 | 35 | 21 | 2.67 |
| HD2377B | Pleiades | 6400 | 75 | 68 | 2.9 |
| HD23584 | Pleiades | 6500 | 85 | 68 | 2.8 |
| HD23351 | Pleiades | 6700 | 80 | 70 | 2.9 |
| HD23608 | Pleiades | 6650 | 110 | 70 | 2.8 |

into account. The rotational velocities were interpolated linearly in between those 2 $T_{eff}$. Above $T_{eff}$ =6900 K, Li is shielded by $g_R$(Li). One could argue that the agreement is quite satisfactory at least for the Hyades. A somewhat larger rotational velocity would be needed to explain the results for UMa.

In Table 1 are shown a number of stars that have large Li abundances and rotational velocities. Are indicated both the measured rotational velocities and the limiting rotational velocity beyond which Li should be strongly depleted by the mechanism just described. These contradict the model just described. They require that the penetration of the convection zone by meridional circulation be at most partial.


## VI. CONCLUSIONS


While gravitational settling explains the presence of a Li abundance gap in F stars, it cannot explain underabundances by more than a factor of about 30 in the Hyades. If larger underabundance factors were confirmed, it may imply that Li has been destroyed by nuclear reactions in at least some of the F stars. It was shown how meridional circulation may then be implied. The blue side of the abundance gap would still be explained by the drop of $g_R$(Li) between $T_{eff}$ =7000 and 6700 K. Observations of N and O may allow to distinguish between the effect of settling and of nuclear burning of Li though more calculations of radiative accelerations are needed to confirm this test. While the meridional circulation model of Tassoul and Tassoul appears to pass the test of abundance anomalies in HgMn and FmAm stars, it may not pass that of the F stars. The difference may come from the deeper convective zones of the F stars which may modify the solutions obtained by Tassoul and Tassoul for purely radiative envelopes. Note that given the observed solar rotational velocity, meridional circulation should have no effect on the Li abundance before the Sun is 2 $10^{10}$ yr old.

Abundance anomalies imply mass loss rates smaller than $10^{-13}$ $M_\odot$ yr$^{-1}$. The Li abundance implies mass loss rates smaller than $10^{-14}$ $M_\odot$ yr$^{-1}$. This contradicts the assumption of Guzik, Willson, and Brunish (1987) that stars with 1 < M < 3 $M_\odot$ lose mass at the rate of $10^{-9}$ to $10^{-8}$ $M_\odot$ yr$^{-1}$.

I thank Paul Charbonneau and Yves Charland for carrying out most of the calculations reported in this paper.

REFERENCES

Abt, H. A., and Levy, S. G. 1985, <u>Ap. J. Suppl.</u>, 59, 229.
Baglin, A., Morel, P., and Schatzman, E. 1985, <u>Astr. Ap.</u>, 149, 309.
Baschek, B., and Searle, L. 1969, <u>Ap. J.</u>, 155, 537.
Boesgaard, A. M. 1976, <u>Ap. J.</u>, 210, 466.
Boesgaard, A. M. 1987, <u>Pub. A. S. P.</u>, in press.
Boesgaard, A. M., Budge, K. G., and Burck, E. E. 1988, <u>Ap. J.</u>, February.
Boesgaard, A. M., and Lavery, R. J. 1986, <u>Ap. J.</u>, 309, 762.
Boesgaard, A. M., and Tripicco, M. J. 1986a, <u>Ap. J. (Letters)</u>, 302, L49.
Boesgaard, A. M., and Tripicco, M. J. 1986b, <u>Ap. J.</u>, 303, 724.
Boesgaard, A. M., and Tripicco, M. J. 1987, preprint.
Burkhart, C., Coupry, M. F., Lunel, M., and Van't Veer, C. 1987, <u>Astr. Ap.</u>, in press.
Cayrel, R., Cayrel de Strobel, G., Campbell, B., and Dappen, W. 1984, <u>Ap. J.</u>, 283, 205.
Charbonneau, P., and Michaud, G. 1987a, submitted for publication.
Charbonneau, P., and Michaud, G. 1987b, in preparation.
Duncan, D. K. 1981, <u>Ap. J.</u>, 248, 651.
Duncan, D. K., and Jones, B. F. 1983, <u>Ap. J.</u>, 271, 663.
Guzik, J. A., Willson, L. A., and Brunish, W. M. 1987, <u>Ap. J.</u>, 319, 957.
Hobbs, L. M. 1984, <u>Ap. J.</u>, 286, 252.
Hobbs, L. M., and Duncan, D. K. 1987, <u>Ap. J.</u>, 317, 796.
Hobbs, L. M., and Pilachowski, C. 1986a, <u>Ap. J. (Letters)</u>, 309, L17.
Hobbs, L. M., and Pilachowski, C. 1986b, <u>Ap. J. (Letters)</u>, 311, L37.
Michaud, G. 1977, <u>Nature</u>, 266, 433.
Michaud, G. 1982, <u>Ap. J.</u>, 258, 349.
Michaud, G. 1985, in <u>Solar Neutrinos and Neutrino Astronomy</u>, Ed. M.L Cherry, K. Lande
    and W. A. Fowler (New York: American Institute of Physics), p. 75.
Michaud, G, 1986, <u>Ap. J.</u>, 302, 650.
Michaud, G., Charland, Y. 1986, <u>Ap. J.</u>, 311, 326.
Michaud, G., Charland, Y., Vauclair, S., and Vauclair, G. 1976, <u>Ap. J.</u>, 210, 447.
Michaud, G., Fontaine, G., and Beaudet, G. 1984, <u>Ap. J.</u>, 282, 206.
Michaud, G., Tarasick, D., Charland, Y., and Pelletier, C. 1983, <u>Ap. J.</u>, 269, 239.
Paquette, C., Pelletier, C., Fontaine, G., and Michaud, G. 1986, <u>Ap. J. Suppl.</u>, 61, 177.
Pilachowski, C. A., and Hobbs, L. M. 1987, preprint.
Schatzman, E. 1969, <u>Astr. Ap.</u>, 3, 331.
Schatzman, E. 1977, <u>Astr. Ap.</u>, 56, 211.
Schatzman, E., Maeder, A., Angrand, F., and Glowinski, R. 1981, <u>Astr. Ap.</u>, 96, 1.
Spite, F., and Spite, M. 1982, <u>Astr. Ap.</u>, 115, 357.
Spite, M., Maillard, J.-P., and Spite, F. 1984, <u>Astr. Ap.</u>, 141, 56.
Tassoul, J.-L., and Tassoul, M. 1982, <u>Ap. J. Suppl.</u>, 49, 317.
Tassoul, M., and Tassoul, J.-L. 1983, <u>Ap. J.</u>, 271, 315.
Tassoul, M., and Tassoul, J.-L. 1984, <u>Ap. J.</u>, 279, 384.
Thévenin, F., Vauclair, S., and Vauclair, G. 1986, <u>Astr. Ap.</u>, 166, 216.
Tomkin, J., Lambert, D. L., and Balachandran, S. 1985, <u>Ap. J.</u>, 290, 289.
Van't Veer-Menneret, C., Coupry, M. F., and Burkhart, C. 1985, <u>Astr. Ap.</u>, 146, 139.
Vauclair, G., Vauclair, S., and Michaud, G. 1978a, <u>Ap. J.</u>, 223, 920.
Vauclair, S., Vauclair, G., Schatzman, E., and Michaud, G. 1978b, <u>Ap. J.</u>, 223, 567.
Wolff, S. C., and Preston, G. W. 1978, <u>Ap. J. (Suppl.)</u>, 37, 371.

OBSERVED LITHIUM ABUNDANCES AS A TEST OF STELLAR INTERNAL STRUCTURE

Sylvie Vauclair - Observatoire Midi-Pyrénées - France

The "lithium gap" observed in the Hyades and other galactic clusters by Ann Boesgaard and her collaborators (Boesgaard and Tripicco 1986, Boesgaard 1987, Boesgaard, Budge and Burck 1987) gives a challenge to theoreticians. Indeed a good fit between the theoretical results and the observations will give, a clue for our understanding of the stellar internal structure and evolution.

A theoretical explanation of the "lithium gap" by gravitational and radiative diffusion has been proposed by Michaud 1986. In G type stars, the convection zone is too deep for gravitational settling to take place : the density at the bottom of the convection zone is so large that the diffusion time scale exceeds the age of the star. Increasing the effective temperature leads to a decrease of the convection zone, and consequently to a decrease of the diffusion time scale. In F stars it becomes smaller than the stellar age, leading qualitatively to a lithium abundance decrease as observed. When the convection zone is shallow enough, the radiative acceleration on lithium becomes important as lithium is in the hydrogenic form of li III (while it is a bare nucleus, li IV, deeper in the star). This radiative acceleration may prevent lithium settling for hotter F stars. This is a very attractive explanation, which leads to a minimum of the lithium abundance nearly at the place where it is observed in effective temperature. However it suffers from some difficulties : the theory predicts an increase of the lithium abundance larger than normal in the hottest F stars, which is not observed, and the predicted minimum lithium abundance is one or two orders of magnitude higher than the minimum observed in the Hyades. The former may be overcome if mass loss occurs in these stars (Michaud 86). Let us focus on the latter.

The radiative acceleration on a given element, through a bound-bound transition, may be written :

$$g_R = \frac{1}{m} \frac{N_{i,n}}{N} \int_0^\infty \sigma_{i,n}(v) \frac{\phi v \, dv}{c} \tag{1}$$

where m is the mass of the considered element, $N_{i,n}/N$ the fraction of the element in the lower level of the line, $\sigma_{i,n}(v)$ the transition section and $\phi_v \, dv$ the available photon flux.

With the diffusion approximation, a lorentz profile for the line, and after integration over v, $g_R$ becomes :

$$g_R = \frac{1}{mN} \frac{8\pi^2 k^3}{3h^2 c^3} T^2 (-\frac{dT}{dr}) \frac{z^4 e^z}{(e^z - 1)^2} \frac{\Delta/2}{\sqrt{\frac{\kappa_c}{\kappa_L}(\frac{\kappa_c}{\kappa_L} + 1)}} \tag{2}$$

with $z = \dfrac{hv}{kT}$ and $\kappa_L = N_n \dfrac{\pi e^2}{m_e c} \dfrac{f}{\pi} \dfrac{2}{\Delta}$

where f is the oscillator strength of the line and $\Delta/2$ the half width. $\kappa_c$ is the monochromatic opacity due to all the opacity sources except the considered line.

For an unsaturated line ($\kappa_L << \kappa_c$), (2) may be transformed into :

$$g_R = \frac{1.6 \times 10^{-4}}{A} \frac{N_{i,n}}{N} f \frac{z^4 e^z}{(e^z-1)^2} \frac{T_e^4}{T} \frac{R^2}{r^2} \frac{\overline{\kappa}}{\kappa_c} \tag{3}$$

where $\overline{\kappa}$ is the Rosseland mean opacity, $T_e$ the effective temperature, T the local temperature, R the stellar radius, r the local radius

The ratio $\overline{\kappa}/\kappa_c$ which appears in $g_R$ represents the fact that the radiative acceleration through one line strongly depends on the other sources of opacity at the same frequency. Up to now, the radiative accelerations have been computed with the approximation $\kappa_c \approx \overline{\kappa}$. However if, for example, a line of an abundant element sits at the same place as the Ly $\alpha$ lithium resonance line, the radiative acceleration on lithium may be strongly decreased.

A table of the important lines which may blend the lithium and beryllium resonance lines has been given in Vauclair 1987. This table is not exhaustive as this part of the spectrum is not well known. Also the atomic parameters of these lines are very uncertain. The influence of these lines on the lithium and beryllium radiative accelerations is shown on fig. 1. It seems too small to account for the discrepancy between the theoretical and observed minima of Li abundances. However these results are _very uncertain_ as this part of the spectrum is not well known. I would like to emphasize the urgent need of good atomic parameters for these far UV lines.

Another explanation of the lithium gap in the Hyades could be found in terms of turbulent diffusion and nuclear destruction. Turbulence is definitely needed to explain the lithium abundance decrease in G stars. If this turbulence is due to the shear flow instability induced by meridional circulation (Baglin, Morel, Schatzman 1985, Zahn 1983), turbulence should also occur in F stars, which rotate more rapidly than G stars. Fig. 2 shows a comparison between the turbulent diffusion coefficient needed for lithium nuclear destruction and the one induced by turbulence. Li should indeed be destroyed in F stars : This effect gives an alternative scenario to account for the Li gap in the Hyades. The fact that Li is normal in the hottest observed F stars could be due to their slow rotation.

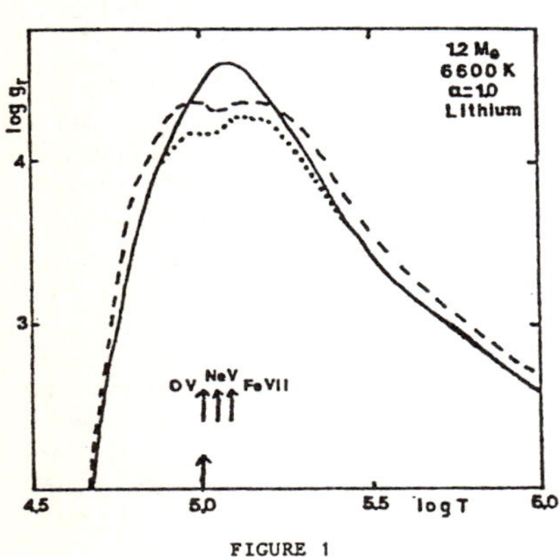


FIGURE 1

Radiative acceleration on Li in a 1.2 M$_\odot$ star.

— $\kappa_\upsilon = \overline{\kappa}$

... $\kappa_\upsilon = \overline{\kappa}$ + opacity of blendings lines

--- $\kappa_\upsilon = \overline{\kappa}/2$ + opacity of blending lines

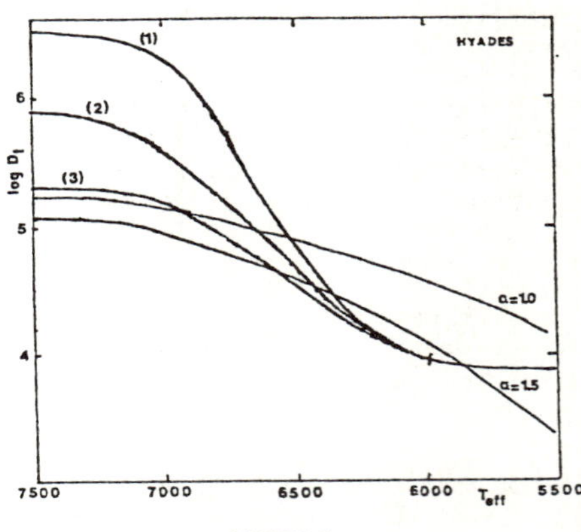


FIGURE 2

Turbulent diffusion coefficients.

-curves labelled α =1.0 and α=1.5 : D$_T$ needed for Li nuclear destruction ( α= mixing length parameter)
-curves (1),(2),(3): D$_T$ induced by rotation (Zahn 1983). these curves are adjusted for Li nuclear destruction in solar type stars.
(1): present rotation law; (2): assuming Endal and Sofia 1981 braking; (3): assuming Skumanitch 1972 braking.

## References


Baglin,A., Morel,P., Schatzman,E. 1985, Astron. Astrophys. _149_, 309
Benz,W., Mayor,M. Mermilliod,J.C. 1984, Astron. Astrophys. _138_, 93
Boesgaard,A.M., Tripicco,M.J. 1986, ApJ _302_, L49
Boesgaard,A.M., Budge,K.G., Burck,E.E. 1987, preprint
Boesgaard,A.M. 1987, P.A.S.P., in press
Boesgaard,A.M. 1987, ApJ, in press
Endal,A.S., Sofia,S. 1981, Ap. J. _243_, 625
Michaud,G. 1986, ApJ _302_, 650
Skumanitch, A. 1972, Ap. J. _171_, 565
Vauclair,S. 1987, proceedings of the IAU Symposium 132 "The impact of very high S/N spectroscopy on stellar physics".
Zahn,J.-P. 1983, in "Saas Fee advanced courses in Astrophysics".

## ON THE DISTRIBUTION OF THE LITHIUM ABUNDANCE IN NORMAL LATE-TYPE GIANTS


Liia Hänni
Tartu Astrophysical Observatory
202444 Tõravere, Estonian SSR, USSR


The abundance of lithium in stellar atmospheres presents an important observational constraint to the hydrodynamical models of the outer layers of stars. It can be considered as a cumulative measure of the extent of matter exchange between surface and deeper layers during the stellar evolution.

From the observed large scatter of lithium abundances in evolved stars it follows that the efficiency of mixing has been highly variable from one object to another. At present, it seems to be difficult to find any satisfactory explanation to the lithium abundances of individual stars. We suppose that at this stage the statistics of lithium abundances in different types of stars can give some insight into the character of mixing processes operating in stars. In this report some observational results about the distribution of the lithium abundances in normal late-type giants are presented.

The observations of the lithium resonance line at $\lambda 6708.8$ Å were carried out with a SIT vidicon detector attached to the coudé spectrometer of the 1.5 m telescope of the Tartu Astrophysical Observatory. The sample of stars observed consists of 70 K0 - K5 and 75 M0 - M4 giants. A set of spectra of K giants with different strengths of lithium resonance doublet is shown in Fig. 1.

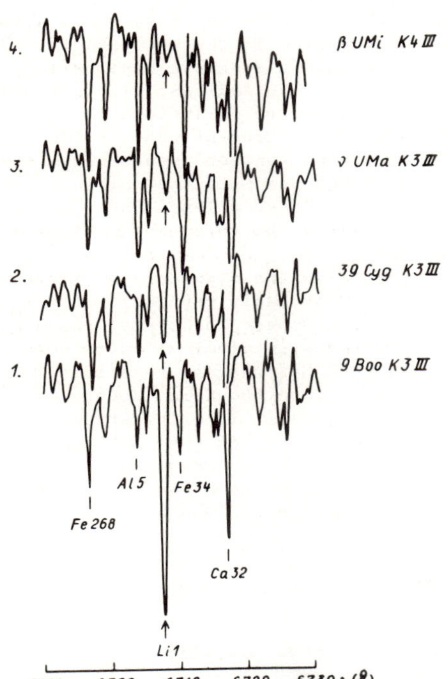


Fig. 1. A set of the spectra of K giants with different strength of the lithium resonance line.

Comparatively low spectral resolution ($\Delta\lambda \sim 0.6$ Å) and difficulties with the intensity calibration did not warrant any detailed abundance analysis of our spectral material. Instead, the stars with similar spectral types were divided into four groups according to the strength of the lithium line. The groups may be described as follows (see Fig. 1): 1 - the lithium line is extraordinarily strong, its intensity exceeds that of the nearby CaI line $\lambda 6717.7$ Å; 2 - the lithium line is strong; 3 - the lithium line is weak, but its contribution to the blend is still noticeable; 4 - the line is undetectable in our spectra.

In Fig. 2 the frequency distribution of lithium line strengths is shown for three intervals of spectral classes. Our sample of K0 - K1 giants mainly consists of the objects with undetectable or weak lines of lithium. Due to the high degree of the ionization of lithium in these stars, the line can be detected only when the lithium abundance exceeds $\log N_{Li} \sim 1.0$ (in the scale of $\log N_H = 12.0$). This estimate is based on the common stars of this study and the work by Lambert et al. (1980), where the abundance of lithium has been derived by the method of spectrum synthesis using high-quality spectra. Only one star out of 20 early K giants surveyed by us was found to possess a relatively strong line of lithium - BS 5361 (K0III). Certainly, our sample of K0 - K1 giants is too small to guarantee the detection of extreme cases of lithium line strengths.

In the case of K2 - K5 giants as well as M0 - M4 giants again more than 60% of the stars have no detectable lithium resonance line, although the detection limit lies at much lower abundance values. The constancy of the relative number of the stars with an undetectable lithium line despite different detection limits seems to be noteworthy. It may imply that about 63 - 65% of field late-type giants independently of their position on the red giant branch, really have a very low abundance of lithium. Lithium deficiency of these stars, most probably, originates from the period of main sequence evolution.

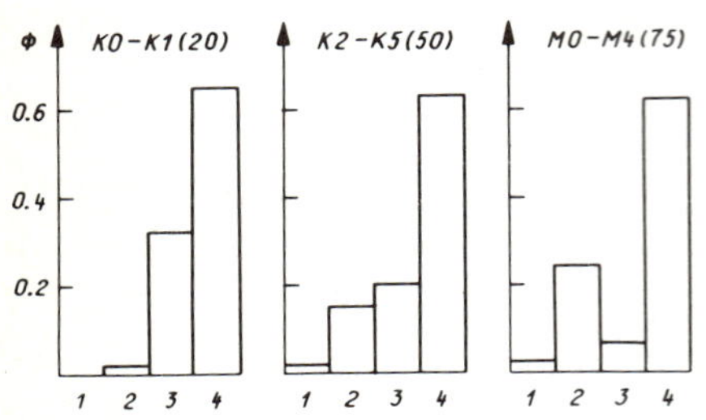



Fig. 2. Frequency distribution of the lithium resonance line strengths in red giants. The number of observed stars is indicated in the brackets.

The next feature that deserves attention in Fig. 2, is bimodal frequency distribution of lithium line strengths for M0 - M4 giants. Here we get some hint at the existence of a gap in the distribution of the lithium abundances of the evolved stars. The results of a more precise abundance analysis of 25 M giants by Luck and Lambert (1982) and 10 M giants by Hänni (1983) also tend to show a discontinuity in the lithium abundances within the sample of stars with similar effective temperature.

About 25% of M giants - the stars of group 2 - show a definite lithium line in their spectra. As indicated by common objects of the present survey and of the two above-mentioned studies, the lithium abundances of these stars are clustering around the abundance value of log $N_{Li}$ ∿ 0.0. They form an extension to the sequence of G - K giants of declining surface lithium abundance with advancing spectral type defined by the stars with a detectable and strong line of lithium. The decrease of the surface lithium content may be attributed to the increasing convective dilution as a star evolves upwards along the red giant branch. The starting value of the lithium abundance of the evolved stars with a still measurable lithium content would have been high. Most probably, these stars have left the main sequence with the initial atmospheric abundance of lithium which, according to abundant observational data, is close to the cosmic abundance of log $N_{Li}$ ∿ 3.0. As evidenced by the theory of stellar evolution, the early M-type giants should have been experienced the maximum amount of mixing caused by the growth of the convective envelope in the stage of the first red giant branch. Therefore, the abundance of log $N_{Li}$ ∿ 0.0 probably represents the final value of the lithium abundance after the first dredge-up phase for stars, which have become red giants with undepleted surface lithium content. It gives us an estimate of the reduction factor of the lithium abundance equal to 1000, whereas theoretical calculations by Iben (1967a, b) predict the factor no more than 60. Some doubt may be expressed about the validity of standard models of the stellar interior structure.

Finally, I would like to stress that the bulk and quality of the lithium abundance data for red giants need to be considerably raised for an adequate statistical analysis. The main task of the present report was to provoke some interest in the problem concerning the distribution of lithium abundances in evolved stars.


REFERENCES
Hänni, L. 1983, *Tartu Astrophys. Obs. Preprint* **A-2,**
Iben, I. 1967a, *Astrophys. J.,* **147,** 624.
Iben, I. 1967b, *Astrophys. J.,* **147,** 650.
Lambert, D.L., Dominy J.F., and Sivertsen S. 1980, *Astrophys. J.,* **235,** 114.
Luck, R.E., and Lambert D.L. 1982, *Astrophys. J.,* **256,** 189.

ABUNDANCES IN COOL EVOLVED STARS

Catherine A. Pilachowski
National Optical Astronomy Observatories, Kitt Peak National Observatory
PO Box 26732, Tucson, AZ  85726-6732,  USA


Nature has filled the upper right quadrant of the Hertzsprung-Russell diagram with more varieties of peculiar stars and odd chemical compositions than even our most speculative observers and theorists could dream up.  To bring some structure to this vast subject I will categorize the phenomena we observe according to our model of stellar evolution, dividing the stars among the first ascent of the giant branch and the core-helium burning phase, the asymptotic giant branch (double shell-burning) phase, and the post-AGB and pre-planetary nebula stars.  The types of stars found in these three groups are summarized below.

| Warm Giants: | AGB Stars: | Post-AGB Stars: |
|---|---|---|
| Ba II Stars | M Stars | R CrB Stars |
| Early Carbon Stars | MS Stars | Hydrogen-Deficient Carbon Stars |
| CH Stars | S Stars | RV Tauri Stars |
| Subgiant CH Stars | N Stars | W Vir Stars |
| Weak G-band Stars | SC Stars | SRd Variables |
| Li-Rich Giants | J Stars | |

Two dominant themes run throughout the evolution of late type star compositions: the abundances of the isotopes of carbon, nitrogen, and oxygen, and the abundances of the metals heavier than the iron peak - the neutron capture elements usually associated with the s-process.  In addition to these elements, the abundance of lithium can also be a distinguishing characteristic of some groups, and can be used to interpret possible origins for some of these peculiar stars.


The Warm Giants


Most samples of warm giants in the literature are comprised of primarily low mass, old disk stars that fall either on their first ascent of the red giant branch or in the core-helium burning clump immediately following core helium ignition.  The abundances of several classes of peculiar warm giants, as well as normal K giants and the Sun, are summarized in Table 1.  Looking first at the comparison of the abundances in normal K giants with the predictions of stellar evolution theory for the first dredge-up (Iben and Renzini 1984), we see generally

**Table 1: Abundances in the Warm Giants**

| | C/O | $^{12}C/^{13}C$ | C/N | $^{16}O/^{17}O$ | $^{16}O/^{18}O$ | [Fe/H] | [s/Fe] | binary? |
|---|---|---|---|---|---|---|---|---|
| Solar | 0.56 | 89 | 4.8 | 2630 | 490 | 0.0 | 0.0 | --- |
| Normal K Giants | 0.30 | 6 - 20 | 1.15 | 500 | 500 | solar | solar | no |
| 1st Dredge Up | 0.4 | 20 - 30 | 1.9 | 2630 - 250 | 600 - 1000 | no change | no change | --- |
| Ba II Stars | 0.76 | 25 | 2.1 | 300 | 500 | -0.5 - 0.1 | +0.7 | yes |
| Mild Ba II Stars | 0.30 | 20 | 1.3 | >100 | >100 | ~solar | +0.2 | yes? |
| Ba-C Stars | C-rich | --- | --- | --- | --- | --- | enhanced | ? |
| Subgiant CH Stars | ≤1.0 | 10 - >40 | --- | --- | --- | -0.35 | +0.7 | yes |
| CH Stars | C-rich | 6 | --- | --- | --- | <-0.5 | +1.4 | yes |
| CH-like Stars | C-rich | (6) | --- | --- | --- | old disk? | +0.3 | ? |
| Early R Stars | 1.6 | <10 | 2.2 | --- | --- | -0.2 | +0.2 | no |
| Weak G-band Stars | 0.04 | 4 - 10 | 0.08 | --- | --- | 0.0 | ~solar | no |
| Li-Rich Stars | 0.5 - 1.0 | 22 | 1.5 - 4 | --- | --- | -0.6 - 0.3 | ~solar | ? |


Dominy 1984
Harris and Lambert 1984
Harris et al. 1985a
Kovacs 1985
Krishnaswamy and Sneden 1984
Lambert 1985
Lambert and Dominy 1980
Lambert and Ries 1981

Lambert and Sawyer 1984
Luck 1982
Luck and Bond 1982
Luck and Sneden 1986
McClure 1985
Smith 1984
Sneden 1983
Sneden and Bond 1976

Sneden et al. 1981
Sneden and Pilachowski 1984
Tomkin et al. 1984
Wallerstein and Sneden 1982
Wannier 1985

good qualitative agreement.  Material left over from main-sequence CN-cycle hydrogen burning is mixed with relatively pristine material on the stellar surface, reducing the carbon abundance slightly, enhancing the nitrogen abundance, and raising the $^{13}$C abundance.  The oxygen abundance remains constant for all practical purposes, since stellar interior temperatures are too low for the ON cycle to operate, except that a small amount of $^{16}$O is converted to $^{17}$O.

The uncertain distances, luminosities, masses, and evolutionary states of field giants confuse our attempts to verify theoretical predictions in detail, but a recent study of the carbon and nitrogen abundances in M67 giants by Brown (1987) offers an excellent observational test.  Brown determined carbon and nitrogen abundances for giants from $3.7 > M_V > 0.9$; his C/N ratios are plotted versus absolute magnitude in Figure 1.  The first dredge-up begins at $M_V=3.5$, but it is fully complete by a luminosity of $M_V \simeq 3.0$.  The giants of M67 are able to complete the first dredge-up more quickly than expected from theoretical calculations, and the change in the C/N ratio is larger than predicted, as well.  The field K giants in Table 1 also show lower C/N ratios than predicted by first dredge-up calculations.  Brown offers two hypotheses to explain these results: a) that a real stellar envelope becomes fully convective at lower luminosity than predicted, or b) that the CN-cycled material lies closer to the surface than expected, either due to mass loss of the outer layers of the star or due to a greater extent of the CN-cycled region in the interior.

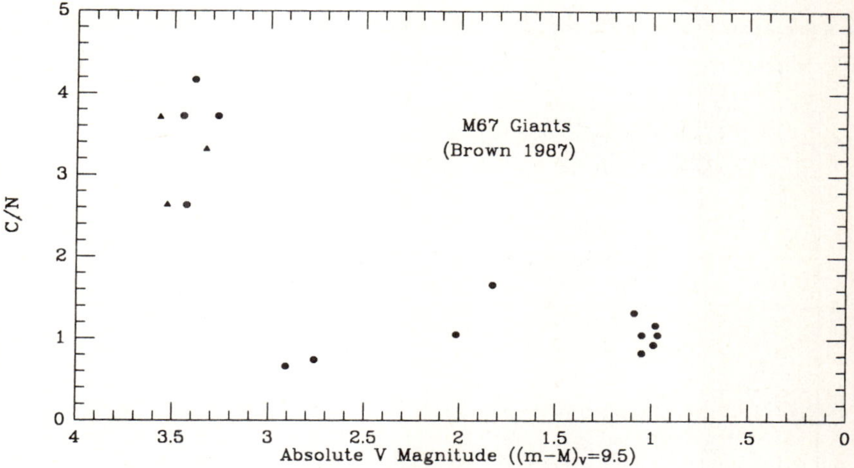


Figure 1 - The C/N ratio in M67 giants vs. absolute magnitude.

The mass loss alternative is contradicted by the detection of lithium in low mass giants, but support for more extensive CN-cycling in the interior is provided by measurements of the $^{12}$C/$^{13}$C ratio in field stars:  the well-known Arcturus problem

that has been haunting us since carbon isotope ratios were first measured.
Theoretical models of the first dredge-up reduce the carbon isotope ratio from the
solar value of about 90 to approximately 20-30. Standard models are unable to
produce carbon isotope ratios lower than this. Many K giants, however, have $^{12}C/^{13}C$
as low as 7 - 10 (Lambert and Ries 1981), approaching values appropriate for CN-
cycle equilibrium. Metal-deficient giants achieve carbon isotope ratios as low as 4
(Sneden et al. 1986). Theorists have addressed this problem by tweaking the
standard models to increase the amount of mixing (c.f. Dearborn et al. 1976).
Brown's M67 data eliminate some of these alternatives, specifically those invoking
mixing during the helium core flash. Models which partially mix the stellar
material either on the main sequence or before the first ascent of the giant branch,
perhaps through meridional circulation currents or turbulent diffusion, have been
the most successful at reproducing the observed composition changes at the first
dredge-up.

The compositions of several varieties of peculiar warm giants are also listed in
Table 1. These stars differ from normal G and K giants in a variety of ways:
lithium is unusually high (or low); carbon is enriched through triple-$\alpha$
nucleosynthesis or depleted through CN-cycle processing; and/or s-process elements
are enhanced. The origin of these groups of peculiar warm giants has been a mystery
for decades. Their relative rarity in the Galaxy suggests unusual circumstances are
required to produce peculiar giants. While their compositional anomalies are
reminiscent of double-shell flashes on the asymptotic giant branch, the luminosities
of these warm giants are much too low for them to have undergone helium shell

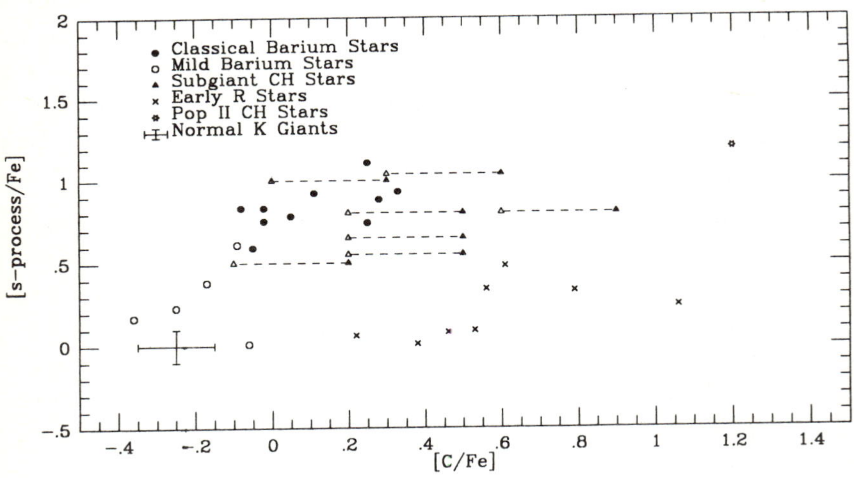


Figure 2 - S-process enrichment vs. carbon enrichment in peculiar warm giants. The
expected locus of the subgiant CH stars following the first dredge-up is indicated
by open triangles. The barium stars, mild barium stars, subgiant CH stars, and CH
stars all seem to follow a common relationship.

flashes.  Other explanations must be sought, and three fundamental classes of
hypotheses are often invoked: a) mixing at the helium core flash, b) mass transfer
from a binary companion, and c) diffusion.  Since so many of the groups are now
known to be binary stars (McClure 1985), a mass transfer hypothesis seems likely.

The barium stars, and their colleagues the mild barium stars, the marginal barium
stars, the semi-barium stars, and the barium-carbon stars, are G and K giants with
an excess of carbon, presumably due to the triple-$\alpha$ process, normal abundances of
other CNO nuclides, and enhancements of the heavy s-process species Sr-Y and Ba, La,
etc.  Recent authors have concluded that the differences among the sub-groups of
barium stars are mostly in degree.  A plot of the relative s-process enrichment vs.
carbon enrichment for a mixed sample of these stars (suggested by Lambert 1985) is
shown in Figure 2.  Data are taken from several authors listed under Table 1.  The
mild or marginal barium stars follow the same relation between s-process enrichment
and carbon enrichment as do the classical barium stars, suggesting a common origin.

Aside from the obvious carbon and s-process enhancements in the barium stars,
their compositions are similar to those of other G and K giants.  The oxygen and
carbon isotope ratios and the lithium (Pinsonneault et al. 1984), nitrogen and
oxygen abundances are normal, suggesting red giant evolution has proceeded
normally.  In a recent study of oxygen isotope ratios in barium stars, Harris,
Lambert, and Smith (1985a) argue that if the carbon excess in the barium stars
resulted from helium burning in the star itself, either from a core helium flash or
from the third dredge-up between helium shell flashes on the AGB, then the star must
also have mixed up material from the $^{17}$O peak and below, decreasing the surface
$^{16}$O/$^{17}$O ratio to less than 200.  The observed oxygen isotope ratio establishes that
the barium stars are unlikely to have produced the excess carbon (or the excess s-
process elements) themselves.  Harris et al. conclude that the carbon and heavy
elements must have been produced elsewhere, presumably in a more massive, evolving
companion, and been dumped onto the surfaces of the barium stars.

The chemical composition of the subgiant CH stars is superficially different from
that of the barium stars.  While they show similar enhancements of s-process
elements relative to iron, the carbon isotope ratio varies widely among the subgiant
CH stars, ranging from 10 to >40 for most of the sample, and C/O <1 in all cases
(Sneden 1983).  In Figure 2, the subgiant CH stars appear to follow a different
relationship than the barium stars, but if they are allowed to evolve through a
standard first dredge-up event, they will move to the left to overlap the barium
stars.  The barium stars and the subgiant CH stars may in fact be related.  Since
the CH stars are also binaries (McClure 1985), a mass transfer explanation for their
origin is tempting, although Luck and Bond (1982) offer an alternative at the helium
core flash.

The CH stars and the CH-like stars are included in Table 1 for completeness, but we don't know much about them.  The CH stars are metal-deficient, Population II giants enriched in carbon and s-process elements.  They may be enriched in $^{13}$C (Lambert 1985).  They appear to be binaries (McClure 1985).  The CH stars are probably Pop II barium stars, and to accept their designation as Pop II barium stars.  Not much is known about the CH-like stars of Yamashita (1972, 1975), which are Pop I, carbon-rich giants.  At low spectral resolution the $\lambda$4554 line of Ba II is enhanced.  These stars may be related to the barium or barium-carbon stars, but we need more information about their compositions, and especially about their binary status.

The early R stars were the subject of an excellent study by Dominy (1984).  They differ from the barium stars in that the s-process elements are not enhanced, they contain significant excess $^{12}$C, with C/O ratios of typically 1.6, and evidence of substantial CN-cycle processing, with $^{12}$C/$^{13}$C ratios of <10, and N/Fe ratios of a factor of 2-3 above normal K giants.  Oxygen is not depleted, however, so the high abundance of carbon (C/O>1) cannot be attributed to the CNO cycle operating near equilibrium.  The early R stars are not binaries, unlike the barium stars.  Dominy suggests that these stars are a genuine example of mixing at the core helium flash.

Finally we come to two groups of peculiar warm giants which share unusually high abundances of lithium.  Some of the weak G-band stars contain lithium as high as the initial "cosmic" abundance of 3.0 found in young main sequence stars (Lambert and Sawyer 1984).  Carbon is extremely deficient (a factor of 10 relative to normal K giants), the carbon isotope ratios are very low (from 4 to 10), and nitrogen is enhanced such that the sum C+N is constant.  The weak G-band stars are not binaries (Tomkin et al. 1984).  The material on the surfaces of the weak G-band stars has clearly been subjected to CN-cycle processing, which would certainly have quickly destroyed the original lithium.  It may be necessary to invoke diffusion to explain the unusual lithium abundances (Lambert and Sawyer 1984) and possibly internal mixing on the main sequence to provide the extreme CN-cycle processing.  The lithium rich G and K giants are extremely rare, and little quantitative information is available.  They appear to be otherwise essentially normal, but with lithium abundances up to the "cosmic" limit.


The Stars on the Asymptotic Giant Branch


The stars on the AGB are difficult to understand not only because so many groups of peculiar stars are present, but also because so many pathological stars occur within each class.  Many AGB stars show evidence of recent nucleosynthesis, from the presence of technetium (Merrill 1952, Little-Marenin and Little 1979) and $^{93}$zirconium (Zook 1985).  The AGB stars have exhausted helium in their cores, and

## Table 2:  Abundances in the AGB Stars

| | C/O | $^{12}$C/$^{13}$C | [N/Fe] | $^{16}$O/$^{17}$O | $^{16}$O/$^{18}$O | [C+N+O/Fe] | [s/Fe] |
|---|---|---|---|---|---|---|---|
| Solar | 0.56 | 89 | 0.0 | 2630 | 490 | 0.0 | 0.0 |
| K Giants | 0.30 | 6-20 | +0.5 | 500 | 500 | 0.0 | solar |
| Early M Giants | 0.45 | ~13 | +0.4 | 160-1100 | ~500 | ~0.0 | solar |
| MS Stars | 0.64 | ~30 | +0.4 | 900-3000 | 1100-4700 | >0.0 | 0 - 0.6 |
| S Stars | 0.81 | ~38 | +0.5 | 500-2400 | 1300-5000 | +0.2 | +0.65 |
| SC Stars | ~1.0 | 5-53 | +1.3 | 160-400 | ≥300 | 0.5 | ++ |
| N Stars | 1.1 | ~50 | 0.0 | 550-4100 | 700-2400 | (~0.2) | 1.0-2.0 |
| J Stars | ~0.9 | 3-5 | 0.0 | 350-850 | uncertain | (~0.2) | solar |

Boesgaard 1970
Catchpole 1982
Catchpole and Feast 1971
Dominy and Wallerstein 1987
Dominy et al. 1986
Fujita and Tsuji 1965, 1977

Harris et al. 1987
Harris et al. 1985b
Hinkle and Scarlach 1985
Johnson et al. 1982
Lambert 1985
Lambert et al. 1986

Smith and Lambert 1985, 1986
Torres-Peimbert and Wallerstein 1966
Tsuji 1985, 1986
Utsumi 1986
Wannier 1985

produce energy through two burning shells of hydrogen or helium surrounding an inert carbon core.  Above a luminosity of $M_V \sim -3$, this configuration is unstable, leading to thermal pulses (Schwarzschild and Härm 1965) which may induce mixing of the convective envelope with the layers between or below the hydrogen and helium burning shells.  For stars of mass greater than about 4-5$M_\odot$ a dredge-up event which penetrates the hydrogen burning shell has been identified with similar effect as the first dredge-up.  A series of "third" dredge-up events occur for low and intermediate mass stars during the time between shell flashes.  These events bring up helium-burned material enriched in $^{12}$C and (probably) the s-process elements.

A selected set of the abundances in AGB giants is given in Table 2.  The determination of abundances in AGB giants is a difficult problem, and isotopic ratios are known with much greater certainty than the abundances of individual elements.  Indeed, progress in determining the compositions of cool giants is due to the relatively recent availability of high resolution IR spectra (in particular the KPNO 4M FTS) and to improved models for cool stars.  To complicate the problem, cool giants are usually variable stars, displaying shock components and multiple velocity structure in their spectra.  The compositions of the early M giants resemble the K giants, and in reasonable accord with the predictions of stellar evolution theory, But among the more luminous and more evolved stars, the story becomes more interesting.  The MS stars begin to show signs of enrichment of triple-$\alpha$ $^{12}$C with an increase in the C/O and the $^{12}$C/$^{13}$C ratios.  The nitrogen abundance remains unchanged.  The sum of C+N+O begins to increase above the solar value, and enhancements of s-process elements appear.  Some MS or S stars have large nitrogen excesses (c.f. RS Cnc and HR 8714 in Smith and Lambert 1986), which could result if these stars are more massive than most MS stars, and have undergone an extensive second dredge-up event.  The relatively low luminosity of most MS stars would suggest that most of them may be low mass stars.

The MS stars blend smoothly into the S stars, with small increases in the $^{12}$C abundance, the C/O ratio, the sum C+N+O, and the abundance of the s-process elements.  The nitrogen abundance remains the same as for normal M giants, indicating no further CN-cycle processing.  The $^{12}$C abundance in MS and S stars is approximately double what it is in normal M giants.  If the mass of the convective envelope is about 0.4$M_\odot$, then roughly $3 \times 10^{-3} M_\odot$ of carbon must be added by the third dredge-up.  Only 3-6 thermal pulses are required; only 2-3 pulses are needed to produce the s-process enhancement.  The MS and S stars are probably in the early phases of double shell flashing (Smith and Lambert 1986).  The carbon and s-process element abundances in MS and S stars are in quantitative agreement with the predictions of third dredge-up calculations.

The SC stars may represent the onset of a new phenomenon, CN-cycle processing of material in the convective envelope.  Dominy et al. (1986) have determined C, N, and

O abundances and carbon and oxygen isotope ratios for a sample of SC stars.
Unfortunately metal (Fe) abundances are not available and an unambiguous
interpretation of their results depends on knowing the metal abundance. Their
nitrogen abundances also depend very sensitively on the C/O ratio. The $^{12}$C and the
s-process abundances, and the C/O ratio have increased above values seen in the MS
and S stars. If the low oxygen abundances reported by Dominy et al. represent the
original metal content of the stars, the nitrogen abundances are enriched, which
suggests further CN-cycle processing of the fresh carbon introduced between thermal
pulses. This explanation yields a general excess of C+N+O over the original
composition of about a factor of three, requiring >5 thermal pulses. An alternative
explanation, proposed by Dominy et al., is that the oxygen abundance is low due to
ON-processing. They invoke successive $\alpha$ captures on $^{14}$N to deplete the nitrogen and
produce $^{22}$Ne. The most luminous carbon stars in the Magellanic Clouds are probably
SC or CS stars (formerly classified as J stars, see Utsumi 1988), suggesting that
this phase of evolution occurs near the tip of the AGB. These stars may be part of
the cycle defined by the MS, S, and N-type carbon stars, with the addition of CN
cycle processing of $^{12}$C-rich material in more massive examples.

Our understanding of the nature and origins of the carbon stars has increased
dramatically in recent years, motivated by the discovery and study of so many carbon
stars in the Magellanic Clouds. A comprehensive study of the abundances of C, N,
and O in N- and J-type carbon stars in the Galaxy was provided by Lambert et al.
(1986). Most carbon stars have $1.0 < C/O < 1.6$, and typical values of 1.1. The
carbon isotope ratios of the N-type stars range from 30 to nearly 100. While
results are ambiguous, oxygen, nitrogen and the metals may be slightly sub-solar.
Like the MS and S stars, the N-type carbon stars have oxygen isotope ratios much in
excess of theoretical predictions, and of their K and M giant predecessors (Harris
et al. 1987). Utsumi (1985) reports that the s-process elements are enriched by
factors of 10-100. Abundances in the N-type carbon stars are generally consistent
with advanced thermal pulsation models, and with an advancing evolutionary sequence
M-MS-S-C. The high $^{12}$C/$^{13}$C ratios reflect the addition of triple-$\alpha$ carbon. The
absence of stars with high C/O ratios may be because stars with too much carbon
produce copious graphite grains which ultimately shroud the stars' light. The C/O
ratio in the heavily obscured carbon star IRC +10° 216, while very uncertain, seems
to be >1.7 (McCabe et al. 1979).

The J-type, or $^{13}$C-rich stars offer a greater mystery. They differ form N-type
stars in having much lower carbon isotope ratios (as low as 3.2, at the CN-cycle
equilibrium limit), and they lack enhancements of s-process elements (neglecting
such anomalies as WZ Cas). Oxygen and nitrogen may be deficient, although this
conclusion will remain uncertain until we really can establish the abundances of the
iron peak elements. $^{12}$C/O is generally less than unity, but ($^{12}$C+$^{13}$C)/O is slightly
greater than unity. Oxygen isotope ratios in J stars are lower than in N stars, and

are similar to the values in K giants and early M stars. Utsumi (1988) reports that the genuine J-type stars in the Magellanic Clouds are of relatively low luminosity. The origins of the J-type carbon stars pose a special problem. The absence of s-process enhancements probably eliminates a thermal-pulse mechanism. Harris et al. (1987) argue that no form of the CNO-cycle, including hot bottom convection, can account for the composition of the J stars because it is difficult to get C/O>1 and $^{12}C/^{13}C\sim3-5$ without also producing a high $^{14}N$ abundance. They are lead to explosive hydrogen burning as the only recourse, and suggest that for low enough temperatures at the helium core flash, explosive hydrogen burning can achieve low carbon isotope ratios without a nitrogen excess. Such an event may also produce enough $^{12}C$ to raise C/O above unity. The J-type carbon stars would then be post-helium core flash, rather than AGB, stars. The low oxygen isotope ratios and absence of s-process enhancements support this view. Lloyd Evans suggests that the J stars may form an evolutionary sequence with the early and late R stars.

While oxygen isotopes of the K giants and early M giants fulfill the predictions of the first dredge-up rather well, the AGB giants contain far less of the heavy oxygen isotopes than predicted. Several explanations have been proposed and dismissed by Harris et al. (1985b). The enhancement of $^{16}O$ by the reaction $^{12}C(\alpha,\gamma)^{16}O$ is unlikely because we don't see enormous oxygen excesses this explanation would require. The $^{18}O$ cannot be destroyed by convective mixing on the main sequence (with temperatures too low to create $^{17}O$) because the oxygen isotopes of the K giants are as expected following the first dredge-up. Destroying the $^{17}O$ and $^{18}O$ during helium burning would require processing the entire convective envelope, a difficult prospect. Harris et al. are left to invoke explosive nucleosynthesis at the helium core flash or during helium shell flashing to destroy the heavy oxygen isotopes.

The question of the oxygen isotopes is further complicated by the apparent correlation of the isotope ratios with the neutron exposure parameter (as defined by Cowley and Downs 1980) for MS, S, and N-type carbon stars. If read, this correlation must suggest that the envelope is depleted in $^{17}O$ and $^{18}O$ during the AGB lifetime, and in proportion to the number of shell flashing events endured. Depletion of species in the whole envelope by the same mechanism that adds new $^{12}C$ and s-process elements is difficult. To compound the mystery, the oxygen isotope ratios in the SC stars are consistent with the dredge-up predictions, so any depletion mechanism should not operate in these stars.

The compositions of the AGB stars lead to an evolutionary progression up the asymptotic giant branch and through thermal pulsations of the helium burning shell from the M giants to the MS, S, and N-type carbon stars, and finally to such objects as IRC +10° 216. The SC stars and the J stars do not yet fit smoothly into this sequence.

The Post-AGB Stars


Two subgroups of warmer, luminous supergiants offer candidates for the post-AGB phase of stellar evolution - the group of R CrB stars and hydrogen-deficient carbon stars, and the Population II SRd variables (the 89 Her stars), RV Tauri variables, and the W Virginis stars.

The R CrB variables and related hydrogen-deficient carbon stars show strong carbon features (except CH) and very weak Balmer lines.  The R CrB stars are surrounded by circumstellar dust shells, and continue to eject puffs of new circumstellar material on time scales of a month or two;  mass loss rates of order $10^{-6} M_O$ $yr^{-1}$ are reported by Walker 1986.  The compositions of the R CrB stars and hydrogen-deficient carbon stars are were compiled by Lambert (1986).  Hydrogen is extremely deficient in both groups ($H/He \sim 10^{-5}-10^{-6}$), and C/Fe is enriched by typically an order of magnitude over the solar ratio.  C/O is typically 2 (more than is measured in the AGB stars, and [N/Fe]$\sim$ 1. R CrB itself contains a strong lithium line, indicating a lithium abundance [Li/Fe] near the cosmic value.  S-process elements are not enriched (U Aqr is a noted exception).  $^{12}C/^{13}C$ is high, typically >50. These stars appear to be nearly exposed stellar cores whose surface abundances contain not only nitrogen enriched material from CNO-cycle processing, but also $^{12}C$ from helium burning.  Very little of the original hydrogen envelope can remain.

Several scenarios for creating such stars have been proposed.  Lambert (1986) discussed the possibility of an explosion at the helium core flash ejecting the hydrogen envelope.  The remnant would probably be a helium star which would eventually evolve to the region of the R CrB stars.  $^{12}C$ would be produced in the core flash.  Webbink (1984) and Iben and Tutukov (1985) proposed that extreme helium stars might also be formed by the merger of two white dwarfs.  The third model is that the R CrB stars are the result of continued mass loss on the AGB driven by instabilities in the envelope (helium shell flashes).  A helium and carbon rich star would result if most or all of the envelope were driven off.  The high Li/Fe ratios in at least some R CrB stars probably supports an AGB origin.

The Population II SRd variable supergiants (89 Her stars), the RV Tauri stars, and the W Virginis stars may qualify as halo post-AGB stars.  Their compositions were summarized recently by Bond and Luck (1987b) for the 89 Her stars.  Some of these stars show [C/Fe] excesses comparable to the R CrB stars, and the 89 Her stars also show nitrogen enhancements, which indicate the presence of hydrogen burning products at the stellar surface.  Small oxygen enhancements are within the range of those seen in Population II stars.  Stars with more extreme CNO abundances also show enhanced sulfur abundances.  Bond and Luck speculate from the depressed neutron capture element abundances that these stars were originally very metal poor, but are now somewhat hydrogen depleted through mass loss and hydrogen burning.  Analysis of

IRAS observations of RV Tauri stars show that these objects have recently (i.e.
within the last 500 years) significantly decreased their mass loss rates from a
level near $10^{-5} M_\odot$ yr$^{-1}$ Jura (1986).  These stars may have recently undergone an
episode of very rapid mass loss, as one might expect for stars in the last stages of
AGB star evolution.  Jura speculates that the RV Tauri stars will become planetary
nebulae if they evolve to high temperatures to photo-ionize the surrounding
circumstellar material before it dissipates.


Summary


   Great progress in observational programs and theoretical studies in recent years
has provided a structure within which we can understand some of the processes which
create the peculiar red giants.  The origin of most groups of the peculiar warm
giants can be explained through some form of mass transfer from a more massive and
more evolved companion.  The compositions of the AGB stars can be understood through
the mechanism of mixing from the thermal pulses of stars with helium and hydrogen
burning shells.  The peculiar AGB stars form a sequence M-MS-S-(SC)-C which is
consistent with this picture.  Many of the peculiar supergiants can be understood
within the context of post-AGB evolution.

   Many fundamental problems remain to be solved, and much of the basic abundance
data required to guide us are still missing.  Some important problems yet before
include a) the question of internal mixing on the main sequence to account for the
abundance changes seen in the first dredge-up.  This problem may have application to
the question of the lithium rich giants and the weak G-band stars as well.  b) The
origin of the early R stars.  If they do arise from a violent helium-core flash, we
may learn something about the physics of that event.  c) We need to obtain much more
basic data about the other groups of peculiar warm giants, such as the CH stars and
the CH-like stars, and the lithium-rich and weak G-band stars.  Are they binaries?
What about their compositions?  d) How do the M, MS, S, and C stars actually
accomplish the mixing required between the helium shell flashes to modify their
surface compositions?  Can we reproduce their detailed abundance evolution?  Can we
distinguish among models of the s-process environment using the pattern of
enrichment for different species?  We need to obtain much more accurate (and
reliable) CNO, iron and heavy element abundances than currently available, and for
many more stars and more species.  e) What happens to the heavy oxygen isotopes
during the AGB?  Are they destroyed, and how?  f) How do the SC stars and the J-type
carbon stars fit into the general scheme of things?  Their compositions don't match
our predictions very well.  For the J-type carbon stars, both shell flashes and hot
bottomed convection appear to be ruled out.  g) What happens to stars at the end of
their AGB evolution - do they finally lose their envelopes, and how?  How do they

evolve to planetary nebulae?  Are the R CrB stars and the RV Tauri and 89 Her stars
involved?

The last five years has seen great steps forward in our understanding of the
evolution of late type giants stars.  This Colloquium happens at the right time for
us to step back to assess what we have learned and to figure out where to to from
here.  I anticipate that with the development of new IR detectors, and the
construction of new high resolution IR spectrographs with much fainter limiting
magnitudes than now available, we will see much more progress in the next five
years.


A. M. Boesgaard 1970, Ap. J., 161, 1003.
H. E. Bond and R. E. Luck 1987b, "Proceedings of the ESO Workshop on Stellar
    Evolution and Dynamics in the Outer Halo of the Galaxy," April, 1987; Garching.
J. A. Brown 1987, Ap. J., 317, 701.
R. M. Catchpole and M. W. Feast 1971, M.N.R.A.S., 154, 197.
C. R. Cowley and P. L. Downs 1980, Ap. J., 236, 648.
D. S. P. Dearborn, P. E. Eggleton, and D. N. Schramm 1976, Ap. J., 203, 455.
J. F. Dominy 1984, Ap. J. Suppl., 55, 27.
J. F. Dominy and G. Wallerstein 1987, Ap. J., 317, 810.
J. F. Dominy, G. Wallerstein, and N. B. Suntzeff 1986, Ap. J., 300, 325.
Y. Fujita and T. Tsuji 1965, Publ. Dom. Astro. Obs., 12, 339.
Y. Fujita and T. Tsuji 1977, P.A.S.J., 29, 711.
K. H. Hinkle and W. W. G. Scharlach 1985, in "Cool Stars with Excesses of Heavy
    Elements," M. Jaschek and P. C. Keenan, eds.; Dordrecht: Reidel, pp. 255-261.
M. J. Harris and D. L. Lambert 1984, Ap. J., 285, 674.
M. J. Harris, D. L. Lambert, K. H. Hinkle, B. Gustafsson, and K. Eriksson 1987, Ap.
    J., 316, 294.
M. J. Harris, D. L. Lambert, and V. V. Smith 1985a, Ap. J., 292, 620.
M. J. Harris, D. L. Lambert, and V. V. Smith 1985b, Ap. J., 299, 375.
I. Iben and A. Renzini 1984, Phys. Letters, 105, 329.
I. Iben and A. V. Tutukov 1985, Ap. J. Suppl., 58, 661.
H. K. Johnson, G. T. O'Brien, and J. C. Climenhaga 1982, Ap. J., 254, 175.
M. Jura 1986, Ap. J., 309, 732.
N. Kovacs 1985, Astron. Ap., 150, 232.
K. Krishnaswamy and C. Sneden 1984, P.A.S.P., 97, 407.
D. L. Lambert 1985, in "Cool Stars with Excesses of Heavy Metals," M. Jaschek and P.
    C. Keenan, eds.; Dordrecht: Reidel; pp. 191-219.
D. L. Lambert 1986, in "Hydrogen Deficient Stars and Related Objects," K. Hunger, D.
    Schonberner, and N. K. Rao, eds.; Dordrecht: Reidel; pp. 127-150.
D. L. Lambert and J. F. Dominy 1980, Ap. J., 235, 114.
D. L. Lambert, B. Gustafsson, K. Eriksson, K. H. Hinkle 1986, Ap. J. Suppl., 62,
    373.
D. L. Lambert and L. M. Ries 1981, Ap. J., 248, 228.
D. L. Lambert and S. R. Sawyer 1984, Ap. J., 283, 192.
R. E. Luck 1982, P.A.S.P., 94, 811.
R. E. Luck and H. E. Bond 1982, Ap. J., 259, 792.
R. E. Luck and C. Sneden 1986, P.A.S.P., 98, 320.
I. Little-Marenin and S. Little 1979, A. J., 84, 1374.
T. Lloyd Evans 1986, M.N.R.A.S., 220, 723.
E. M. McCabe, R. Connon Smith, and R. E. S. Clegg 1979, Nature, 281, 263.
R. D. McClure 1985, in "Cool Stars with Excesses of Heavy Metals," M. Jaschek and P.
    C. Keenan, eds.; Dordrecht: Reidel, pp. 315 - 330.
P. W. Merrill 1952, Ap. J., 116, 21.
M. Parthasarathy, C. Sneden, and E. Bohm-Vitense 1984, P. A. S. P., 96, 44.
M. H. Pinsonneault, C. Sneden, and V. V. Smith 1984, P.A.S.P., 96, 239.
M. Schwarzschild and R. H_*:arm 1965, Ap. J., 142, 855.

V. V. Smith 1984, Astron. Ap., 132, 326.
V. V. Smith and D. L. Lambert 1985, Ap. J., 294, 326
V. V. Smith and D. L. Lambert 1986, Ap. J., 311, 843.
C. Sneden 1983, P.A.S.P., 95, 745.
C. Sneden and H. Bond 1976, Ap. J., 204, 810.
C. Sneden, D. L. Lambert, and C. A. Pilachowski 1981, Ap. J., 247, 1052.
C. Sneden and C. A. Pilachowski 1984, P.A.S.P., 96, 38.
C. Sneden, C. A. Pilachowski, and D. A. VandenBerg 1986, Ap. J., 311, 826.
J. Tomkin, C. Sneden, and P. L. Cottrell 1984, P.A.S.P., 96, 609.
S. Torres-Peimbert and G. Wallerstein 1966, Ap. J., 146, 724.
T. Tsuji 1985, in "Cool Stars with Excesses of Heavy Elements," M. Jaschek and P. C.
    Keenan, eds., Dordrecht: Reidel, pp. 295-300.
T. Tsuji 1986, Astron. Ap., 156, 8.
K. Utsumi 1985, in "Cool Stars with Excesses of Heavy Elements," M. Jaschek and P.
    C. Keenan, eds., Dordrecht: Reidel, pp. 243-247.
K. Utsumi 1988, this volume.
H. J. Walker 1986, in "Hydroden Deficient Stars and Related Objects," K. Hunger, D.
    Schonberner, and N. K. Rao, eds.; Dordrecht: Reidel; pp. 407-419.
G. Wallerstein and C. Sneden 1982, Ap. J., 255, 577.
P. G. Wannier 1985, Proceedings of the ESC Workshop on "Production and Distribution
    of C, N, O Elements," Garching, pp. 233- 247.
R. F. Webbink 1984, Ap. J., 277, 355.
Y. Yamashita 1972, Ann. Tokyo Astron. Obs., 13, 169.
Y. Yamashita 1975, P. A. S. J., 27, 325.
A. C. Zook 1985, Ap. J., 289, 356.

# CHEMICAL PECULIARITIES, MASS LOSS, AND FINAL EVOLUTION OF AGB STARS IN THE MAGELLANIC CLOUDS


P.R. Wood
Mount Stromlo and Siding Spring Observatories
Private Bag, Woden P.O., A.C.T. 2606
Australia


## I.  INTRODUCTION

The Magellanic Clouds are sufficiently close that evolved stars which exhibit chemical peculiarities and the effects of mass loss can be readily observed.  Such objects include carbon stars, S stars, long-period variables, OH/IR stars and planetary nebulae. Because of the relatively well-known distances of the Magellanic Clouds, the intrinsic luminosities of these objects can be accurately determined, in contrast to the situation in the Galaxy where the great majority of asymptotic giant branch (AGB) stars occur in the field population.  In this review, observations of AGB stars in the Magellanic Clouds will be discussed with particular reference to those features which can shed light on mass loss and chemical peculiarities resulting from stellar evolution.

## II.  CHEMICAL PECULIARITIES IN AGB STARS

*a. Mechanisms for producing abundance peculiarities*

There are four processes which can produce chemical peculiarities in AGB stars: the first, second and third dredge-up events and envelope (or hot-bottom) burning.

The first and second deedge-up events occur prior to the AGB phase when the star moves from core hydrogen or helium burning, respectively, on to the red giant branch.  At this time, envelope convection extends downward in mass into regions which have undergone some CN cycle processing with the result that material enhanced in $^{14}N$ and $^{4}He$ and deficient in $^{12}C$ is mixed into the envelope.  The first and second dredge-up events have been studied in detail by Becker and Iben (1980).

The third dredge-up occurs as a result of the envelope convection extending downward in mass during a helium shell flash, with the result that the products of both the CNO cycle and the 3-$\alpha$ process are brought to the stellar surface (Iben 1975).  As a result of the third dredge-up, the stellar surface is enriched in $^{12}C$ and s-process elements.

The final process which can alter the surface abundances of AGB stars is envelope or hot-bottom burning (Scalo, Despain and Ulrich 1975; Iben 1975).  In the more massive and luminous AGB stars, envelope convection during the evolutionary phase between helium shell flashes may extend deep enough that the CN cycle is active at the bottom of the convective region.  As a result, the entire envelope may be slowly cycled through the CN cycle resulting in the conversion of $^{12}C$ to $^{14}N$ and $^{1}H$ to $^{4}He$.

A detailed review of the above processes is given by Iben and Renzini (1983).  Here only effects of the third dredge-up and envelope burning will be considered.

*b. Observations of the third dredge-up*

An excellent observational demonstration of the operation of the third dredge-up is exhibited by the intermediate age clusters in the Magellanic Clouds where, going from lower to higher luminosity on the AGB we pass through the sequence of spectral types M to S to C (Bessell, Wood and Lloyd-Evans 1983; Lloyd Evans 1984).  Low on the AGB where there have been few or no helium shell flashes, very little 3-$\alpha$ processed material has been dredged up and the AGB stars have essentially normal abundances (apart from the effects of the first and second dredge-up events) and are of spectral type M (or K).  However, at the tip of the sequence of oxygen-rich stars in these clusters we find stars whose spectra show strong bands of the s-process element Zr (these stars are MS or S stars).  Model atmosphere calculations by Brett and Bessell (1987) indicate that these strong bands are not due to a temperature effect but are due to real abundance enhancements of Zr.  Thus these stars provide direct evidence for the operation of the third dredge-up in oxygen-rich AGB stars.

At luminosities above the MS and S stars in the Magellanic Cloud clusters, we find cool (N type) carbon stars in which the third dredge-up has increased the C/O ratio to >1. There have been many studies of the carbon stars in the Magellanic Clouds; a good review and comparison of this work with theory is given by Iben and Renzini (1983), with some more recent work being found in Lattanzio (1986).

*c. Indirect evidence for envelope burning in AGB stars*

It is well known that the carbon stars in the Magellanic Clouds have an upper limit to their luminosity of $M_{bol} = -6$ (e.g. Mould and Aaronson 1987).  On the other hand, it is also known that oxygen-rich AGB stars in the LMC and SMC extend in luminosity right up to the AGB limit at $M_{bol} = -7.1$ (Wood, Bessell and Fox 1983; Hughes and Wood 1987).  Why are none of these stars carbon stars?  It seems that, in the Magellanic Clouds, there is considerable indirect evidence that envelope burning is preventing the formation of carbon stars with $M_{bol} < -6$ by converting $^{12}C$ dredged-up during helium shell flashes into $^{14}N$.

The first piece of evidence is the finding of Richer *et al* (1979) that the brightest carbon stars in the LMC have strong bands of $^{13}C$ indicating that a significant amount of CN cycling has been occurring in the envelopes of these stars.  From this result we might conclude that the brightest carbon stars are undergoing envelope burning at a rate which is fast enough to produce a significant amount of $^{13}C$ but not fast enough to convert much of the dredged-up $^{12}C$

to $^{14}$N and so prevent C/O exceeding 1. In the more luminous AGB stars, the envelope burning process works fast enough to prevent carbon star formation.

The second piece of evidence for envelope burning in upper AGB stars in the Magellanic Clouds comes from examination of the spectra of SMC long-period variables (LPVs) which are on the AGB and have $M_{bol} < -6$. The pulsation (current) masses of these stars are ~3.5 $M_\odot$ while the core masses are ~1 $M_\odot$, giving an envelope mass of ~2.5 $M_\odot$ (Wood, Bessell and Fox 1983). The O/H and C/O ratios in the gas from which these stars formed were ~$10^{-4}$ and ~0.14, respectively (Dufour 1983). Applying the enrichment/depletion factors for the first and second dredge-ups given by Becker and Iben (1979), we find that C/O $\approx 0.1$ at the beginning of the thermally pulsing AGB phase. Brett and Bessell (1987) have compared the spectra of the above SMC AGB stars with spectra computed from model atmosphere calculations. They found that it was not possible to tell directly from the spectra whether envelope burning had been operating in a thermally pulsing AGB star. The important quantity determining the nature of the stellar spectrum (particularly the strength of TiO bands) is the amount of $^{16}$O not tied up in CO; this quantity can be adjusted to any desired value by altering the relative amounts of dredge-up and envelope burning. Although large increases in $^{14}$N abundance gave rise to stronger CN bands, the effect was difficult to disentangle from variations in the C/O ratio. However, Brett and Bessell (1987) were able to deduce that the AGB stars they looked at have C/O $\approx 0.7$. Assuming that there has been no envelope burning on the AGB, this would imply that ~0.0018 $M_\odot$ of $^{12}$C has been dredged up since the beginning of the thermally pulsing phase. From the calculations of Iben (1977), this amount of dredge-up should occur in only 18 shell flashes, corresponding to only ~$2\times10^4$ years on the AGB during which $M_{bol}$ would have increased by only ~0.02 mag. This is such a brief interval in magnitude and time on the AGB (given that the mass loss rates for optically visible LPVs such as these are $<10^{-5}$ $M_\odot$ yr$^{-1}$ e.g. Knapp and Morris 1985) that it is hard to believe that the stars could not stay there for 50% longer, by which time there would have been sufficient dredge-up to convert the star to a carbon star. A simple way to allow these stars to remain on the AGB as oxygen-rich stars for a time interval sufficient for their magnitudes to increase by a few tenths or for mass loss to dissipate their envelopes is to have envelope burning occur in them.

In summary, we have two pieces of indirect evidence for the operation of the envelope burning mechanism in upper AGB stars, but no direct measurement of the $^{14}$N abundance in these stars. The importance of the operation of the envelope burning mechanism is that it provides a method of producing primary nitrogen.


III. MASS LOSS FROM MAGELLANIC CLOUD AGB STARS

Direct measurements of mass lass rates from AGB stars in the Magellanic Clouds have not yet been made, although the stellar wind has been observed in a number of OH/IR stars.  However, deductions about the total amount of mass lost from AGB stars in the Magellanic Clouds can be made from observations of long-period variables, the tips of AGB sequences in clusters, AGB luminosity functions, and planetary nebulae.  These aspects of mass loss will now be discussed.

*a.  Long-period variables*

The long-period variables (LPVs) in the Magellanic Clouds provide some of the most definitive evidence available for large-scale mass loss from AGB stars.  Termination of the AGB during the LPV phase is clearly demonstrated in Figures 6 and 7 of Wood, Bessell and Fox (1983) (WBF) who show that the optically visible LPVs in the Magellanic Clouds "disappear" from the AGB with a considerable amount of mass still left in their hydrogen-rich envelopes.  This disappearance is presumably due to the formation of a thick circumstellar shell which hides the central star from view.  WBF find that stars initially less massive than ~5 $M_\odot$ eject their envelopes to become planetary nebulae, while the more massive AGB stars reach the AGB limit, corresponding to the core mass attaining the Chandrasekhar limiting mass of 1.4 $M_\odot$, and explode as supernovae (this type of supernova has been designated Type $1^1/2$ by Iben and Renzini 1983).  The most massive planetary nebulae produced in this picture is ~2.1 $M_\odot$, while the Type $1^1/2$ supernovae have hydrogen envelopes of ~2.1 to 7 $M_\odot$ at explosion.

Given the current prominance of SN1987A, it is interesting note that there are many LPVs in the LMC which are core helium burning supergiants with $M_{bol} = -8$, close to the luminosity of the progenitor of SN1987A (this is, of course, much more luminous than the AGB limit).  The main sequence mass of this progenitor star must have been about 20 $M_\odot$ (see contributions in these proceedings), whereas the pulsation masses indicated in Figure 6 of WBF indicate that in the red supergiant stages, LPVs with $M_{bol} = -8$ have masses in the range ~7 to ~20 $M_\odot$.  If these pulsation masses are at all reliable (see WBF) then they indicate that there has been considerable mass loss from massive red supergiants in the LMC.  SN1987A should therefore be surrounded by a shell containing >10 $M_\odot$ of mass loss material, although the distance of this shell from the supernova is very uncertain.

Deep searches for LPVs in the LMC (Wood, Bessell and Paltoglou 1985; Reid Glass and Catchpole 1987) show that the LPV population is dominated by objects of age a few times $10^9$ years which, at the time of their exit from the AGB, have envelope mass ~0.35 $M_\odot$, core mass ~0.65 $M_\odot$ and $M_{bol} = -5$.  These objects are presumably the immediate precursors of most of the planetary nebulae in the Magellanic Clouds, and their properties will be compared below with those of the planetary nebulae.

*b.  AGB tips and AGB luminosity functions*

By observing the luminosities of the brightest AGB stars in clusters of different age in the Magellanic Clouds, the final AGB luminosity can be found as a function of age or initial mass; a recent summary of this work is given by Mould and Aaronson (1987).  For clusters of age $2-3 \times 10^9$ years, the final AGB luminosity is $M_{bol} = -5$, in agreement with the results for LPVs (although the brightest cluster stars generally are not *known* variables). However, in the youngest of the rich clusters with AGB stars, which have main sequence turnoff masses of ~5 $M_\odot$, the most luminous AGB stars have $M_{bol} = -6$. Similarly, Reid and Mould (1985) have obtained luminosity functions for AGB stars away from the bar of the LMC which they interpret as indicating that there seems to be a limit of $M_{bol} = -6$ for AGB stars.

However, there are two sets of observations which indicate that AGB stars do indeed reach the AGB limit $M_{bol} = -7.1$. Firstly, the results of WBF and Reid, Glass and Catchpole (1987) for the LPVs show that such stars do reach the AGB limit. Secondly, a luminosity function for red stars in the bar of the LMC obtained by Hughes and Wood (1987) shows that AGB stars extend to $M_{bol} = -7.1$, although there is a steep fall-off with luminosity above $M_{bol} = -6$. One explanation for the rapid fall-off in the number of AGB stars above $M_{bol} = -6$ is provided by Wood and Faulkner (1986) who show that envelope ejection will occur at this point in all but the most massive AGB stars due to the luminosity of the star exceeding the Eddington limit at the base of the hydrogen-rich envelope during the surface luminosity peak of a helium shell flash.

*c. OH/IR and IR stars in the Magellanic Clouds*

The IRAS point source catalog has provided many candidate OH/IR stars in the LMC, and to a lesser extent, in the SMC. Elias, Frogel and Schwering (1986) have found that two of the LMC IRAS sources are red supergiants with $M_{bol} \sim -9$ and thick circumstellar shells. Wood, Bessell and Whiteoak (1986) showed that one of these objects was an OH/IR star with 1612 MHz maser emission. In a further study of IRAS sources in the LMC, Wood *et al* (1987a) have found a further 4 AGB stars in the LMC with thick circumstellar shells, with three of these objects being OH maser sources. An interesting feature of these observations is that the expansion velocities of the stellar winds giving rise to the mass loss can be derived from the 1612 MHz line profiles. In the LMC AGB stars, the expansion velocities are ~12 km s$^{-1}$, only half the value in similar Galactic OH/IR stars. Wood, Bessell and Whiteoak (1986) argue that this is due to the different metal abundances in the LMC and the Galaxy.

*d. Planetary Nebulae*

The masses of planetary nebulae give a direct measure of the amount of rapid mass loss at the end of the  AGB phase of evolution. Masses for planetary nebulae have been recently derived by Wood *et al* (1987b), Meatheringham, Dopita and Morgan (1987), and Barlow (1987), using either observed angular diameters or electron densities, together with H$\beta$ fluxes. The results of these studies show that planetary nebulae of small size and high density have small ionized masses; the central stars of such objects have presumably only recently left the AGB. There is an evolutionary sequence from these small, dense nebulae to larger, less dense nebulae. The maximum mass of ionized material found in the nebulae is close to the value of ~0.35 $M_\odot$ predicted as the mass ejected from the dominant population of intermediate age LPVs in the Magellanic Clouds.

The luminosities of the central stars derived from Hβ fluxes (Wood *et al* 1987b) are all less than $M_{bol} = -5$, the luminosity at which the LPVs of age $2\text{-}3 \times 10^9$ years leave the AGB. This is as expected, as there are a number of processes (such as dust absorption in the nebula, being optically thin) which will act to reduce the *derived* bolometric luminosities of the central stars.

The combined planetary nebula and OH/IR star data allow one to derive a mean mass loss rate for the "superwind" which leads to planetary nebula formation in the stars of age $2\text{-}3 \times 10^9$ years in the LMC. The total nebula mass is $\sim 0.35\ M_\odot$: the nebula becomes fully ionized at a radius $R \sim 0.1\mathrm{pc}$: this is assumed to occur when the volume occupied by superwind material has become fully ionized: with a superwind expansion velocity of 12 km s$^{-1}$, the time from beginning of the superwind to full ionization of the nebula is thus $\sim 8000$ years: hence, the mean mass loss rate during the superwind mass loss phase is $\sim 0.35/8000 = 4 \times 10^{-5}\ M_\odot$ yr$^{-1}$. This value is similar to values derived for Galactic stars (e.g. Renzini 1980). Finally, it is worth noting that the time of 8000 years is roughly the time taken for a planetary nebula nucleus of $0.65\ M_\odot$ (the mass of a typical Magellanic Cloud planetary nebula nucleus as deduced from the LPV data) to evolve from $T_{eff} = 10^4$ K to $\sim 10^5$ K (near the blue end of its evolution) where it can ionize the surrounding nebula (Wood and Faulkner 1987). Hence, it appears that the time taken for the planetary nebula nucleus to evolve from the AGB to $T_{eff} = 10^4$ K is quite small ($<$ a few thousand years).


REFERENCES

Barlow, M.J. 1987, M.N.R.A.S., **227**, 590.

Becker, S.A. and Iben, I. 1980, Ap.J., **237**, 111.

Bessell, M.S., Wood, P.R. and Lloyd Evans, T. 1983, M.N.R.A.S., **202**, 59.

Brett, J.M. and Bessell, M.S. 1987, Astr. Ap., submitted.

Dufour, R.J. 1983, in IAU Symposium 108, "Structure and Evolution of the Magellanic Clouds", eds. S. van den Bergh and K.S. de Boer (Reidel), p.353.

Elias, J.H., Frogel, J.A. and Schwering, P.B.W. 1986, Ap.J., **302**, 675.

Hughes, S.M.G. and Wood, P.R. 1987, Proc. Astr. Soc. Australia, in press.

Iben, I. 1975, Ap.J., **196**, 525.

Iben, I. 1977, Ap.J., **217**, 788.

Iben, I. and Renzini, A. 1983, Ann. Rev. Astr. Ap., **21**, 271.

Knapp, G.R. and Morris, M. 1985, Ap.J., **292**, 640.

Lattanzio, J.C. 1986, Ap.J., **311**, 708.

Lloyd Evans, T. 1984, M.N.R.A.S., **208**, 447.

Meatheringham, S.J., Dopita, M.A. and Morgan, D. 1987, Ap.J., submitted.

Mould, J.R. and Aaronson, M. 1987, Ap.J., **303**, 10.

Reid, N., Glass, I.S. and Catchpole, R.M. 1987, M.N.R.A.S., in press.

Reid, N. and Mould, J.R. 1985, Ap.J., **299**, 236

Renzini, A. 1980, in "Physical Processes in Red Giants", eds. I. Iben and A. Renzini (Reidel), p.431.

Richer, H.B., Olander, N. and Westerlund, B.E 1979, Ap.J., **230**, 724.

Scalo, J.M., Despain, K.M. and Ulrich, R.K. 1975, Ap.J., **196**, 805.

Wood, P.R., Bessell, M.S. and Fox, M.W. 1983, Ap.J., **272**, 99.

Wood, P.R., Bessell, M.S., Hughes, S.M.G., Whiteoak, J.B., Gardner, F.F. and Otrupcek, R.E. 1987a, in preparation.

Wood, P.R., Bessell, M.S. and Paltoglou, G. 1985, Ap.J., **290**, 477.

Wood, P.R., Bessell, M.S. and Whiteoak, J.B. 1986, Ap.J. Letters, **306**, L81.

Wood, P.R. and Faulkner, D.J. 1986, Ap.J., **307**, 659.

Wood, P.R., Meatheringham, S.J., Dopita, M.A. and Morgan, D.H. 1987b, Ap.J., **320**, 178.

Nucleosynthesis and Mixing in Low- and Intermediate-Mass AGB Stars


David Hollowell and Icko Iben, Jr.
Astronomy Dept., University of Illinois
Urbana, Illinois
61801 U. S. A.


ABSTRACT.  The existence of carbon stars brighter than $M_{bol}=-4$ can be understood in terms of dredge up in thermally pulsing asymptotic giant branch (AGB) stars.  As a low- or intermediate-mass star evolves on the AGB, the large fluxes engendered in a helium shell flash cause the base of the convective envelope to extend into the radiative, carbon-rich region, and transport nucleosynthesis products to the stellar surface.  Numerical models indicate that AGB stars with sufficiently massive stellar envelopes can become carbon stars via this standard dredge-up mechanism.  AGB stars with less massive stellar envelopes can become carbon stars when carbon recombines in the cool, carbon-rich region below the convective envelope.
Neutron capture occurs on iron-seed nuclei during a shell flash, and the products of this nucleosynthesis are also carried to the stellar surface.  The conversion of $^{22}$Ne into $^{25}$Mg can initiate neutron capture nucleosynthesis in large-core mass AGB stars, but only if these stars can survive their large mass loss rates.  The current estimates of nuclear reaction rates do not allow for appreciable neutron capture nucleosynthesis via the $^{22}$Ne source in lower mass AGB stars.  The carbon recombination that induces dredge up in AGB stars of small envelope mass, however, also induces mixing of $^1$H and $^{12}$C in such a way that ultimately a $^{13}$C neutron source is activated in these stars.  The $^{13}$C source can provide an abundant supply of neutrons for the nucleosynthesis of both light and heavy elements.  While the existence of neutron-nucleosynthesis products in AGB stellar atmospheres can be understood qualitatively in terms of an active neutron source, the combination of nuclear reaction theory and evolutionary models has yet to provide quantitative agreement with stellar observations.


I.   INTRODUCTION


     In this review we wish to discuss how observations of AGB stars can be used to determine the manner in which heavy elements are created during a thermal pulse, and how these heavy elements and carbon are transported to the stellar surface.  In particular we wish to study how the periodic hydrogen and helium shell burning above a degenerate carbon-oxygen (C-O) core forms a neutron capture nucleosynthesis site that may eventually account for the observed abundance enhancements at the surfaces of AGB stars.  In section II we discuss the nucleosynthesis provided by stellar evolution models (for a general review see [1]).  In section III we discuss the isotopic abundances provided by nucleosynthesis reaction network calculations (see [2, 3]).  In section IV we discuss how observations of AGB stars can be used to discriminate between the neutron capture nucleosynthesis sources (see [4]).  And in section V we note some of the current uncertainty in this work.


II.  AGB EVOLUTION


     A thermally pulsing AGB star spends ~80% of each pulse burning hydrogen in a thin shell above a hydrogen-depleted core.  The H-burning shell remains at a radius ~0.01 R$\odot$ and produces the AGB surface luminosity (~$10^4$ L$\odot$) as it burns outward (in

mass) through the star.  The byproducts of CNO burning in this shell (principally .
$^4$He, with some $^{14}$N) are "dumped" onto the temporarily dormant helium shell ($L_{H_e}$ ~
10 L☉) that surrounds the degenerate C-O core.  Although the hydrogen-burning shell
is not an active site of neutron capture nucleosynthesis, it is where raw materials
for this nucleosynthesis, such as $^4$He and $^{14}$N, are created.  This $^4$He will be fuel
for the following thermal pulse and the $^4$He and $^{14}$N may be reactants in the neutron
capture nucleosynthesis reactions.

As the helium-rich shell below the hydrogen-burning shell becomes more
massive, the temperature and density at the base of this shell increase until $3\alpha$
burning begins.  Due to the compact, thin-shell nature of this burning, a thermal
runaway develops [5]; the maximum temperature in the helium-burning shell can reach
$250 - 400 \cdot 10^6$ K and the luminosity of the shell can reach $10^7 - 10^8$ L☉.  The high
temperatures and large luminosities cause a convective shell to form in the He-rich
region.  This hot convective shell first forms in the lower, He-burning regions and
the temperature at the base of the shell, $T_{csb}$, is initially $\approx 100 \cdot 10^6$ K.  The
convective shell then grows inward with $T_{csb}$ increasing to $250 - 400 \cdot 10^6$ K.  It
also grows outward and engulfs almost all of the products of hydrogen burning left
behind by the advancing hydrogen-burning shell.

It is this hot convective shell that can be an active site of neutron capture
nucleosynthesis in a thermally pulsing AGB star [6].  Convection mixes the
nucleosynthesis raw materials for $\alpha$ capture and for neutron producing reactions to
the hot base of the convective shell.  The material mixed to the base contains
heavy elements that originally were in the stellar envelope (perhaps in a solar
system distribution), as well as heavy elements from previous thermal pulses
(perhaps in a neutron-rich distribution).  Processed material is simultaneously
mixed away from the shell base to cooler outer regions of the shell, and this
material contains $\alpha$-burning byproducts and a rearranged heavy element distribution
(if neutron capture nucleosynthesis has occurred).

When the temperature at the base of the convective shell becomes $\geq 250 \cdot 10^6$ K,
the $^{14}$N in the convective shell is rapidly converted to $^{22}$Ne by two $\alpha$ captures and
a $\beta^-$ decay.  If the base temperature is in excess of $300 - 350 \cdot 10^6$ K, the Coulomb
barrier between $^{22}$Ne and $^4$He can be overcome and the $^{22}$Ne$(\alpha,n)^{25}$Mg reaction acts as
a neutron source in the convective shell.  Extant evolutionary models, however,
suggest that only the high-core mass ($M_{core} \geq 0.8$ M☉, $M_{bol} \leq -6.0$) AGB stars can
attain these high convective shell temperatures long enough to produce a sufficient
number of neutrons for substantial neutron capture nucleosynthesis.  Models of low-
core mass AGB stars ($M_{bol} \geq -6.0$) do not strongly activate the $^{22}$Ne source since
their convective shells are too cool.  A possible source of neutrons in the low-
metallicity, low-core mass AGB stars is the $^{13}$C$(\alpha,n)^{16}$O reaction [7].  We will
shortly discuss how $^{13}$C can be formed in an AGB star just after a thermal runaway.

As the energy from the pulse escapes from the hot convective shell (now C and
He rich), the region between the H and He shells becomes radiative.  As the pulse
energy flows through the base of the convective envelope, the convective envelope
moves inward in mass.  If the luminosity is large enough (i.e. if the AGB core mass
is large enough) the convective envelope will extend into the region that was
previously occupied by the hot convective shell [6, 8, 9].  The convective envelope
then mixes the nucleosynthesis byproducts (particularly $^{12}$C, and perhaps neutron
irradiated material) to the stellar surface in the "classical" dredge-up mechanism.

When this dredge-up phase occurs, the extinct hydrogen shell (which marks the
discontinuity between the H-rich envelope and the C- and He-rich interior) has
expanded to ~ 1 R☉ due to the thermal pulse heating.  During this phase the carbon
nuclei will begin to recombine with free electrons in those regions of the AGB star
that are cooling to temperatures $\leq 10^6$ K [10, 11].  In AGB stars with a large
envelope ($M_{env}$ ~ 1 M☉) temperatures below $10^6$ K only occur in the H-rich, C-poor,
convective envelope, and carbon recombination has little effect.  In AGB stars with
a small envelope ($M_{env}$ ~ 0.1 M☉) these temperatures will be found in the C- and He-
rich interior just below the H-rich envelope.  In these stars carbon recombination
will increase the local opacity, due to the bound-bound and bound-free transitions
of carbon and a convective region forms that mixes hydrogen downward to the
carbon/helium rich region and mixes carbon upward toward the hydrogen-rich region.
Our calculations show that if enough carbon is transported upward into sufficiently
cool regions, the convective envelope will mix with these recombination regions and

carbon will be dredged up to the surface.  Independent of whether or not carbon is
mixed to the stellar surface, the hydrogen mixed downward will form a region $\sim 10^{-4}$
M$\odot$ in size which is .1% H (by mass) in a $\sim$ 20% C, 80% He mixture.

In all AGB stars, as the dredge-up phase ends and as helium burning (now in a
radiative structure) above the C-O core decreases, the C/H discontinuity will
contract to $\sim$ 0.01 R$\odot$.  Hydrogen will begin to burn again, and will continue to
burn until conditions are ready for the next thermal pulse.  If a low-envelope-mass
AGB star has a "pocket" of .1% hydrogen and 20% carbon as previously described, the
pocket will also contract and the hydrogen will burn with $^{12}$C in the CNO cycle.  It
will, however, only burn to form $^{13}$C as there is too little $^{1}$H and too much $^{12}$C for
further CNO burning.  In this way a low-envelope-mass AGB star forms a pocket of
$^{13}$C which eventually will be engulfed by a hot convective shell during the
following thermal pulse.  Since the pulse temperature is always high enough
(>150$\cdot$10$^{6}$ K) to overcome the Coulomb repulsion between $^{13}$C and $^{4}$He nuclei, the
$^{13}$C$(\alpha,n)^{16}$O reaction will readily convert this $^{13}$C matter into $^{16}$O and neutrons -
the only requirement is that the AGB star produce a sufficient quantity of $^{13}$C.


III. AGB NEUTRON CAPTURE NUCLEOSYNTHESIS


While stellar evolution models describe neutron production, the calculations
must be supplemented with nucleosynthesis calculations to determine what abundances
of heavy elements the $^{13}$C or $^{22}$Ne neutron sources will produce.  While analytic
theory can approximate the production of heavy elements [3], numerical modelling of
$\sim$ 500 isotopes is used for detailed comparison of stellar evolution theory and
observation.  In general, the destruction of an isotope between Fe and Bi occurs
due to that isotope capturing a neutron or $\beta^-$ decaying, while the creation of an
isotope will be due to the neutron capture or $\beta^-$ decay of lighter elements.

The neutron capture rate in the AGB convective shell is primarily dependent
upon the neutron density $N_n$.  Higher neutron densities tend to build up the
neutron-rich isotopes.  The $\beta^-$ decay rate of an unstable isotope can, in some
instances, be sensitive to the temperature at the base of the convective shell,
$T_{csb}$.  Higher temperatures mean greater excitation of low-lying nuclear levels from
which $\beta^-$ decay may proceed much more rapidly than from the nuclear ground state.
In addition, the lifetime of the convective shell describes how long the
nucleosynthesis will occur — a larger shell lifetime allows more light elements to
be transmuted into heavy elements.  In this review we will consider how $N_n$ and $T_{csb}$
differ between the $^{13}$C and $^{22}$Ne neutron sources, and how this difference affects
the heavy element nucleosynthesis.

The $^{22}$Ne source, as we recall, operates at a significant rate only in AGB
stars of large core mass.  These stars have high maximum shell temperatures and
hence have more rapid neutron production and large neutron densities [12, 13, 14].
Since the $^{22}$Ne reaction rate is temperature dependent, $N_n$ will vary as the shell
temperature varies.  A gradual temporal change in $N_n$ during a pulse will be found
in models with a gradual temporal change in $T_{csb}$ during a pulse.  The maximum shell
temperature during a thermal pulse and the maximum value of $N_n$ during a thermal
pulse are both functions of core mass.  For an AGB star with a 0.8, 1.0, or 1.2 M$\odot$
C-O core, the maximum neutron density in the convective shell would be $10^8$, $10^{10}$,
or $10^{11}$ n/cm$^3$ respectively, and thus the $^{22}$Ne source can provide an s-process
nucleosynthesis environment or an environment intermediate between those which
produce s- and r-process distributions [16].  Of course, high maximum shell
temperatures can, in many instances, accelerate the conversion of nuclear neutrons
to nuclear protons by $\beta^-$ decays, and thus with the $^{22}$Ne source there is a coupling
between the neutron capture nucleosynthesis and the $\beta^-$ decay of the isotopes.

The $^{13}$C source liberates neutrons as soon as the $^{13}$C pocket is mixed into the
hot convective shell (typically at $T_{csb} \approx 150 \cdot 10^6$ K).  The release of neutrons is
essentially immediate, depending only upon the abundance of $^{13}$C in the shell and
not on shell temperature.  The abundance of $^{13}$C in the shell is a function of the
small amount of $^{1}$H that was mixed into a $^{12}$C-rich region (which occurred $\sim 10^5$
years earlier) and thus $N_n$ from the $^{13}$C source is mixing dependent, not temperature

dependent.  This $^{13}$C source provides a "pulse" of neutron irradiation that lasts
until all the $^{13}$C is incorporated into the convective shell (~ 1 year).  After
this, no more neutrons are released, and $N_n$ drops rapidly as neutrons are captured
on the elements.  However, after neutron production from the $^{13}$C source has ceased,
the $\beta^-$ decay rates that control the conversion of nuclear neutrons to nuclear
protons become temperature sensitive when the pulse temperatures approach their
maximum values ($250-300 \cdot 10^6$ K in low-core mass stars).  Hence, for the $^{13}$C source
there is not the same kind of coupling between neutron release and $\beta^-$ decay as
occurs with the $^{22}$Ne source.  Our calculations show the $^{13}$C source always produces
a neutron density ~ $10^{12}$ n/cm$^3$ in the convective shell, and thus the $^{13}$C source
always provides a nucleosynthesis environment intermediate between the classical s-
process and r-process environments.


IV.  AGB OBSERVATIONS


The existence of radioactive $^{99}$Tc seen in many AGB stars is strong evidence
that AGB stars are active sites of neutron capture nucleosynthesis.  With a half
life of $< 2 \cdot 10^5$ years, the technetium must have been created and mixed to the
stellar surface within the last ~ $10^6$ years [4, 15, 17, 18] (a time equal to or
longer than a thermal pulse time).  Although the lifetime of $^{99}$Tc is only 10 years
at the $350 \cdot 10^6$ K at which the $^{22}$Ne source operates [19], it has been found [20]
that $^{99}$Tc can always be made by the $^{22}$Ne source.  The coupling between $N_n$ and $T_{csb}$
always allows more $^{99}$Tc to be created by neutron capture nucleosynthesis than can
be destroyed by $\beta^-$ decay.  The $^{13}$C source, on the other hand, operates at
relatively low temperatures, where $^{99}$Tc always decays more slowly than neutrons are
captured.  We note, however, that neutron capture nucleosynthesis and $\beta^-$ decay are
not tightly coupled in the $^{13}$C source, and thus the $^{99}$Tc created in the convective
shell at an early, cool period of a thermal pulse (via the $^{13}$C source) may be
destroyed by $\beta^-$ decay at a later, hotter portion of a pulse.
From the observations of $^{99}$Tc in Mira type AGB stars and from assumed period-
luminosity relations [21] for such stars, it is concluded that $^{99}$Tc may be observed
in AGB stars as dim as $M_{bol} \approx -4.0$ ($M_{core} \approx 0.6$ M$\odot$).  As current AGB evolution
models show that the $^{22}$Ne neutron source is only active in AGB stars with $M_{bol} \leq$
$-6.0$ ($M_{core} \geq 0.8$ M$\odot$), one concludes that the $^{13}$C source must be the
nucleosynthesis source in the low-core mass objects.
If the observed absence of $^{99}$Tc lines in a significant percentage of C stars
[21] is not due to the difficulties in Tc line identification, the absence must be
due to a lack of atmospheric Tc.  This would imply that all of the surface material
in these evolved AGB stars was irradiated at a time long ago compared with the $^{99}$Tc
decay lifetime.  If the $^{13}$C source is active in these stars, then the thermal pulse
temperatures may have become high enough to insure the destruction of any $^{99}$Tc that
is created during the $^{13}$C source neutron capture nucleosynthesis.  Another
possibility is that during dredge up the convective envelope may reach deep enough
into the star to mix the C- and He-rich pocket (containing .1% hydrogen) to the
surface, creating the observed C star.  Since the pocket is destroyed, the $^{13}$C
neutron source will not exist, and no $^{99}$Tc will be formed during the following
thermal pulse.  Finally, independent of the nucleosynthesis properties of C stars,
some AGB models do show that dredge up can "turn off" during an advanced AGB stage
[8], such that neutron irradiated material is no longer transported to the surface.
In this case the surface abundance of Tc (brought to the surface during previous
dredge-up episodes) would drop as the envelope $^{99}$Tc slowly $\beta^-$ decays.
While observations of $^{99}$Tc suggest the occurrence of neutron capture
nucleosynthesis, isotopic observations can be used to probe more deeply the
conditions in the hot convective shell during this nucleosynthesis.  In particular,
the observations of the ZrO bands in S stars can be used to study the neutron flow
through the Zr isotopes, $^{90}$Zr $- ^{96}$Zr.  The existence of $^{93}$Zr (with a half life of
~$10^6$ years) at the surface of these stars confirms that they contain an active site
of neutron capture nucleosynthesis in their interiors.  Abundance measurements of
the stable isotope $^{96}$Zr have been used to determine the physical conditions that

existed during neutron capture nucleosynthesis.  The A=90 to 94 isotopes of Zr will capture neutrons and form $^{95}$Zr, but $^{95}$Zr is unstable and decays with a lifetime of $\approx 2$ months.  Only when neutron irradiation occurs with a high neutron flux ($N_n > 10^{10}$ n/cm$^3$) will the $^{95}$Zr capture a neutron and form the stable $^{96}$Zr before $\beta^-$ decaying.  For lower neutron densities ($N_n < 10^{10}$ n/cm$^3$) the $^{95}$Zr will $\beta^-$ decay before any $^{96}$Zr can be formed.  Since observations suggest little or no $^{96}$Zr exists in these S stars [4, 22, 23], the nucleosynthesis source must be a mild one.  Malaney [13] finds that the $^{13}$C source and the high-core mass $^{22}$Ne source always produce large amounts of $^{96}$Zr, and only a low-core mass $^{22}$Ne source ($M_{core} \approx 0.8$ M$\odot$, $N_n \sim 10^8$ n/cm$^3$) is able to match the low $^{96}$Zr abundance.

In order to match the observed overabundance of atomic zirconium with the relatively small enhancement of atomic zirconium that a low core mass $^{22}$Ne source provides, the observed stars must have a small envelope mass - we may be seeing almost pure irradiated matter in the envelope [13].  If the stellar surface consists largely of irradiated matter, and if the $^{22}$Ne$(\alpha,n)^{25}$Mg reaction was the source of neutrons, one would expect the $^{25}$Mg (or $^{26}$Mg, the neutron capture daughter) to be enhanced relative to $^{24}$Mg at the stellar surface [24, 25].  However no significant $^{25}$, $^{26}$Mg enhancements are seen in S stars [26], which suggests that the $^{22}$Ne source is not responsible for the Zr enhancement.  Unlike the $^{22}$Ne source, the daughter nuclei of the $^{13}$C source could never be noticed in an AGB atmosphere.  The increase in $^{16}$O abundance from the $^{13}$C$(\alpha,n)^{16}$O reaction is always negligible compared to the increase from the $^{12}$C$(\alpha,\gamma)^{16}$O reaction.


## V.   DISCUSSION


Why do some observations, when compared with theoretical calculations, suggest $^{22}$Ne is the active neutron source, and other observations suggest $^{13}$C is the source?  If a change in reaction theory were to allow the $^{22}$Ne source to produce neutrons at a lower temperature, all low-mass, low-luminosity AGB stars (independent of metallicity and envelope mass) could be active sites of neutron capture nucleosynthesis.  This would require, however, a modification in our understanding of light element nucleosynthesis, which must explain why $^{25}$Mg and $^{26}$Mg are not enhanced in some AGB stars.  Similarly, if further stellar evolution calculations show AGB stars with $M_{bol} = -4.0$ develop convective shells that are hot enough to activate the $^{22}$Ne source, we will still have to understand how heavy elements can appear neutron enriched, while $^{25}$, $^{26}$Mg does not appear enriched.  The $^{13}$C source, on the other hand, is dependent upon the stellar mixing, and hence our current understanding of this source may simply reflect our use of the mixing-length theory of convection in 1-D, quasistatic modelling.

It should also be realized that we have considered $^{13}$C source nucleosynthesis independently from $^{22}$Ne source nucleosynthesis.  The final element distribution that would result if both sources are active during the same thermal pulse in an AGB star may be different than the distributions created by the sources separately.  We note that, even with current reaction rates, the $^{22}$Ne source is activated at about the 1% level in the convective shell in low-core mass AGB models [8, 28], long after the $^{13}$C source has been exhausted.  Although the effect on elemental distribution of neutron irradiation from a two neutron source model has not been studied, it is known that the time evolution of neutron density does effect the resultant isotopic distribution [3, 27].  It has also been shown [13] that the $^{96}$Zr abundance is only sensitive to the maximum value of $N_n$ during a pulse, and hence the large neutron flux provided by the $^{13}$C source should always produce a large amount of $^{96}$Zr, whether or not the $^{22}$Ne source is active during a thermal pulse.

While significant uncertainty exists in determining the absolute stellar abundances of the heavy elements, the relative isotopic abundance of an atomic species can be determined somewhat independently of the atmospheric modelling uncertainties [4, 29].  Observations of different ZrO bands in a star do produce slightly different surface abundance values, but the uncertainty this introduces into our analysis of stellar nucleosynthesis is small compared to other uncertainties we have already discussed.  In addition, the heavy element

enhancements in the AGB stars may be due to nucleosynthesis that occurred before
the AGB phase (particularly during the helium core flash) [24, 30].  If such pre-
AGB nucleosynthesis occurs, our analyses of these abundances in terms of AGB
evolution may be faulty.

This work has been supported in part by the U.S. National Science Foundation,
grant AST 84-13371, by the U. S. Department of Energy, and by the National Center
for Supercomputing Applications.

REFERENCES

[ 1] Iben, I., Jr. and Renzini, A. 1983, Ann. Rev. Astr. Ap., 21, 271.
[ 2] Mathews, G. J., Takahashi, K., Ward, R. A., and Howard, W. M. 1986, Ap. J.,
     302, 410.
[ 3] Käppeler, F., Beer, H., Wisshak, K., Clayton, D. D., Macklin, R. L., Ward, R.
     A. 1982, Ap. J., 257, 821.
[ 4] Smith, V. V. 1987, preprint, to appear in Proceedings of the ACS Symposium on
     the Origin and Distribution of the Elements.
[ 5] Schwarzchild, M. and Härm, R. 1965, Ap. J., 142, 855.
[ 6] Iben, I., Jr. 1975, Ap. J., 196, 549.
[ 7] Iben, I., Jr. and Renzini, A. 1982, Ap. J. Lett., 263, L23.
[ 8] Iben, I., Jr. 1983, Ap. J. Lett., 275, L65.
[ 9] Lattanzio, J. 1987, Ap. J., 313, L15.
[10] Sackmann, I.-J. 1980, Ap. J. Lett., 241, L37.
[11] Iben, I., Jr. and Renzini, A. 1982, Ap. J. Lett., 259, L79.
[12] Iben, I., Jr. 1977, Ap. J., 217, 788.
[13] Malaney, R. A. 1987, preprint, to appear in Proceedings of the 2nd IAP
     Rencontre on Nuclear Astrophysics.
[14] Howard, W. M., Mathews, G. J., Takahashi, K., and Ward, R. A. 1986, Ap. J.,
     309, 633.
[15] Smith, V. V. and Wallerstein, G. 1983, Ap. J., 273, 742.
[16] Clayton, D. D. 1968, Principles of Stellar Evolution and Nucleosynthesis (U.
     Chicago Press: Chicago), 546.
[17] Merrill, P. W. 1952, Ap. J., 116, 21.
[18] Mathews, G. J., Takahashi, K., Ward, R. A., and Howard, W. M. 1986, Ap. J.,
     302, 410.
[19] Takahashi, K., Mathews, G. J., and Bloom, S. D. 1986, Phys. Rev. C, 33,  296.
[20] Takahashi, K., Mathews, G. J., Ward, R. A. and Becker, S. A. 1986,
     Nucleosynthesis and Its Implications on Nuclear and Particle Physics, ed. J.
     Audouze and N. Mathieu (Reidel: Dordrecht), 285.
[21] Little, S. J., Little-Marenin, I. R., and Bauer, W. H. 1987, Astr. J., 94,
     981.
[22] Zook, A. C. 1978, Ap. J. Lett., 221, L113.
[23] Peery, B. F., Jr., and Beebe, R. F. 1970, Ap. J., 160, 619.
[24] Truran, J. W. and Iben, I., Jr. 1977, Ap. J., 216, 797.
[25] Almeida, J. and Käppeler, F. 1983, Ap. J., 265, 417.
[26] Smith, V. V. and Lambert, D. L. 1986, Ap. J., 311, 843.
[27] Cosner, K., Iben, I., Jr. and Truran, J. W. 1980, Ap. J. Lett., 238, L91.
[28] Becker, S. A. 1981, Physical Processes in Red Giants, ed. I. Iben, Jr. and A.
     Renzini (Reidel: Dordrecht), 141.
[29] Johnson, H. R. 1985, Cool Stars with Excess of Heavy Elements, ed. M. Jaschek
     and P. C. Keenan (Reidel: Dordrecht), 271.
[30] Pilachowski, C. 1987, this volume.

ABUNDANCES IN J-TYPE CARBON STARS

Kazuhiko Utsumi

Department of Astronomy, Faculty of Integrated Arts and
Sciences, Hiroshima University, Hiroshima 730, Japan.

I) Introduction

It has been found by Utsumi(1985a,b) that in J-type carbon stars of C4-5 and
WZ Cas(C9,2J Li), abundances of s-process elements with respect to Fe are nearly
normal, while in normal carbon stars of C5-8, heavy metals are overabundant by factors
of 10-100, and rare-earth elements are overabundant by a factor of about 10.

In the MK system, most J-type stars are classified as C4-5,4-5 stars which show
very strong $C_2$ and CN bands. Yamashita(1972,1975) classified many C7-9 stars most of
which are CS or SC stars. His classification of C7-9J stars is mainly based on
$C^{12}C^{13}(0,1)$ band at 6168 A, $C^{13}N(4,0)$ band at 6260 A, and LiI 6708 A line. In most of
C7-9 stars, lines of s-process elements are greatly enhanced. It is a question if in
all J-type stars abundances of s-process elements are nearly normal or not.

II) Observation

We tested the classification of J-type stars using the spectra obtained at the
Okayama Astrophysical Observatory, in the region between 4400 A and 6800 A with a
dispersion of about 13.5 A/mm. Table 1 shows the stars studied together with the spectral type by the Harvard system and by Yamashita, and the $C^{13}$ index by Yamashita.

From the spectra between 6100 A and 6300 A, it is found that in all C7-9 stars
it appears as if $C^{13}$ features are very strong with low dispersions, because $C_2$ and CN
bands are weak, and strong low-excitation atomic lines coincide in chance with $C^{13}$
features. Table 2 shows the strong atomic lines which coincide in chance with these
molecular bands. It is better to use $C_2(1,0)$ bands to find J-type stars.

Table 1

| Star | Spectral Type | $C^{13}$ | Star | Spectral Type | | $C^{13}$ |
|---|---|---|---|---|---|---|
| HD 16115 | R3; C2,3J | 4 | R CMi | | C7,1(J)e | 4 |
| HD 19557 | R5; C4,5J | 5 | FU Mon | | C8,0(J) | 4 |
| UV Cam | R8; C4,5J | 4 | | | | |
| HD 52432 | R5; C4,5J | 5 | WZ Cas | N1p; | C9,2J Li | 4 |
| HD 79319 | R4; C4,4J | 4 | WX Cyg | N3e; | C9,2J Li | 5 |
| HD 168227 | R5; C4,5J | 5 | | | | |
| | | | U Hya | N2; | C6,3 | 3+ |
| Y CVn | N3; C5,5J | 5 | RR Her | K5ep; | C8,2e | 3+ |
| RY Dra | N4; C4,5J | 5 | RS Cyg | N0pe; | C8,2e | 4 |
| | | | U Cyg | Np; | C8,2e | 4 |

Table 2

| Wavelength | Molecular Band | Wavelength | Atomic Line |
|---|---|---|---|
| 6102 A | $C_{12}^{12}C_{12}^{13}(1,3)$ | 6102.72 A | CaI 3 |
| 6122 | $C_{12}^{12}C_{12}^{12}(1,3)$ | 6122.22 | CaI 3 |
| 6168 | $C^{12}C^{13}(0,2)$ | 6170.34 | VI 20 |
| | | 6162.18 | CaI 3 |
| 6191 | $C_{12}^{12}C^{12}(0,2)$ | 6191.73 | YI 3 |
| 6206 | $C_{12}^{12}N(4,0)R_1$ | ------- | ----- |
| 6260 | $C^{13}N(4,0)R_1^1$ | 6258.96 | ScI 3 |
| | | 6258.62 | VI 19 |
| | | 6261.23 | VI 20 |

Among C7-9 stars studied, only WZ Cas certainly, and WX Cyg probably belong to
J-type stars. Both stars belong to Li stars which show unusually strong LiI 6708 A
line. The other C7-9 stars in which lines of s-process elements are greatly enhanced
are found not to belong to J-type stars. Hot J-type star HD 16115 is found to be
certainly classified as a J-type star. It has been found that in HD 16115 abundances
of s-process elements with respect to Fe are nearly normal.

III) J-type stars in the Local Group galaxies

Richer et. al.(1979,1981,1983) made photometry and spectroscopy of many carbon
stars in the Large Magellanic Cloud and other Local Group galaxies. They constructed
the (I, R-I) color-magnitude diagram for these carbon stars. They also took spectra
of these cabon stars, and classified them on the C-classification system as defined
by Yamashita. They found two kinds of J-type stars, one is of high-luminosity stars
with weak CN bands, and the other is of low-luminosity stars with strong CN bands.

We believe that luminous $C^{13}$-rich stars in the Local Group galaxies are not
J-type but C7-9 stars including CS and SC stars, and "true" J-type stars are low-
luminosity stars with very strong $C^{13}$ bands, and Li stars like WZ Cas.

It may be said that in all J-type stars abundances of s-process elements with
respect to Fe are nearly normal.

References


Utsumi, K., 1985a, Proc. Japan Acad., 61, 193.

Utsumi, K., 1985b, Cool Stars with Excesses of Heavy Elements, ed. M. Jaschek and
P. C. Keenan, p 243.

Yamashita, Y., 1972, Ann. Tokyo Astr. Obs., 2nd series, 13, no. 3.

Yamashita, Y., 1975, Ann. Tokyo Astr. Obs., 2nd series, 15, no. 1.

Richer, H. B., Olander, N., and Westerlund, B. E., 1979, Astrophys. J., 230, 724.

Richer, H. B., 1981, Astrophys. J., 243, 744.

Richer, H. B., and Westerlund, B. E., 1983, Astrophys. J., 264, 114.

THE ABUNDANCE OF OXYGEN IN M92 GIANTS


Catherine A. Pilachowski
National Optical Astronomy Observatories
Kitt Peak National Observatory*
PO Box 26732, Tuscon, AZ   85726-6732,   USA


Studies of the carbon and nitrogen abundances in metal poor giants have generally supported the models of Sweigart and Mengel (1979) which indicate the action of meridional circulation in CN-processing metal poor red giant envelopes.  Carbon <u>et al</u>. (1982) find the trends in the carbon and nitrogen abundances in M 92 stars to be at least qualitatively consistent with CN-cycle processing of the envelope, but stars at all phases of giant branch evolution have unusual, and unexplained, carbon and nitrogen abundances.  The oxygen abundance is the missing piece of information which may explain what's going on.  Sweigart and Mengel's models suggest that ON-processing might also occur in the most metal poor giants.  The sample of giants in M 92 for which Carbon <u>et al</u>. have provided carbon and nitrogen abundances is ideal for the determination of oxygen abundances to look for the effects of ON-processing.  Once the oxygen abundances are known, the sum C+N+O can be computed to see if it is constant along the giant branch or varies from star to star.


Spectra of the $\lambda6300$ line of [O I] were obtained for 6 stars at the top of the M92 giant branch with the KPNO 4M Telescope and echelle spectrograph, using a CCD detector with high resolution (0.22Å) and high signal-to-noise (>50).  The resolution is sufficient to separate the Sc II feature 0.4Å longward of the [O I] line.  Synthetic spectra were calculated assuming an oxygen abundance of Log $\varepsilon(O)=6.67$ ([O/H]=[Fe/H]=-2.2).  A carbon abundance of [C/Fe]=-1.0, consistent with the results of Carbon <u>et al</u>. was assumed for all 6 stars.  Model atmosphere parameters of Carbon <u>et al</u>. were adopted for the synthetic spectrum calculation.  The oxygen, carbon, and nitrogen abundances are given in Table I, and log $\varepsilon(O)$ is plotted vs. log $\varepsilon(N)$ in Figure 1.


These results indicate that the sum of the carbon, nitrogen, and oxygen abundances in bright M92 giants remains constant within observational uncertainty, despite large variations in the abundance of nitrogen from star to star.  The constancy of C+N+O suggests that the abundance differences are due to variation in the degree of mixing, and not to initial differences in the amount of C+N+O.


*Operated by the Association of Universities for Research in Astronomy, Inc., under contract with the National Science Foundation.

Table I - The Abundance of Oxygen in M92 Giants

| Star | $T_{eff}$ | log g | Log W/λ | [C/Fe] | [N/Fe] | [O/Fe] | Log (C+N+O) |
|---|---|---|---|---|---|---|---|
| III-13 | 4210K | 0.66 | 19mÅ | -1.0 | +0.3 | +0.2 | 6.98 |
| VII-18 | 4230K | 0.71 | ≤6mÅ | -1.2 | +1.0 | ≤0.4 | ≤6.93 |
| III-65 | 4340K | 0.94 | 15Å | -1.2 | +0.5 | +0.1 | 6.96 |
| XII-8 | 4510K | 1.17 | ≤12mÅ | -1.0 | -0.3 | ≤0.2 | ≤6.98 |
| V-45 | 4530K | 1.22 | 12mÅ | -1.0 | +0.0 | +0.2 | 6.99 |
| II-70 | 4580K | 1.36 | ≤12mÅ | -1.0 | -0.3 | ≤0.2 | ≤6.98 |

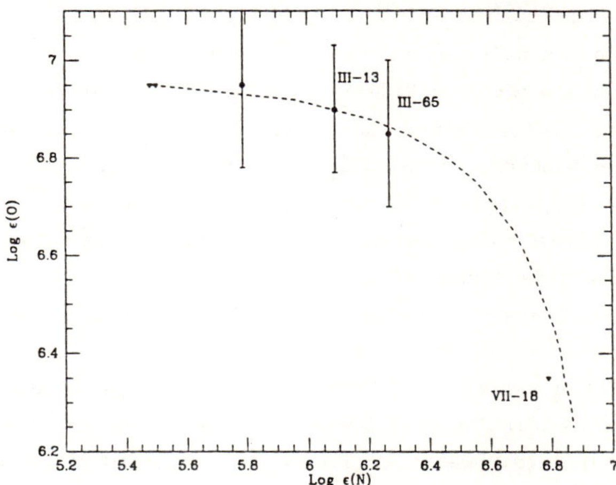


Figure 1 - Log ε(O) vs. Log ε(N) for M92 giants. Upper limits are shown as triangles. A curve of constant N+O is indicated.

The abundances of nitrogen and oxygen are anti-correlated (see especially the pair of stars III-13 and VII-18, which have very similar atmospheric parameters, but very different oxygen and nitrogen abundances). The anti-correlation of oxygen and nitrogen suggests that the ON-cycle has been active to modify the abundances, starting at about $M_V = -1.5$. A curve of constant C+N+O, with oxygen being converted to nitrogen, is drawn in Figure 1 for comparison to the data. The original oxygen abundance in M92 was probably Log ε(O)=6.98, or [O/Fe]= +0.2, and [C+N+O/Fe]= +0.03 ± 0.05.


Carbon, D. F., Langer, G. E., Butler, D., Draft, R. P., Suntzeff, N. B., Kemper, E., Trefzger, C. F., and Romanishin, W. 1982, Ap. J. Suppl., 49, 207.
Sweigart, A. and Mengel, J. 1979, Ap. J. 229, 642.

# A SEARCH FOR GALACTIC CARBON STARS

Hideo Maehara

Tokyo Astronomical Observatory, Mitaka, Tokyo 181, Japan

## 1. Carbon star as a probe

The carbon star is one of the best probes for the galactic study;

(1) it is intrinsically bright ($M_{bol} = -2$ to $-6$) especially in the red and infrared wavelength regions,

(2) it has spectral features readily detectable on objective prism plates due to their strong carbon molecular bands,

(3) it is an evolved star distributed abundantly ($\sim$1 star per square degree) along the galactic plane.

We can detect it in the Galaxy up to several kpc from the sun on objective prism plates of the Schmidt telescope.

## 2. Observation

We have been making survey observations of faint cool carbon stars using the Kiso 105-cm Schmidt telescope. Kodak IN and 103aF plates are respectively taken behind the 4-degree objective prism (700 $\text{Åmm}^{-1}$ at H$\alpha$) for the detection and for the spectral classification. $V$-band plates are utilized to obtain the position and the brightness of the stars detected.

The survey areas are distributed along the northern galactic plane. Seven fields in the Cassiopeia region ($l = 115°$ to $133°$) and eight fields in the Taurus-Auriga-Gemini region ($l = 170°$ to $188°$) have been observed and processed up to now (Maehara and Soyano 1987a,b).

## 3. Detection

The detection is made on hypersensitized IN plates covering $\lambda\lambda$ 6900 – 8800 Å, where strong CN bands exist in carbon stars' spectra. The limiting magnitude of our detection is around I=13 mag, and about 1 mag fainter than that of the catalog compiled by Stephenson (1973).

In total, 210 and 125 carbon stars are respectively detected in the 180 square-degree area of the Cassiopeia region, and in the 200 square-degree area toward the galactic anticenter direction (figure 1). The mean surface number density is around 1 star per square degree at $l \sim 0°$, and half at $l \sim 5°$.

## 4. Spectral classification

Bright carbon stars in the Cassiopeia region are classified into the C classification system (Yamashita 1972, 1975) using 103aF plates ($\lambda\lambda$4500 – 6800 Å). Six criteria are extracted from spectral tracings of standard stars. Fifty nine stars are classified into the 2-dimensional $C$ system with the temperature $T$ and the carbon abundance $A$ subclasses to an accuracy of $\pm 1$ subclass.

Carbon stars earlier than C3 cannot be well detected due to the weakness of $C_2$ Swan bands. A majority of the stars belong to C4 and C5 stars with high carbon abundance, while there are a few C8 and C9 stars.

## 5. Space distribution

The space distribution in the Cassiopeia region is estimated on the basis of the spectral classification and the absolute magnitudes determined by Mikami (1975). It is shown that the carbon stars are distributed over the galactic plane without strong concentration onto the (Perseus) arm. The number ratio of C4 – C5, C6 – C7, and C8 – C9 stars is nearly $1 : 0.4 : < 0.1$.

The mean space density of cool carbon stars is about 100 stars $kpc^{-3}$ in the Cassiopeia region up to 3 kpc from the sun. The density in the anticenter region is fairly less than that of the Cassiopeia region. The space distribution of cool carbon stars is different from region to region, and the ratio to M-type giant stars is likely to be correlated with the metal abundance distribution in the Galaxy.

## References


Maehara, H. 1985, *Publ. Astron. Soc. Japan*, **37**, 333.
Maehara, H., and Soyano, T. 1987a, *Ann. Tokyo Astron. Obs., 2nd Ser.*, **21**, 293.
Maehara, H., and Soyano, T. 1987b, *Ann. Tokyo Astron. Obs., 2nd Ser.*, **21**, 423.
Mikami, T. 1975, *Publ. Astron. Soc. Japan*, **27**, 445.
Stephenson, C. B. 1973, *Publ. Warner Swasey Obs.*, **1**, No.4.
Yamashita, Y. 1972, *Ann. Tokyo Astron. Obs., 2nd Ser.*, **13**, 169.
Yamashita, Y. 1975, *Ann. Tokyo Astron. Obs., 2nd Ser.*, **15**, 47.


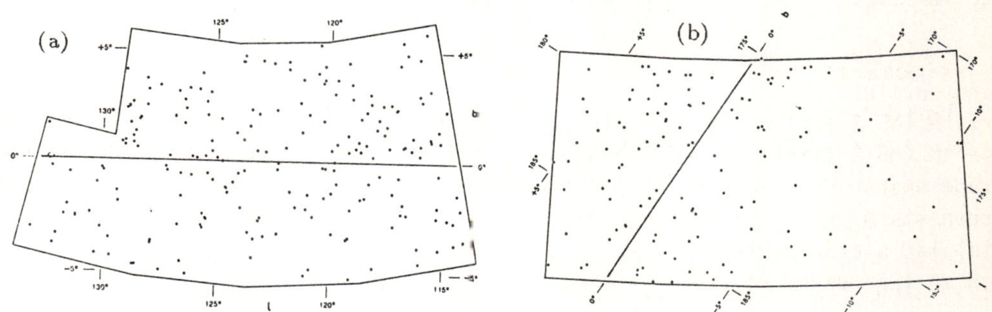

Fig. 1. Carbon stars detected in the surveyed areas.
The coordinates are given in the galactic longitude and latitude.
(a) Cassiopeia region
(b) Taurus-Auriga-Gemini region

# SURFACE DISTRIBUTION OF M STARS WITH DIFFERENT IRAS COLOUR


Keiichi ISHIDA
Tokyo Astronomical Observatory, University of Tokyo
Mitaka, Tokyo 181, Japan
and
Mazlan OTHMAN
Physics Department, National University of Malaysia
Bangi, Malaysia



The surface distribution of M stars is studied by differentiating them according to whether they show a circumstellar dust shell (CS) or not. Analysis shows that galactic latitudinal and longitudinal distributions are not determined by spectral subclasses alone. The study also indicates that the M type stars with CS have higher intrinsic luminosities in the K band than those without CS. The M stars used in the study are obtained from the Two Micron Sky Survey catalogue (IRC) which is an unbiased sample with respect to the interstellar extinction. The CS feature is identified by the ratio of flux densities at 12 and 25 $\mu$m in the IRAS point source catalog.


Table 1. Spectral composition of stars without and with CS.

| spectral type | number of stars | | |
| --- | --- | --- | --- |
| | without CS | with CS | total |
| K and earlier | 1318 | 36 | 1354 |
| early M (M5 & earlier) | 1650 | 614 | 2264 |
| late M (M6 & later) | 325 | 1117 | 1442 |
| S type stars | 31 | 40 | 71 |
| carbon stars | 130 | 58 | 188 |
| total IRC stars identified with an IRAS point source | 3454 | 1865 | 5319 |

SiO Isotope Emissions from Late-Type Stars

N. Ukita and N. Kaifu

Nobeyama Radio Observatory, Tokyo Astron. Obsevatory
Minamisaku, Nagano 384-13, Japan

Abstract
   The $J=2-1$, $v=0$ emissions of $^{28}SiO$, $^{29}SiO$, and $^{30}SiO$ from three late-type stars
were simultaneously observed with the Nobeyama 45-m telescope in Jaunuary 1987.
The relative intensities of $(^{29}SiO)$ / $(^{30}SiO)$ were measured to be 2.4 for
$\chi$ Cyg, 1.5 for NML Tau, and 2.9 for V1111 Oph.  These values are lower limits
for the relative isotope abundance of $(^{29}Si)$ / $(^{30}Si)$ , and are larger than
the terrestrial value of 1.51.  Recent theoretical studies suggest that
hydrostatic nucleosysnthesis and supernova explosions result in smaller values
of $(^{29}Si)$ / $(^{30}Si)$ than solar.  These models cannot explain the observed
excess of $^{29}Si$.

## 1.  Introduction
Observations of isotopic abundances provides information on the nucleosynthesis
operating in the compact core of stars and supernova explosions and on the
chemical evolution of the Galaxy.  The CNO nuclides in late-type stars are
affected by freshly synthesized core material brought up by "dredge-up" events.
On the other hand, the Si isotopes are involved in later phases of nuclear
burning, a narrow span of the red giant lifetime before planetary nebulae or
supernovae.  Therefore relative abundances of Si isotopes we observe remain
unchanged from those of interstellar matter from which a star was formed.

## 2.  Observations and Results
The $J=2-1$, $v=0$ emissions of $^{28}SiO$, $^{29}SiO$, and $^{30}SiO$ from V1111 Oph, NML Tau, and
$\chi$ Cyg were observed with the Nobeyama 45-m telescope in Jaunuary 1987.  At 86
GHz, the half-power beamwidth was 20", and the aperture efficiency 0.37.  The
three lines were simultaneously observed using two receivers with instantaneous
bandwidths of 2 GHz and 0.5 GHz to reduce errors in relative intensities due to
pointing and intensity calibration errors.  The intensity scale reported here
is the antenna temperature T  , corrected for atmospheric and ohmic losses.
   Figure 1 shows line profiles of NML Tau.  NML Tau shows spike components at
Vlsr=38 km/s in the spectra of $^{28}SiO$ and $^{29}SiO$.  We excluded this velocity range

in the following analysis. Table 1 summarizes integrated intensities and relative line intensities for these stars, together with those for IRC+10215 observed with the IRAM 30-m telescope.

## 3. Discussion

The $^{28}$SiO emissions show parabolic line profiles, suggesting that they are optically thick thermal emissions. Optically thin emissions should display rectangular line profiles. The line profiles of the $^{29}$SiO and $^{30}$SiO emissions seems to be rectangular, at most intermediate between them. The optical thickness of the line usually leads to underestimates of the abundances of the most abundant isotopes. Therefore we concentrate on the two minor isotopes.

The relative intensities of $[^{29}$SiO$]$ / $[^{30}$SiO$]$ were measured to be 2.9 for V1111 Oph, 1.5 for NML Tau, and 2.4 for $\chi$ Cyg, giving lower limits for the relative isotope abundance of $[^{29}$Si$]$ / $[^{30}$Si$]$ . Our mesurements indicate that the relative isotope abundances of $[^{29}$Si$]$ / $[^{30}$Si$]$ vary from star to star and that they are equal to or larger than the terrestrial value of 1.51.

A model of hydrostatic nucleosysnthesis(Thielemann and Arnett, 1985) suggests that $^{29}$Si and $^{30}$Si are mainly produced in He, C, Ne burning, and that a smaller relative abundance of $[^{29}$Si$]$ / $[^{30}$Si$]$ than solar is expected. Recent models of supernova explosion indicate that expected relative abundance of $[^{29}$Si$]$ / $[^{30}$Si$]$ is 0.1 for type II supernovae(Nomoto et al., 1984) and 0.8 for type II supernovae(Woosley and Weaver, 1986). These models cannot explain the observed excess of $^{29}$Si.

Refernces

Nomoto. K., Thielemann, F.-K., and Yokoi, K., 1984, Astrophys. J.,286, 644.
Thielemann, F.-K., and Arnett, W.D., 1985, Astrophys. J.,295, 604.
Woosley, S.E. and Weaver, T.A., 1986, Ann. Rev. Astron. Astrophys, 24, 205

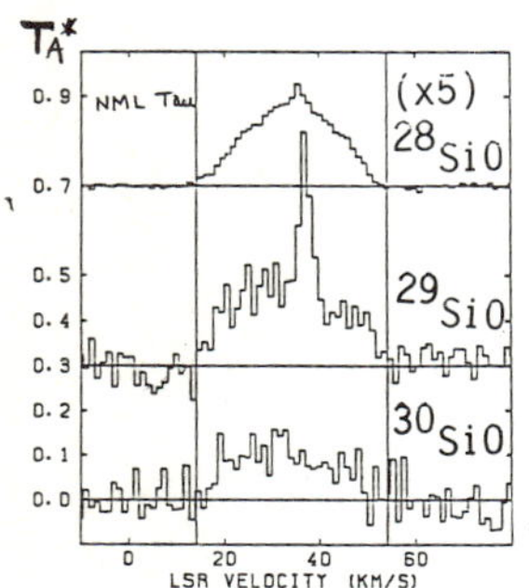


Table 1. Line Intensities

| Source | Type | $^{28}$SiO (K km/s) | $^{29}$SiO (K km/s) | $^{30}$SiO (K km/s) | Relative Intensity $[^{29}$SiO$]$ / $[^{30}$SiO$]$ |
|---|---|---|---|---|---|
| $\chi$ Cyg | S | 8.6+0.10 | 2.74+0.17 | 1.13+0.17 | 2.4+0.6 |
| NML Tau | M | 11.7+0.10* | 2.74+0.17* | 1.88+0.17* | 1.5+0.2 |
| V1111 Oph | M | 10.9+0.10 | 2.31+0.16 | 0.81+0.14 | 2.9+0.8 |
| IRC+10216 | C | 36.0+1.0 | 3.09+0.26 | 1.80+0.11 | 1.68+0.20 # |
| Terrestrial | | | | | 1.51 |

* integrated intensities for 4<V<35 km/s
\# measurements with the IRAM 30-m telescope, Kahane et al.(1987, preprint)

# DETECTION OF WATER MASER EMISSION FROM A CARBON STAR V778 CYGNI


Y. Nakada, H. Izumiura, T. Onaka, and O. Hashimoto
Department of Astronomy, Faculty of Science, University of Tokyo,
Bunkyo-ku, Tokyo 113, Japan


N. Ukita, and S. Deguchi
Nobeyama Radio Observatory, Tokyo Astronomical Observatory, University of Tokyo,
Nagano 384-13, Japan


and


T. Tanabé
Tokyo Astronomical Observatory, University of Tokyo
Mitaka, Tokyo 181, Japan



Infrared spectra of evolved stars are generally dominated by the radiation from their circumstellar shells. M stars are characterized by the 10 μm emission feature from silicate dust grains, while C stars by the 11 μm SiC band. However, some C stars have been found to show the 10 μm feature indicating the oxygen-rich property of their circumstellar dust (Willems and de Jong 1986, Little-Marenin 1986).

In order to investigate the gas phase chemistry of the circumstellar envelopes around these peculiar objects, we have observed radio molecular lines of $H_2O$, SiO, HCN, and CO towards three of them BM Gem (C5, 4J), V778 Cyg (C4, 5J), and EU And (C4, 4).

Water maser emission has been detected towards V778 Cyg (Fig. 1) in the present observation. It is another evidence of the oxygen-rich nature of the circumstellar envelope around V778 Cyg, because water vapor has always been found around M stars. Thermochemical calculations also support the idea that the water molecules are present in the oxygen-rich environment but not in the carbon-rich one. The positional coincidence of the water maser with the optical star was found to be less than 0.5 arcsec (Deguchi *et al.* 1987), (Fig. 2). Unfortunately no other emission has been found, though Benson and Little-Marenin (1987) observed a water maser in EU And. Our negative detection of maser emission in EU And was probably due to the intensity variation of the object.

Our observation has confirmed the oxygen-rich chemistry of the circumstellar gas around the carbon-rich star V778 Cyg. The radial vleocity of the maser peak ($-17$ km s$^{-1}$) is red-shifted by only 2 km s$^{-1}$ relative to the optical photospheric velocity ($-19$ km s$^{-1}$). Therefore, the expansion velocity of the maser emission career in V778 Cyg seems 2 to 10 km s$^{-1}$, a moderate value contrary to the case of EU And.

The interpretation for these peculiar objects has not been established. There are two possibilities though. One is that we are observing the transition stage of an evolved star from M- to C-type with a remnant circumstellar shell expelled in the preceeding M giant stage. The difficulty in this hypothesis is that the transition time will be too short to be observed. The other possibility is that an invisible M star which is bright in the infrared forms the binary system with a visible C star. The accurate observation of the radial velocity variation is necessary to see if this hypothesis is correct or not.


## REFERENCES


Benson, R. J., and Little-Marenin, I. R. 1987, *Ap. J. (Letters)*, **316**, L37.

Deguchi, S. *et al.* 1987, *Ap. J.*, in press.

Little-Marenin, I. R. 1986, *Ap. J. (Letters)*, **307**, L15.

Willems, F.J., and de Jong, T. J. 1986, *Ap. J. (Letters)*, **309**, L39.


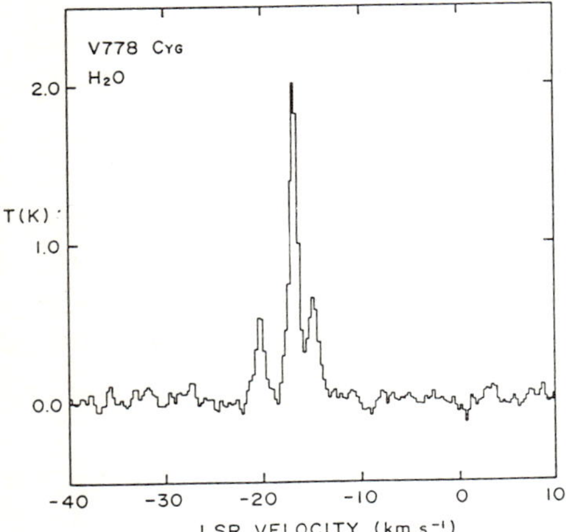


Fig.1. Spectrum of the H$_2$O maser emission in V778 Cyg.

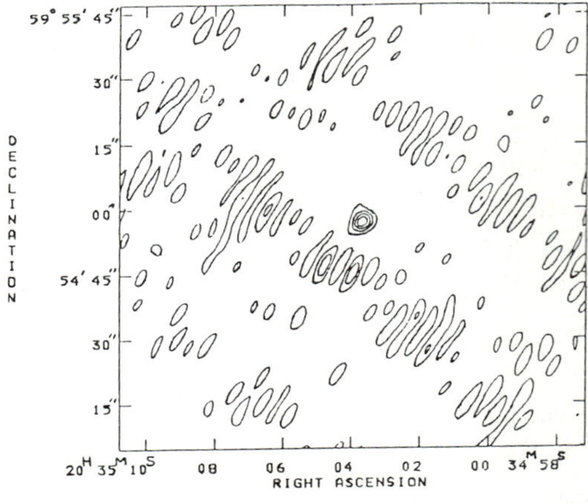

Fig.2. A map of water maser emission toward V778 Cyg. The contour levels are −0.2, 0.2, 0.4, 0.6, and 0.8 normalized to the peak flux 5.9 Jy/beam.

CHEMICAL AND EXPANSION PROPERTIES OF COMPACT PLANETARY
NEBULAE IN THE GALACTIC ANTI-CENTER REGION

Shin'ichi Tamura
Astronomical Institute, Tohoku University
Aobayama, Sendai, Japan  980

A spectroscopic diagnosis on M1-5, M1-9, K3-66, and K3-67 is presented.  Our sample were chosen from the catalogue of radial velocities
(Schneider et al 1983) by the reason of not only their kinetic peculiarity, but also apparently compact images.  Main purpose is to analyze the
chemical properties which should give us an information about galactic
chemical abundance distribution in the direction of galactic anti-center
region based upon kinetic peculiarity.  Another one is to study on an
expansion characteristics by which we can recognize intrinsically compact
planetary nebulae in young phase.

Spectroscopic observations were made at the Okayama Astrophysical
Observatory (Shibata and Tamura 1985), the Steward Observatory, and Lick
Observatory (Tamura and Shaw 1987).

(1) Due to the chemical abundance diagnostic criteria (Peimbert 1983),
    M1-9 and K3-67 seem to have evolved from their massive progenitors
    in the relation between He/H versus N/O (Figure 1).

(2) K3-66 belongs chemically and kinematically to Population II, while
    K3-71 is a rather high-ionization planetary, and may have Population
    II kinematics.

(3) From the analyses of emission line profiles (Figure 2), K3-66 should
    be in distant site from us and M1-5 seems to be intrinsically compact
    planetary nebula.

(4) Due to the line profiles and the estimation of HI mass (Schneider
    et al 1987), there is still a possibility M1-9 is the really compact
    planetary nebula like M1-5.


## References

Kaler, J. B. 1983, Ap. J., **271**, 188.
Peimbert, M. 1983, in IAU Symposium 103, Planetary Nebulae, ed. by
    D. Flower (Dordrecht: Reidel), p. 233.
Schneider, S. E., Terzian, Y., Purgathofer, A., and Ferinotto, M. 1983,
    Ap. J. Suppl., **52**, 399.
Schneider, S. E., Silverglate, P. R., Altschuler, D. R., and Giovanardi,
    G. 1987, Ap. J., **314**, 572.
Shibata, K. and Tamura, S. 1985, Publ. Astron. Soc. Japan, **37**, 325.
Tamura, S. and Shaw, R. A. 1987, Publ. Astron. Soc. Pacific, in press.

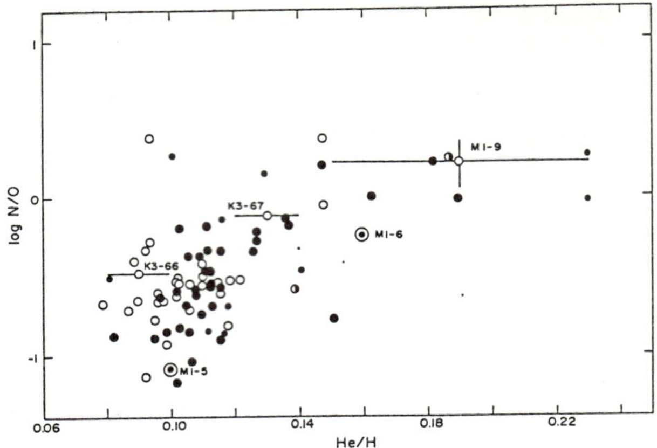

Fig. 1. He/H versus log N/O of M1-6 and M1-9 from Shibata and Tamura (1985) (as labeled), and of K3-66 and K3-67 from Tamura and Shaw(1987) superposed on Kaler's(1983) survey. Filled symbols: Population I; open symbols: Population II.

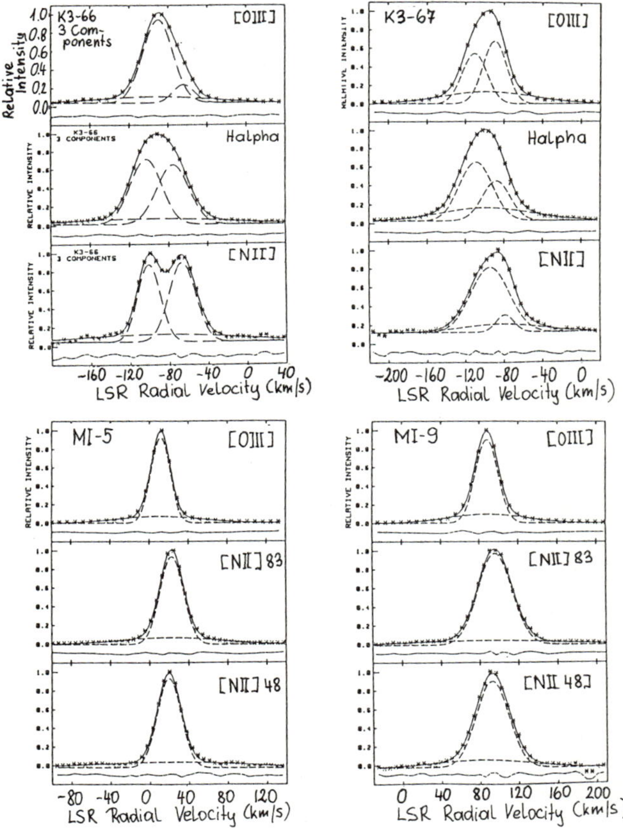

Fig. 2. Emission line profiles of M1-5, M1-9, K3-66, and K3-67. Crosses and real lines are observational results, broken lines are Gaussian components.

# HELIUM FLASHES THROUGH THE NCO REACTION


K. Arai and K. Kaminisi
Department of Physics, Kumamoto University, Kumamoto 860
M. Hashimoto
Max Planck Institut für Astrophysik, Garching bei München D-8046
and  K. Nomoto
Department of Earth Science and Astronomy, University of Tokyo
Meguro-ku, Tokyo 153


## 1.  Introduction

It is generally accepted that the helium flash occurs when the $3\alpha$ reaction commences in the degenerate helium core of low mass stars. In this core, original CNO isotopes have been converted into $^{14}N$ and the electron Fermi energy becomes large enough to approach the threshold energy for e-capture on $^{14}N$. Hence Kaminisi et al. (1975) have pointed out that in these circumstances the $^{14}N(e^-,\nu)^{14}C(\alpha,\gamma)^{18}O$ (NCO) reaction may play an important role for igniting the helium flash.

We, therefore, examine the effects of the NCO reaction on the evolution of low mass stars.  A key ingredient of the NCO reaction is that the density reaches the threshold for e-capture ($\rho_{th} \simeq 10^6$ g cm$^{-3}$). Evolutionary sequences are presented for the cases of accreting helium white dwarfs (Hashimoto et al. 1986) and a 0.7 $M_\odot$ , Population II star ascending the giant branch.

## 2.  Evolution of accreting helium white dwarfs

An initial model has the mass M = 0.3 $M_\odot$ and the chemical abundances $X(^4He)$ = 0.9879 and $X(^{14}N)$ = 0.0121. In three cases, A, B and C, the accretion rates are $\dot{M}$ = $10^{-8}$, $10^{-9}$ and $3 \times 10^{-10}$ $M_\odot$ yr$^{-1}$.

The evolutionary paths of the central density and temperature are plotted by the solid lines in Figure 1 for cases A-C.  The dashed lines show the paths for the same accretion rates as for cases A-C but with the NCO reaction switched off. The dotted lines denote the ignition curves for the $3\alpha$ and NCO reactions. Note that the NCO reaction dominates over the $3\alpha$ reaction to heat up the core.  The inclusion of the NCO reaction leads to the ignition of the helium flash at considerably lower density.

In case A, once $\rho_c$ reaches $\rho_{th}$, produced $^{14}C$ quickly burns to $^{18}O$. The core halts further contraction. The helium flash is initiated by the NCO reaction.  In cases B and C,  the condition $\rho_c > \rho_{th}$ is realized.  Then $^{14}N$ is completely converted into $^{14}C$ in the central region.  However, $^{14}C$ has not yet been processed into $^{18}O$, because $T_c$ is too low for the $\alpha$-capture to occur.  The central density continues to increase until the helium flash is triggered by $^{14}C$ burning.

3. Evolution of a low mass star ascending the giant branch

Referring to a specified sequence of a 0.7 $M_\odot$ star presented by Sweigart and Gross (1978), we calculate the red giant sequence from the subgiant branch to the onset of the helium flash. The initial composition is $X(^4He) = 0.999$ and $X(^{14}N) = 0.001$.

The evolutionary track followed by the center and the temperature profiles against density are shown by the solid lines in Figure 2. The numerals attached to the lines denote the time before the onset of the flash in units of $10^6$ yr. The center does not reach the NCO ignition curve before the $3\alpha$ reaction ignites at the site of the maximum temperature. The NCO reaction does not change the existing evolutionary models, as was pointed out by Spulak (1980), except that a considerable amount of $^{18}O$ is produced in the central region.

4. Concluding remarks

It is shown that the original CNO isotopes in the core of a low mass star are converted into $^{18}O$ during the helium flash. We note that the NCO and the subsequent $^{18}O(\alpha,\gamma)^{22}Ne(\alpha,n)^{25}Mg$ reactions can be a prolific source of neutrons for the synthesis of s-process elements.

References


Hashimoto, M., et al. 1986, Ap. J., **307**, 687.
Kaminisi, K., et al. 1975, Prog. Theor. Phys., **53**, 1855.
Spulak, R. G., Jr. 1980, Ap. J., **235**, 565.
Sweigart, A. V., and Gross, P. G. 1978, Ap. J. Suppl., **36**, 405.


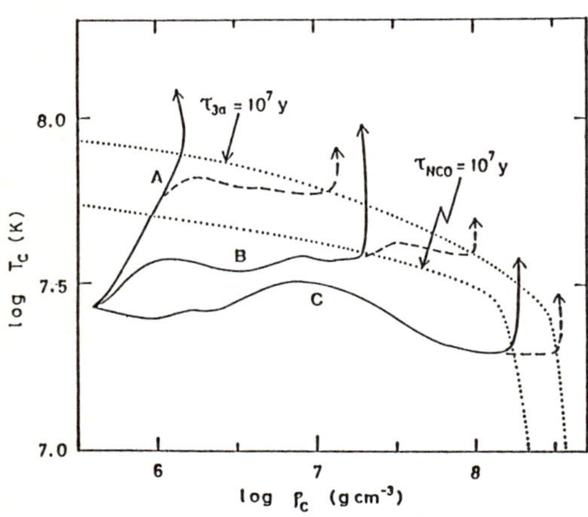


Fig. 1. Evolutionary tracks of accreting helium white dwarfs.

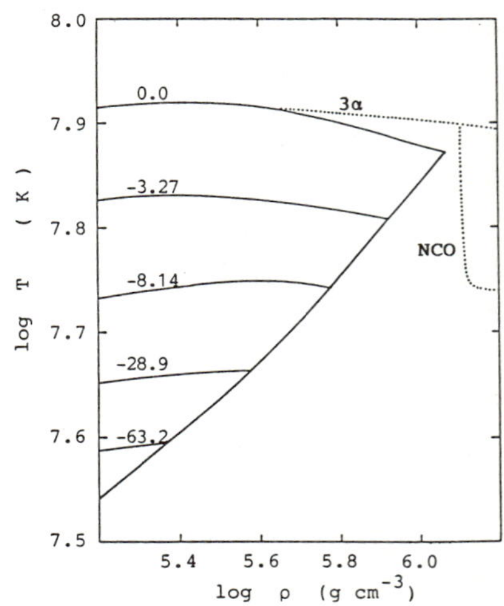


Fig. 2. Evolutionary track of a 0.7 $M_\odot$ star.

ABUNDANCES IN HOT EVOLVED STARS

D. Husfeld
Universitäts-Sternwarte München
Scheinerstr. 1, D-8000 München 80, Germany

## 1. INTRODUCTION

In the upper left corner of the HR-diagram various stars have been found that are clearly not members of the hydrogen main sequence. The majority of them lie to the left and below the main sequence, indicating that they are highly evolved stars close to the extinction of their thermonuclear power source and hence to rapid cooling towards the white dwarf domain. The implication is that a "short" time ago they were red giants and as such experienced phenomena like mixing, dredge-up and heavy mass loss. Therefore, one expects the hot evolved stars to display the consequences of the afore mentioned processes that are up to now not satisfactorily understood.

However, the hot evolved stars do not form a homogeneous group. The most prominent division is into Central Stars of Planetary Nebulae (CSPN) and those stars that are apparently not surrounded by nebulosity. Further obvious differences exist in spectral type and chemical composition (as deduced from medium and high resolution spectra). Consequently, a unique progenitor for all kinds of hot evolved stars appears to be very unlikely and accurate stellar parameters are needed to compare these stars to predictions of stellar evolution theory.

This paper summarizes the results of analyses of highly evolved stars with spectral type B or hotter, namely sdB, sdOB and sdO types, CSPN and extremely helium-rich stars. It does not consider white dwarfs since their chemical surface composition is apparently governed by diffusion processes and accretion of interstellar material (Wesemael, 1979; Vauclair et al., 1979; Wesemael and Truran, 1982) and is not linked to their past evolution. Section 2 deals with the positions of the hot evolved stars in the ($\log T_{eff}$-$\log g$) plane and their helium to hydrogen ratios. Metal abundances are considered in section 3 and comparisons of stellar evolution calculations with the available data are performed in section 4.

59

## 2. THE (LOG $T_{eff}$-LOG G) DIAGRAM AND THE HE ABUNDANCE

The quantities determined directly by the spectroscopic analysis as performed for hot stars are effective temperature $T_{eff}$, surface gravity g and element abundances. Of course, this is not sufficient to place a star in the HR diagram. This is possible only with further knowledge of either luminosity, radius, mass or distance of the star. However, uncertainties in these quantities (which are usually much larger than the uncertainties in $T_{eff}$ and g) directly translate into the HR diagram. On the other hand, theoretical evolutionary tracks can be easily expressed in terms of $T_{eff}$ and g without loss of precision. It is therefore good practice to discuss the results of spectroscopic analyses directly in a (log $T_{eff}$-log g) diagram as we shall do in this paper.

A detailed description of the analysis method to obtain $T_{eff}$, log g and the He/H ratio has been given elsewhere (see the references in Groth et al. (1985), Méndez et al. (1985), Heber (1987)) and will not be repeated here. A few important points, however, should be indicated. In the case of the B type stars, effective temperatures are low enough to bring the peak of the stellar flux distribution within - or close to - the range of the IUE satellite. $T_{eff}$ is, therefore, most accurately determined by fitting model atmosphere fluxes to low resolution IUE spectrograms and visual photometric data. The surface gravity is then determined using the widths of the hydrogen lines that are subject to the linear Stark broadening effect. Model calculations for B stars can usually be made under the assumption of LTE.

On the other hand, O stars are too hot for a deduction of $T_{eff}$ from their continuum slope in any reliable way. Instead, one has to use model predictions concerning ionization equilibria and related line strengths (commonly the HeI/HeII ratio) whereas log g is again determined from the wings of the broad H and HeII lines. All calculations must account for non-LTE effects even for the relatively high densities in the atmospheres of sdO stars (Kudritzki, 1979). It should be pointed out that the construction of non-LTE model atmospheres is still somewhat unsatisfactory because no realistic opacities of elements heavier than helium can be included due to limitations imposed by the available algorithms and computers. The situation will hopefully change in the near future when approaches like the one of Werner (1986) are used. Nevertheless, existing non-LTE model atmospheres are able to produce H-He line profiles accurately even when compared with spectrograms with S/N ≈ 100 (see for example Heber et al., 1987a,b).

Figure 1 summarizes the results of all available fine analyses of hot evolved stars. It is important to note that most of these stars are not known to be members of binary systems. We find that all sdB and sdOB stars are drastically depleted in helium. A well-defined border line at $T_{eff}$ = 42000K separates these stars from the classical or "compact" sdOs which cluster at $T_{eff}$ ≈ 50000K, have surface gravities log g ≥ 5 and are systematically enriched in helium (y:= $N_{He}/(N_H+N_{He})$ ≈ 0.5). Some members of the class of extremely helium-rich sdOs (no traces of hydrogen in the spectrum) are located in the same area of the (log $T_{eff}$-log g) plane as the sdOB and classical sdO stars, falling on either side of the border-line between them.

Recently, another subtype of sdO stars has been found at significantly lower gravities and higher temperatures thus placing these stars among the CSPN (Husfeld, 1986; Husfeld et al., 1986; Heber et al., 1987a). As they have higher luminosities than the classical sdOs, we call this subtype the "luminous" sdOs. For this subtype helium abundances range from roughly normal to extremely high (y > 0.9).

The bulk of the analyzed CSPN is hotter than 60000K, covering a wide range in gravity. They are clearly separated from the classical sdOs and the sdB/sdOB stars. Most of them have normal helium abundances. However, some cases of intermediate and extreme enrichments as well as depletion of helium have been found too. (Note that the helium-poor CSPN are almost at a DA white dwarf stage).

## 3. METAL ABUNDANCES

Up to now, the sdB/sdOB stars, the classical sdOs and the extremely helium-rich luminous sdOs have been analyzed for the most important (and accessible) metal abundances. The analyses usually require extensive non-LTE line formation calculations to solve the statistical equilibrium in detailed model atoms simultaneously with the radiative transfer equations for all relevant frequencies. With the advent of computer codes based on modern powerful solution algorithms (Auer and Heasley, 1976; Werner and Husfeld, 1985) it has now become possible to test (and eventually remove) approximations necessary in older computations. This and the availability of improved atomic data make the non-LTE predictions more reliable, and obstacles in obtaining accurate abundance determinations come now mainly from the observational side where high-quality spectra are needed to identify and to measure weak

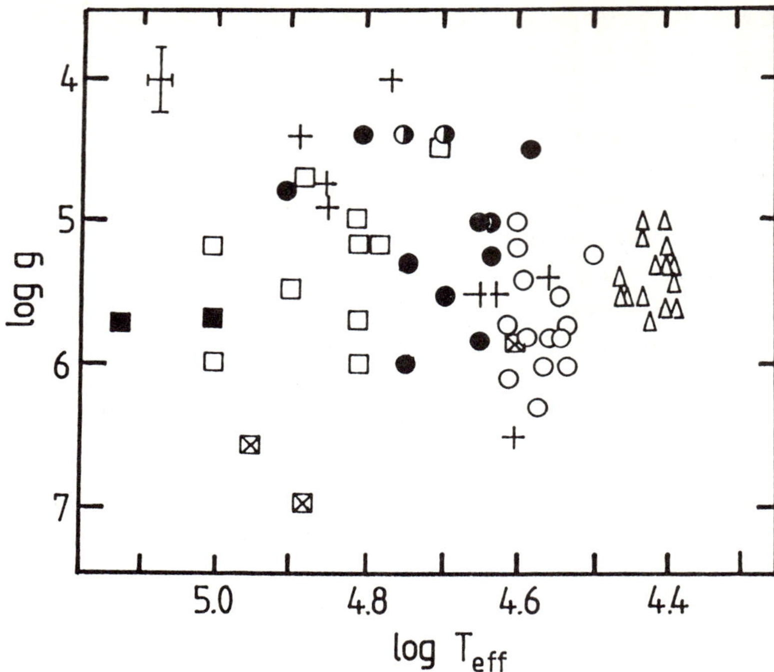


Figure 1: Positions of spectroscopically analyzed hot evolved stars in the ($\log T_{eff}$-log g) plane. Symbols denote type and He abundance. $\Delta$: sdB ($y \ll 0.09$); O: sdOB ($y \ll 0.09$); ◐: sdO ($y \approx 0.09$); ●: sdO ($y \approx 0.5$); +: sdO ($y \approx 1.0$); □: CSPN ($y \approx 0.09$); ■: CSPN ($y \gtrsim 0.5$); ⊠: CSPN ($y \approx 0.01$). The cross in the upper left corner indicates typical ranges of uncertainty in $\log T_{eff}$ and log g.

Table 1: Logarithms of mass fractions for abundant elements in four classical sdO stars and in the sun. Col. 6 gives theoretical predictions from hydrogen fusion in the CNO cycle at $T = 2 \cdot 10^7$ K (Caughlan, 1964). Solar values are from Holweger (1979).

|    | HD 49798 | BD+75°325 | HD 127493 | HD 128220B | sun    | CNO cycle |
|----|----------|-----------|-----------|------------|--------|-----------|
| H  | -0.70    | -0.85     | -0.85     | -0.36      | -0.152 | -0.85     |
| He | -0.10    | -0.07     | -0.07     | -0.25      | -0.55  | -0.07     |
| C  | -3.7     | -3.7      | -4.1      | -2.7       | -2.40  | -4.1      |
| N  | -1.6     | -1.9      | -2.1      | -2.2       | -3.01  | -1.9      |
| Si | -3.0     | -3.1      | -3.0      |            | -3.16  |           |

lines $(W_\lambda = 20$-$30$ mÅ) of subordinate transitions and/or less abundant ions. An abundance derived from several multiplets is usually accurate to 0.2-0.3 dex.

The abundances derived for some classical sdO stars - given as logarithm of the mass fractions - are compared with solar values in Table 1. Besides the notorious increase in the He/H ratio a consistent depletion of carbon and an enrichment of nitrogen is evident whereas silicon is essentially unaltered. From this pattern it might be suspected that these stars display the products of hydrogen burning in the CNO cycle. That this hypothesis is also in quantitative agreement with the observations can be seen from the last column in Table 1 where theoretical predictions at a plasma temperature of $2*10^7$K (Caughlan, 1964) are given.

Similarly, Table 2 presents the abundances in some sdB/sdOB stars. Remarkable here is the strong depletion of He and Si that can only be understood qualitatively as the result of gravitational settling. Carbon is also strongly underabundant with respect to the sun whereas nitrogen is roughly solar. The latter abundance could be explained in the framework of gravitational settling if a previous nitrogen enrichment due to the CNO cycle is assumed.

Table 3 summarizes the results of an analysis of four extremely helium-rich luminous sdOs (Husfeld, 1986; Husfeld et al., in preparation). Here, only upper limits for the hydrogen abundances can be given as no traces of this element can be found in the spectra. Consequently, helium appears as the most abundant element. Significantly overabundant are also carbon (with one exception: LSE 263) and nitrogen. Silicon is effectively unaltered. This abundance pattern compares well with the abundances found in the extreme helium stars of spectral type B (given in col. 6 of Table 3). However, it should be stressed that the carbon depletion in LSE 263 makes this star a peculiar object in its class.

Finally, we mention the central star of NGC 246. This star is known to display CIV as the strongest absorptions in the blue spectrum and broad shallow lines of HeII; hydrogen can not be detected (Heap, 1975; Husfeld, 1986). Analysis by Husfeld (1986) revealed that this CSPN is extremely hydrogen-deficient ($n_H/n_{He} < 0.1$) and strongly overabundant in carbon (30% by number). Nitrogen is not present, neither in the blue nor in the UV spectrum, but a solar abundance of this element cannot be excluded.

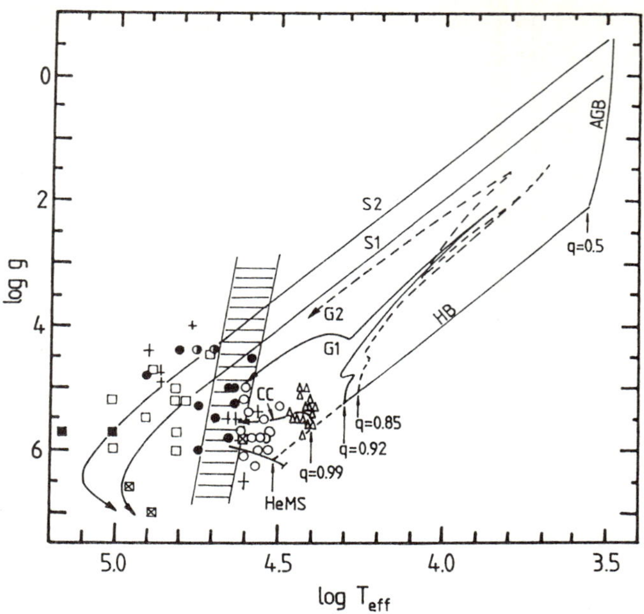


Figure 2: Comparison between observations and evolutionary tracks applicable for sdB, sdOB and classical sdO stars. HB is the horizontal branch with core mass $M_{core}$ = 0.475 $M_O$, on which some values of q = $M_{core}/M_{total}$ are indicated. CC denotes the track of Caloi and Castellani (1985), the tracks G1 and G2 are from Gingold (1976). Post-AGB tracks applicable for CSPN are from Schönberner (1983; S1: M = 0.546 $M_O$, S2: M = 0.565 $M_O$). The hatched strip marks the area in which Groth et al. (1985) have found photospheric convection. Stellar symbols are the same as in Fig. 1.


Table 2: Logarithms of mass fractions for abundant elements in four sdB/sdOB stars.

|     | Feige 110 | HD 149382 | Feige 66 | LB 2459 | sun    |
|-----|-----------|-----------|----------|---------|--------|
| H   | -0.05     | -0.07     | -0.03    | -0.005  | -0.152 |
| He  | -0.96     | -0.85     | -1.12    | -1.93   | -0.55  |
| C   | <-7.9     | -4.4      | -4.5     | -4.4    | -2.40  |
| N   | -3.1      | -2.9      | -2.7     |         | -3.01  |
| Si  | <-6.8     | <-7.7     | <-7.8    | -4.2    | -3.16  |

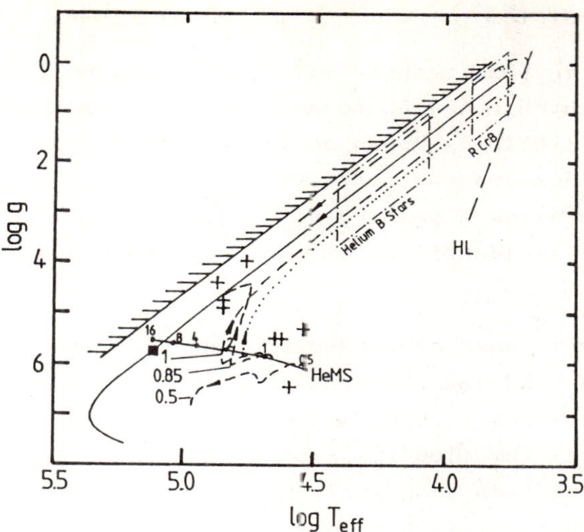


Figure 3: Comparison between observations and evolutionary tracks applicable for extremely helium-rich stars. The helium main sequence (HeMS) is labelled with stellar masses, HL is the Hayashi limit. The hatched line indicates the Eddington limit for pure helium composition. Evolutionary tracks are from Paczynski (1971; dashed lines labelled with $M/M_O$), Schönberner (1977, $M = 0.7\ M_O$, full drawn line) and Law (1982; $M = 1\ M_O$, dotted line). Stellar symbols are the same as in Fig. 1.

Table 3: Logarithm of mass fractions of abundant elements in four extremely helium-rich luminous sdOs, in the sun and in the Helium B stars (mean value derived from Heber, 1983).

|     | LSE 153 | LSE 259 | LSE 263 | BD+37°442 | sun | Helium B stars |
|-----|---------|---------|---------|-----------|-----|----------------|
| H   | <-1.7   | <-2.0   | <-1.7   | <-1.7     | -0.15 | <-3.5 |
| He  | -0.05   | -0.05   | -0.05   | -0.05     | -0.55 | -0.05 |
| C   | -1.3    | -1.2    | -3.6    | -1.6      | -2.40 | -1.7  |
| N   | -2.1    | -1.5    | -1.8    | -2.5      | -3.01 | -2.5  |
| Mg  | -2.5    | <-1.8   | <-2.8   |           | -3.23 | -3.0  |
| Si  | -2.9    |         |         | -2.9      | -3.16 | -3.1  |

## 4. EVOLUTIONARY STATUS

The discussion in this chapter will be restricted to those classes of hot stars for which metal abundances are available and described in section 3. Too little is known at present about the extremely helium-rich compact sdOs and the moderately helium-rich and helium-normal luminous sdOs to allow a meaningful consideration. CSPN, on the other hand, are clearly identified as post-AGB objects and do not need re-discussion.

Groth et al. (1985) have addressed the question as to why the sdB/sdOB stars on the one hand and the classical sdOs on the other display such remarkably different abundance patterns despite their close neighbourhood in the ($\log T_{eff}$-$\log g$) diagram. In their discussion they identified the sdB/sdOB stars with objects on the extended horizontal branch (EHB), i.e. with $q:=M_{core}/M_{total} \geq 0.95$. Because of the small envelope mass such stars populate the blue end of the horizontal branch (HB) close to the helium main sequence where surface gravities are high and gravitational settling reduces the surface abundances of all elements heavier than hydrogen. The remaining hydrogen mass, however, is too small to re-ignite and the sdB stars evolve directly into the sdOB domain and then quickly towards the blue, leaving little chance to observe them at temperatures $T_{eff} \geq 45000K$. An evolutionary sequence with q = 0.99 from Caloi and Castellani (1985) is plotted in Figure 2.

Greenstein and Sargent (1974) proposed that a helium-poor sdOB star would turn into a helium-rich sdO once it reaches a sufficiently high temperature to doubly ionize helium. The resultant convection zone should then reverse (or perhaps overcompensate) the preceding effect of gravitational settling. However, as Groth et al. (1985) have shown, no convection zone develops at photospheric depths when helium is already depleted. Therefore, an evolutionary link between sdOB and sdO stars can be ruled out. Instead, the conclusion is that the precursors of sdO stars must have had larger envelope masses (q < 0.95) on the HB to be located there at lower gravities. According to evolutionary calculations by Gingold (1976) such stars make a redward excursion after leaving the HB but do not ascend the AGB. It is presently still an open question of how CNO-processed material is brought to the surface (mass loss? convective mixing? helium shell flashes?) and at what time this happens. In any case, when these stars with high helium abundance evolve towards the blue and reach $T_{eff} \approx 42000K$ photospheric

convection  sets in and impedes gravitational settling  (see  Fig. 2).
Therefore, the sdO stars remain helium-rich even at higher gravities.

Presently,  no  definite conclusions about the evolutionary status  of
the  extremely  helium-rich stars and the way in which they have  lost
their  hydrogen envelope can be drawn although several scenarios  have
been  proposed.  In the white dwarfs (WD) merger scenarios of  Webbink
(1984)  and Iben and Tutukov (1985) it is assumed that a close  binary
system  - consisting of a CO- and a He-WD - contracts due to  gravitational  radiation  and  that finally both white dwarfs  merge  into  a
helium star that evolves up the AGB. Alternatively, one might - conti-
nuing  somewhat the discussion of Groth et al.  (1985) concerning  the
origin  of the classical sdOs - imagine that the precursors of  helium
stars are slightly more massive than the classical sdO precursors  and
therefore evolve further up the AGB. The stronger mass loss would then
lead  to an even more pronounced depletion of hydrogen.  However,  our
understanding of the mass loss phenomena in the red giant phase is too
poor  to decide whether it is really possible to remove nearly all  of
the hydrogen without triggering the (still unknown) PN ejection mechanism.  A third possibility has been studied by Iben et al.  (1983)  in
which  a  post-AGB star already in or close to the white  dwarf  stage
suffers  a  final shell flash consuming most or all of  the  remaining
hydrogen.  For  a very short time it becomes a cool supergiant  again,
before  it finally evolves - this time as a helium star - through  the
CSPN  domain until helium burning ceases and the star ultimately  ends
as  a  (DB-?)  WD.  In this case many of the luminous sdOs  should  be
surrounded by a very diluted PN ejected when the central star left the
AGB for the first time.  The evolution of stars with a C/O core and  a
helium envelope from the AGB through the domains of RCrB, helium B and
luminous  sdO  stars  has  been studied by Schönberner (1977)  (see
Fig. 3).  It  would  nicely  explain the agreement of  abundances  and
luminosities  in most of the extremely helium-rich luminous sdO and  B
stars.  However,  post-AGB evolution cannot explain the extreme helium
stars  in  the compact sdO/sdOB domain.  Instead,  it is  tempting  to
identify  these  as  stars on or close to  the  helium  main  sequence
(HeMS).  Even  the  luminous helium sdOs could be produced  by  direct
evolution  from  the HeMS without redward excursion if masses  of  the
order  of 0.85 $M_O$ are assumed (Paczynski,  1971).  However,  it is not
clear how the HeMS can be populated.  Nevertheless,  the diversity  of
possible evolutionary histories offers an explanation of the inhomoge-
nity  of  the luminous sdO star class (i.e.  varying He/H  ratios  and
carbon abundances).

In future work,  more information about inhomogenities in the class of
luminous sdOs should be collected.  Similarly, more data (particularly
metal  abundances)  are needed for the extremely helium-rich  sdOs  at
high  gravities.  Finally,  we urgently need _quantitative_  predictions
from   the  above-mentioned  evolutionary  scenarios  to  be  able  to
discriminate between them.


## ACKNOWLEDGEMENTS

I wish to thank Rolf-Peter Kudritzki and Ulrich Heber who  contributed
much  to  this review - directly and indirectly.  Also,  I  gratefully
acknowledge support by the Deutsche Forschungsgemeinschaft (DFG) under
grants Hu 39/21-1 and Ku 474/9-2 and by the Local Organizing Committee
of this colloquium.


## REFERENCES

Auer, L.H., Heasley, J.N.: 1976, Astrophys. J. _205_, 165
Caloi, V., Castellani, V.: 1985, private communication
Caughlan, G.R.: 1964, Astrophys. J. _141_, 688
Gingold, R.A.: 1976, Astrophys. J. _204_, 116
Greenstein, J.L., Sargent, A.I.: 1974, Astrophys. J. Suppl. Ser. _28_, 157
Groth, H.G., Kudritzki, R.P., Heber, U.: Astron. Astrophys. _152_, 107
Heap, S.R.: 1975, Astrophys. J. _196_, 195
Heber, U.: 1983, Astron. Astrophys. _118_, 39
Heber, U.: 1987, Proc. IAU Coll. 95 "The Second Conference on Faint
     Blue Stars", Davis press, in press
Heber, U., Hunger, K., Werner, K.: 1987a, Proc. IAU Coll. 132 "The
     Impact of Very High S/N Spectroscopy on Stellar Physics", in
     press
Heber, U., Kudritzki, R.P., Groth, H.G.: 1987b, in preparation for
     Astron. Astrophys.
Holweger, H.: 1979, Proc. XXII Colloque International D'Astrophysique,
     Université de Liege, p. 117
Husfeld, D.: 1986, Ph.D. thesis, University of Munich
Husfeld, D., Heber, U., Drilling, J.S.: 1986, Proc. IAU Coll. 87
     "Hydrogen-Deficient Stars and Related Objects", eds. K. Hunger,
     D. Schönberner, N.K. Rees, Reidel/Dordrecht, p. 353
Iben, J., Jr., Kaler, J.B., Truran, J.W., Renzini, A.: 1983, Astro-
     phys. J. _264_, 605

Iben, J, Jr., Tutukov, A.V.: 1985, Astrophys. J. Suppl. Ser. 58, 661

Kudritzki, R.P.: 1979, Proc. XXII Colloque International D'Astrophy-
    sique, Université de Liege, p. 295

Law, W.-Y.: 1982, Astron. Astrophys. 108, 118

Méndez, R.H., Kudritzki, R.P., Simor, K.P.: 1985, Astron. Astrophys.
    142, 289

Paczynski, B.: 1971, Acta Astronomica 21, 1

Schönberner, D.: 1977, Astron. Astrophys. 57, 437

Schönberner, D.: 1983, Astrophys. J. 272, 708

Vauclair, G., Vauclair, S., Greenstein, J.L.: 1979, Astron. Astrophys.
    80, 79

Webbink, R.F.: 1984, Astrophys. J. 277, 355

Werner, K.: 1986, Astron. Astrophys. 161, 177

Werner, K., Husfeld, D.: 1985, Astron. Astrophys. 148, 417

Wesemael, F.: 1979, Astron. Astrophys. 72, 104

Wesemael, F., Truran, J.W.: 1982, Astrophys. J. 260, 807

# EVOLUTIONARY HELIUM AND CNO ANOMALIES IN THE ATMOSPHERES AND WINDS OF MASSIVE HOT STARS


*Nolan R. Walborn*
Space Telescope Science Institute*
3700 San Martin Drive
Baltimore, Maryland 21218


## ABSTRACT


The ubiquitous evidence for processed material in the atmospheres, winds, and circumstellar ejecta of massive stars will be reviewed. A broad array of normal and peculiar evolutionary stages is considered, up to and including Type II supernova progenitors. The quantitative analysis of these spectra is difficult, and until recently for the most part only qualitative or approximate results have been available. However, several important current programs promise reliable abundance determinations, which will enable detailed comparisons with recent evolutionary calculations. A significant emerging result is that the morphologically normal majority of both hot and cool supergiants may already display an admixture of CNO-cycle products in their atmospheres. It may become possible in this way to identify blue supergiants returning from the red supergiant region, as appears to have been the case for the SN 1987A progenitor.


## I. INTRODUCTION

In this review I wish to develop two themes. First, that the evidence for processed material at the surfaces of evolving massive stars is virtually everywhere; in fact it is now reasonable to hypothesize that all massive stars display at least the products of hydrogen burning prior to their finales as Type II (or Ib?) supernovae. Second, that current progress both in the quantitative observation of abundance anomalies, and in their computation from theoretical evolutionary models, promises an imminent and significant synthesis. Consistent with the nature of the observational evidence, this review will be concerned essentially with effects of the CNO cycle, that is the enhancement of helium and nitrogen at the expense of hydrogen, carbon, and oxygen, with one exception (the WC stars). A wide range of specific topics is addressed, encompassing the OBN/OBC and related objects, including mass-transfer binaries; morphologically normal OB stars; the Luminous Blue Variables, including Eta Carinae; briefly, the Wolf-Rayet stars and late-type supergiants; a survey of the pertinent evolutionary calculations; and finally, Type II supernova progenitors, including that of SN 1987A. Evolutionary chemical anomalies will be seen to exist in the stellar atmospheres, winds, and/or ejected circumstellar material of all these classes of objects.

## II. OBN/OBC STARS AND RELATED OBJECTS

### A. Background

Two nitrogen-deficient, early-B supergiants described exactly twenty years ago by the Jascheks (1967) were the first known cases of CNO anomalies in OB absorption-line spectra. The first examples of the inverse phenomenon, OB spectra with enhanced nitrogen, were reported by Walborn (1970), and the OBN/OBC classification dichotomy was introduced shortly thereafter (Walborn 1971). While this notation has mnemonic value, it does incur the subtle disadvantage of suggesting, by analogy with the WR classification, a very advanced evolutionary state for the OBC stars. Such an implication was not intended, since the OBC stars, all supergiants, have hydrogen

---


* Operated by AURA, Inc. under contract with NASA.

lines of normal strength and in fact, as further discussed in the next section, the most likely interpretation is that they are *less* evolved than their morphologically normal relatives. A general review of the OBN/OBC categories and their possible interpretations was presented by Walborn (1976). Walborn and Panek (1985) showed that the optical anomalies are reflected in the ultraviolet stellar-wind features as well, significantly extending the ionization range over which they are seen in a given spectrum, and supporting their interpretation in terms of abundance effects (see also Walborn, Nichols-Bohlin, and Panek 1985).

## B. Current Programs

Two important current programs are contributing to alleviate the scarcity of quantitative abundance determinations for the OBN/OBC stars. Schönberner *et al.* (1987) have analyzed the optical spectra of four OBN and related dwarf or giant objects, finding the very significant result that not only the N/C ratios but also the He/H are larger than normal, thus establishing beyond a reasonable doubt that CNO-cycled material is indeed being observed. Wollaert, Lamers, and de Jager (1987) have compared equivalent widths in the ultraviolet spectra of fourteen OBN/OBC stars of all luminosity classes to those of normal stars, providing limits on the abundance anomalies by means of a curve-of-growth analysis.

## C. Individual Objects

### 1. HD 93840

This star is a new, extreme BN supergiant, the discovery of which I find quite pleasing. As described above, the OBN/OBC spectral anomalies were first detected optically, and correlated anomalies were later found to exist in the ultraviolet. These latter effects have now in turn permitted the discovery of the BN nature of HD 93840, through the careful analysis of Savage and Massa (1987). The UV spectrum shows a very strong N V P Cyg profile, with little or no corresponding C IV emission, which is highly abnormal (Walborn and Nichols-Bohlin 1987). Savage and Massa classified the star as BN1 II-Ib on the basis of the UV spectrum and previous optical types, while N. Houk classified it B1/2 Iab/b in the Michigan HD Catalogue. Its galactic latitude is +11°, in keeping with the relatively high latitudes of the OBN class (Walborn 1970; Bolton and Rogers 1978; Bisiacchi, López, and Firmani 1982). Further investigation of the optical spectrum is planned for the coming season.

### 2. HD 157038

This interesting supergiant was described optically as related to the OBN category, because while silicon and carbon appear normal, the strength of nitrogen would correspond to a higher temperature and that of magnesium to a lower (Walborn 1976, 1980). A quantitative optical analysis has been performed by Lennon and Dufton (1986), revealing enhanced nitrogen and helium abundances, which led them to suggest that this object may be evolving blueward following a red supergiant phase. Recently, J. Nichols-Bohlin and I in collaboration with D. Lennon and colleagues at Belfast have observed the 1200–1900 Å spectrum of HD 157038 at high resolution in a very long exposure with the International Ultraviolet Explorer. The stellar-wind features correspond well to a normal B3 Ia spectrum, which is perhaps not surprising since they are mostly due to C and Si at that type. A detailed analysis will be undertaken by Lennon *et al.*

## D. Mass-Transfer Binaries

It is possible that all OBN objects are a result of either active or previous mass transfer in close binary systems (Bolton and Rogers 1978). Two exceptionally interesting active systems, both of β Lyrae type, are HD 72754 and HD 163181. The optical spectrum of the former, including the nitrogen anomaly, was extensively discussed by Thackeray (1971), and a UV analysis is promised by de Freitas Pacheco and Codina Landaberry (1986). The anomalous optical CNO spectrum of HD 163181 was discovered by Walborn (1972) and analyzed by Kane, McKeith, and Dufton (1981). A brief description of the UV spectrum has been given by Hutchings and van Heteren (1981). In collaboration with J. Nichols-Bohlin, I have inspected the IUE high-resolution data and find an anomalously large ratio of N V/C IV stellar-wind features, together with some indications of spectral variability.

Significantly, Balachandran, Lambert, Tomkin, and Parthasarathy (1986) have recently established that the spectrum of $\beta$ Lyrae itself reveals the presence of CNO-cycle products. Its spectral type is too late for ready morphological detection of CNO anomalies in the blue-violet, but the quantitative analysis has shown that it is physically related to the BN binaries. Furthermore, Peters and Polidan (1984) have found evidence for extreme carbon deficiency in the high-temperature accreting material of a number of Algol-type binary systems. Hence it may be that not only are all OBN objects mass-transfer binaries, but many other mass-transfer binaries are related to the OBN category.

## III. MORPHOLOGICALLY NORMAL OB STARS

### A. Supergiants

All of the O9-B0 supergiants in the Orion Belt ($\delta$, $\epsilon$, and $\zeta$ Orionis) and in NGC 6231 (Scorpius OB1) have systematically nitrogen-deficient spectra, relative to the majority of supergiants of the same types (Walborn 1976). A prior result of nitrogen deficiency among the NGC 6231 main-sequence members, which would have favored an explanation in terms of initial abundances at formation, has recently been withdrawn (Brown et al. 1986b). An interpretation of the supergiant anomaly in terms of evolutionary processes is therefore indicated. The hypothesis proposed by Walborn (1976) was that the morphologically normal majority of OB supergiants actually display some fraction of CNO-cycled material at their surfaces, while the rarer, nitrogen-weak OBC objects such as those in the two associations, are relatively less evolved and have physically normal (i.e. main-sequence) surface abundances. This hypothesis receives substantial support from the very important current results of Bohannan, Voels, Abbott, and Hummer (1987), who find normal helium abundances in $\delta$ and $\zeta$ Orionis, but enhanced helium in the morphologically normal O9.5 Ia standard $\alpha$ Camelopardalis. These determinations corroborate the pioneering study by Baschek, Kodaira, and Scholz (1972), who found that the atmosphere of the morphologically normal O9.5 Iab supergiant HD 188209 contains an admixture of 20–40% of CNO-cycled material, relative to that of $\zeta$ Orionis. Kudritzki et al. (1987) have derived significant overabundances of helium and nitrogen in the LMC supergiants Sanduleak $-65°21$ (O9.5 Ia) and $-68°41$ (B0.5 Ia), and suggested that they are post-red supergiants; the spectra of these two objects are definitely morphologically normal, as one can easily see by comparing the high-quality digital data of Fitzpatrick (1987) to the spectrograms of OBN supergiants with similar types shown by Walborn (1971, 1972). Similarly, both Kudritzki, Simon, and Hamann (1983) and Bohannan, Abbott, Voels, and Hummer (1986) have found an enhanced helium abundance in $\zeta$ Puppis using state-of-the-art model-atmosphere techniques. Bohannan et al. (1987) have also suggested that the systematically smaller gravity effect shown by P. Conti's helium equivalent-width measures in O-type spectra, as compared to that for hydrogen, could be due to compensation by higher helium abundances in the supergiants. While quantitative analysis for a larger sample is certainly desirable before definitive conclusions are drawn, nevertheless the hypothesis that all OB supergiants mix CNO-cycle products into their atmospheres must now be considered very seriously.

### B. Main-Sequence Stars

CNO anomalies in main-sequence stars naturally suggest origins in initial conditions or mass transfer. However, Lyubimkov (1977, 1984) has presented surprising, detailed evidence for trends of increasing helium and nitrogen abundances with age in main-sequence and giant OB stars. Such an effect would present a new challenge for interpretation. On the other hand, the recent results of Brown et al. (1986a, b) do not appear to show these trends. Currently, D. R. Gies and D. L. Lambert are obtaining new, high-quality observational material for a welcome, independent investigation of this important question.

## IV. LUMINOUS BLUE VARIABLES

This term is now used to encompass the Hubble-Sandage, S Doradus, and P Cygni variables, as well as $\eta$ Carinae as an extreme case, for all of which there is an emerging unity of phenomenology and interpretation. These objects lie along the Humphreys-Davidson (1979) limit in the HR diagram, which evidently corresponds to an instability leading to episodic shell ejection and curtailing further redward evolution of the most massive stars. It is probable that this phenomenon represents the transition from OB to WN stages, at least for masses between about 50 and 100 $M_\odot$.

### A. HS, S Dor, P Cyg Stars

Evidence for an enhanced nitrogen abundance and/or a deficiency of carbon and oxygen in P Cygni itself was presented by Luud (1967), and in the related southern object AG Carinae by Caputo and Viotti (1970). Lamers, de Groot, and Cassatella (1983) discussed P Cygni as an object evolving between red supergiant and WN stages. The $\lambda4600$ regions in the beautiful and remarkably similar spectra of the HS stars AF Andromedae in M31 and Variable B in M33 shown by Kenyon and Gallagher (1985) are dominated by P Cyg profiles of the N II multiplet at 4601.5, 4607.2, 4613.9, 4621.4, 4630.5, and 4643.1 Å, contrary to statements by the authors. These spectra are essentially identical to those of P Cygni and AG Carinae (in 1976) themselves. The spectrum of the LMC S Dor-type variable HDE 269006 at minimum shown by Wolf, Appenzeller, and Stahl (1981) has the same multiplet dominating in absorption. More detailed analyses of all these spectra are likely to prove rewarding and demonstrate significant overabundances of nitrogen. Enhanced nitrogen abundances have also been found in the probably related objects HD 38489 in the LMC by Shore and Sanduleak (1983), and S18 in the SMC by Shore, Sanduleak, and Allen (1987).

I have been particularly interested in a category of LMC objects now designated Ofpe/WN9, which combine high-excitation Of characteristics with lower-excitation emission and P Cyg spectral features (Walborn 1977). One of these objects, HDE 269858 (=R127) has subsequently undergone an outburst followed in detail by Stahl et al. (1983), Stahl and Wolf (1986a), and Wolf, Stahl, Smolinski, and Cassatella (1987), and it is now regarded as the hottest known S Dor variable. An important result of high-resolution observations of these objects with the Cerro Tololo 4-meter echelle spectrograph was the discovery of velocity-doubled forbidden lines, implying nonspherical circumstellar nebulae, for which an overabundance of nitrogen and an underabundance of oxygen were found (Walborn 1982). Stahl (1987) has resolved two of these shells by direct imagery, while Stahl and Wolf (1986b) have found spectroscopic evidence for similar structures around a wider range of related LMC objects, including HDE 269006 and S Doradus itself.

Recently, Leitherer and Zickgraf (1987) have detected the circumstellar shell of P Cygni by direct imaging, finding evidence for an abnormally large [N II]/H$\alpha$ ratio. Two other objects which are probably galactic counterparts of the LMC Ofpe/WN9 stars are HD 148937 (Walborn 1973) and AG Carinae (Stahl 1986); both have spectacular ejected circumstellar nebulae. Leitherer and Chavarría (1987) have established a nitrogen overabundance in the HD 148937 nebula (NGC 6164–6165). The AG Carinae nebula has been described by Thackeray (1977) and Viotti, Cassatella, Ponz, and Thé (1987); it may well be nitrogen-rich, but an abundance analysis remains to be done.

### B. Eta Carinae

This famous object may be considered the most spectacular LBV, on the basis of its luminosity and outburst magnitude. Its location in a giant H II region is also atypical, and its mass is probably substantially greater than 100 $M_\odot$. From the fact that no such object is observed in 30 Doradus, despite the much larger early O and WR populations, one may infer that it represents a very short-lived phase. Because of the spectroscopic continuity among the O3 If and WN-A stars in the Carina Nebula (Walborn 1974), I suspect that Eta Carinae is a *post*-WNL object. An important step toward understanding its nature was the discovery that the expanding knots in its complex outer shell are nitrogen- and helium-rich (Davidson, Walborn, and Gull 1982; Davidson, Dufour, Walborn, and Gull 1986). These knots have simpler optical and ultraviolet nebular spectra than the denser inner regions, displaying five ionization stages of N and none of C or O. The helium overabundance has been nicely corroborated by the infrared line analysis of Allen, Jones, and Hyland (1985). The compositions and velocities of the knots are highly reminiscent of the quasi-stationary flocculi in Cassiopeia A, believed to be circumstellar material ejected by the progenitor star, making Eta Carinae a strong presupernova candidate.

## V. WOLF-RAYET STARS

I shall cite only briefly the most significant recent results bearing on the chemical compositions of these key objects, since they are the subject of other papers at this meeting which promise further advances. The WR stars are probably the first to have revealed evolutionary abundance anomalies, but unfortunately the same extravagance which facilitated their discovery has obstructed quantitative analysis. A new idea which has the markings of a breakthrough toward a definitive model is that of very high underlying surface temperatures, which may be capable of ionizing and driving the envelopes radiatively (Cherepashchuk, Eaton, and Khaliullin 1984; Pauldrach, Puls, Hummer, and Kudritzki 1985; D. Abbott, private communication).

The status of WR abundance analyses through IAU Symposium 99 was very well summarized by Willis (1982); see also Linda J. Smith and Willis (1982, 1983). Since then, additional information on the He/H ratios has been provided by Conti, Leep, and Perry (1983), while substantial progress on modeling the He spectra has been contributed by Hamann and Schmutz (1987 and references therein) and by Hillier (1987a, b). The WC stars are special, as the only well-defined class of massive stars showing us the products of helium burning. Currently, improved determinations of their C/He ratios are being made by Lindsey F. Smith and Hummer (1987) from near infrared spectra, and by Torres (1987) optically.

A number of WN stars also have circumstellar ring nebulae, for which abundance determinations are of obvious relevance. Nitrogen and helium enrichments have been demonstrated in several of them by Talent and Dufour (1979), Contini and Shaviv (1980), and Kwitter (1981, 1984).

## VI. LATE-TYPE SUPERGIANTS

This review is about hot massive stars, but at least up to about 50 $M_\odot$ they are either destined to become or have already been late-type supergiants. Hence for completeness, and as a preliminary to the final two sections below, a brief survey of the impressive evidence for CNO-cycle products in the atmospheres of the latter is also of interest. An enhancement of nitrogen and/or deficiency of carbon, and in some cases anomalous isotope ratios of C and O, have been observed in the following objects and classes: Cepheids and related nonvariable supergiants (Luck and Lambert 1981, 1985); Canopus (Desikachary and Hearnshaw 1982); Betelgeuse and Antares (Harris and Lambert 1984; Lambert, Brown, Hinkle, and Johnson 1984); and Polaris (Luck and Bond 1986).

## VII. EVOLUTIONARY CALCULATIONS OF SURFACE ABUNDANCES

An extensive array of theoretical calculations of evolving surface abundances is available for comparison with the observations discussed in the previous sections. Intermediate-mass stars up to 11 $M_\odot$ have been considered by Becker and Iben (1979, 1980) and Becker and Cox (1982); the mechanism is convective dredge-up, and the results are applicable to the Cepheids and related supergiants. For massive stars, stellar-wind mass loss becomes the key mechanism; results for 15–150 $M_\odot$ single stars have been computed and in several cases compared with observations by Stothers and Chin (1979); Noels, Conti, Gabriel, and Vreux (1980); Gabriel and Noels (1981); Noels and Gabriel (1981); Brunish and Truran (1982a, b); Doom (1982a, b); Dearborn and Blake (1984); Pylyser, Doom, and de Loore (1985); Prantzos, Doom, Arnould, and de Loore (1986); and Langer and El Eid (1986). Rotationally induced mixing has been considered by Maeder (1987b). Results for mass-transfer binaries are presented by Vanbeveren and Doom (1980); Vanbeveren (1982, 1983); and Doom and de Grève (1983).

The most detailed charting of evolving surface abundances in the HR diagram for massive stars has been carried out by Maeder (1983; 1987a) and Maeder and Meynet (1987). In principle, very fine comparisons with the observations are enabled; the precision of the latter is now the limiting factor, but that situation is currently improving as discussed above. An enormous amount of theoretical information is made usefully accessible by this work, about which you will hear more in the following talk. I wish to emphasize only a few points here. Stars with initial masses between 12 and about 20 $M_\odot$ will explode as red supergiants, after revealing CNO-cycle products in their cool atmospheres. Stars with masses between about 20 and 50 $M_\odot$ *return* from the red supergiant region, displaying CNO-cycled material during their blueward evolution, and becoming blue-supergiant or possibly WR supernova progenitors. This history most likely describes SN 1987A. Finally, stars more massive than about 50 $M_\odot$ reveal processed material in their atmospheres soon after leaving the main sequence, and they never reach a red supergiant stage in agreement with observation (Humphreys and Davidson 1979). Thus all supernova progenitors are expected to have CNO-cycle products in their atmospheres or circumstellar environments, except perhaps for those very massive objects which reach a WC stage. It should be noted in passing that there may also be an *upper* mass limit for the latter, not yet reflected in these models, since few or no WC stars are observed in giant H II regions (see also Langer 1987).

## VIII. TYPE II SUPERNOVA PROGENITORS

The observations are in excellent accord with the above expectation: in every instance for which information is available, there is evidence for CNO-cycled material at the surfaces or in the vicinities of massive supernova progenitors. The classical case is the Cassiopeia A remnant, which consists of nitrogen-rich, relatively slow-moving "quasi-stationary flocculi", and oxygen-rich, fast-moving knots; the former have been interpreted as circumstellar material ejected by the progenitor, while the latter are interior debris from the explosion itself (Peimbert and van den Bergh 1971; Chevalier and Kirshner 1978; Lamb 1978). Recently, fast-moving nitrogen-rich knots have been discovered at the periphery of the Cas A remnant, which are consistent with explosive ejection from the stellar surface (Fesen, Becker, and Blair 1987). Puppis A has now been shown to have a dichotomous composition/kinematical structure of nitrogen-rich, slow and oxygen-rich, fast material entirely analogous to that of Cas A (Winkler and Kirshner 1985). Nitrogen-rich filaments have also been found in W50, the remnant associated with SS433 (Kirshner and Chevalier 1980 and references therein). The remnant of Kepler's supernova may provide yet another example (van den Bergh and Kamper 1977; Dennefeld 1982; Bandiera 1987). It would be of considerable interest to search for nitrogen-rich features near other young SNR.

But the evidence is not limited to remnants; nitrogen-rich material has in fact been detected in three Type II SN in action. Fransson *et al.* (1984) derived a very large N/C ratio from the ultraviolet spectrum of 1979C in M100 shortly after maximum. Niemela, Ruíz, and Phillips (1985) observed a WN-like optical spectrum from 1983K in NGC 4699 in unprecedented pre-maximum coverage.

And now there is SN 1987A in the Large Magellanic Cloud for our delight and edification. So far as is known, the best spectroscopic observations of the progenitor star, Sanduleak $-69°202$ (Walborn, Lasker, Laidler, and Chu 1987), consist of 460 Å/mm and 110 Å/mm (at H$\gamma$) objective-prism plates obtained at the European Southern Observatory, the B3 I classification being based upon the latter (Rousseau *et al.* 1978). González *et al.* (1987) have investigated the lower-resolution material in comparison with another, normal B3 Ia object in the LMC observed similarly, finding that a blend of N II and He I lines is considerably stronger in the progenitor spectrum. Unfortunately, the interpretation is ambiguous, since the effect could be due either to a chemical anomaly or to a slightly earlier spectral type (say B2) of Sk $-69°202$. The higher-resolution material confirms that the features in question are stronger in the latter (L. Prévot, private communication), but without resolving the composition/temperature ambiguity as yet; further investigation of these important spectrograms is planned. Meanwhile, definitive evidence for nitrogen-rich material associated with SN 1987A is provided by the current far-ultraviolet, narrow emission-line spectrum reported by Wamsteker, Gilmozzi, Cassatella, and Panagia (1987) and Kirshner *et al.* (1987). This spectrum most likely arises in circumstellar material being excited by the outburst, and it is extremely similar to those of the $\eta$ Carinae knots (Davidson, Walborn, and Gull 1982; Davidson, Dufour, Walborn, and Gull 1986). Hence, the direct observation of Type II supernovae confirms the presence of CNO-cycled material in or around the progenitors, as predicted by both the observations and theory of massive stars reviewed here.

Thanks to Dorothy Whitman for her consistently outstanding manuscript support.

## REFERENCES

Allen, D. A., Jones, T. J., and Hyland, A. R. 1985, *Ap. J.* **291**, 280.

Balachandran, S., Lambert, D. L., Tomkin, J., and Parthasarathy, M. 1986, *M.N.R.A.S.* **219**, 479.

Bandiera, R. 1987, *Ap. J.* **319**, 885.

Baschek, B., Kodaira, K., and Scholz, M. 1972, *Ap. Letters* **12**, 227.

Becker, S. A. and Cox, A. N. 1982, *Ap. J.* **260**, 707.

Becker, S. A. and Iben, I. Jr. 1979, *Ap. J.* **232**, 831.

__________. 1980, *Ap. J.* **237**, 111.

Bisiacchi, G. F., López, J. A., and Firmani, C. 1982, *Astr. Ap.* **107**, 252.

Bohannan, B., Abbott, D. C., Voels, S. A., and Hummer, D. G. 1986, *Ap. J.* **308**, 728.

Bohannan, B., Voels, S. A., Abbott, D. C., and Hummer, D. G. 1987, I.A.U. Symposium 132.

Bolton, C. T. and Rogers, G. L. 1978, *Ap. J.* **222**, 234.

Brown, P. J. F., Dufton, P. L., Lennon, D. J., Keenan, F. P., and Kilkenny, D. 1986a, *Astr. Ap.* **155**, 113.

Brown, P. J. F., Dufton, P. L., Lennon, D. J., and Keenan, F. P. 1986b, *M.N.R.A.S.* **220**, 1003.

Brunish, W. M. and Truran, J. W. 1982a, *Ap. J.* **256**, 247.

__________. 1982b, *Ap. J. Suppl.* **49**, 447.

Caputo, F. and Viotti, R. 1970, *Astr. Ap.* **7**, 266.

Cherepashchuk, A. M., Eaton, J. A., and Khaliullin, Kh. F. 1984, *Ap. J.* **281**, 774.

Chevalier, R. A. and Kirshner, R. P. 1978, *Ap. J.* **219**, 931.

Conti, P. S., Leep, E. M., and Perry, D. N. 1983, *Ap. J.* **268**, 228.

Contini, M. and Shaviv, G. 1980, *Astr. Ap.* **88**, 117.

Davidson, K., Dufour, R. J., Walborn, N. R., and Gull, T. R. 1986, *Ap. J.* **305**, 867.

Davidson, K., Walborn, N. R., and Gull, T. R. 1982, *Ap. J. (Letters)* **254**, L47.

Dearborn, D. S. P. and Blake, J. B. 1984, *Ap. J.* **277**, 783.

de Freitas Pacheco and Codina Landaberry 1986, *Rev. Mexicana Astron. Astrof.* **12**, 184.

Dennefeld, M. 1982, *Astr. Ap.* **112**, 215.

Desikachary, K. and Hearnshaw, J. B. 1982, *M.N.R.A.S.* **201**, 707.

Doom, C. 1982a, *Astr. Ap.* **116**, 303.

__________. 1982b, *Astr. Ap.* **116**, 308.

Doom, C. and de Grève, J. P. 1983, *Astr. Ap.* **120**, 97.

Fesen, R. A., Becker, R. H., and Blair, W. P. 1987, *Ap. J.* **313**, 378.

Fitzpatrick, E. L. 1987, I.A.U. Symposium 132.

Fransson, C., Benvenuti, P., Gordon, C., Hempe, K., Palumbo, G. G. C., Panagia, N., Reimers, D., and Wamsteker, W. 1984, *Astr. Ap.* **132**, 1.

Gabriel, M. and Noels, A. 1981, *Astr. Ap.* **94**, L1.

González, R., Wamsteker, W., Gilmozzi, R., Walborn, N., and Lauberts, A. 1987, ESO SN1987A Workshop.

Hamann, W.-R. and Schmutz, W. 1987, *Astr. Ap.* **174**, 173.

Harris, M. J. and Lambert, D. L. 1984, *Ap. J.* **281**, 739.

Hillier, D. J. 1987a, *Ap. J. Suppl.* **63**, 947.

__________. 1987b, *Ap. J. Suppl.* **63**, 965.

Humphreys, R. M. and Davidson, K. 1979, *Ap. J.* **232**, 409.

Hutchings, J. B. and van Heteren, J. 1981, *Pub. A.S.P.* **93**, 626.

Jaschek, M. and Jaschek, C. 1967, *Ap. J.* **150**, 355.

Kane, L. G., McKeith, C. D., and Dufton, P. L. 1981, *M.N.R.A.S.* **194**, 537.

Kenyon, S. J. and Gallagher, J. S. III 1985, *Ap. J.* **290**, 542.

Kirshner, R. P. and Chevalier, R. A. 1980, *Ap. J. (Letters)* **242**, L77.

Kirshner, R., Sonneborn, G., Cassatella, A., Gilmozzi, R., Wamsteker, W., and Panagia, N. 1987, *I.A.U. Circ.*, No. 4435.

Kudritzki, R. P., Groth, H. G., Butler, K., Husfeld, D., Becker, S., Eber, F., and Fitzpatrick, E. 1987, ESO SN1987A Workshop.

Kudritzki, R. P., Simon, K. P., and Hamann, W.-R. 1983, *Astr. Ap.* **118**, 245.

Kwitter, K. B. 1981, *Ap. J.* **245**, 154.

__________. 1984, *Ap. J.* **287**, 840.

Lamb, S. A. 1978, *Ap. J.* **220**, 186.

Lambert, D. L., Brown, J. A., Hinkle, K. H., and Johnson, H. R. 1984, *Ap. J.* **284**, 223.

Lamers, H. J. G. L. M., de Groot, M., and Cassatella, A. 1983, *Astr. Ap.* **123**, L8.

Langer, N. 1987, *Astr. Ap.* **171**, L1.

Langer, N. and El Eid, M. F. 1986, *Astr. Ap.* **167**, 265.

Leitherer, C. and Chavarría K., C. 1987, *Astr. Ap.* **175**, 208.

Leitherer, C. and Zickgraf, F.-J. 1987, *Astr. Ap.* **174** 103.

Lennon, D. J. and Dufton, P. L. 1986, *Astr. Ap.* **155**, 79.

Luck, R. E. and Bond, H. E. 1986, *Pub. A.S.P.* **98**, 442.

Luck, R. E. and Lambert, D. L. 1981, *Ap. J.* **245**, 1018.

__________. 1985, *Ap. J.* **298**, 782.

Luud, L. S. 1967, *Soviet Astron.* **11**, 211

Lyubimkov, L. S. 1977, *Astrophysics* **13**, 71.

__________. 1984, *Astrophysics* **20**, 255.

Maeder, A. 1983, *Astr. Ap.* **120**, 113.

__________. 1987a, *Astr. Ap.* **173**, 247.

__________. 1987b, *Astr. Ap.* **178**, 159.

Maeder, A. and Meynet, G. 1987, *Astr. Ap.* **182**, 243.

Niemela, V. S., Ruíz, M. T., and Phillips, M. M. 1985, *Ap. J.* **289**, 52.

Noels, A., Conti, P. S., Gabriel, M., and Vreux, J.-M. 1980, *Astr. Ap.* **92**, 242.

Noels, A. and Gabriel, M. 1981, *Astr. Ap.* **101**, 215.

Pauldrach, A., Puls, J., Hummer, D. G., and Kudritzki, R. P. 1985, *Astr. Ap.* **148**, L1.

Peimbert, M. and van den Bergh, S. 1971, *Ap. J.* **167**, 223.

Peters, G. J. and Polidan, R. S. 1984, *Ap. J.* **283**, 745.

Prantzos, N., Doom, C., Arnould, M., and de Loore, C. 1986, *Ap. J.* **304**, 695.

Pylyser, E., Doom, C., and de Loore, C. 1985, *Astr. Ap.* **148**, 379.

Rousseau, J., Martin, N., Prévot, L., Rebeirot, E., Robin, A., and Brunet, J. P. 1978, *Astr. Ap. Suppl.* **31**, 243.

Savage, B. D. and Massa, D. 1987, *Ap. J.* **314**, 380.

Schönberner, D., Herrero, A., Becker, S., Eber, F., Butler, K., Kudritzki, R. P., and Simon, K. P. 1987, *Astr. Ap.,* in press.

Shore, S. N. and Sanduleak, N. 1983, *Ap. J.* **273**, 177.

Shore, S. N., Sanduleak, N., and Allen, D. A. 1987, *Astr. Ap.* **176**, 59.

Smith, L. F. and Hummer, D. G. 1987, preprint.

Smith, L. J. and Willis, A. J. 1982, *M.N.R.A.S.* **201**, 451.

__________. 1983, *Astr. Ap. Suppl.* **54**, 229.

Stahl, O. 1986, *Astr. Ap.* **164**, 321 and **170**, 197.

__________. 1987, *Astr. Ap.* **182**, 229.

Stahl, O. and Wolf, B. 1986a, *Astr. Ap.* **154**, 243.

__________. 1986b, *Astr. Ap.* **158**, 371.

Stahl, O., Wolf, B., Klare, G., Cassatella, A., Krautter, J., Persi, P., and Ferrari-Toniolo, M. 1983, *Astr. Ap.* **127**, 49.

Stothers, R. and Chin, C.-W. 1979, *Ap. J.* **233**, 267.

Talent, D. L. and Dufour, R. J. 1979, *Ap. J.* **233**, 888.

Thackeray, A. D. 1971, *M.N.R.A.S.* **154**, 103.

__________. 1977, *M.N.R.A.S.* **180**, 95.

Torres, A. V. 1987, preprint.

Vanbeveren, D. 1982, *Astr. Ap.* **105**, 260.

__________. 1983, *Astr. Ap.* **119**, 239.

Vanbeveren, D. and Doom, C. 1980, *Astr. Ap.* **87**, 77.

van den Bergh, S. and Kamper, K. W. 1977, *Ap. J.* **218**, 617.

Viotti, R., Cassatella, A., Ponz, D., and Thé, P. S. 1987, *Astr. Ap.,* in press.

Walborn, N. R. 1970, *Ap. J. (Letters)* **161**, L149.

__________. 1971, *Ap. J. (Letters)* **164**, L67.

__________. 1972, *Ap. J. (Letters)* **176**, L119.

__________. 1973, *A. J.* **78**, 1067.

__________. 1974, *Ap. J.* **189**, 269.

__________. 1976, *Ap. J.* **205**, 419.

__________. 1977, *Ap. J.* **215**, 53.

__________. 1980, *Ap. J. Suppl.* **44**, 535.

__________. 1982, *Ap. J.* **256**, 452.

Walborn, N. R., Lasker, B. M., Laidler, V. G., and Chu, Y.-H. 1987, *Ap. J. (Letters)* **321**, L41.

Walborn, N. R. and Nichols-Bohlin, J. 1987, *Pub. A.S.P.* **99**, 40.

Walborn, N. R., Nichols-Bohlin, J., and Panek, R. J. 1985, *International Ultraviolet Explorer Atlas of O-Type Spectra from 1200 to 1900 Å*, NASA RP-1155.

Walborn, N. R. and Panek, R. J. 1985, *Ap. J.* **291**, 806.

Wamsteker, W., Gilmozzi, R., Cassatella, A., and Panagia, N. 1987, *I.A.U. Circ.*, No. 4410.

Willis, A. J. 1982, I.A.U. Symposium 99, p. 87.

Winkler, P. F. and Kirshner, R. P. 1985, *Ap. J.* **299**, 981.

Wolf, B., Appenzeller, I., and Stahl, O. 1981, *Astr. Ap.* **103**, 94.

Wolf, B., Stahl, O., Smolinski, J., and Cassatella, A. 1987, *Astr. Ap.,* in press.

Wollaert, J. P. M., Lamers, H. J. G. L. M., and de Jager, C. 1987, *Astr. Ap.,* in press.

THE POINT ON THE THEORETICAL CHANGES OF SURFACE
CHEMISTRY DURING MASSIVE STAR EVOLUTION

André Maeder
Geneva Observatory, CH-1290 Sauverny


# 1. INTRODUCTION

For main sequence stars, the central nuclear processing generally has no effect on surface abundances. Later in the evolution, the newly synthetized elements may be revealed at the stellar surface by processes such as mass loss, convective dredge-up, overshooting, diffusion, rotational and tidal mixing, etc. The changes of CNO abundances are the most conspicuous and the easiest to observe spectroscopically; some abundance ratios like C/N, O/N may undergo changes by more than $10^2$. On the whole, surface chemistry is a most powerful diagnostics of stellar evolution, model assumptions and nuclear cross sections.

# 2. MAPPING OF CNO ABUNDANCES IN THE HR DIAGRAM

For massive stars, the peeling-off by stellar winds (1) contributes to big changes of chemical abundances (2, 3, 4, 5, 6, 7). Several of the quoted works present the time evolution of surface chemistry and reference is made to them for detailed information. Figs. 1 and 2 illustrate the changes of the $^{12}C/^{14}N$ and $^{16}O/^{14}N$ ratios along the evolutionary tracks in the HR diagram for models (7) with moderate overshooting, i.e. with an overshooting distance $d_{over}$=0.25 $H_p$ (8). Similar data for models without (6) or with considerable overshooting (5) exist, and such mappings can also be made for other interesting chemical ratios. Along the tracks, we distinguish the following abundance sequence:

-1. Initial cosmic abundances: C/N=4, O/N=9 in mass fraction (as always here).

-2. Intermediate abundances: for initial $M \geq 60$ $M_\odot$ the transition from a cosmic to an equilibrium C/N ratio ($\cong 10^{-2}$) is abrupt near the end of the main sequence. For O/N, the transition is more progressive. Changes of minor species due to the Ne-Na, Mg-Al cycles also occur (9, 5). For $M \leq 40$ $M_\odot$, changes of abundances only occur from the red supergiant (RSG) stage and onwards, convective dilution in large envelopes making these changes rather limited and stepwise. A second plateau may be produced if the intermediate fully convective zone appears at the stellar surface. Many other changes of abundances occur like the strong decrease in $^3$He, $^{15}$N and $^{18}$O.

-3. CNO equilibrium: it can be reached by massive stars in the Luminous Blue Variable (LBV) stage and by WN stars. The changes in H and He contents are rather smooth and cases of CNO equilibrium with H (LBV and WNL stars) and without H (WNE stars) may be distinguished (see also §6).

-4. Products of partial He-burning: they are visible only in WC and WO stars. C and O appear prominently at the stellar surfaces. The appearance of these elements is predicted to be quite fast because there is usually a large chemical discontinuity at the edge of the He-burning core (cf. 3).

On the whole, the smaller the initial mass, the earlier the above itinerary through the abundance sequence is stopped. Also, a star with overshooting behaves to first order like a more massive star without overshooting.

## 3. O-STARS:  DO SOME EVOLVE HOMOGENEOUSLY?

Models with initial masses M<80 $M_\odot$ and no overshooting keep their original abundance during main sequence (MS) evolution (6). Above this mass, models predict CNO equilibrium only near the end of the MS phase. For an overshooting of $d_{over}$=0.25 $H_p$, this limit is about 60 $M_\odot$ (cf. Figs. 1 and 2). On the whole, both these results agree with the observations that most MS O-stars have solar abundances (e.g. 10, 11, 12).

The subgroups of OBN and OBC show N or C enhancement (13). Do the OBC have just the original unmodified abundance, while the so-called normal O-stars already exhibit evidence of mild CNO processing? This is an attractive suggestion (14);  to be fully confirmed it would need detailed abundance analyses for a large group of O-stars in young clusters. Four ON stars, analysed in detail (12), show C/N, O/N ratios and He contents typical of advanced CNO processing. Quite unexpectedly, two of these ON stars lie close to the zero-age sequence.

Suggestions have been made that ON stars result from mass transfer in close binaries (13, 15), from mass loss by stellar winds (16), or from convective overshooting (17). The models in Figs. 1 and 2, which include overshooting and mass loss at the observed rates, clearly show that these two processes are unable alone to account for

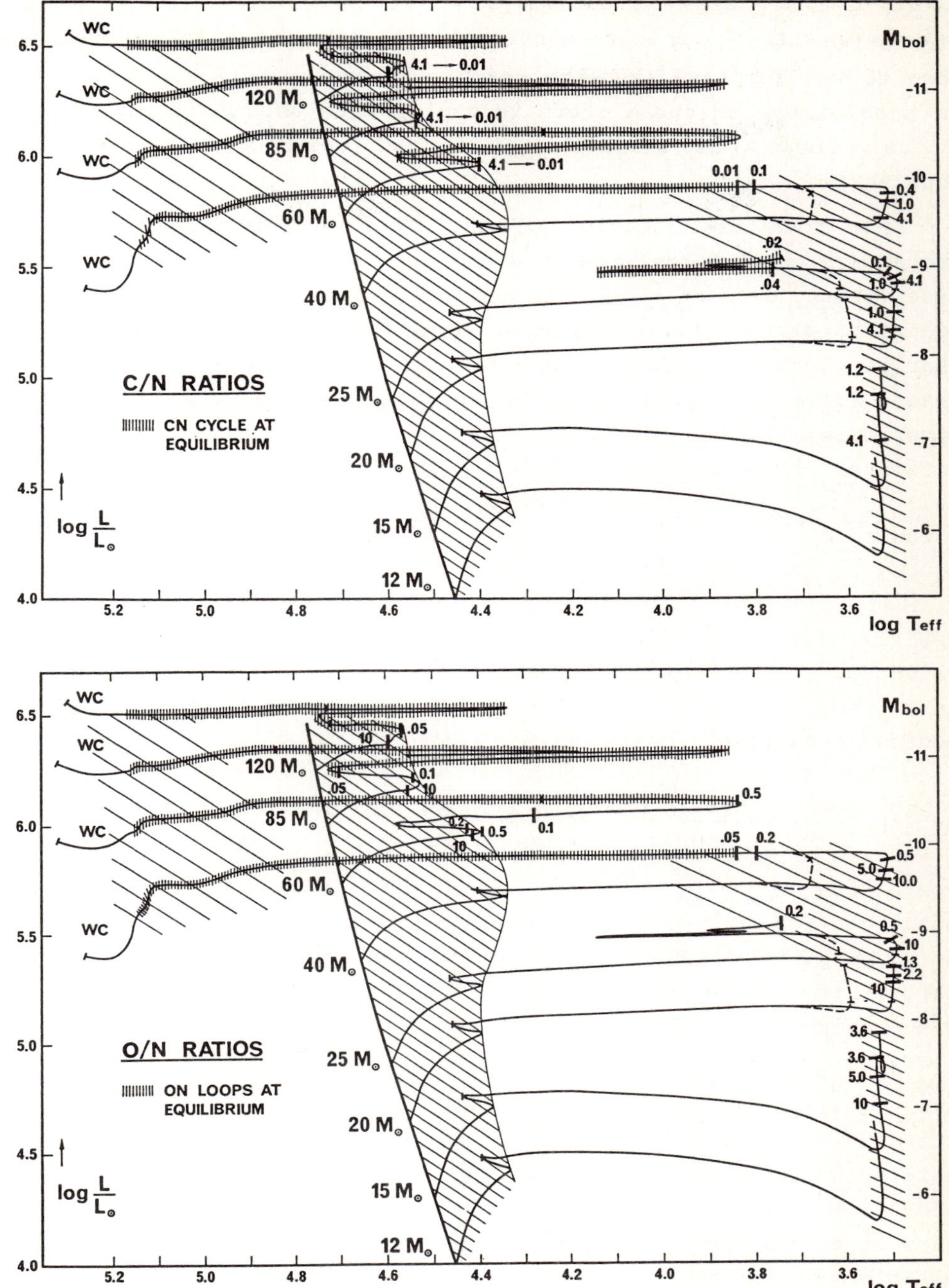


Figs. 1 and 2: The values of the abundance ratios $^{12}C/^{14}N$ and $^{16}O/^{14}N$ along the evolutionary tracks (7).

the ON stars close to the zero-age sequence. This is especially interesting since the occurrence of N-rich stars as blue stragglers close to the zero-age sequence seems quite general (e.g. 18, 19, 20). The possibility of these stars belonging to a second star generation in the clusters is unlikely in view of their very high chemical peculiarity.

From both the location of the ON blue stragglers in the HR diagram and their large N-enhancement, it has been suggested (19) that these stars evolve close to chemical homogeneity. As possible mechanism, distorsions either due to rotation or tidal interaction in binaries could induce baroclinic instabilities (21), i.e. instabilities occurring when the surfaces of constant pressure and temperature do not coincide. Thus matter can move freely in a horizontal direction and create a two-dimensional turbulence. At small scales, the cascade of turbulence produces a three-dimensional diffusion (21). Evolutionary models including such a diffusive mixing induced by rotation indicate (19) that full mixing may occur in cases of high rotation. The relative role of rotation, tidal mixing and mass transfer in the formation of ON blue stragglers is still to be evaluated. In any case, the homogeneous evolution of a fraction of O-stars would have great implications for nucleosynthesis and galactic chemical evolution.

## 4. BLUE SUPERGIANTS IN RELATION WITH SN 1987 A

Stars in the location of LBV (luminous blue variable), i.e. blue supergiants with $M_{bol} < -10$, are predicted to exhibit CNO equilibrium abundances (cf. Fig. 1 and 2), whether or not overshooting is present. The observations (22) for $\eta$ Carinae and the models agree, which confirms the evolutionary status of this intriguing object as a post-MS supergiant.

Blue supergiants on the first redwards tracks are predicted to have normal cosmic abundances, whether or not overshooting is present. According to the models (6, 7), the blue supergiants exhibiting CN equilibrium and ON intermediate values should be in a post red supergiant (RSG) stage. (This statement could be revised if such stars have undergone some substantial diffusion). Interestingly, the pulsation properties of blue supergiants on the first and second crossing are also different (23).

There are very few detailed abundance analyses for blue supergiants. A study for two stars indicates (24) one star with normal abundance and the other with marked evidence of CNO processing.

Two blue supergiants in the LMC also show (25) CNO processed elements. These few results show that at least some blue supergiants are in a post RSG stage. This is quite interesting in relation with the large N/C and N/O ratios shown by UV spectra (26, 27) of SN 1987 A, which also support the idea that the SN precursor was in a post RSG stage. Analyses for a large number of blue supergiants are necessary to give the percentage of stars on the blue- and redwards crossing. A list of candidate blue supergiants in young clusters has been selected (6).

## 5. RED SUPERGIANTS:  TEST OF NUCLEAR CROSS SECTIONS

RSG are predicted to exhibit CNO processed elements with various rates of dilution according to initial mass. In a given area of the HR diagram, the predicted C/N and O/N ratios present some significant scatter (cf. Figs. 1 and 2). Thus great care has to be taken about conclusions drawn from the comparisons between models and observations. The M1.5 Iab stars $\alpha$ Ori and $\alpha$ Sco have C/N, $^{12}C/^{13}C$ and O/N ratios characteristic of CNO processing (28, 29). The observed values indicate more advanced nuclear processing than predicted; however care has to be taken regarding these conclusions, for the reasons given above.

The comparisons for $^{16}O$, $^{17}O$, $^{18}O$ in the two mentioned red supergiants show (6) rough agreement between observations and models regarding the isotopes $^{16}O$ and $^{18}O$. For $^{17}O$ the predicted abundance is too high by a factor of about 25. A discussion of the various intervening reactions suggests (6) that the reaction rate of $^{17}O(p,\alpha)^{14}N$ has to be pushed to its higher resonant limit (factor f=1).

## 6. WR STARS:  DIFFERENT SENSITIVITY OF WN AND WC STARS TO MODEL PHYSICS

Various comparisons (30, 2, 31, 32, 5, 6, 7) of models and observations confirm that the sequence of WN, WNE, WC and WO is consistent with a progression in the exposition of nuclearly processed materials:

    WNL (WN6-WN9):  H, He, N
    WNE (WN2-WN6):  He, N, no H left
    WC            :  He, C, O, no H and no N left
    WO            :  same as WC, but with larger O/C

Some exceptions to this scheme exist. Moreover, the range of initial masses does not seem to be identical for stars of various subtypes.

For WN stars, the typical abundance ratios are $C/He = (1.8-3.9) \cdot 10^{-4}$, $N/H = (1.3-1.7) \cdot 10^{-2}$, $C/N = (1.1-2.9) \cdot 10^{-2}$ and $C/O = 0.25-1.3$ in mass fraction (7). These ratios depend very little on initial masses, on the exact mass loss rates, on overshooting etc. They are essentially model-independent as normally expected for equilibrium ratios. All model results agree well with observations, particularly for the C/N ratio, which is the most accurate observationally. This is not a success for the models, it just means that the abundances in WN stars are not a very constraining test for the model assumptions such as mixing. However, the above agreement implies that the cross sections for CNO burning are correct, which in itself is an essential result for stellar evolution.

During the WC stage, the chemical abundances change quite a lot. since materials which are more and more processed are progressively revealed at the stellar surface. The values of C/He and O/C at the beginning of the WC phase are very sensitive to overshooting: the larger the cores, the lower the initial C/He and O/C ratios (e.g. 7). One has $C/He = 0.9-3$, $0.3-3$, $0.1-2.5$ for models with no, moderate and large overshooting respectively (e.g. 6, 7, 5).

There have recently been many new observations for C/He in WC stars (33, 34), which give C/He ratios in the range 0.4-2.4. The value of the lower boundary is in favour of models with moderate overshooting, a conclusion which is also supported by cluster sequences in the HR diagram and the number ratio of WN and WC stars (7).

IRAS data for NeIII at 15.8 $\mu$ give a Ne/He ratio of 0.05 in mass fraction (35), in agreement with model predictions (5, 6, 7). However, from ground-based data for NeIII at 12.8 $\mu$, a ratio of $Ne/He = 0.005$ has been obtained (36). Who is right? If this last result is true, it raises the interesting question in which form are the ashes of CNO elements. If most $^{22}Ne$ has been converted into $^{25,26}Mg$ (which is normally not predicted), this would imply that through the reaction $^{22}Ne (\alpha, n) \, ^{25}Mg$, WC stars may synthetize more s-elements than expected (cf. also 5).


## REFERENCES


1. Chiosi, C., Maeder, A. 1986, Ann. Rev. Astron. Astrophys. 24, 329
2. Noels, A., Gabriel, M. 1981, Astron. Astrophys. 101, 215
3. Maeder, A. 1983, Astron. Astrophys. 120, 113
4. Greggio, L. 1984, in *Observational Tests of the Stellar Evolution Theory*, IAU Symp. 105, eds. A. Maeder and A. Renzini, p. 329

5. Prantzos, N., Doom, C., Arnould, M., de Loore, C.  1986, Astrophys. J. _304_, 695

6. Maeder, A.  1987, Astron. Astrophys. _173_, 287

7. Maeder, A., Meynet, G.  1987, Astron. Astrophys. _182_, 243

8. Mermilliod, J.C., Maeder, A.  1986, Astron. Astrophys. _158_, 45

9. Dearborn, D.S.P., Blake, J.B.  1985, Astrophys. J. _288_, L21

10. Kane, L., McKeith, C.D., Dufton, P.L.  1980, Astron. Astrophys. _84_, 115

11. Dufton, P.L., Kane, L., McKeith, C.D.  1981, Mon. Not. R. Astr. Soc. _194_, 85

12. Kudritzki, R.P.  1985, in _Production and Distribution of C, N, O Elements_, ESO Workshop, eds. I.J. Danziger, F. Matteucci, K. Kjär, p. 277

13. Walborn, N.R.  1976, Astrophys. J. _205_, 419

14. Walborn, N.R.  1987, this meeting

15. Bolton, C.T., Rogers, G.L.  1978, Astrophys. J. _222_, 234

16. Dearborn, D.S.P., Eggleton, P.P.  1977, Astrophys. J. _213_, 448

17. Doom, C.  1982a, Astron. Astrophys. _116_, 303, 308

18. Schild, H., Berthet, S.  1986, Astron. Astrophys. _162_, 369

19. Maeder, A.  1987, Astron. Astrophys. _178_, 159

20. Mathys, G.  1987, Astron. Astrophys., in press

21. Zahn, J.P.  1983, in _Astrophysical Processes in Upper MS Stars_, 13th Saas-Fee Course, eds. B. Hauck and A. Maeder, Geneva Obs., p. 253

22. Davidson, K., Dufour, R.J., Walborn, N.R., Gull, T.R.  1984, in _Observational Tests of the Stellar Evolution Theory_, IAU Symp. 105, eds. A. Maeder and A. Renzini, p. 261

23. Lovy, D., Maeder, A., Noels, A., Gabriel, M.  1984, Astron. Astrophys. _133_, 307

24. Lennon, D.J., Dufton, P.L.  1986, Astron.Astrophys. _155_, 79

25. Kudritzki, R.P.  1987, in _ESO Workshop on SN 1987 A_, ed. J. Danziger, in press.

26. Cassatella, A.  1987, in _ESO Workshop on SN 1987 A_, ed. J. Danziger, in press

27. Kirshner, R.P.  1987, this meeting

28. Harris, M.J., Lambert, D.L.  1984, Astrophys. J. _281_, 739

29. Lambert, D.L., Brown, J.A., Hinkle, K.H., Johnson, H.R.  1984, Astrophys. J. _284_, 223

30. Vanbeveren, D., Doom, C.  1980, Astron. Astrophys. _87_, 77

31. Smith, L.J., Willis, A.J.  1983, Astron. Astrophys. Suppl. _54_, 229

32. Conti, P.S., Leep, E.M., Perry, D.N.  1983, Astrophys. J. _268_, 228

33. Smith, L.F., Hummer, D.G.  1987, Astrophys. J., in press

34. Torres, A.V.  1987, Astrophys. J., in press

35. Van der Hucht, K.A., Olnon, F.M.  1985, Astron. Astrophys. _149_, L17

36. Barlow, M.J.  1987, preprint

# WHITE DWARF SEISMOLOGY : INVERSE PROBLEM OF g-MODE OSCILLATIONS


HIROMOTO SHIBAHASHI and TAKASHI SEKII
Department of Astronomy, Faculty of Science, University of Tokyo,
Bunkyo-ku, Tokyo 113, Japan

STEVEN KAWALER
Center for Solar and Space Research, Yale University
260 Whitney Aven., P.O. Box 6666, New Haven, CT 06511, U.S.A.


Since light variability in white dwarfs was first discovered twenty years ago, eighteen DA white dwarfs, several pulsating DB white dwarfs, and hotter pre-white dwarfs have so far been found to be pulsating variables. The most conspicuous characteristics of pulsations in these stars are that they seem to consist of multiple g-modes of nonradial oscillations. Attention should be paid to multiplicity of modes. Stimulated by the success of helioseimology, a research field called 'asteroseismology', in which we may probe the internal structure of stars by means of observations of their oscillations, is going to develop. How well such a seismological approach succeeds is dependent on how many modes are observed in each of stars. Since the number of modes of an individual pulsating white dwarf is larger than those of other types of pulsating stars but for the Sun, the seismological study may be the most promising as to the white dwarfs. In fact, by applying the asymptotic relations among eigenfrequencies of high order g-modes with low degree, the degree $l$, and the radial order $n$, Kawaler(1987a,b,c) succeeded to get some constraints on the physical quantities of some of pulsating white dwarfs.

The eigenfrequencies of g-modes are mainly governed by the Brunt-Väisälä frequency distribution in the star. Therefore, we can, in principle, obtain some information about the Brunt-Väisälä frequency distribution by means of a seismological approach. The Brunt-Väisälä frequency distribution in a white dwarf is related with the entropy gradient, and depends on the cooling rate, the past nuclear reaction, mass loss, convection, diffusion of elements, degeneracy, and so on. None of these processes has so far been well understood. Therefore, seismology may provide a unique tool to investigate these elementary processes. The mathematical procedure to diagnose the Brunt-Väisälä frequency distribution in a white dwarf has been outlined by Shibahashi(1986). This method is a variant of an inversion method to probe the sound velocity distribution in the Sun from the p-mode oscillations of the Sun ( Shibahashi 1987, Sekii and Shibahashi 1987 ). In this paper, we perform some numerical simulation to examine the validity of this method.

The wave equation for g-mode oscillations is, in some limiting cases, reduced to a form similar to the Schrödinger equation in quantum mechanics, which is written as

$$d^2 v/dr^2 + N^2(r)r^{-2}[l(l+1)/\omega^2 - \Xi_l(r)]v = 0. \tag{1}$$

Here, $v$ denotes an eigenfunction, $\omega$ means the eigenfrequency, $l$ is the degree of the spherical harmonic function which describes the pattern of the mode on the stellar surface, and $\Xi_l(r)$ is the 'gravity wave potential' which consists of the inverse square of the Brunt-Väisälä frequency, $N(r)$, multiplied by $l(l+1)$ and an $l$-independent part;

$$\Xi_l(r) = l(l+1)/N^2(r) + \Theta(r). \tag{2}$$

The first term in the right hand side of equation (2) dominates over $\Theta(r)$ in the inner part of a white dwarf, while $\Theta(r)$ becomes large in the outer part. Based on the WKB method, the quantization rule leads to

$$(n + \varepsilon)\pi = \int_{r_1}^{r_2} [l(l+1)/\omega^2 - \Xi_l]^{1/2}N/r \ dr, \tag{3}$$

where $n$ is the radial order of the mode and $r_i(i=1,2)$'s are the turning points at which $\Xi_l(r_i) = l(l+1)/\omega^2$. Strictly speaking, the quantization rule gives only a relation between discrete eigenvalues $l(l+1)/\omega^2$ and the corresponding integers $l$ and $n$. But we extend this relation to nonintegers $n$ by interpolation and treat equation (3) as if it were a continuous function of $l(l+1)/\omega^2$ and $l$. If we can identify each of observed modes, then we can regard equation (3) as an integral equation giving $N/r \ dr/d\Xi_l$, since, for a fixed value of $l$, the left hand side of equation (3) is a function of $l(l+1)/\omega^2$. The solution gives the distance between two turning points measured with the gravity wave velocity :

$$s(\Xi_l,l) \equiv \int_{r_1}^{r_2} N/r \ dr = 2 \int_{[l(l+1)/\omega^2]_{min}}^{\Xi_l} \partial n/\partial[l(l+1)/\omega^2] \ [\Xi_l - l(l+1)/\omega^2]^{-1/2}d[l(l+1)/\omega^2]. \tag{4}$$

Here, the lower limit of the integral region corresponds to the minimum of the gravity wave potential, and it is obtained by extrapolating eigenvalues to $n=0$. Once we get solutions (4) for several values of $l$, we obtain

$$d[1/N(r)]/d\ln r_1 = (2l+1)/[2l(l+1)]\cdot(\partial s/\partial l)^{-1} \tag{5}$$

by differentiating solutions (4) with respect to $l$ since the inner turning point of $r_1$ is approximately given by

$$N^2(r_1) = \omega^2. \tag{6}$$

The right hand side of equation (5) is evaluated at a given $l(l+1)/\omega^2$, and then by using equation (6), we should regard equation (5) as an equation to give $d[1/N(r_1)]/d\ln r_1$ as a function of $N(r_1)$. By using a reasonable range of $l$, we eventually obtain the Brunt-Väisälä frequency distribution in the white dwarf.

To apply the present method to practical cases, we have to perform the mode identification prior to the inversion. It is a quite difficult work, because we cannot resolve the stellar disk image so that the available observational data do not include any direct information of spatial patterns of modes except for the solar case. So, the methods for mode identification is an important subject in asteroseismology. Probably, fitting the observed period spectrum to the asymptotic formula for periods, $l$, and $n$ (e.g. Tassoul 1980) will be useful for the mode identification. But, at the moment, we suppose the modes are well identified, and we examine how well the Brunt-Väisälä frequency distribution in a white dwarf is inferred from the oscillation data in an ideal case in order to see the validity of the inversion method. To do so, we use a pre-white dwarf model calculated by Kawaler(1986), which is a $0.60\,M_\odot$ star with a luminosity of $100\,L_\odot$, and its theoretically calculated eigenfrequencies of 95 g-modes with $1 \le l \le 10$ and $1 \le n \le 10$. Figure 1 shows the eigenmodes in the $\{l, [l(l+1)/\omega^2]^{1/2}\}$-diagram, which corresponds to the so-called $(k,\omega)$-diagram in helioseismolgy. The range of the degree $l$ is wider than the really detectable modes, but we dare to use even high degree modes such as $l > 4$ because our present purposes are to develop the inversion method and to explore the possibility of the asteroseismological study of white dwarfs in an ideal case. Figure 2 shows the result of the inversion; the thin curve indicates the true Brunt-Väisälä frequency of the model as a function of $\int_0^r N/r\, dr$ (lower scale), and the thick curve indicates the Brunt-Väisälä frequency distribution obtained by using the present method. As seen in this figure, the solution reproduces well the real distribution. Since the inner turning points of g-modes with $l \le 10$ of the present model are at the deep interior, the outer part of the model is not inferred at all. In a case of cool DA white dwarfs, the outer part will be reproduced rather than the inner part.

## References


Kawaler, S. 1986, Ph.D. thesis, University of Texas at Austin.
Kawaler, S. 1987a, in *Stellar Pulsation*, ed. A. N. Cox, W. M. Sparks, and S. G. Starrfield (Springer, Berlin), p.367.
Kawaler, S. 1987b, in *IAU Symp. No.123, Advances in Helio- and Asteroseismolgy*, ed. J. Christensen-Dalsgaard (Reidel, Dordrecht), in press.
Kawaler, S. 1987c, in *IAU Colloq. No. 95, The Second Conference on Faint Blue Stars*, in press.
Sekii, T. and Shibahashi, H 1987, in *Stellar Pulsation*, ed. A. N. Cox, W. M. Sparks, and S. G. Starrfield (Springer, Berlin), p.322.
Shibahashi, H. 1986, in *Hydrodynamic and Magnetohydrodynamic Problems in the Sun and Stars*, ed. Y. Osaki (University of Tokyo, Tokyo), p.195.
Shibahashi, H. 1987, in *IAU Symp. No.123, Advances in Helio- and Asteroseismology*, ed. J. Christensen-Dalsgaard (Reidel, Dordrecht), in press.
Tassoul, M. 1980, *Astrophys. J. Suppl.*, **43**, 469.


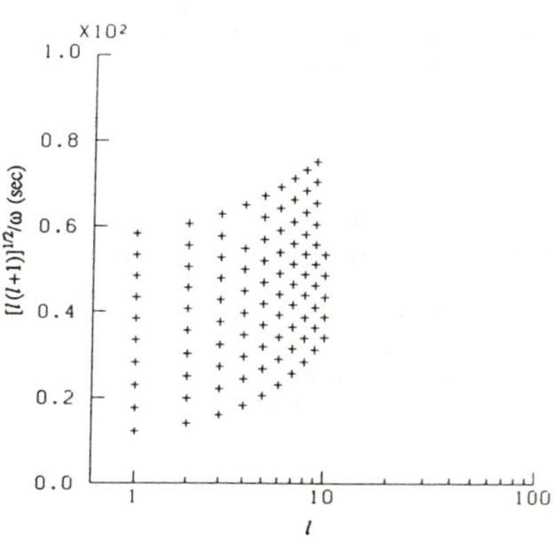

Fig.1. The eigenmodes used in the present paper. The ordinate and the abscissa indicate $l$ and $[l(l+1)]^{1/2}/\omega$, respectively.

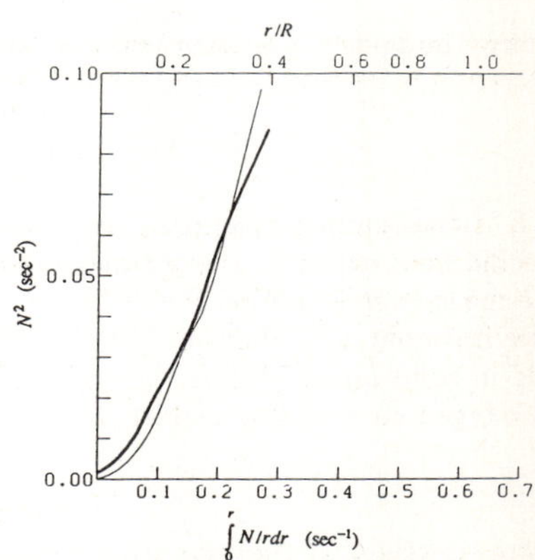

Fig.2. The Brunt-Väisälä frequency distribution in the model (thin curve) and the inverted result (thick curve). The upper scale of the abscissa indicates $r/R$ as a reference.

ance" mechanism being the separation

# THE WHITE DWARF LUMINOSITY FUNCTION AND THE PHASE DIAGRAM
## OF THE CARBON–OXYGEN DENSE PLASMA


E. Garcìa–Berro[1], M. Hernanz[1,2], R. Mochkovitch[3], J. Isern[1,4]

1 — Departament de Fisica de la Terra i del Cosmos, Universitat de Barcelona , Spain
2 — Departament de Fisica, ETSEIB, Universitat Politecnica de Catalunya, Spain
3 — Institut d'Astrophysique du CNRS, Paris, France
4 — Instituto de Astrofisica de Andalucia, CSIC, Spain



**Abstract.** We show that the theoretical white dwarf luminosity function depends very much on the assumed phase diagram for the carbon–oxygen dense plasma. Since it is still very uncertain, we compare the two possible extreme cases of complete miscibility and complete separation of carbon and oxygen in solid phase. In the latter case we find that the paucity of low luminosity — $\log(L/L_\odot) \leq -4.5$ — white dwarfs can be explained by the formation of an oxygen core, which releases a large amount of gravitational energy and slows down the cooling rate.


## 1. Introduction.

The paucity of low luminosity white dwarfs [1], if not a selection effect, can have different origins : (i) uncertainties in the envelope opacity [2], (ii) the presence of some "delay" mechanism which can decrease the cooling rate or simply (iii) an age of the Galactic disk which does not exceed $10^{10}$ years [3]. In this contribution, we consider case (ii), the "delay" mechanism being the separation of carbon and oxygen during crystallization of the white dwarf interior. We use the phase diagram of the carbon–oxygen plasma proposed by Stevenson [4] where either pure carbon or pure oxygen freezes out, depending on the composition of the liquid mixture compared to an entectic having a carbon mass fraction $X_c=0.6$.

## 2. Method.

The energy budget of a cooling white dwarf can be written as follows :

$$L + L_\nu = -\int_0^{M_{wd}} C_v \frac{dT}{dt} + \ell \frac{dM_{sol}}{dt} + L_{grav},\qquad(1)$$

where $L$ is the luminosity and $L_\nu$ are the neutrino losses ; $C_v, \ell, M_{wd}$ and $M_{sol}$ are respectively the specific heat capacity at constant volume, the latent heat, the total mass of the white dwarf and the mass in solid phase.
The gravitational contribution to the luminosity, $L_{grav}$, is negligible if carbon and oxygen are miscible in solid phase. When they separate, denser solid oxygen accumulates at the star center and an oxygen core is progressively formed. The white dwarf slightly contracts and then

$$L_{grav} = e_{grav} \frac{dM_{ox}}{dt},\qquad(2)$$

where $M_{ox}$ is the mass of the oxygen core and $e_{grav}=10^{14}$ erg.g$^{-1}$ for a 0.6 M$_\odot$ white dwarf with equal mass fractions of carbon and oxygen [5]. This value of $e_{grav}$ represents about ten times the latent heat so that $L_{grav}$ now strongly affects the cooling rate.

## 3. Results and discussion.

In Figure 1, we present the luminosity functions obtained with the assumptions of total miscibility (full line) and total separation (dashed line) for an age of the Galactic disk of 15 $10^9$ years. These luminosity functions take into account the increase of the vertical scale height over the plane of the disk with the age of the objects. This geometrical effect [2] nearly suppress the bump at $L \approx 10^{-4} L_\odot$ which would have been produced by the strong decrease of the cooling rate in the case of total separation [3].

It can be seen that the low luminosity cut-off of the luminosity function can be reproduced, even for a large age of the Galactic disk, if carbon and oxygen separate.

The total separation we have considered may be however an extreme case and the time delay introduced by a partial separation would naturally be smaller. The uncertainty on the carbon-oxygen phase diagram then appears to be a main difficulty for a reliable determination of the age of the Galactic disk using the white dwarf luminosity function.

Figure 1

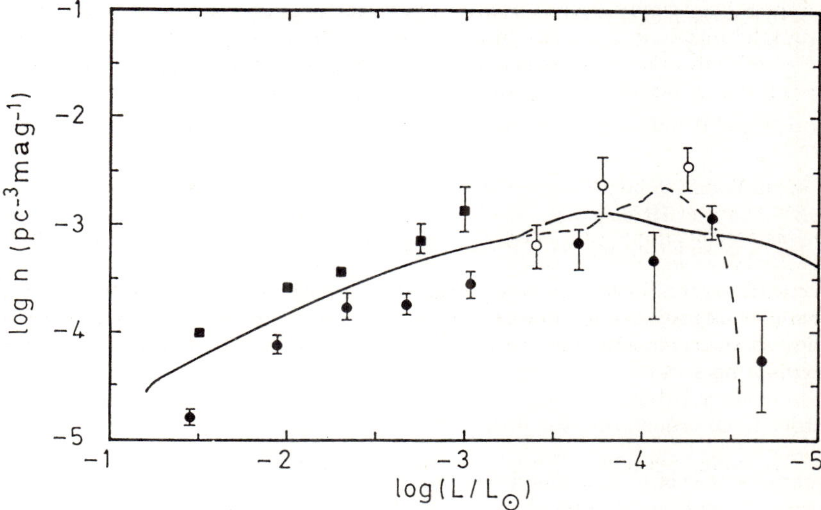




References

[1] Liebert, J. : 1980, Ann. Rev. Astron. Astrophys. 18, 363
[2] Iben I., Jr, Tutukov, A.V. : 1984, Astrophys. J. 282, 615
[3] Winget, D.E., Hansen, C.J., Liebert, J., Van Horn, H.M., Fontaine, G. , Nather, R.E., Kepler, S.L., Lamb, D.Q. : 1987, Astrophys. J. 315, L77
[4] Stevenson, D.J. : 1980, J. Phys. Suppl. N°3 41, C2-61
[5] Garcìa-Berro, E., Hernanz, M., Mochkovitch, R., Isern, J. : 1987, to be published in Astron. Astrophys.

# Wolf-Rayet stars as a diagnostic of internal mixing processes in massive mass losing stars


N. Langer

Universitäts-Sternwarte Göttingen, Geismarlandstraße 11,

D–3400 Göttingen, F.R.G.


Massive stars ($M_{ZAMS} \gtrsim 30\,M_\odot$) develop during their observable hydrostatic evolutionary phases — i.e. central H- and He-burning — three different large scale convective zones, which are: **1)** The H-burning convective core, **2)** the intermediate convective shell (ICZ) above the hydrogen shell source, which forms at time of hydrogen exhaustion, and **3)** the He-burning convective core. The spatial extent of these convective regions, wherein the chemical structure is rapidly homogenised, can be predicted from theory only with a large uncertainty. Different assumptions on the efficiency of these mixing processes in stellar evolution calculations lead to quite different evolutionary pictures for massive stars, especially regarding their Wolf-Rayet (WR) phases. On the other side, many observational data concerning WR stars became available in recent years. For this reason, we attempt to perform a comparison of theoretical evolutionary sequences with observed properties of WR stars in order to derive restrictions on the efficiency of the three mixing processes mentioned above.

**1)** Effects of greatly enlarged H-burning convective cores are (cf. Langer and El Eid, 1986; Prantzos et al., 1986):

- A reduced width of the main sequence band and the avoidance of the domain of the Luminous Blue Variables (LBVs) in the HR diagram.

- The formation of very massive (luminous) WR stars of types WNE and WC.

Both points disagree with observations: The observed main sequence width requires only a moderate core mass increase (cf. Mermilliod and Maeder, 1986), the LBVs exist, and very massive WNE and WC stars are not observed (cf. references in Langer, 1987; Doom, 1987). Evolutionary calculations without overshooting avoid both discrepancies. We conclude that convective overshooting is not very efficient in massive H-burning stars, but that the Schwarzschild-criterion may be a fair approximation in order to determine the size of the convective core.

**2)** To predict the spatial extent of ICZs is especially complicated, since they establish in regions of varying mean molecular weight, and the problem of semiconvection is encountered (cf. Langer et al., 1983; 1985). Langer (1987) argued, that in order for massive stars to terminate their nuclear evolution as a WR star of type WNL, the mass of the ICZ is required to exceed a critical value of $M_{ICZ} \geq M_{WR} \cdot \tau_{He} \simeq 10-20\,M_\odot$. There are several arguments in favour of massive stars exploding as WNL stars: As mentioned above, very massive stars probably do not evolve into WNE or WC stars. Furthermore, there exist supernovae (SNe)/SN remnants (SNRs) possibly originating from a WNL precursor, the most well known example being Cas A (cf. El Eid and Langer, 1986; Fesen et al., 1987). But also the SNe 1961v (Utrobin, 1984) and 1986j (Rupen et al., 1987) are suspected to originate from very massive precursors which still contain hydrogen in their outer layers, i.e. from WNL stars. Therefore we conclude, that the mass of the ICZ in very massive stars exceeds $10\,M_\odot$.

We note that hydrogen shell burning and consequently an ICZ in very massive stars develop only if convective overshooting during central H-burning is small or negligible. This is an additional argument supporting our conclusion of point 1).

**3)** Convective core overshooting during central He-burning should be much less efficient compared to that in the H-burning phase, since mostly during central He-burning the convective core is

""

growing with time, which consequently leads to the formation of a huge molecular weight barrier on top of it.

An enlarged He-burning convective core has two consequences for the WR stages of massive stars:

- It leads to more extreme surface compositions in WC stars, i.e. to smaller surface helium concentrations but higher C- and O-abundances. The surface helium mass fraction $Y$ can be roughly estimated to be larger than a certain value, depending on the mass of the convective core $M_{cc}$: $Y \geq (M_{WC} - M_{cc})/(\dot{M}_{WC} \cdot \tau_{He})$, where $M_{WC}$ is the mass of the WC star, $\dot{M}_{WC}$ its average mass loss rate, and $\tau_{He}$ its He-burning lifetime.

- It leads to a larger upper limit for the mass of WC stars. Such limit exists since WR stars of higher mass have larger envelope masses $M_{env} = M_{WR} - M_{cc}$ and shorter He-burning lifetimes, but presumably no larger mass loss rates. The mass limit is determined by $M_{env}(M_{WR}) = M_{WR} - M_{cc} \geq \dot{M}_{WR} \cdot \tau_{He}$.

Concerning the first point, observations are not yet sufficiently accurate in order to draw conclusions (cf. Torres, 1987). For the second point, we can conclude at least that no overshooting is consistent with observations, since it leads to a maximum mass for WC stars of $\sim 60\,M_\odot$, which is consistent with the absence of very massive WC stars mentioned above (cf. also: Langer and Kiriakidis, 1988).

## Summary


- The convective core size in very massive H-burning stars may well be approximated by the Schwarzschild criterion.
- In sufficiently massive stars the extension of the intermediate convection zone exceeds $10-20\,M_\odot$.
- The convective core size in massive He-burning stars can hardly be restricted by observations. However, the absence of very massive WC stars is consistent with the case of no overshooting at all.


Theoretical evolutionary sequences taking into account the above points lead to the following scheme, which is basically consistent with recent papers of Schild and Maeder (1984), Langer (1987), and Doom (1987):
most massive stars $\rightarrow$ WNL $\rightarrow$ SN
very massive stars $\rightarrow$ WNL $\rightarrow$ WCE $\rightarrow$ SN
massive stars $\rightarrow$ WNE $\rightarrow$ WCL $\rightarrow$ SN
less massive stars $\rightarrow$ WNE $\rightarrow$ SN.


The author gratefully acknowledges a travel grant of the Deutsche Forschungsgemeinschaft.


## References


Doom, C.: 1987, Astron. Astrophys. Letters **182**, L43

El Eid, M.F., Langer, N.: Astron. Astrophys. **167**, 274

Fesen, R.A., Becker, R.H., Blair, W.P.: 1987, Astrophys. J. **313**, 378

Langer, N.: 1987, Astron. Astrophys. Letters **171**, L1

Langer, N., Sugimoto, D., Fricke, K.J.: 1983, Astron. Astrophys. **126**, 207

Langer, N., El Eid, M.F., Fricke, K.J.: 1985, Astron. Astrophys. **145**, 179

Langer, N., El Eid, M.F.: 1986, Astron. Astrophys. **167**, 265

Langer, N., Kiriakidis, M.: 1988, this volume

Mermilliod, J.C., Maeder, A.: 1986, Astron. Astrophys. **158**, 45

Prantzos, N., Doom, C., Arnould, M., de Loore, C.: 1986, Astrophys. J. **304**, 695

Rupen, M.J., v. Gorkom, J.H., Knapp, G.R., Gunn, J.E., Schneider, D.P.: 1987, Astron. J. **94**, 61

Schild, H., Maeder, A.: 1984, Astron. Astrophys. **136**, 237

Torres, A.V.: 1987, Astrophys. J., in press

Utrobin, V.P.: 1984, Astrophys. Space Sci. **98**, 115

# A systematic study of stellar models for C/O-rich Wolf-Rayet stars

N. Langer and M. Kiriakidis

Universitäts-Sternwarte Göttingen, Geismarlandstraße 11
D–3400 Göttingen, F.R.G.

A grid of homogeneous stellar models for Wolf-Rayet stars in the mass range from 1 to $60\,M_\odot$ has been computed. For each stellar mass, stars with eight different sets of chemical compositions — from pure helium stars with (Y,C,O)=(1,0,0) to extreme helium poor stars with (Y,C,O)=(0.02,0.11,0.87) — have been calculated in order to investigate the dependence of the stellar structure on the stellar mass and chemical composition. Modern input physics adapted to the exotic chemical composition of the Wolf-Rayet stars has been incorporated in the models, including effects of partial recombination of helium, carbon, and oxygen, and detailed opacity tables for 20 different combinations of the (Y,C,O)-abundances. Furthermore, we estimated the effect of the intense, partly optically thick Wolf-Rayet winds on their apparent effective temperature, using the formalism of de Loore et al. (1982, IAU-Symp. **99**, 53) with the parameters: $\dot{M} = 3 \cdot 10^{-5}\,M_\odot\,yr^{-1}$, $\beta = 2$, and $v_\infty = 2000\,km\,s^{-1}$. The main results may be summarized as follows:

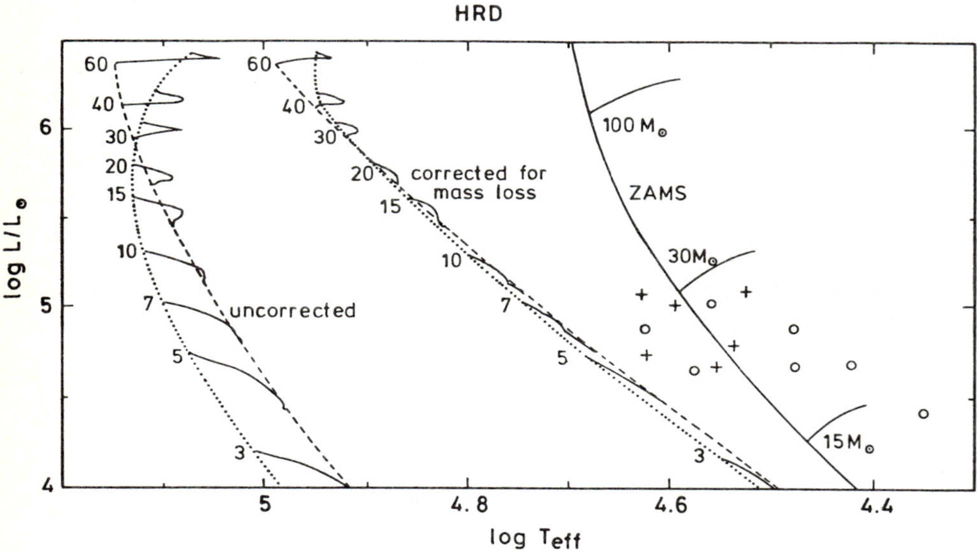


**Figure 1.** *Lines of constant mass and varying chemical composition for the computed Wolf-Rayet models of 3, 5, 7, 10, 15, 20, 30, 40, and $60\,M_\odot$ in the HR diagram (solid lines). The pure helium stars are connected through a dashed line, while the extreme helium poor stars are connected through a dotted line. Also the HRD positions after applying a correction for the partly optically thick stellar wind on the effective temperature are shown. Furthermore, the theoretical zero age main sequence (ZAMS) is indicated, together with schematic evolutionary tracks for stars of 15, 30, and $100\,M_\odot$. The crosses and circles correspond to HRD positions of observed WNE and WC stars, respectively, according to Smith and Willis (1983, Astron. Astrophys. Suppl. **54**,229).*

**a)** Influence of the chemical composition

At high masses ($30 - 60\,M_\odot$), for a given stellar mass the models become progressively cooler and increase the radius at almost constant luminosity when the helium mass fraction decreases. This means that very massive WC stars are found to the right of their WNE counterparts (which are represented here by the pure He-stars) in the HR-diagram — at least if the stellar wind shifts the HRD positions in both cases in a similar way (cf. Fig. 1). At low mass the situation is different: For a given stellar mass, the surface temperature is increasing at approximately constant radius with decreasing helium abundance, and therefore also the luminosity is increasing.

**b)** Influence of the stellar mass

For low helium concentrations, the effective temperature of the WC models is increasing with decreasing mass in the range $60\,M_\odot \geq M_{WC} \gtrsim 15\,M_\odot$, and decreasing for $M_{WC} \lesssim 15\,M_\odot$, while for high helium concentration the effective temperature always decreases for decreasing mass (cf. Fig. 1). However, after applying a correction to the effective temperature taking into account the stellar mass loss we find in general a strong decrease of the effective temperature with decreasing mass

The following conclusions emerge:

- At high stellar mass ($30\,M_\odot \lesssim M_{WR}$) the chemical composition affects the surface temperature but not the luminosity of the star. For lower masses both quantities are influenced.
- For WR stars less luminous than $\sim 3 \cdot 10^5\,L_\odot$ (i.e. $M_{WR} \lesssim 20\,M_\odot$) no strict mass-luminosity relation exists.
- The incorporation of the effect of the intense WR mass loss on the apparent effective temperature is absolutely necessary in order to overcome the large gap between HRD positions of observed and computed WR stars.
- The stellar mass is the main parameter which determines the effective temperature of the WC stars, indicating that the so called earlier types (WCEs) correspond to higher stellar masses than the late WC stars (WCLs).
- If it is confirmed that WC stars in general have luminosities less than $\sim 10^5\,L_\odot$ (cf. also Schmutz et al., this volume) this would mean that WC stars have as surprisingly low masses as $M_{WC} \lesssim 10\,M_\odot$.

Detailed results will be published in a paper now in preparation for Astronomy and Astrophysics.

N. L. gratefully acknowledges a travel grant of the Deutsche Forschungsgemeinschaft.

OVERABUNDANCE OF SODIUM IN THE ATMOSPHERES OF MASSIVE
SUPERGIANTS AS A POSSIBLE MANIFESTATION OF NeNa CYCLE

A.A.Boyarchuk[1], P.A.Denisenkov[2],
I.Hubeny[3], V.V.Ivanov[2], I.Kubát[3],
L.S.Lyubimkov[1], N.A.Sakhibullin[4]

1 - Crimean Astrophysical Observatory, USSR
2 - Leningrad University, USSR
3 - Astronomical Institute, CSSR
4 - Kazan' University, USSR

Studies of chemical composition of the atmospheres of F-K supergiants have revealed the existence of two specific peculiarities, deficit of carbon and excess of nitrogen. Anomalous abundances of C and N in yellow supergiants is explained by mixing of surface material with CNO-processed material of stellar interiors (the first dredge-up) . Somewhat unexpectedly, F-K supergiants were found to show one more general chemical peculiarity: overabundance of sodium. Fig.1 shows the available values of $[\mathrm{Na/H}]=\log[N(\mathrm{Na})/N(\mathrm{H})]_* -\log[N(\mathrm{Na})/N(\mathrm{H})]_\odot$ as summarized by Boyarchuk and Lyubimkov(1983). Different symbols refer to data of various authors. Fig.1 shows that sodium overabundance increases with the decrease of surface gravity. For small log g it reaches $[\mathrm{Na/H}] \sim$ $\sim 1$. Recently Sasselov(1986) has suggested that this correlation is essentially a consequence of a relation between $[\mathrm{Na/H}]$ and a mass of supergiant.

All the estimates of $[\mathrm{Na/H}]$ presented in Fig.1 are found from the LTE analysis of subordinate and not resonance lines of NaI. To estimate the role of non-LTE effects on inferred values of $[\mathrm{Na/H}]$ for several F-

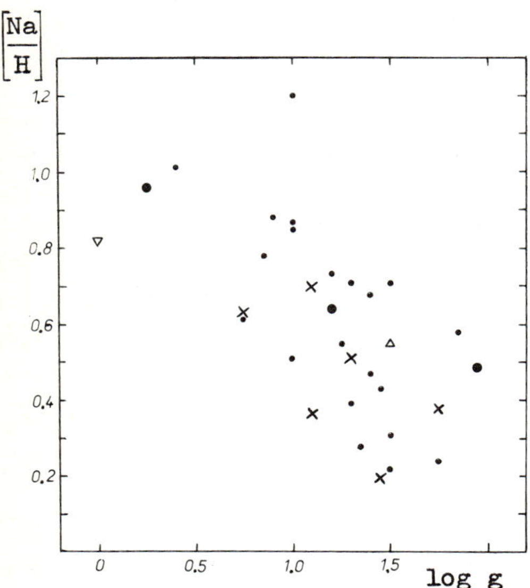

Figure 1. Sodium overabundance versus surface gravity of F-, G- and early K-supergiants

supergiants we have calculated the non-LTE level populations of NaI and the equivalent widths $W_\lambda$ of the most important NaI lines. Nineteen levels of NaI and the ground state of NaII were taken into account in calculations. Two different computer codes were used. They were independently developed in the Astronomical Institute of the Czechoslovac Academy of Sciences and in Kazan' University. The results obtained using these two codes are in excellent agreement. See Boyarchuk et al. (1987) for details. The calculations showed that the differences between LTE and non-LTE $W_\lambda$ values of NaI subordinate lines do not exceed 10%. The corresponding differences in Na abundances are less than 0.1 dex. Only for the most massive supergiants ($\log g \sim 0$) the LTE sodium abundance may be overestimated up to 0.2 dex.

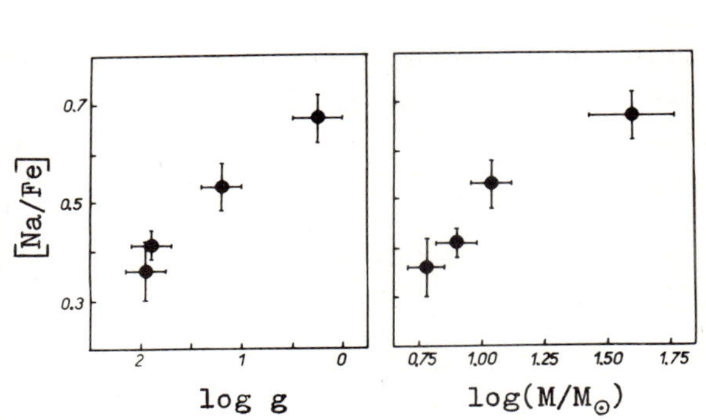

Figure 2. Non-LTE sodium overabundance for F-type supergiants $\alpha$UMi, $\alpha$ Car, $\gamma$ Cyg and $\rho$ Cas plotted against surface gravity and mass.

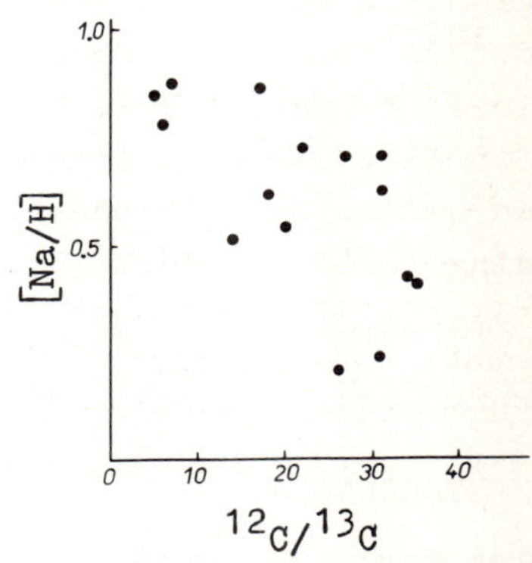

Figure 3. Sodium overabundance versus $^{12}C/^{13}C$.

Among those stars for which we have performed non-LTE calculations are $\alpha$UMi, $\gamma$ Cyg, $\rho$ Cas and $\alpha$Car. For these four F-supergiants detailed and accurate measurements of $W_\lambda$ values of sodium lines were known. For $\alpha$ Car the observed values of $W_\lambda$ were taken from literature, while for the three other stars the spectra were obtained at the Crimean Astrophysical Observatory (with dispersion 4, 6 and 8 Å/mm). In Fig.2 non-LTE [Na/Fe] values for these four stars are plotted versus **g** and **M**. One may conclude that non-LTE analysis also shows overabundance of Na correlated with **g** and **M**.

Earlier two of us (Boyarchuk and Lyubimkov, 1983) have suggested that the observed overabundance of Na in F-K supergiants is a manifestation of NeNa cycle

$$^{20}Ne(p,\gamma)^{21}Na(\beta^+\nu)^{21}Ne(p,\gamma)^{22}Na(\beta^+\nu)^{22}Ne(p,\gamma)^{23}Na(p,\alpha)^{20}Ne \ .$$

This cycle operates in the interiors of main sequence stars parallel to the CNO cycle. It produces nuclei of $^{23}$Na which together with the products of CNO cycle at a later phase of evolution are dredged up.

Detailed analysis of the kinetics of this cycle has shown (Denisenkov and Ivanov, 1987) that the reaction $^{22}$Ne$(p,\gamma)^{23}$Na, due to the presence of recently discovered resonance at $E_r$=30 KeV, is fast enough to provide a mechanism of (5-6)-fold sodium enrichment of central regions of core hydrogen burning (MS) stars with $M \gtrsim 1.5$ $M_\odot$ due to $^{22}$Ne burning (assuming Cameron's (1982) isotopic abundance of Ne). Lifetime of $^{22}$Ne is markedly longer than the time needed for $^{12}$C/$^{13}$C to reach its equilibrium value. Hence, if surface layers of supergiants (and red giants) are enriched in material processed deeply enough in core regions on MS phase, one should expect a correlation between [Na/H] and $^{12}$C/$^{13}$C. Such correlation does exist (Fig.3: $^{12}$C/$^{13}$C and [Na/H] values by Luck, 1977). Low abundance of C in Na-rich atmospheres shown in Fig.3 is a manifestation of Na production on hydrogen rather than helium burning phase. The data given in Fig.2 and 3 indicate that mixing is more effective with the increase of mass.

To summarise: <u>sodium is to be added to those few elements, whose atmospheric abundance is a probe of stellar interiors.</u>

REFERENCES


Boyarchuk,A.A., Hubený,I., Kubát,J., Lyubimkov,L.S., Sakhibullin,N.A. 1987, Astrofizika, in press.

Boyarchuk,A.A.,and Lyubimkov,L.S. 1983, Izv. Krymsk. Asrofiz. Obs., v.66, p.130.

Cameron,A.G.W. 1982, in Essays in Nuclear Astrophysics, ed. C.Barnes, D.D.Clayton, and D.N.Schramm, Cambridge Univ. Press, p.23.

Denisenkov,P.A., and Ivanov,V.V. 1987, Pis'ma Astron. Zh., v.13,p.520.

Luck,R.E. 1977, Astrophys.J., v.218, p.752; v.212, p.743.

Sasselov,D.D. 1986, Publ. Asron. Soc. Pacific, v.98, p.561.

# Excitation of Non-Radial Oscillations by Overstable Convection in Differentially Rotating Massive Main Sequence Stars


*Umin Lee*

Department of Astronomy, University of Tokyo, Tokyo 113, Japan



**Summary**.

We investigated overstable convective modes of differentially rotating massive main sequence stars. It is examined that the overstable convective modes in a rapidly rotating convective core may excite envelope non-radial oscillations by the resonance coupling between them. Let us denote by $q$ the ratio of the angular velocity of the rapidly rotating core to that of the envelope. Then, it is found that as the ratio $q$ increases, the order of the non-radial $g$ modes resonantly coupled with the convective mode becomes lower, so that the excited $g$ modes come to have shorter periods in an inertial frame. It is also found that when the separation of the frequencies of the consecutive envelope $g$ modes is wider than the resonance width, there occurs the alternation between the strong and the weak resonances, i.e., the resonantly coupled mode becomes overstable or neutral according to the strength of the resonance.

Our equilibrium model is a main sequence of $10M_o$, which has a convective core of $3.25M_o$. The model is assumed to rotate with velocity of about 110 km/sec at the equator. Modes investigated are with $-m=l=2$. In this case, the shortest excited mode we obtain has periods of about a half day, in which the ratio $q$ is about 1.5. We may conclude that if the differential rotation is assumed, the excitation by the overstable convection may give the widely applicable explanation to the early type star pulsations.



**References**

Lee, U. & Saio, H. 1986, *Mon. Not. R. astr. Soc.*, **221**, 365.

Lee, U. & Saio, H. 1987, *Mon. Not. R. astr. Soc.*, **224**, 513.

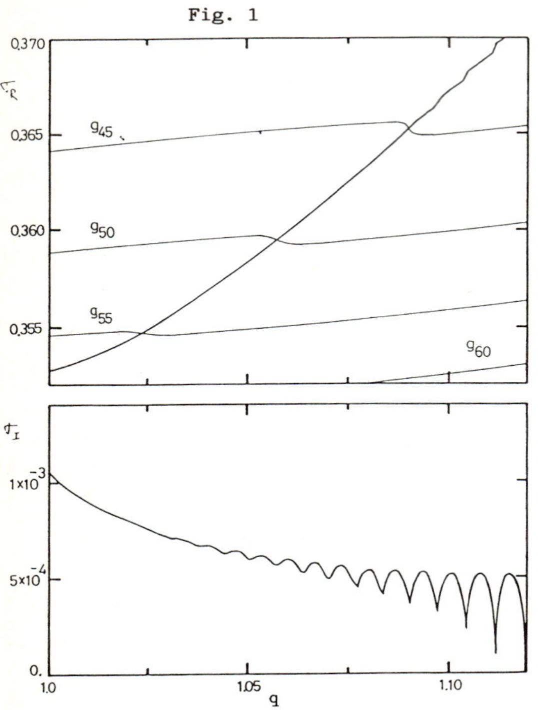



Fig. 1

Complex eigen-frequencies of overstable convective modes coupled with envelope g modes as a function of the ratio q (thick curves). Real frequencies of high order envelope g modes are also given (thin curves). Resonance couplings find themselves in the wavy features in the imaginary part of the frequency.

Fig. 2

Complex eigen-frequencies of the convective mode coupled with envelope g modes as a function of the ratio q. Real frequencies of the envelope g modes are also given. The thick curves denote the overstable modes and the thin curves neutral modes. The convective mode becomes overstable (complex freq.) or neutral (real freq.) according to the strength of the resonance.

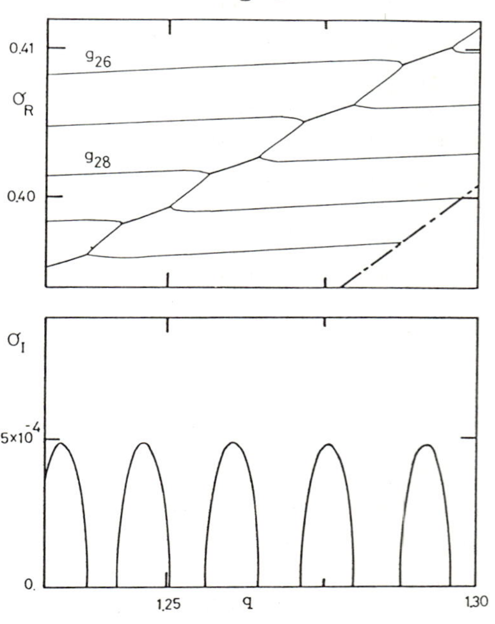

# Dredge-up by Sound Wave Emission from a Convective Core

Masa-aki Kondo

Senshu Universrity
Higashi-mita, Tama-ku, Kawasaki-shi, Kanagawa, 214 Japan

Concerning the scsattering of OB stars in the HR diagram (Humphry 1980), the effects of overshooting of convective core (Maeder 1984), mass loss (cf. Chiosi and Maeder 1986), and generous stability criterion of semi-convection (Stothers and Chin 1976) have been discussed. Here, we will note the dredge up effect is caused by the sound waves emitted from a convective core.

The sound mode of nonradial oscillation, with the spherical harmonics $Y_{\ell m}(\theta,\phi)$ and the frequency $\omega$, can exist in the propagation zone, where the bottom boundary locates at the position of $\omega = L_\ell \ [\equiv \sqrt{\ell(\ell+1)}c_s/r]$, and the upper boundary does near the photosphere. Here, $L_\ell$ is called as the Lamb frequency, and $c_s$ is the sound velocity.

In early type stars, the bottom boundary penetrates into convective core (Osaki 1975). Accordingly, convective motion of eddies excites sound waves, as in the case of acoustic noise emmision from imcompressible turbulence, shown by Lighthill (1978). Since the frequency of excited waves is higher than the Brunt-Väisälä frequency at the photosphere, the waves are not trapped, but running outward (cf. Unno et al 1979).

If the random displacement of generated wave is denoted by $\xi$, it causes first order fluctuation of mean molecular weight $\mu_1$, given by $-\xi_r \partial \mu_0/\partial r$, where $\mu_0$ is the unperturbed distribution of mean molecular weight. Then, the nonlinear coupling between $\xi$ and $\mu_1$ produces the convective effect for the mean molecular weight, originated from the term of $(v_1 \cdot \nabla)\mu_1$, where $v_1 = \partial \xi/\partial t$.

If we take time average over shorter period than an evolutional time scale and over spherical angular average, we obtain the evolutional equation for the averaged mean molecular weight $\bar{\mu}$;

$$\frac{\partial \bar{\mu}}{\partial t} + (v_{ev} + v_{wind})\frac{\partial \bar{\mu}}{\partial r} = -C(r)\frac{\partial \bar{\mu}}{\partial r} \quad ,$$

where
$$C(r) = -<(v_1 \cdot \nabla)\xi_r>,$$

and $v_{ev}$ and $v_{wind}$ means the velocity field of evolutionally secular change and the interior component of stellar wind. It should be noted the diffusion term of $<v_r \xi_r>\partial^2 \bar{\mu}/\partial r^2$ vanishes in this sound mode case, because $v_r$ and $\xi_r$ are in orthogonal phase to each other. The effect of diffusion has been considered by Schatzman (1977), in the case of late typestars, and by Langer

99

et al (1985), in the case of semiconvection zone of massive stars.

Now, it is proved after some manupilation of sound modes that

$$C(r) = \frac{1}{\Gamma_1 p_0} <p'v_r>,$$

where $p_0$ is the unperturbed pressure, $p'$ a pressure perturbation of sound mode, and $\Gamma_1$ the specific heat ratio. Consequently, the convective coefficient C is proportional to the acoustic power $<p'v_r>$, which has been fully considered in the case of the isothermal atmosphere (Stein 1967). In this problem, the exciting region for sound waves locates from the bottom boundary of propagation zone to the edge of convective core. This region is wider for higher frequency waves. However, energy densities of exciting eddies decrease in the high frequency case.

The mode of $\ell = m$ has the largest amplitude, which is determined by the forth order correlation of turbulent velocity $<u_i u_i u_i' u_i'>$, and there are monopole, dipole and quadrupole emmisions for each spherical harmonics, with regard to the radial direction (cf. Unno 1964). The mode of low $\ell$ contributes to the monopole emmision, but that of high $\ell$ to the latter ones, as the same as in the isothermal case.

Now, strong stellar wind blows on the photosphere of early type stars (cf. de Jager 1980), so that mass outward flow $v_{wind}$ up to 1 cm/sec exists in the edge of core. The precise consideration is required to determine whether the convective velocity of C is more effective for dredge-up than $v_{wind}$ or not. Quantative results will be shown in the seperate paper.

## References


Chiosi, C. and Maeder, A., 1986, Ann. Rev. Astron. Astrophys., 24, 329.

Humphry, R.M., 1970, Astrophys. Letters, 6, 1.

de Jager, C., 1980, The Brightest Stars, (Reidel, Dordrecht).

Langer, N., El Eid, M.F., and Fricke, K.J., 1985, Astron. Astrophys.,145, 179.

Lighthill, M.J., 1978, Waves in Fluids, (Cambridge Univ. Press, Cambridge) p57.

Maeder, A., 1984, in Observational Tests of the Stellar Evolution Theory, ed. Maeder and Renzini, (Reidel, Dordrecht), p299.

Osaki, Y., 1975, Publ. Astron. Soc. Japan, 27, 237.

Schatzman, E., 1977, Astron. Astrophys., 56, 211.

Stein, R.F., 1967, Solar Physics, 2, 385.

Stothers, C. and Chin, C., 1976, Astrophys. J., 204, 472.

Unno, W., 1964, Transaction I.A.U. XIIB, (Academic Press, New York), p555.

Unno, W., Osaki, Y., Ando, H. and Shibahashi, H., 1979, Nonradial Oscillations of Stars, (Univ. of Tokyo Press, Tokyo), p140.

# II. Mass-Losing Stars in Different Stages of Evolution

*Chair: A. Maeder and P. Wood*

**STELLAR MASS LOSS AND
ATMOSPHERIC INSTABILITY**

Cornelis de Jager and Hans Nieuwenhuijzen

Astronomical Observatory and Laboratory for Space Research
Beneluxlaan 21, 3527 HS  UTRECHT, the Netherlands

**Abstract:** A review is given of rate of mass-loss values $\dot{M}$ in the upper part of the Hertzsprung-Russell diagram. Near the luminosity limit of stellar existance $\dot{M}$ = $-10^{-4}$ $M_{\odot}$ $yr^{-1}$. Episodical mass loss in bright variable super- and hypergiants does not significantly increase this value. For Wolf-Rayet stars the rate of mass loss is larger by a factor 140 than for non-evolved stars with the same $T_{eff}$ and L; for C stars this factor is ten. This can be explained qualitatively. Rotation appears hardly to influence the rate of mass loss except for $v_{rot}$-values close to the break-up velocity. This is in accordance with theory. We suggest the existence of a Red Supergiant Branch; along that branch mass loss is virtually independent of luminosity. Stellar winds along the upper limit of stellar existence are mainly due: to <u>radiation pressure</u> for hot supergiants ($\gtrsim$ 10 000 K); to <u>turbulent pressure</u> for cool supergiants (3000-10 000 K), and to <u>dust-driven and pulsation-driven</u> winds for cooler stars. The turbulent pressure may originate in large-scale stochastic motions as observed in Alpha Cyg. Episodical mass loss, as observed in P Cyg, HR 8752 and other Very Luminous Variables may be due to occasional violent stochastic motions, resulting in a shock-driven episodical mass-loss component.

## 1. <u>Mass loss of chemically not evolved stars</u>

Values for the rate of mass-loss $-\dot{M}$ from stars are mostly derived from the following data:
- middle-ultraviolet resonance line profiles (C IV, Si IV and others);
- profiles of subordinate lines, like Hα, mainly in the visual spectral range;
- infrared continuum photometric data (assumed due to free-free emission);

102

- microwave continuum data (free-free emission);
- infrared molecular lines, mostly C-components;
- microwave maser lines.

A few more $\dot{M}$-values were derived from other sources.

Although the various methods are based on very different approaches, the intercomparison of $\dot{M}$-values for the same stars demonstrated that the various methods yield values for the rate of mass loss that do not differ <u>systematically</u>. Also, the average scatter of the data per method is about the same: 0.45 dex. On the basis of these findings an interpolation formula has been derived giving $-\dot{M}$ as a function of $\log T_{eff}$ and $\log(L/L_\odot)$ for the whole upper part of the Hertzsprung-Russell diagram (De Jager et al., 1987); cf. Figure 1. Although one would expect that in such a representation over a broad $(T,L)$ domain the accuracy of the adaption would be less than in interpolation formulae restricted to smaller parts of the Hertzsprung-Russell diagram, this appears not to be the case: other representations have also a one-sigma scatter of 0.5 dex per determination or show even larger scattering.

The choice of the parameters may be a point for discussion. A $(T,L)$-representation is essentially an $(R,L)$-representation. It does not appear difficult, though, to add a third parameter, such as the stellar mass $M$, and this investigation is under way by the present authors.

An $(R,M,L)$-representation is, however, not physically better founded than an $\cdot(R,L)$ or $(T,L)$-representation. As Vardya (1987) showed: such representations, like virtually all those published so far (cf. Table 1) are essentially numerically- (not physically-) based interpolation formulae, because the constant A in the representation

$$\dot{M} = A.L^{\alpha} M^{\beta} R^{\gamma}, \tag{1}$$

is for most of the representations of Table 1 not dimension-less, and therefore it only represents an approximated constant zero-order value of a function $A(L,M,R)$. If one wishes A to be without dimensions, then the solution of eq (1) is:

$$\dot{M} = A \ (L \ M^2/R^2)^{1/3} \ , \tag{2}$$

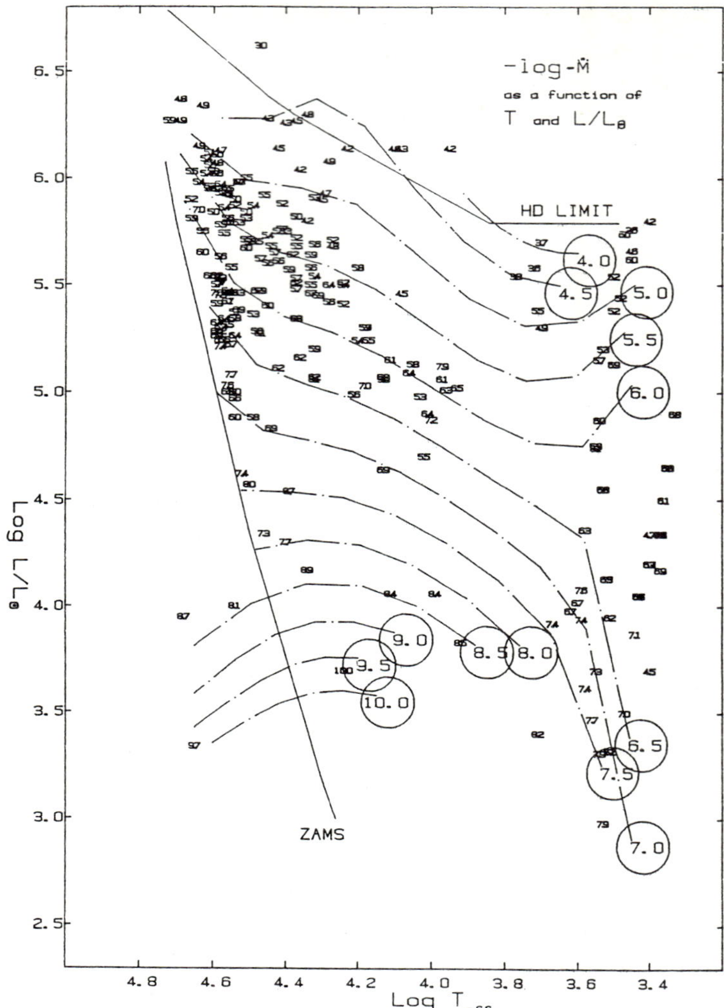


Figure 1: Mass loss over the Hertzsprung-Russell diagram. The numbers give values of −log (−Ṁ) for individual stars, to one decimal. The lines are interpolation lines according to a formula given by De Jager et al. (1987).

**Table 1. Interpolation formulae for the rate of mass loss
(partly after Vardya, 1984)**

| Reference | $\dot{M}$ x const. |
|---|---|
| McCrea (1962) and Reimers (1975) | $L\,R\,M^{-1}$ |
| Abbott et al. (1980) | $L^{1.8}$ |
| Chiosi (1981) | $L^{0.72}(R/M)^{5/2}$ |
| ibid | $L^{5/4}(R/M)^{13/8}$ |
| ibid | $L^{2}(R/M)^{7/2}$ |
| Andriesse (1979) and Chiosi (1981) | $L^{3/2}(R/M)^{9/4}$ |
| Lamers (1981) | $L^{1.42}R^{0.61}M^{-0.99}$ |
| Garmany et al. (1981) | $L^{1.75}$ |
| Vardya (1984) | $L^{815}(R/M)^{9/10}$ |
| ibid | $L^{7/4}(R/M)^{9/8}$ |
| ibid | $L^{2}(R/M)^{3/2}$ |
| De Jager et al. (1987) | $\phi(T_{eff},L)$ |
| Nieuwenhuijzen and De Jager (1988) | $f(T_{eff},L,v_{rot})$ |

but since that representation appears not to fit to the observed data,
Vardya (1985) proposed after some attempts:

$$\dot{M} = (A/(G^{1/2}c^{2}))L^{2}(R/M)^{3/2}, \qquad (3)$$

where G and c are the gravitation constant and the speed of light, respectively.  The formula (3) has not yet been applied to the $\dot{M}$-values
over the whole HR-diagram.

The uppermost part of the Hertzsprung-Russell diagram is of particular
interest since the stars in that area are apparently close to their
limit of existence, which is shown by their stochastic variability,
pulsations, large rate of mass loss and occasional episodic mass loss.
The curve above which no stars appear to exist is called the Humphreys-
Davidson limit (Humphreys and Davidson 1979; De Jager, 1980); cf.
Figure 2. Stars close to that limit exhibit many of the properties listed above. In that area one also finds the Luminous Blue Variables,
which are stars that erratically expell a large amount of mass. At some
distance from the star the gas condenses into dust particles and thus
the star becomes reddened. Sometimes the expelled gas is optically

thick enough to shift the star's photosphere outward, thus lowering the
star's effective temperature and changing the spectral type, while the
bolometric luminosity remains constant: the star's position then undergoes horizontal excursions in the HR-diagram. Well-known examples of
LBV's are S Dor, R 127 and P Cyg.

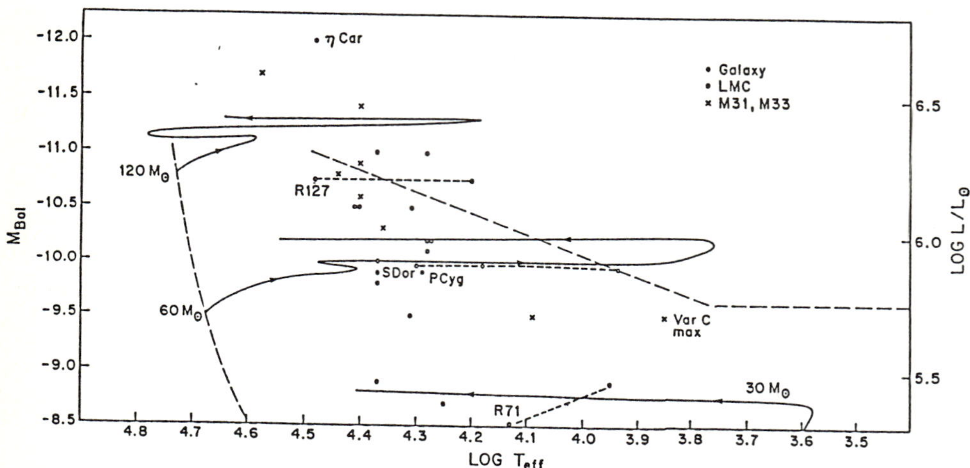


Figure 2: Stars and evolutionary tracks near the Humphreys-Davidson
limit. From Humphreys (1987).

But such behaviour is not really restricted to the Luminous Blue Variables. Humphreys (1987) described a cool star ("variable A") that shows
the same behaviour, and so does the cool hypergiant HR 8752 (Piters et
al., 1987): here an episodical mass ejection started around 1968; the
star obtained a later spectral type; the expelled gas remained detectable till 1980-1982. It would make sense to include such variables in
the sample and to speak just of Very Luminous Variables, hence adding
the word "Very" and deleting "Bright".

It is sometimes claimed or assumed that the episodic mass loss of stars
near the HD limit is so large that its contribution would significantly
increase the average (over the centuries) rate of mass loss, over the
quiet-star's value. But that viewpoint seems hard to maintain for it
would demand much larger or more frequent episodic mass loss events
than actually observed. We therefore suggest to take $\dot{M} = -10^{-4}$ $M_\odot yr^{-1}$
along the Humphreys-Davidson limit as the present best value.

106

## 2. Influence of atmospheric chemical composition on the rate of mass loss

It is remarkable that He-rich stars appear to have a higher rate of mass loss than stars with solar-type atmospheres with the same $T_{eff}$- and L-values: The Wolf-Rayet stars have, on the average, $\dot{M}$-values that are 140 times larger than the values for corresponding O and B type stars (De Jager et al., 1987). This may be due to the fact that WR stars, with their large Helium abundance, are relatively closer to their Eddington limit than the most luminous O-type stars.

The C stars are another case: their average mass loss is slightly more than 10 times the value for solar-type stars with the same atmospheric parameters. This must indicate that C-star mass loss is dust-driven, for the driving efficiency of Carbon dust particles is about ten times the value for silicates (Sedlmayr, private comm. 1987).


## 3. Rotation and mass loss

Vardya (1985) has published an interesting diagram suggesting a strong dependence of the rate of mass loss on (projected) stellar rotation $v_R \sin i$. Nieuwenhuijzen and De Jager (1988) could confirm his result for a larger material. But, as also shown by the latter authors, that result is certainly not correct, physically speaking. For, both $\dot{M}$ and $v_R \sin i$ vary more or less monotonically over the HR diagram, and both quantities tend to increase for increasingly luminous stars. This explains why a plot of $\dot{M}$ against $v_R \sin i$ has to show correlation although the two phenomena are perhaps physically hardly correlated. As theoretical predictions by De Grève et al. (1972), Poe and Friend (1986), Friend and Abbott (1986), and Pauldrach et al. (1986) have suggested: $\dot{M}$ increases only by a few tens of percent for an increase in $v_R$ by a factor of ten. It is only close to the critical equatorial (or: breakup) rotational velocity that $\dot{M}$ increases quicker. These theoretical predictions were confirmed by Nieuwenhuijzen and De Jager (1988) in a differential analysis of 140 stars in which it was attempted to avoid running into the trap of a quasi correlation.

The Be-stars need special mention. The $\dot{M}$-values derived from UV resonance line profiles are by about a factor 100 smaller than those found
from infrared continuum measurements. This can be explained if we
accept, following Lamers and Waters (1987), that the UV data refer to
mass flow from the high-latitude parts of the stellar surface, while
the IR data give the mass flow from the stars' equatorial discs. Apparently, the mass flow from Be stars comes essentially from the equatorial discs, while only about one percent of the contribution comes from
the high-latitude parts. It appears also that the observed mass flux
from the discs is somewhat higher than the values from a theoretical
prediction by Poe and Friend (1986).


## 4. Mass loss from red stars; the Red Supergiant Branch

The diversity of groups of stars in the extreme red part of the HR diagram is reflected in the fact that the $\dot{M}$ data in that region hardly
allow for a smooth numerical-mathematical representation. Clearly, the
common assumption that $\dot{M} \sim L$ is certainly unjustified here. We mentioned already the C-stars, with their rate of mass loss about ten
times that of other stars at the same location in the HR diagram. The
explanation for these large values: dust-driven winds involving carbon
particles implies that the mass loss of other stars at the same location as the C stars is also dust-driven, via silicates, which have ten
times lesser efficiency. Gail and Sedlmayr (1987) and Sedlmayr (these
Proceedings) have shown that the mechanism of dust-driven winds works
for $T < 3000$ K, and high luminosities ($L/L_\odot \gtrsim 4$).
Many of the red stars are pulsating and/or show irregular or semi-
regular variations of brightness and radial velocity. For the Mira
stars the mechanism of pulsation-(shock-)driven mass loss has been proposed (Wood and Cahn, 1977; Hill and Willson, 1979), but it appears
difficult to make quantitative predictions.

HR-diagrams of our or of other galactic systems (Humphreys and
Davidson, 1984) show in the red a branch of supergiants, definitely
differing from the Asymptotic Giant Branch, because they are much
brighter. The lower part of this branch is marked by stars like $\subset$ Sco
and $\alpha$ Ori; at its upper part is the famous object VY CMa. The branch
has an inclination of -7 in the (log $T_{eff}$, log L)-diagram, suggestive

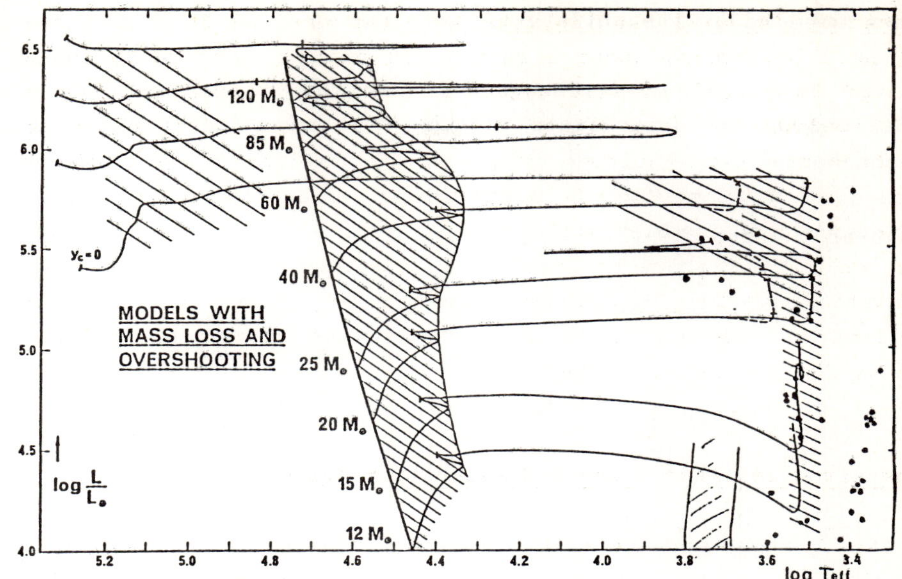


Figure 3: The upper part of the HR diagram with evolutionary tracks calculated by Maeder and Meynet (1987). The branch of dots in the red part is the proposed Red Supergiant Branch. Lower to the right (at $\log T_{eff} \approx 3.4$) is the uppermost part of the Asymptotic Giant Branch. The hatched area near $\log T_{eff} = 3.75$ is the upper part of the Cepheid branch.

of a Hayashi-track. The rate of mass loss is roughly constant along the branch and equal to a few times $10^{-6}$ $M_☉$ $yr^{-1}$, with the exception of VY CMa, however, for which $\dot{M}$ has been determined (De Jager et al., 1987), according to different methods, and with great accuracy:

$$\log(-\dot{M}) = -3.620 \pm 0.047 \quad [M_☉ yr^{-1}].$$

What makes this branch interesting is that it contains more stars than can be expected on the basis of current ideas on stellar evolution. From Maeder and Meynet's (1987) evolutionary calculations it appears that the upper part of this Red Supergiant Branch (as we propose to call it) contains about 3 times more stars than one would expect on the basis of the counted numbers of main sequence O-type stars and evolutionary time schedules. Such high numbers, on the other hand, would rather be expected if stars of about 15 $M_☉$ would climb up, in their evolution, along this Red Supergiant Branch, but so far there is no clear physical basis for supporting this idea.

## 5. <u>**Mass loss and stellar instability of cool stars**</u>

One of us (De Jager, 1984) has suggested that the Humphreys-Davidson
limit is defined by the approximate balance of three accelerations in
stellar atmospheres:

$$g_{grav} + g_{rad} + g_{turb} \approx 0 .$$

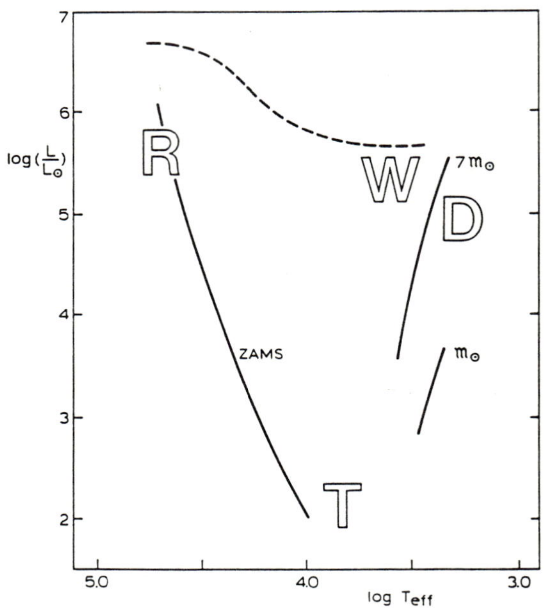


Figure 4: Proposed solar wind mechanisms in the Hertzsprung-Russell
diagram. R = radiation driven winds; W = wave-(turbulence-)
driven; D = dust-driven; T = thermal (coronal) winds.


The classical Eddington criterion is restricted to the first two terms,
and, as Lamers and Fitzpatrick (1987) showed, accounts for the instability in hot stars, with $T_{eff} \gtrsim 10^4$ K. For cooler stars radiative acceleration is ineffective, but turbulent acceleration appears to be able
to balance the gravitation term. The atmospheres of stars closest to
the Humphreys-Davidson limit appear to be strongly turbulent, with
microturbulent velocities equal to or even surpassing the velocity of

sound (Boer et al., 1988). There is considerable dissipation of turbulent energy, which causes transfer of momentum and energy. The consequent heating of hot gas is small, but the momentum transfer causes an outward directed <u>turbulent acceleration</u> (Figure 4).
For stars near the Humphreys-Davidson limit the value of the turbulent acceleration is about equal to that of the gravitational acceleration, but oppositely directed, which explains the instability of cool hypergiant atmospheres (Figure 5).

Microturbulence seldom occurs alone; it is generally driven by larger-scale motions: microturbulence is the high-wavenumber part of the atmospheric <u>spectrum of turbulence</u>. For the stars discussed here the origin of the motion field may be found in pulsations or in convective motions. Such motions have been discovered in $\alpha$ Cyg (Boer et al., 1987): they have up- and downward velocities of 14 km s$^{-1}$ and the

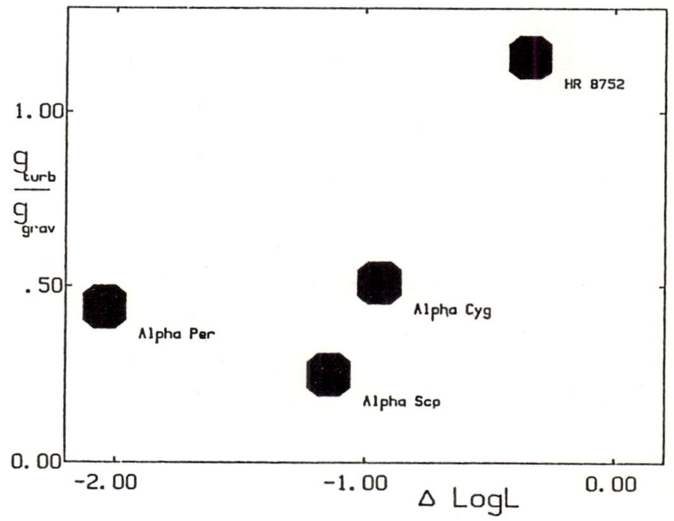


Figure 5: Values of $-g_{turb}/g_{grav}$ for a few well-studied super- and hypergiants suggest an increase of this ratio towards the Humphreys-Davidson limit (Boer et al, These Proceedings).

elements have average diameters of about 30 million km. Whether these motions should be called convection, non-radial or stochastic pulsations is not just a matter of taste: we prefer the last suggestion, because the star is too hot for convection to develop, and too large for having an ordered system of non-radial pulsations.

The concept of stochastic pulsations offers also a natural mechanism to explain <u>episodical mass loss</u> as due to occasionally occurring exceptionally large or rapidly moving pulsation elements. In forwarding this suggestion we realize that its proof should still be given.

## References

Abbott, D.C., Bieging, J.H., Churchwell, E., Cassinelli, J.P.: 1980, Astrophys.J. 238, 196.

Andriesse, C.D.: 1979, Astrophys. Space Sci. 61, 205

Boer, B., De Jager, C., Nieuwenhuijzen, H.: 1987, Astron. Astrophys. submitted.

Boer, B., Carpay, J., De Jager, C., De Koter, A., Nieuwenhuijzen, H., Piters, A., Spaan, F.: 1988, Proc. Int. Astr. Union Coll. 108.

Chiosi, C.: 1981, Astron. Astrophys. 93, 163.

De Grève, J.P., De Loore, C., De Jager, C.: Astrophys. Space Sci. 18, 128.

De Jager, C.: 1980, The Brightest Stars, Reidel, Dordrecht.

De Jager, C.: 1984, Astron. Astrophys. 138, 246.

De Jager, C., Nieuwenhuijzen, H., Van der Hucht, K.A.: 1987, Astron. Astrophys. Suppl. Series, in press.

Friend, D.B., Abbott, D.C.: 1986, Astrophys. J. 311, 701.

Gail, H.P., Sedlmayr, E.: 1987, Astron. Astrophys. 177, 186.

Garmany, C.D., Olson, G.L., Conti, R.S., Van Steenberg, M.E.: 1981, Astrophys. J. 250, 660.

Hill, S.J., Willson, L.A.: 1979, Astrophys. J. 229, 1029.

Humphreys, R.M.: 1987, in H.J.G.L.M. Lamers and C.H.W. de Loore (eds): "Instabilities in Luminous Early type Stars", Reidel, Dordrecht, p.3.

Humphreys, R.M., Davidson, K.: 1979, Astrophys. J. 232, 409.

Humphreys, R.M., Davidson, K.: 1984, Science, 233, 243.

Lamers, H.J.G.L.M.: 1981, Astrophys. J. 245, 593.

Lamers, H.J.G.L.M, Fitzpatrick, E.J.: 1988, Astrophys. J. in press.

Lamers, H.J.G.L.M., Waters, L.B.F.M.: 1987, Astron. Astrophys. in press.

Maeder, A., Meynet, g.: 1987, Astron. Astrophys. 182, 243.

McCrea, W.H.: 1962, Quart. J. Roy. Astron. Soc. 3, 63.

Nieuwenhuijzen, H., De Jager, C.: 1987, Astron. Astrophys. submitted.

Pauldrach, A., Puls, J., Kudritzki, R.P.: 1986, Astron. Astrophys. 164, 86.

Piters, A., De Jager, C., Nieuwenhuijzen, H.: 1987, Astron. Astrophys. in press.

Poe, C.H., Friend, D.B.: 1986, Astrophys. J. 311, 317.

Reimers, D.: 1975, Mem. Soc. Roy. Sci. Liège, 8, 369.

Vardya, M.S.: 1984, Astrophys. Space Sci. 107, 141.

Vardya, M.S.: 1987, Astrophys. J. 299, 255.

Wood, P.R., Cahn, J.H.: 1977, Astrophys. J. 211, 499.

WINDS OF HOT STARS AS A DIAGNOSTIC TOOL OF STELLAR EVOLUTION

R.P. Kudritzki, A. Pauldrach, J. Puls
Institut für Astronomie und Astrophysik der Universität München
Scheinerstr. 1, D-8000 München 80

## I. INTRODUCTION

Modern quantitative spectroscopy of hot stars has two aspects: the analysis of photospheric lines and stellar wind lines. The first one is meanwhile established as an almost classical tool to determine stellar parameters. NLTE model atmosphere and line formation calculations yield $T_{eff}$, log g and abundances with high precision (see recent reviews by Husfeld (this meeting), Kudritzki (1987), Kudritzki and Hummer (1986), Kudritzki, (1985)). The second aspect, however, the quantitative analysis of stellar wind lines is still at its very beginning. For long time the stellar wind lines have been used to determine mass-loss rates $\dot{M}$ and terminal velocities $v_\infty$ only. While these studies were pioneering and of enormous importance, it was also clear that very approximate calculations were done with respect to NLTE ionization and excitation and the radiative transfer in stellar winds. Thus, stellar wind lines could be used only in a more qualitative comparative sense, with no theory behind, which allowed the determination of precise and reliable numbers.

However, during the past few years the situation has dramatically changed. The theory of radiation driven winds has been strongly improved and very detailed and complex multi-level NLTE calculations for stellar winds have become available. The purpose of this paper therefore is to convince that on the basis of this new framework stellar wind lines provide a powerful quantitative diagnostic tool to determine independently radii, luminosities and masses of stars.

## II. THE THEORY OF RADIATION DRIVEN WINDS

Hot stars have an intense radiation field, which by absorption due to UV metal lines leads to an outward accelerating force, which is undoubtedly present. Lucy and Solomon (1970) and Abbott (1979) proved convincingly that this radiation force is sufficient to initialize and to maintain stellar winds. The domain of self-initializing winds predicted by the theory coincides almost perfectly with the occurence of mass-loss in the upper left part of the HR-diagram (see Abbott,

1979). Thus, we can conclude that winds of stars more massive than 20 $M_O$ are basically radiation driven.

The basic dynamical wind quantities are M and $v_\infty$. What does the theory predict with respect to their dependence on the stellar parameters? This has been investigated in the pioneering paper by Castor, Abbott, Klein (1975, "CAK"), who for the first time formulated the theory of radiation driven winds in a selfconsistent way. Besides some crucial simplifying assumptions (see below) the theory in its later version (Abbott, 1982) used a realistic line list of 250000 lines of H to Zn in ionization stages I to VI. The prediction of the theory were:

$$- v_\infty = (\alpha/1-\alpha)^{1/2} \ v_{esc}$$
$$- M \quad L^{1/\alpha}(M(1-L/L_E))^{1-1/\alpha} \tag{1}$$

where the parameter $\alpha$ comes out as the result of the theory and lies between 0.5 and 0.7. As shown in Fig. 1, the predicted proportionality are roughly reproduced by the observations. However, with respect to the proportionality constants the old CAK theory fails. Predicted mass-loss rates are too high by a factor of three (Fig. 2a) and theoretical terminal velocities are too low by a large factor (Fig. 2). Since $v_\infty$ can be measured with high precision from the blue edges of the velocity affected line profiles, the latter discrepancy cast enormous doubts on the theory.

However, recently Pauldrach, Puls and Kudritzki (1986, "PPK") were able to prove that this failure of the theory was caused by one of the crucial approximations made by CAK, namely the "radial streaming approximation". In this approximation the interaction of photospheric photons with the wind plasma is treated as if they were streaming out radially from the stellar surface. This approximation is very poor. Even some stellar radii away from the star the photosphere forms a finite cone angle (Fig. 3), which is crucial for correct treatment of momentum exchange. This "finite cone angle effect" was taken into account by PPK by introducing a correction factor CF to the CAK radiative force $f^{CAK}$.

$$f^{PPK} = CF \ f^{CAK}, \qquad CF = \frac{2}{1-\mu_*^2} \int_{\mu_*}^{1} \left(\frac{\mu^2 \ dv/dr + (1-\mu^2) \ v/r}{dv/dr}\right)^\alpha \mu d\mu \tag{2}.$$

Fig. 4 shows that CF<1 close to stellar surface. As a result M, which is fixed at the critical point close to the stellar surface is significantly reduced. Away from the star we have CF>1. In addition, due to

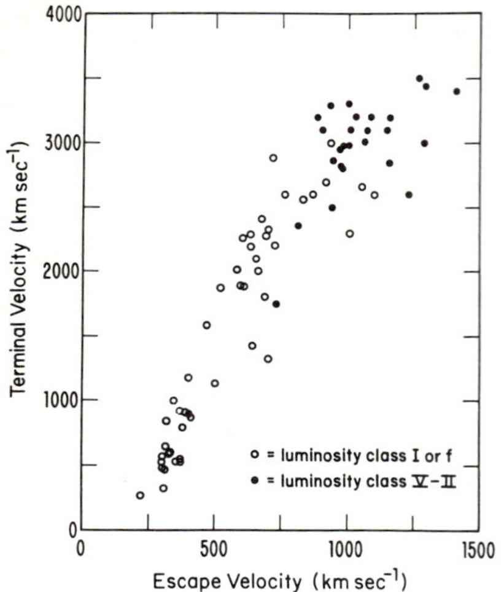

Fig. 1a: The observed relation between terminal velocity $v_\infty$ and photospheric escape velocity $v_{esc}$ (from Abbott, 1982).

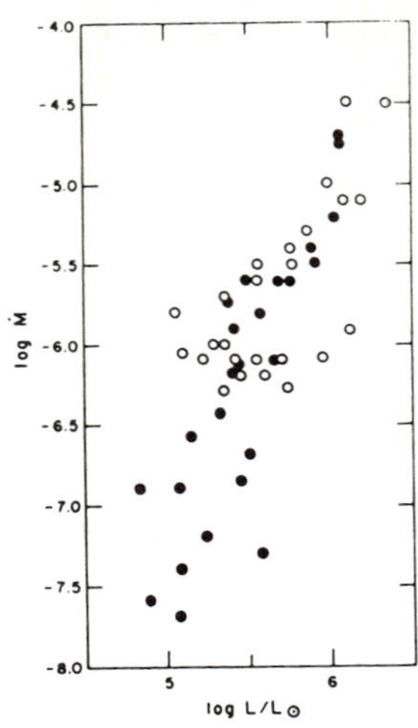

Fig. 1b: The observed relation of log $\dot{M}$ vs. log L, which obeys roughly $\dot{M} \sim L^{1.6}$ (from Garmany and Conti, 1984).

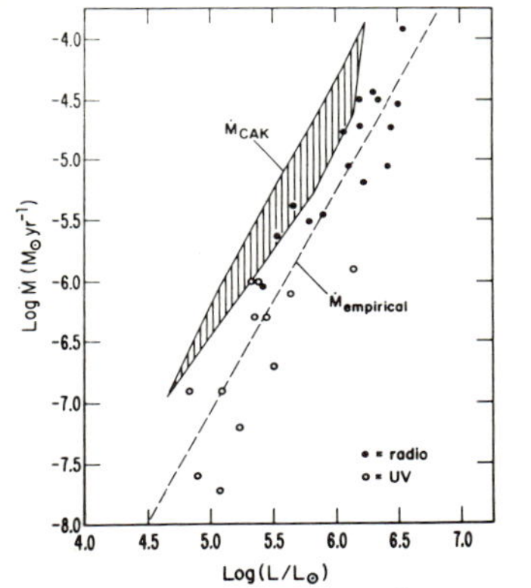

Fig. 2a: Observed log $\dot{M}$ vs. log L (dots) compared with predictions of CAK-theory (hatched area) (from Abbott, 1982).

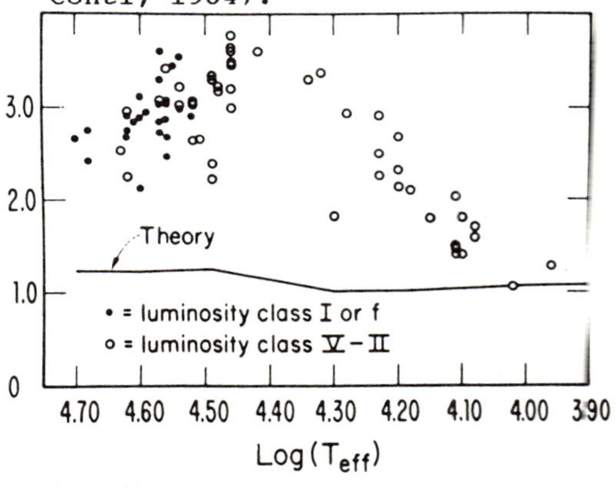

Fig. 2b: $v_\infty/v_{esc}$ as function of $T_{eff}$ for individual stars compared with predictions of CAK-theory (from Abbott, 1982).

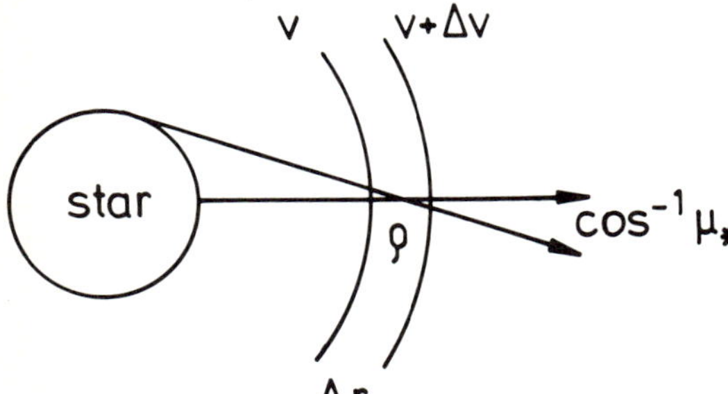



Fig. 3: The photospheric finite cone angle irradiating the expanding stellar wind shell.

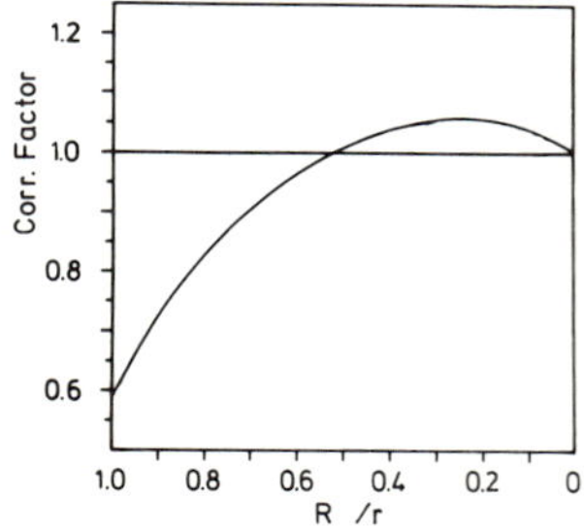

Fig. 4: Finite cone angle correction factor to the radiative force as function of reciprocal radius (from Pauldrach et al., 1986).

the lower $\dot{M}$ the wind density $\rho$ is reduced in these outer layers (relative to the CAK case) and therefore - since $dv^2/dr \approx f/\rho$ - the wind material is much stronger accelerated. Consequently much higher values of $v_\infty$ are obtained. (Similarly results were independently obtained in a recent paper by Friend and Abbott, 1987).

PPK have applied this improved wind theory on a sample of massive OB-stars with well determined wind properties. The results are given in Table 1 and show for given stellar parameters the observed dynamical wind quantities $\dot{M}$ and $v_\infty$ are well reproduced by the theory. This includes even such extreme objects like P Cygni, which has an extremely high mass-loss rate and a very slow wind. The dynamically improved theory of radiation driven winds therefore appears to be highly reliable.

Table 1

| star | spec.type | $T_{eff}$ $(10^3 K)$ | log g (cgs) | log $L/L_\odot$ | $\dot{M}_{obs}$ | $\dot{M}_{calc}$ | $v_\infty^{obs}$ | $v_\infty^{obs}$ |
|---|---|---|---|---|---|---|---|---|
|  |  |  |  |  | $10^{-6} M_O/yr$ |  | km/s |  |
| P Cyg | BIIa | 18.0 | 2.0 | 5.64 | 20-30 | 29 | 400 | 395 |
| $\epsilon$ Ori | B0Ia | 28.5 | 3.25 | 5.91 | 3.1 | 3.3 | 2010 | 1950 |
| $\zeta$ Ori | O9.5I | 30.0 | 3.45 | 5.79 | 2.3 | 1.9 | 2290 | 2274 |
| 9 Sgr | O4(f)V | 50.0 | 4.10 | 5.95 | 4.0 | 4.0 | 3440 | 3480 |
| HD 48099 | O6.5V | 39.0 | 4.00 | 5.40 | 0.63 | 0.64 | 3500 | 3540 |
| HD 42088 | O6.5V | 40.0 | 4.05 | 4.89 | 0.13 | 0.20 | 2600 | 2600 |
| $\lambda$ Cep | O6ef | 42.0 | 3.7 | 5.90 | 4.0 | 5.1 | 2500 | 2500 |

## III. THE EVOLUTION OF MASSIVE STARS AND THE DYNAMICS OF STELLAR WINDS

The observed terminal velocities $v_\infty$ of O-stars provide interesting information about the evolution of massive stars. This is shown in Fig. 5, which is taken from Garmany and Conti (1985) and displays $v_\infty$ vs. $T_{eff}$ for a sample of O-stars (luminosity classes between V and III) in the Galaxy, LMC and SMC. Two striking facts can be read off from Fig. 5:
- the values of $v_\infty$ form an inclined band vs. $T_{eff}$
- the values for LMC and SMC are on the average 500 km/s and 800 km/s deeper than for the Galaxy.

Kudritzki et al. (1987 "KPP") were able to explain both effects in terms of the improved theory of radiation driven winds. They calculated wind models for LMC, SMC and Galaxy using the reduced metal abun-

117

dances for the Clouds

$$z_{LMC} = 0.28 \, z_{Gal}; \qquad z_{SMC} = 0.1 \, z_{Gal}$$

as indicated by the analysis of the HII-region emission line spectra (Dufour, 1984). Table 2 contains the results for a typical O5V star:

Table 2: Calculated wind parameters for a typical O5V-star in Galaxy, LMC, SMC.

|  | $v_\infty$ km/s | $\dot{M}$ $10^{-6}M_o$/yr | $\dfrac{\dot{M}v_\infty c}{L}$ |
|---|---|---|---|
| $z_{Gal}$ | 3350 | 2.12 | 0.66 |
| $z_{LMC}$ | 2900 | 1.35 | 0.36 |
| $z_{SMC}$ | 2435 | 0.72 | 0.16 |

Obviously, both $v_\infty$ and $\dot{M}$ are significantly reduced by the lower metallicity. The effect on $v_\infty$ is of the same order as it is observed, whereas the change in $\dot{M}$ is on the margin to be detectable.

In a next step KPP calculated wind models along evolutionary tracks of O-stars in all three galaxies. The tracks used the same metallicities as above and were calculated by Pylyser et al. (1985). Fig. 6 comprises the major results. The upper part contains simply the evolutionary tracks. However, contrary to the common HR-diagram $v_{esc}$, the escape velocity from the stellar surface, is plotted as function of $T_{eff}$. This "alternative HR-diagram" reveals that during the massive star evolution the $v_{esc}$ values lie within an inclined band of the diagram. When the stars evolve away from the ZAMS, their radii increase and consequently their $v_{esc}$ decrease. However, due to convective mixing, overshooting and mass-loss of the evolution turns back to the blue at roughly constant luminosity. This means that the radii decrease again. However, $v_{esc}$ while mildly increasing does not approach the old values at the ZAMS, since the stars have now already lost a significant fraction of their mass.

In radiation driven wind theory $v_\infty$ is related to $v_{esc}$. Thus we expect $v_\infty$ to form a similar inclined band as $v_{esc}$. This is shown in the lower part of Fig. 6 together with the observed position of massive stars from Fig. 5. The agreement between theory and observation is obvious. We wish to point out here that diagrams of this type also allow to read of present stellar parameters and thus - at least in principle - provide an interesting alternative to the HR-diagram.

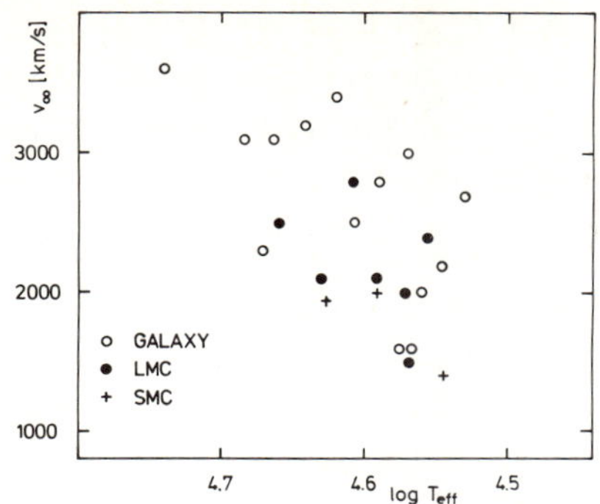

Fig. 5: Observed correlation of $v_\infty$ vs. $T_{eff}$ for O-stars in Galaxy, LMC and SMC (from Garmany and Conti, 1985).

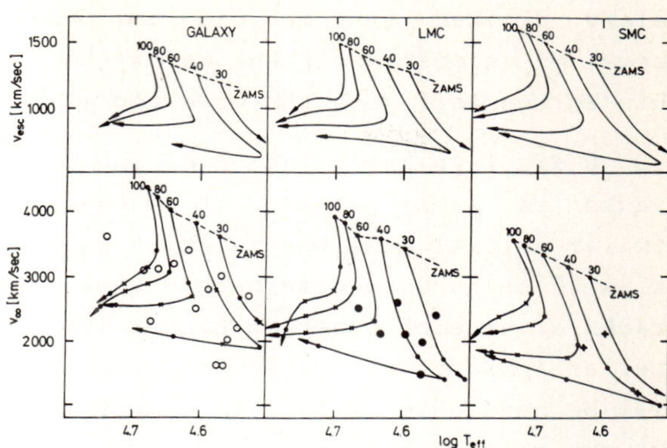

Fig. 6: The "alternative HR-diagram" of massive star evolution. Upper part: Surface escape velocity vs. $T_{eff}$. Lower part: Terminal veolcity vs. $T_{eff}$. The position of observed objects (see Fig. 5) is also shown.

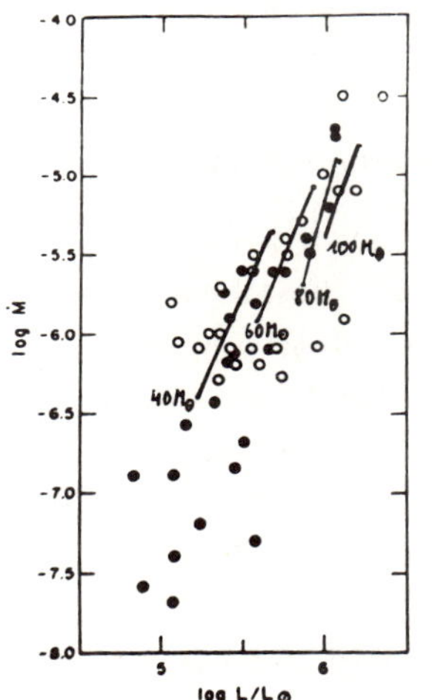

Fig. 9: The observed log L, log $\dot{M}$ relation for galactic O-stars (same as Fig. 16) compared with the theoretical results of Fig. 8.

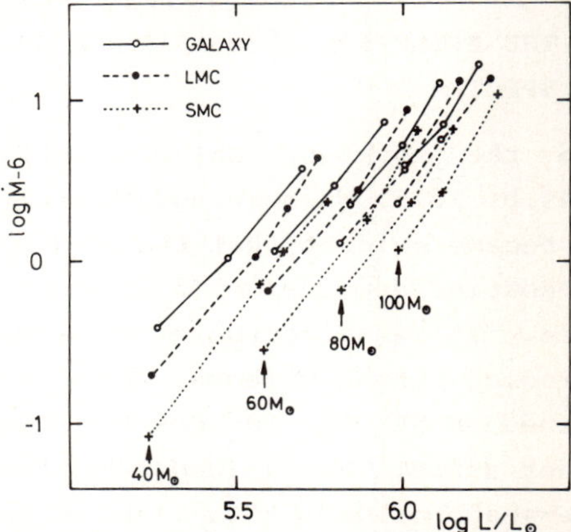

Fig. 8: The log L, log $\dot{M}$ relation along the evolutionary tracks as predicted by radiation driven wind theory.

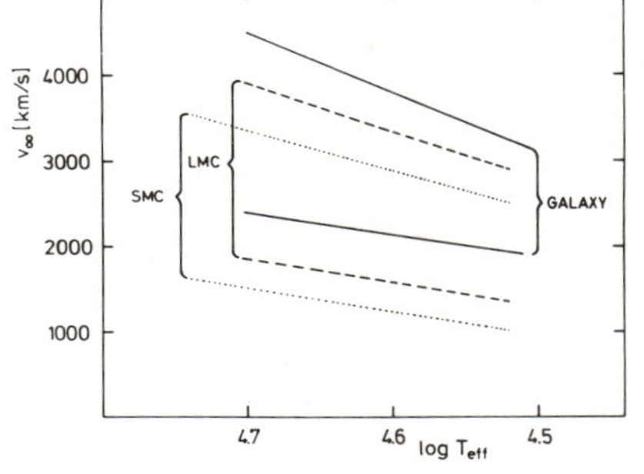

Fig. 7: The ($v_\infty$, log $T_{eff}$)-bands for each galaxy as obtained from the theoretical calculations of Fig. 6.

Fig. 7 shows the envelopes of the $v_\infty$ vs. log $T_{eff}$ tracks for each galaxy plotted into one diagram. Obviously the theory predicts the range of velocities in the Galaxy to lie above the LMC and the SMC. This agrees well with the observed effect in Fig. 5.

Fig. 8 displays the relation of log $\dot{M}$ vs. log L along the evolutionary tracks in each galaxy. It is clearly seen that for each track of similar initial mass the relation $\dot{M}_{Gal} > \dot{M}_{LMC} > \dot{M}_{SMC}$ holds. However, it will be hard to disentangle this effect observationally since the tracks cross each other at higher luminosity. Careful determination of stellar parameters based on detailed photospheric NLTE spectroscopy will be needed for this. (A comparison with observations of stars in our own galaxy is shown in Fig. 9).

Generally, we conclude that the theory of radiation driven winds describes the observed wind dynamics satisfactory. This includes also the dependence on metallicity.

## IV. THE EVOLUTION OF MASSIVE STARS AND THE MORPHOLOGY OF STELLAR WIND SPECTRA

From the work by Walborn and collaborators (Walborn and Panek, 1984a,b, 1985; Walborn and Nichols-Bohlin, 1987; Walborn et al., 1985 it became evident that the appearance of stellar wind spectra shows systematic dependence on luminosity, effective temperature and abundance. It is of course crucial for a reliable stellar wind theory to reproduce these effects. This requires, however, detailed and very refined NLTE calculations for excitation and ionization in the winds of hot stars. For a long period such calculations were carried out in a very approximative way, until just very recently the first realistic calculations became available: Pauldrach (1987) adopted a cool wird $(T_W \approx T_{eff})$ and treated the full multi-level NLTE problem of all relevant elements and ions including electron collisions selfconsistently with the radiation driven wind hydrodynamics. He includes in total 26 elements , 133 ionization stages, 4000 levels, 10000 radiative bound-bound transitions in the rate quations and the ocrrect continuous radiation field for the bound-free rates calculated from the spherical transfer equation. Puls (1987) extended these improvements significantly further by including the "multi-line effects", which arise from the velocity induced line overlap, which causes radiative coupling between different ions and possible multiple momentum transfer from photons to the wind plasma. The first application of this strongly improved theory on the case of the O4f-star $\zeta$ Puppis shows excellent agreement with respect to the dynamical quantities $\dot{M}$ and $v_\infty$. In addition - and this is the really important result - an enormous shift towards higher ionization stages is obtained due to the detailed NLTE treatment. Thus, it was for the first time possible to reproduce by cool wind models the observed high ionization features of OVI, NV, CIV etc. A typical example is given in Fig. 10, which shows how perfectly the observed NV profile in the UV-spectrum of $\zeta$ Puppis can be reproduced by the selfconsistent radiation driven NLTE wind model atmospheres. Note that this is not a profile fit, where $\dot{M}$ and $v(r)$ have been properly adjusted. It is the result of a selfconsistent calculation, which depends only on the choice of the stellar parameters $\log L/L_o$, $T_{eff}$ and $\log g$.

We have now applied calculations of this type along evolutionary tracks for massive stars to investigate whether our wind models do also reproduce the change of spectral morphology during stellar evolution. Fig. 11 displays the investigated parameter domain. Fig. 12 demonstrates the extreme behaviour of the SiIV resonance lines, which exhibit a pronounced luminosity effect both in theory and observation. The physical reason for this effect, which will prove to be of enormous potential for the luminosity classification of extragalactic O-stars using HST, is that in the winds of O-stars most of the silicon is SiV. Supergiants are closer to the Eddington limit and thus according to eq. (1) have denser winds, which leads to stronger recombination towards SiIV. Consequently, the SiIV wind features show up in the supergiants. The behaviour of NV, NIV, CIV can also be reproduced by the models, but this is not shown here for sake of brevity.

The improved stellar wind models are obviously able to describe the observed changes of spectral morphology as function of $T_{eff}$ and L in a proper way. This - after a phase of calibration by detailed quantitative star by star spectral analysis of standards - will render the possiblity to use the ultraviolet wind spectra of hot stars for direct luminosity determination.


V. STELLAR WINDS AND THE EVOLUTION OF CENTRAL STARS OF PN

After the advent of the IUE satellite it became undoubtedly clear that stellar winds are also present in many CSPN (Heap, 1978; Perinotto, 1982). Detailed studies of their wind properties have been carried out by Hamann et al. (1984) and Cerruti-Sola and Perinotto (1985). Here we

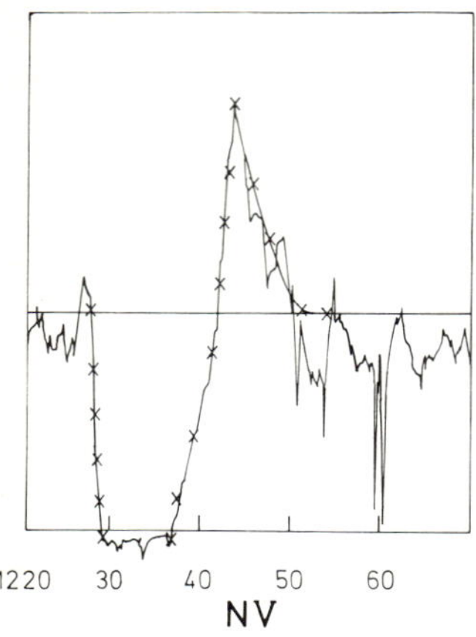



**Fig. 10:** The IUE high resolution profile of NV λ 1240 of ζ Puppis compared with the result from a selfconsistent radiation driven wind model (crosses), which includes full NLTE in all ions and multi line effects (from Puls, 1987).

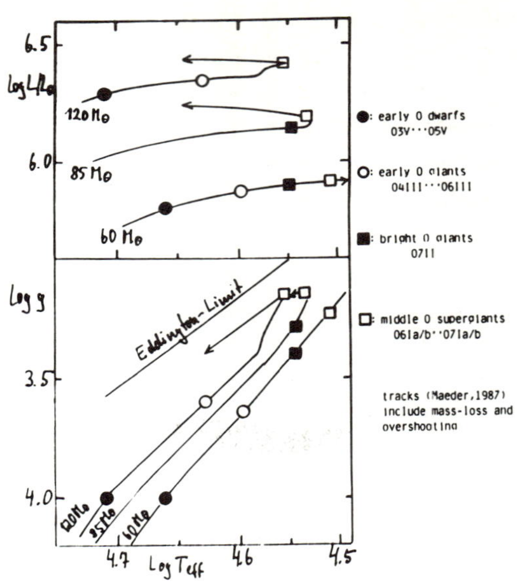



**Fig. 11:** Wind models along evolutionary tracks representing different luminosity classes.

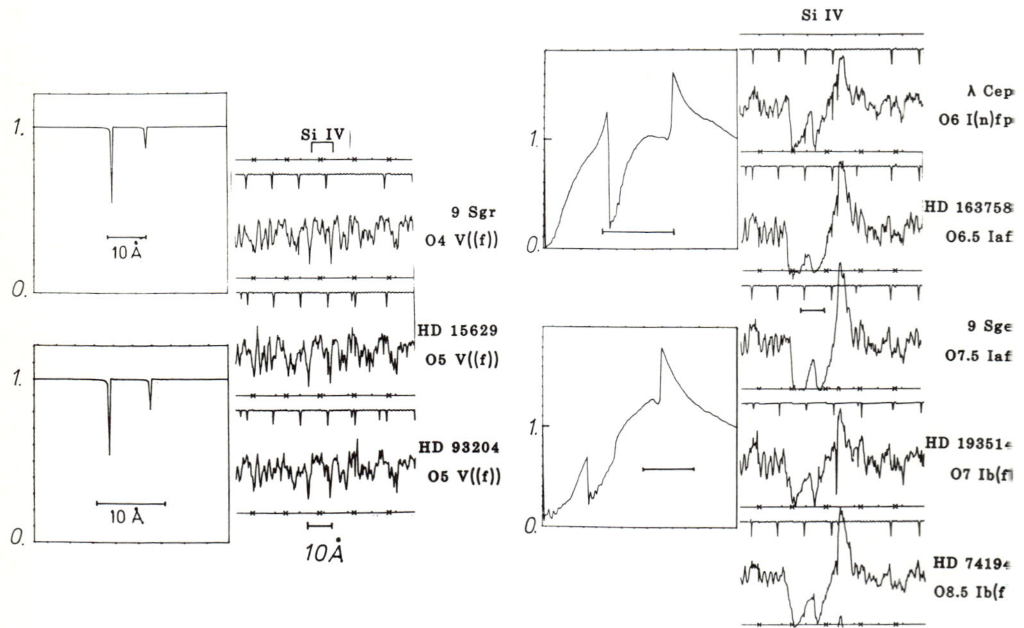



**Fig. 12:** The strong observed luminosity effect of SiIV as reproduced by the wind models. The observations are copied from Walborn et al. (1985). (For discussion see text).

want to concentrate on two striking observational correlations in the case of CSPN. The first one is displayed in Fig. 13, which exhibits the increase of terminal velocity with $T_{eff}$. Fig. 13 strongly points to radiation driven winds being present in the outer layers of CSPN, since in this case the terminal velocity is related to the surface escape velocity. CSPN evolve at constant luminosity towards the blue, so that the escape velocity increases. Consequently, we expect the terminal velocity to increase with $T_{eff}$.

The second observational correlation results from the detailed photospheric quantitative spectroscopy by Méndez et al. (1987), which allows to determine the position of CSPN in the log g, log $T_{eff}$-plane with high precision. By transformation of evolutionary tracks into this plane stellar masses $M/M_O$ and radii $R/R_O$ can be read off directly. The log g, log $T_{eff}$ diagram of CSPN can also be used to investigate how the strength of stellar winds depends on the stellar parameters. This is done in Fig. 14, which reveals clearly that the more massive objects closer to the Eddington limit have observable wind features.

We have now applied our improved radiation driven wind theory also on wind models along post-AGB tracks by Schönberner (1983) and Wood and Faulkner (1986). Fig. 15 shows the calculated relation between terminal velocity and $T_{eff}$ along the evolutionary tracks, including the observed values for CSPN. The result is extremely convincing. It again allows to read off stellar masses directly, and suggests that the masses of CSPN are in a rather narrow range between 0.5 and 0.8 solar masses (Schönberner, 1981; Méndez et al., 1985, 1987). This reveals the power of stellar wind models for the determination of stellar masses.

As indicated in Fig. 14 also the wind features in the optical spectra become significant at higher masses of the CSPN. This is demonstrated by Fig. 16, which shows how HeII 4686 switches from photospheric absorption into wind emission with increasing mass. It is important to test whether radiation driven wind models can reproduce this behaviour.

For this purpose, a new type of "unified model atmospheres" has been developed at the Munich Observatory by R. Gabler (1986) and A. Wagner (1986) in cooperation with J. Puls, A. Pauldrach and R.P. Kudritzki. These NLTE model atmospheres are spherically extended, in radiative equilibrium, and include the density and velocity distribution of radiation driven winds. The spectra of H and He lines are then calcu-

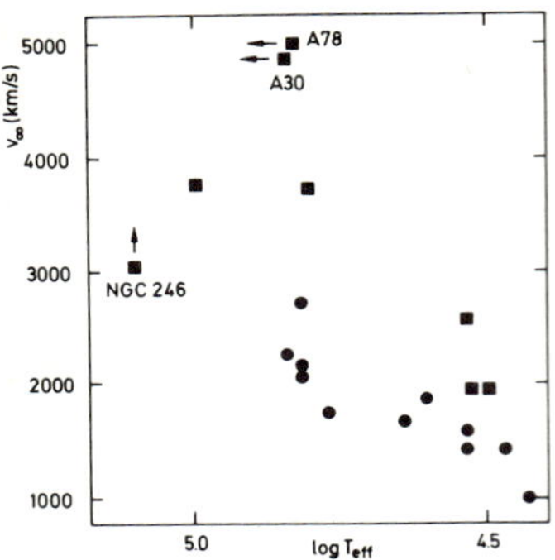

**Fig. 13:** Terminal velocity versus $T_{eff}$ for CSPN. Squares are H-deficient objects; circels, normal Hydrogen abundance.

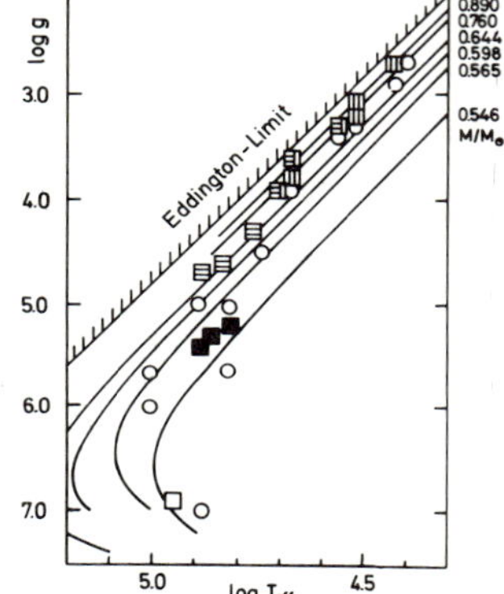

**Fig. 14:** The log g, log $T_{eff}$ diagram of CSPN compared with post AGB-tracks. The presence of winds in the optical and for UV spectra is indicated as follows: IUE wind detections, wind detections by optical spectra, no winds in optical - no IUE high resolution spectra taken, objects with definitely no wind.

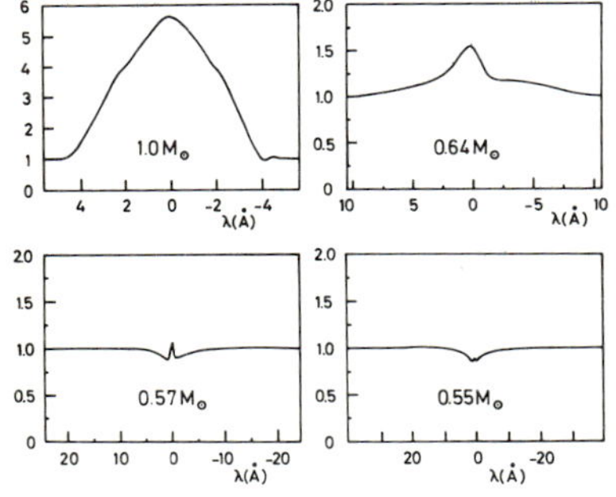

**Fig. 18:** HeII 4686 model profiles at $T_{eff}$ = 50000K for different CSPN masses. Note the enormous emission at one solar mass.

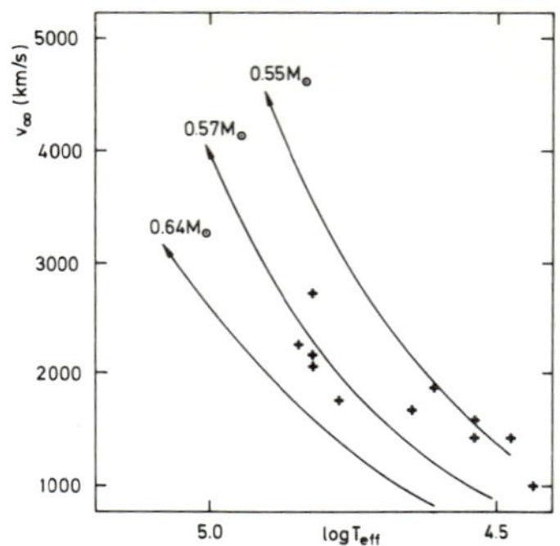

**Fig. 15:** Same as Fig. 13 for CSPN with normal H abundance, but including wind calculations along evolutionary tracks.

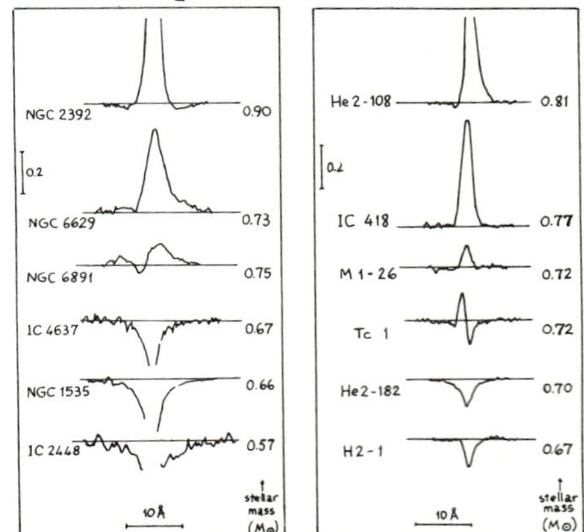

**Fig. 16:** The behaviour of the stellar HeII 4686 as a function of the stellar mass. The left and right panels are for objects with $T_{eff}$ around 50000K and 35000K, respectively.

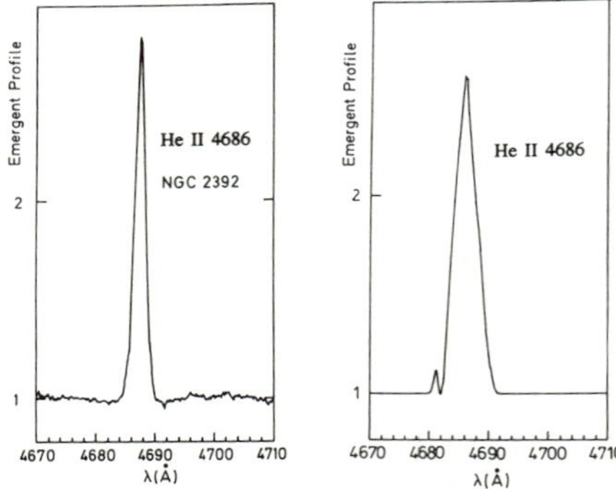

**Fig. 17:** The observed HeII 4686 profile of the high mass CSPN NGC 2392 (left) compared with a model calculation for one solar mass and $T_{eff}$ = 45000K.

lated for these models by detailed NLTE multi-level calculations in the whole atmosphere, thus treating the contribution of subsonic deeper and supersonic outer layers to the emergent line profile in the correct self-consistent unified way, including Stark-effect broadening and velocity fields. We have calculated a sequence of such models for $T_{eff}$ = 50000K and stellar masses equal to 0.55, 0.57, 0.64 and 1 solar masses. The corresponding luminosities of the CSPN (or the gravities or the radii) were obtained from the evolutionary tracks mentioned already in the previous section. In units of the Eddington luminosity $L_E$, we obtain, respectively, $L/L_E$ = 0.06, 0.16, 0.34, 0.74. The more massive object is therefore already rather close to the Eddington limit. Its theoretical HeII 4686 profile should look similar to the one of the central star of NGC 2392, following Méndez et al. (1987). Fig. 17 shows that this is really the case. In addition, the turnover from photospheric absorption to wind emission with increasing $L/L_E$, as observed by Méndez et al., is well reproduced by the theoretical computations displayed in Fig. 18. This demonstrates that the concept of radiation driven winds is very useful also for CSPN. Moreover, it renders the possibility to use the 4686 emission as a powerful luminosity (=mass, =distance)-indicator. In future work this will be tested quantitatively.


## ACKNOWLEDGEMENTS

This work was supported by the Deutsche Forschungsgemeinschaft under grants Ku 474/11-1 and Ku 474/13-1.


## REFERENCES

Abbott, D.C.: 1979, IAU Symp. 83, p. 237

Abbott, D.C.: 1982, Astrophys. J. <u>259</u>, 893

Castor, J., Abbott, D.C., Klein, R.: 1975, Astrophys. J. <u>195</u>, 157

Cerruti-Sola, M., Perinotto, M.: 1985, Astrophys. J. <u>291</u>, 237

Dufour, R.J.: 1984, IAU Symp. 108, p. 353

Friend, D., Abbott, D.C.: 1986, Astrophys. J. <u>311</u>, 701

Gabler, R.: 1986, Diplomarbeit, Universität München

Hamann, W.R., Kudritzki, R.P., Méndez, R.H., Pottasch, S.R.: 1984, Astron. Astrophys. <u>139</u>, 459

Heap, S.R.: 1978, IAU Symp. 83, p. 99

Kudritzki, R.P.: 1985, Proc. ESO Workshop on "Production and Distribution of C,N,O Elements", ed. J. Danziger et al., p. 277 (invited paper)

Kudritzki, R.P.: 1987, "Spectroscopic Constraints on the Evolution of Subluminous O-stars and Central Stars of PN", Proc. of IAU Coll. No. 95 on Faint Blue Stars, ed. Davis Philip, in press

Kudritzki, R.P., Hummer, D.G.: 1986, Proc. IAU Symp. 116, ed. de Loore et al., p. 3

Kudritzki, R.P., Pauldrach, A., Puls, J.: 1987, Astron. Astrophys. 173, 293

Lucy, L.B., Solomon, P.: 1970, Astrophys. J. 159, 879

Méndez, R.H., Kudritzki, R.P., Herrero, A., Husfeld, D., Groth, H.G.: 1987, Astron. Astrophys., in press

Méndez, R.H., Kudritzki, R.P., Simon, K.P.: 1985, Astron. Astrophys. 164, 86

Pauldrach, A.: 1987, Astron. Astrophys. 183, 295

Pauldrach, A., Puls, J., Kudritzki, R.P.: 1986, Astron. Astrophys. 164, 86

Perinotto, M.: 1982, IAU Symp. 103, p. 323

Puls, J.: 1987, Astron. Astrophys. 184, 227

Pylyser, E., Doom, C., de Loore, C.: 1985, Astron. Astrophys. 148, 379

Schönberner, D.: 1981, Astron. Astrophys. 103, 119

Schönberner, D.: 1983, Astrophys. J. 272, 708

Wagner, A.: 1986, Diplomarbeit, Universität München

Walborn, N.R., Panek, R.J.: 1984, Astrophys. J. 280, L27

Walborn, N.R., Panek, R.J.: 1984, Astrophys. J. 286, 718

Walborn, N.R., Panek, R.J.: 1985, Astrophys. J. 291, 806

Walborn, N.R., Nichols-Bohlin, J.: 1987, PASP 99, 40

Walborn, N.R., Nichols-Bohlin, J., Panek, R.J.: 1985, "IUE Atlas of O-Type Spectra from 1200 to 1900Å", NASA Reference Publication 1155

Wood, P.R., Faulkner, D.J.: 1986, Astrophys. J. 307, 659

ATMOSPHERIC PHENOMENA IN ETA CARINAE
AND THE HUBBLE-SANDAGE VARIABLES

R. Viotti, L. Rossi  (Istituto Astrofisica Spaziale, CNR, Frascati, Italy)
A. Altamore, C. Rossi (Istituto Astronomico, Roma University, Italy)
G.B. Baratta (Osservatorio Astronomico, Roma, Italy)
A. Cassatella (IUE Observatory, ESA, Madrid, Spain)

ABSTRACT. From the analysis of the galactic counterparts of the HS variables,$\eta$Car, AG
Car  and  P Cyg we conclude that their atmospheres are near critical physical  conditions  so that small changes of the mass outflow could produce a large  variation  of
the atmospheric structure and nova-like explosions at nearly constant Mbol.

The  Hubble-Sandage  variables (HSV) are bright objects in outer galaxies  which  are
known for their large photometric variability on very long time scales. The HSV's are
among the most luminous stellar objects and may represent a phase of the evolution of
very  massive  stars after having left the main sequence.  The problem is to find  the
origin of their variability and its relation with the stellar parameters and position
in the HR diagram close to the observational Humphreys-Davidson upper boundary. HSV's
are  difficult  to  study because of their apparent faintness.  There are  however  a
number  of  HSV homologues in our Galaxy and in the Magellanic Clouds  which  show  a
similar  behaviour and are commonly considered as the galactic or MC counterparts  of
HSV's.  The best known objects in the Milky Way are $\eta$  Car,  AG Car and P Cyg. In the
following  we shall discuss the main properties of these stars and their evolutionary
implications.

a.  The light curves.  $\eta$_Car is known for its large luminosity variations which took
place the last century.  During 1830-1850 the star was variable between  0 and 2 mag.
Since 1856 . Car gradually faded to the seventh magnitude in about 14 years. Actually
this fading was only apparent. In fact Andriesse et al. (1978) found that the present
bolometric magnitude (=0.0) is close to the estimated bolometric magnitude during the
bright  phase.  This suggests that $\eta$  Car is presently surrounded by a dust  envelope
formed during the fading phase,  which is absorbing most of the optical and UV radiation from the central star.
     The  P Cygni star AG Car has displayed in recent years large  luminosity  varia-
tions  from V=6 to 8,  associated with deep spectral change.  The equivalent spectral
type  was  A at maximum luminosity in 1981,  and became B in 1983 and Of/WN  in  1984
(Viotti et al.  1984;  Stahl 1986). Viotti et al. from an analysis of UV to IR energy
distribution  of AG Car found that the variations occurred at almost  constant  bolometric  magnitude  (about -8.3) in spite of the large visual  luminosity  variations.
Large photometric variations have been observed in luminous P Cygni stars in the LMC.
Also  in  this case it has been found that these variations are caused by flux  redistribution  in  the atmospheric envelope,  whilst the bolometric  luminosity  remains
nearly constant (e.g. Wolf and Stahl 1983).
     This  was  probably  also the case of P Cyg itself which has  been  rather  stable
since around 1780, but previously was found variable with light maxima in 1600-06 and
1639-59 when the star reached the third magnitude and was 'reddish'. Deep minima were
recorded  during 1606-56 and 1659-83 with occasional fadings below visibility (see de
Groot 1969). If the star at maximum was redder than today, then most of the light was
probably  emitted at optical wavelengths,  and its bolometric correction should  have
been  close to zero.  Since according to Lamers et al.(1983) the present BC of P  Cyg
should  be close to -1.6,the magnitude difference between the bright phase (V=3)  and
now  (4.8) is practically equal to the difference in BC's.  The conclusion  is  that
during the 17th century maxima P Cyg probably had the same bolometric luminosity than
now and that no major 'explosion' occurred at that time. The variations between 3 mag
and  5-6  mag should be ascribed to change of the temperature of  the  atmosphere  at

and 5-6 mag should be ascribed to change of the temperature of the atmosphere at constant Mbol and to the related energy redistribution like in the case of AG Car and of the LMC variables discussed above. The deep fadings during 1600 are most likely associated with dust formation like in $\eta$ Car. The following brightening should be the result of the destruction or dilution of the circumstellar dust envelope (Viotti 1987).

b. <u>Chemical composition</u>. The chemical composition of the atmospheres of these objects is difficult to be determined because of their peculiar spectrum. A comparison of the optical spectrum of AG Car with P Cyg and other B and Be stars led Caputo and Viotti (1970) to conclude that carbon and oxygen should be underabundant in the expanding atmosphere of AG Car. This star is surrounded by a ring nebula formed about 4000 years ago and probably containing dust particles (Viotti et al. 1987). A chemical anomaly seems also to be present in $\eta$ Car. Altamore et al. (1986) from the absence of the CIII] and CII] lines in the UV spectrum of $\eta$ Car suggested a low C/N abundance ratio. Also the ejected condensations surrounding the central star appear to be nitrogen rich (Davidson et al. 1982). In $\eta$ Car dust is continuously condensing in the expanding wind (Andriesse et al.1978) and the chemical composition should have an important role in the process of dust formation and growth. Andriesse et al. suggested that the gas condenses in the form of a cluster of silicates. It should be important to investigate the effect of the anomalous chemical composition on the process of dust formation in AG and $\eta$ Car.

To conclude, observations indicate that in the galactic and MC luminous blue variables large variations occur <u>at constant bolometric magnitude</u> associated with change in the structure of the atmospheric envelope and/or to <u>dust formation</u> and destruction. From a study of the light curves of HSVs we conclude that the same should happen in those objects. Probably the atmospheres of these very luminous, high mass losing stars are subject to instabilities whose effects are enhanced by a sudden increase (or decrease) of the line opacity. Thus the star becomes yellower and its visual luminosity increases. Probably an occasional consistent but not necessarily very large increase of the mass loss rate might produce a drastic decrease of the temperature even below the grain condensation temperature followed by dust condensation in the outer atmospheric layers. In this case molecular opacity should play an important role as in M supergiants. This process has been observed in $\eta$ Car and produces a large and sudden luminosity fading like those observed in many HSV's. This fading is only apparent since most of the radiation is emitted in the IR.
     There are many aspects of the HSV phenomenon which should require much future work. One problem is their possible <u>multiplicity</u>. Hofmann and Weigelt (1986) found that $\eta$ Car is (at least) a quadruple system. This has a strong impact on the estimate of the luminosity and mass loss rate of the central star. If the frequency of multiple systems in the luminous blue variables is high (e.g. Mayor and Mazeh 1987), their evolutionary stage as well as the upper boundary of the HR diagram should be reconsidered.

REFERENCES
Altamore, A., Baratta, G.B., Cassatella, A., Rossi, L., Viotti, R.:1986, New Insights in Astrophysics, ESA SP-263, p.303.
Andriesse, C.D., Donn, B.D., Viotti, R.: 1978, MNRAS 185, 771.
Caputo, F., Viotti, R.: 1970, Astron. Astrophys. 7, 266.
Davidson, K., Walborn, N.R., Gull, T.R.: 1982, Ap. J. 254, L47.
de Groot, M.: 1969, Bull. astr. Soc. Netherlands 20, 225.
Hofmann, K.-H., Weigelt, G.: 1986, Astr. Astrophys. 167, L15.
Lamers, H.J.G.L.M., et al.: 1983, Astr. Astrophys. 128, 299.
Mayor, M., Mazeh, T.: 1987, Astron. Astrophys. 171, 157.
Stahl, O.: 1986, Astron. Astrophys. 164, 321.
Viotti, R.: 1987, Instabilities in Luminous Early Type Stars, H.J. Lamers and C. de Loore eds., Reidel, Dordrecht, p.257.
Viotti, R., Altamore, A., Barylak, M., Cassatella, A., Gilmozzi, R., Rossi, C.: 1984, NASA CP 2349, p.231.
Viotti, R., Cassatella, A., Ponz, D., The', P.S.: 1987, Astr. Astrophys. in press.
Wolf, B., Stahl, O.: 1983, The Messenger n.33, p.11.

MODEL OF DUSTY ENVELOPE OF ETA CARINAE


A.B.Men'shchikov, B.M.Shustov, A.V.Tutukov
Astronomical Council of the USSR Academy of Sciences
Moscow, USSR


Eta Carinae is a well-known example of a star with a massive circumstellar nebulous shell. The shell is regarded as a remnant of a great outburst of the star in the last century. $\eta$ Car parameters remain to be a matter of scientific debates: most often investigators revise stellar temperature $T$ , mass loss rate $\dot{M}$ , the velocity of outflow $V$ . We have worked out several numerical evolutionary models of $\eta$ Car envelope (the homunculus) paying attention to the dependence on the mentioned parameters. The dependence on the parameters of grain evolution was also considered. For a representative model we fixed: $V=500$km/s; $M=10^{-1}M_{\odot}/y$ during the first 15 years of the outburst and slowly decreasing in time, $T=20000$K. A coupled two-components hydrodynamical and radiation model transfer problem has been solved.

The theoretical light curve is shown in Fig. 1a together with the observed one (van Genderen A.M., Thé P.S., 1985, Space Sci. Rev., 39, 313). The theoretical curve properly reproduces the dimming timescale and the depth of the observed curve. Theoretical spectrum shown in Fig. 1b displays discrepancies with observations in short waves. The deficit of optical radiation can be explained only by non-uniformity of the dust envelope which increases the contribution of scattering. The slope of far infrared spectrum is due to the adopted extinction tables. Angular distribution of monochromatic brightness (normalized, in arbitrary units) is shown in Fig. 1d.

The analysis of the models with varied parameters gives the following results:

The temperature of the central source should be no less than 20000K (the best fitting at $25-30 \cdot 10^3$K).

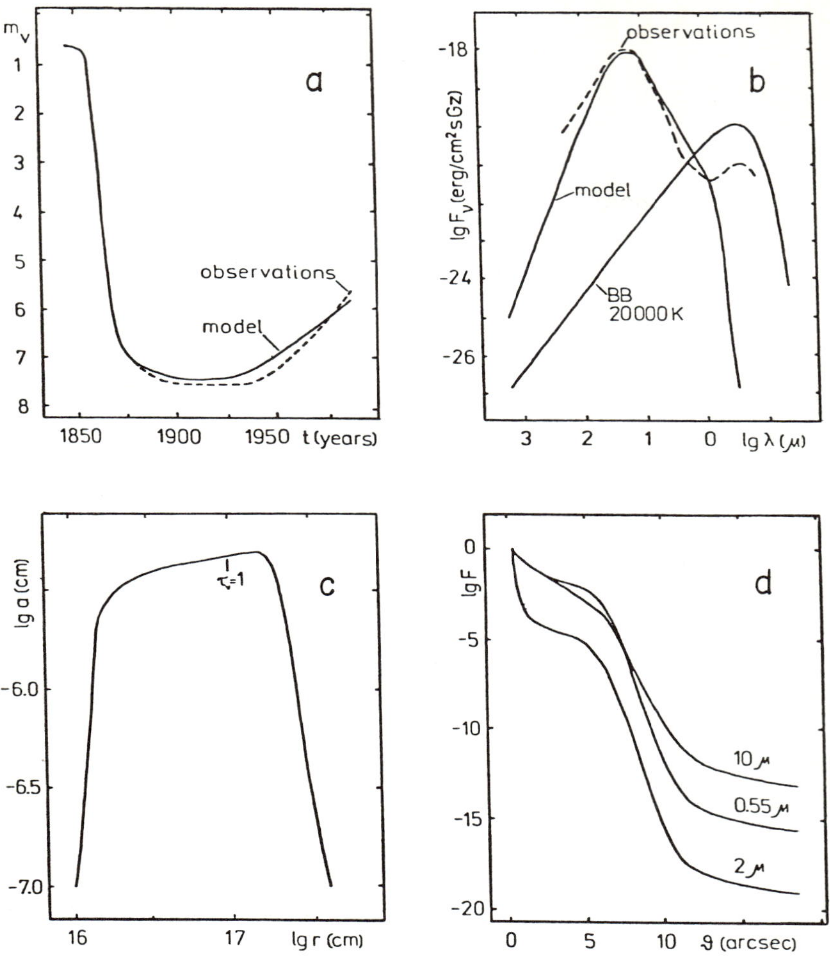


Fig.1. $\eta$ Car model (see explanations in text)

The velocity of the outflow (if constant in time !) should exceed 300-500 km/s.

Mass loss rate should be high during the last 150 years (presumably higher than $3 \cdot 10^{-2} M_\odot/y$).

Dust grains seem to be larger at larger distances, according to the model (Fig. 1c).

TURBULENCE-DRIVEN ATMOSPHERIC INSTABILITY AND
LARGE-SCALE MOTIONS IN SUPER- AND HYPERGIANTS


B. Boer, J. Carpay, A. de Koter, C. de Jager, H. Nieuwenhuijzen,
A. Piters and F. Spaan

Astronomical Observatory and Laboratory for Space Research,
Beneluxlaan 21, 3527 HS  Utrecht, The Netherlands



Abstract: Spectral studies of super- and hypergiants show that the
(outward directed) turbulent acceleration approaches the value of the
gravitational acceleration for the most luminous stars, which makes
their atmospheres unstable.


With the aim to study the influence of microturbulent motions on the
atmospheric stability of super- and hypergiants, such stars are being
studied by us. So far, results are available for the five objects,
listed in the table ( $\zeta_\mu$ = microturbulent velocity; s = sound speed.)

| star | spectrum | $T_{eff}$ (K) | $\Delta \log L$ | $g_{grav}$ (cm s$^{-2}$) | $<\zeta_\mu>/s$ | $<g_{turb}>$ (cm s$^{-2}$) |
|---|---|---|---|---|---|---|
| Alpha Per | F5 Ib | 6500 | − 2 | − 63 | 0.75 | + 24 |
| Alpha Sco | M1.5 Iab | 3600 | − 1.1 | − 1 | 0.5 | + 0.2 |
| Alpha Cyg | A2 Iae | 9200 | − 0.9 | − 26 | 1.0 | + 7 |
| HD 80077 | B2 Ia$^+$ | 17000: | 0 | − 87 | 1.0: | |
| HR 8752 | GO-5 Ia$^+$ | 4200 | − 0.3 | − 1 | 1.1 | + 2 |

From photometric data in the visual and infrared spectral regions and
from literature, the values of the photospheric parameters $T_{eff}$ and
$\log g_{grav}$ were derived. Since the work by Blackwell and Shallis (1977)
it is known that $T_{eff}$ is thus found with high accuracy, particularly
if use is made of photometric data in the near-infrared. The
luminosities were always taken from literature.

The equivalent widths of some 30 well-observed Fraunhofer lines
were then used - with the help of appropriate photospheric models -

for the determination of chemical abundances and of the microturbulent velocity component $\zeta_\mu$. Since the average depth of line formation for different lines normally varies over a large range of optical depths we thus get the variation of microturbulence with height z. The (outward directed) turbulent acceleration $g_t$ is then found with

$$g_t = \frac{\alpha}{\rho} \frac{dP_t}{dz} \qquad (1)$$

where $\rho$ is the density, and $\alpha$ depends on the spectrum of turbulence. We usually took $\alpha = 0.5$. Although the sample is still small, the table shows that $\zeta_\mu/s$ as well as $-g_t/g_{grav}$ tend to approach or surpass unity when $\Delta$ log L approaches zero. Here, $\Delta$ log L is the difference between the stellar luminosity and the upper limit of stellar existence (the Humphreys-Davidson limit). This result is a - preliminary - confirmation of the hypothesis (De Jager, 1984) that the instability of cool hypergiant atmospheres is due to dissipation of turbulent energy, leading to an effective acceleration near to zero:

$$g_{grav} + g_{rad} + g_{turb} \approx 0, \qquad (2)$$

where $g_{rad}$ is negligible in cool supergiants.
A study of high-resolution UV spectra of Alpha Cyg (Boer et al., 1988) has shown that the large-scale (macroturbulent) velocity field peaks at plus and minus 14 km s$^{-1}$, indicating the presence of strong supersonic motions in the atmosphere. A comparison with average radial velocities of the whole star shows that at any time there must be about 30 such elements present on the disk; this yields an average element diameter of appr. $30 \times 10^6$ km. One should call such motions convection were it not that theory does not predict convective motions in a star as hot as $\alpha$ Cyg. Alternative suggestions are non-radial or stochastic pulsations. In any case, it seems obvious that the observed microturbulent motions in Alpha Cyg originate in such large-scale motion fields.

References:

Blackwell, D.E., Shallis, M.J.: 1977, Monthly Not. R. Astron. Soc. 180, 177.
Boer, B., De Jager, C., Nieuwenhuijzen, H.: 1988, Astron. Astrophys. (submitted 1987).
De Jager, C.: 1984, Astron. Astrophys. 138, 246.

QUANTITATIVE SPECTROSCOPY OF WOLF-RAYET STARS


W. Schmutz

Institut für Theoretische Physik und Sternwarte der Universität Kiel

Olshausenstrasse, D-2300 Kiel 1, Federal Republic of Germany



## 1. INTRODUCTION

Advances in theoretical modeling of rapidly expanding atmospheres in the past few years made it possible to determine the stellar parameters of the Wolf-Rayet stars. This progress is mainly due to the improvement of the models with respect to their spatial extension: The new generation of models treat spherically-symmetric expanding atmospheres, i.e. the models are one-dimensional. Older models describe the wind by only one representative point. The older models are in fact 'core-halo' approximations. They have been introduced by Castor and van Blerkom (1970), and were extensively employed in the past (cf. e.g. Willis and Wilson, 1978; Smith and Willis, 1982). First results from new one-dimensional model calculations are published by Hillier (1984), Schmutz (1984), Hamann (1985), Hillier (1986), and Schmutz et al. (1987a); more detailed results are presented by Schmutz and Hamann (1986), Hamann and Schmutz (1987), Hillier (1987a,b), Wessolowski et al. (1987), Hillier (1987c) and Hamann et al. (1987). These results demonstrate that the step from zero- to one-dimensional calculations is essential. The important point is that the complicated interrelation between NLTE-level populations and radiation field is treated adequately (Schmutz and Hamann, 1986; Hillier, 1987). For this interrelation it is crucial to model consistently not only the line-formation region, but also the layers where the continuum is emitted. In fact, it is the core-halo approximation that causes the one-point models to fail in certain aspects.


## 2. MODEL CALCULATIONS

Presently, there are two different codes that are able to model adequately Wolf-Rayet atmospheres: the model of the Kiel group and Hillier's model. The two models are completely different in their technical approach how the NLTE-problem is solved (Hamann, 1986; Hillier, 1987a). Nevertheless, both models yield essentially the same results - an encouraging fact. Both models are founded on similar physical assumptions: The velocity law is predefined analytically and the temperature stratification is determined assuming radiative equilibrium. Up to now, the Kiel-calculations are for a pure helium atmosphere, while Hillier (1987c) includes in addition hydrogen as well as nitrogen and carbon in an approximate way.

The model-input parameters are $T_*$, $R_*$ and $\dot{M}$; the informations provided by the calculations which can be compared with the observations are the emergent line profiles and the continuum fluxes.


## 3. EFFECTIVE TEMPERATURE AND $T_*$

In a spherically-extended stellar atmosphere, the reference radius has to be defined to which the effective temperature is referred. This may be, e.g., $R_{2/3}$, the radius where the Rosseland optical depth becomes 2/3. However, in order to compare the temperature resulting from a spectral analysis with the predictions of the stellar evolution calculations, the temperature should be referred to a fictitious "hydrostatic" radius. The model calculations provide the radius at the base of the model-wind, termed "core radius", $R_*$, and the temperature referred to this core-radius, $T_*$. The core-radius, $R_*$, is the best available approximation to the hydrostatic radius. It has to be admitted, however, that if the continuum optical depth is of the order unity at layers where the wind has reached a velocity which is a considerable fraction of its terminal velocity, the core radius results from an inward extrapolation of the velocity field into regions that are not accessible to the observations. Thus, the derived core radius depends on the adopted velocity law. Luckily, it turns out that for most (about 70%) of the Wolf-Rayet stars the difference between $R_{2/3}$ and $R_*$ is small, i.e. the resulting differences between $T_{eff}(R_{2/3})$ and $T_*$ are comparable to the uncertainties of the derived temperature. Moreover, test calculations for a Wolf-Rayet star with an extended atmosphere (WR136 - Wessolowski et al., in preparation) showed that the resulting stellar parameters $R_*$ and $T_*$ do not depend dramatically on the adopted velocity law. Hence, the stellar parameters $T_*$ and $R_*$ can be compared with the results of the stellar evolution theory without introducing a large ambiguity.


## 4. DIAGNOSTIC OF SPECTRAL LINES

It can be shown that all models for which the parameters $\dot{M}$, $R_*$ and $V_{inf}$ combine to the same value of the parameter $p^W$, yield, to a good degree of approximation, the same (helium-) line equivalent widths. The parameter $p^W$ is defined by

$$p^W = \frac{\dot{M}}{V_{inf} \, R_*^{3/2}} \tag{1}$$

This homologous-atmosphere parameter may be termed "wind density", though its physical dimension is not a density. **The wind density and the effective temperature are the fundamental parameters of a Wolf-Rayet spectrum** - analog to $T_{eff}$ and g in stars with static atmospheres. In other words, for Wolf-Rayet atmospheres the wind density plays the role which has the gravity in static atmospheres. In contrast to the gravity in static atmospheres, however, the wind density cannot be linked with other fundamental stellar parameters. The obvious reason is that the mass-loss mechanism acting in Wolf-Rayet stars is not identified. In the cases when radiation pressure is the driving force of the wind, the wind density is linked to the ratio of radiation forces to gravity, and is hence, a function of mass, radius and luminosity. A minor difference to the gravity-analog in static atmospheres is, that the wind density is only an approximate parameter: i.e. models with the same parameter-pair, $T_*$, $p^W$, are not exactly homologous. E.g., for the He I equivalent widths, $p^W$ depends on a slightly different power of $R_*$, $p^W(\text{He I}) \sim R_*^{-5/3}$, than for the He II lines, $p^W(\text{He II}) \sim R_*^{-3/2}$ (cf. Eq. 1).


## 5. SPECTRAL CLASSIFICATION

The Wolf-Rayet classification is a one-dimensional system. However, as outlined in Sect. 4, their spectra depend on two parameters. Therefore, a two-dimensional classification clearly has to be introduced, e.g. as already suggested by Hiltner and Schild (1966) and refined by Walborn (1974). For later use in this paper, we divide the Wolf-Rayet stars into four classes:

WNE-A    WNE-B    WNL    WC.

Tentatively, we employ as criterion to separate "A" from "B" class the strength of the He II line at 5412Å: if its equivalent width is less than 40Å, the star is assigned to the "A" class, otherwise to the "B" class.


## 6. WOLF-RAYET ANALYSIS

6.1. APPROXIMATE ANALYSIS — We call an analysis as approximate if only the minimum information is employed which is needed to derive the stellar parameters. The minimal information consists on the equivalent widths of one He I and one He II line and the absolute visual magnitude. The latter value allows to split up the wind density parameter, $p^W$, into the radius, $R_*$, and the mass-loss rate, $\dot{M}$. In Fig. 1 we demonstrate this procedure for the star WR138. Two sets of contour-lines are drawn. These lines correspond to the three observational quantities and mark the loci where theoretical and observed values agree. Each set belongs to a different mass-loss rate: $\dot{M} = 10^{-4.4}$ (full drawn) and $10^{-4.0}$ $M_\odot$/yr (dashed). For neither of the two sets a well defined intersection point results. But by interpolating the contour-lines

135

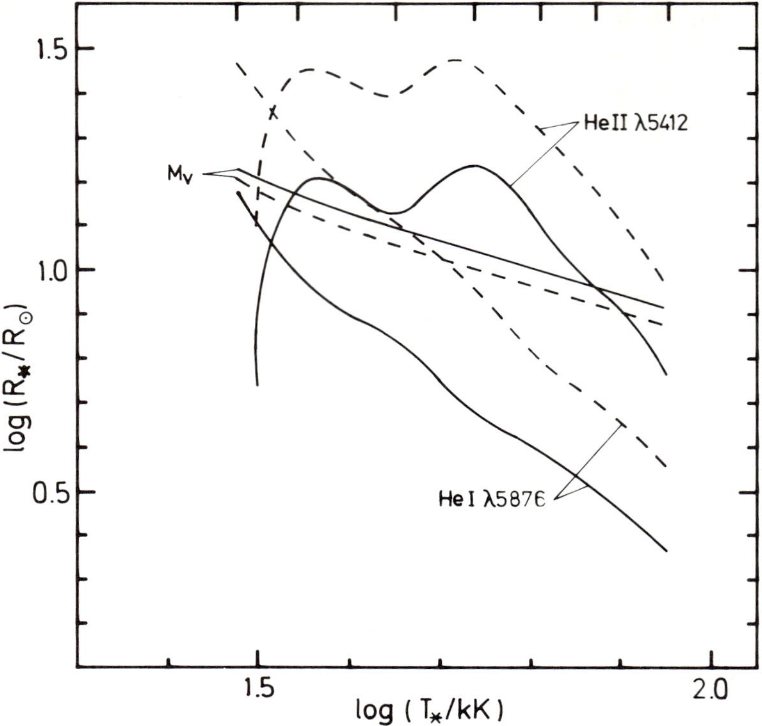


Figure 1.  Fit diagram for WR138, see text (Sect. 6.1)

for a mass-loss rate of about $10^{-4.2}$ $M_\odot$/yr,  all three lines would join in a single point - the fit point.  The model-grid on which the drawn contour-lines are based is calculated with a final wind expansion velocity of 2500 km/s,  while in the case of WR138 the helium line widths correspond to 1500 km/s.  Thus, a model with $T_* = 33$kK, $R_* = 15$ $R_\odot$ and $\dot{M} = 10^{-4.5}$ $M_\odot$/yr reproduces the three observed quantities of WR138 simultaneously.


6.2.  FINE ANALYSIS  —  The accuracy of an analysis may be improved if more than the minimum information is employed.  This may be done by fitting the observed line profiles of several helium lines.  Though the additional helium lines do not provide more information on the stellar parameters,  they can be used as consistency check. More lines help to define the continuum level in crowded spectral regions and to detect line blends.  It also improves the confidence to the model calculations if redundant information is reproduced correctly.  E.g., after the He I λλ4471,5876 and He II λ5412 lines of the WN5+abs-A star WR138 have been fitted carefully (Schmutz et al.,  1987b),  further helium lines are reproduced automatically (Fig. 2).  An other example of a fine analysis can be found in Hamann et al.  (1987) for the WN5-B star

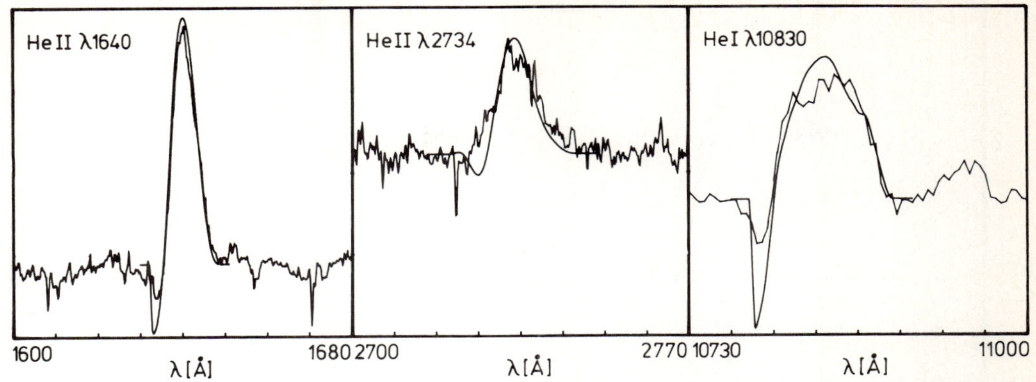


Figure 2. Comparison of calculated helium line profiles to the observed spectrum of the WN5+abs-A star WR138. These lines were not employed in the spectral analyses of this star, but are reproduced automatically by the final model obtained only by fitting the lines He I $\lambda\lambda$4471,5876 and He II $\lambda$5412 (Schmutz et al., 1987b)

WR6. In the case of this analysis, the distance to the star is not known and, therefore, its radius, mass-loss rate and luminosity can not be determined. Note that the known radio flux does not provide new information. Both methods, the analysis of the spectral lines and the radio-method, need the knowledge of the distance in order to derive the mass-loss rate. Unluckily, both methods depend on the distance with the same power of the distance, namely $d^{3/2}$! In the spectral analysis this factor enters by formula 1 in combination with a linear dependence of the stellar radius on the adopted distance; in the radio-method it is contained in the formula of Wright and Barlow (1975).


## 7. STELLAR PARAMETERS OF THE WOLF-RAYET STARS

Approximate analyses as described in Sect. 6.1. have been performed for 30 galactic Wolf-Rayet stars. The resulting stellar parameters $T_*$, L and $\dot{M}$ are given graphically in the Figure 1 of Schmutz et al. (1987c, these proceedings). In the same contribution the resulting parameters of the Wolf-Rayet stars are discussed briefly. As a check, models with the derived parameters are calculated for some stars and the resulting line profiles are compared with the observations. Two such comparisons are shown in Fig. 3. Dispite of the different classification of the two stars, WN5 and WN3, their helium lines are reproduced by two models which agree in their temperature $T_* = 90$kK. The different spectra emergent from the two models result from the different wind densities. Note that not all WN5 stars are as hot as the example of Fig. 3. An analysis of WR6 (Hamann et al., 1987) yields 60kK and for WR138 a temperature of only 33kK is found. This clearly demonstrates the need of a two-dimensional classification.

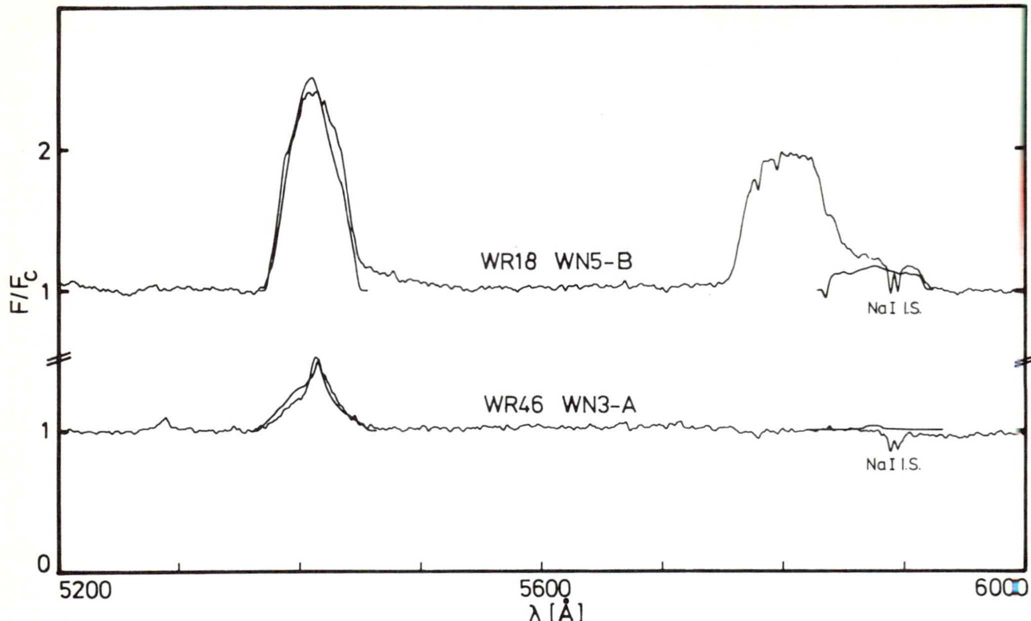


Figure 3.  Rectified spectra of WR46,  WN3pec-A,  and WR18,  WN5-B. Superimposed are theoretical line profiles of models with mass-loss rates log $(\dot{M}/M_\odot/yr)$ = -4.1 (top) and -4.9 (botom) and terminal velocities of 2100 and 2500 km/sec,  respectively. Otherwise, both models have the same parameters $T_\star$ = 90kK and $R_\star$ = 2.5$R_\odot$

## 8. DISCUSSION

In Fig. 4 the resulting stellar parameters of the Wolf-Rayet stars are compared with evolutionary tracks (Maeder and Meynet, 1987). It is obvious that the predictions of the evolutionary calculations do not match the stellar parameters of the Wolf-Rayet stars. We are not competent enough to judge,  whether it is possible to achieve a better agreement by adjusting the free parameters of the evolution theory.  However, we note that the disagreement could be due to the expection that the **majority** of the Wolf-Rayet stars should belong to the "pealed onion" scenario.  There are,  in fact, some Wolf-Rayet stars which are placed in the expected region of the HR-diagram. This holds especially for the hot WNE-A stars which are distinct from the other Wolf-Rayet stars in an important aspect:  Their relatively low mass-loss rates of about $10^{-5}M_\odot$/yr) could be explaind by radiation pressure,  provided that the stars are close enough to the Eddington limit (Pauldrach et al.,  1985). For all the other Wolf-Rayet stars,  particularly for the WC stars,  an alternative scenario could apply,  concerning their evolutionary status as the mass-loss mechanism.  Among the numerous possibilities proposed,  the "rotationally induced mixing" (Maeder,  1982), recently explored quantitatively (Maeder,  1987), looks very promisingly. The dashed

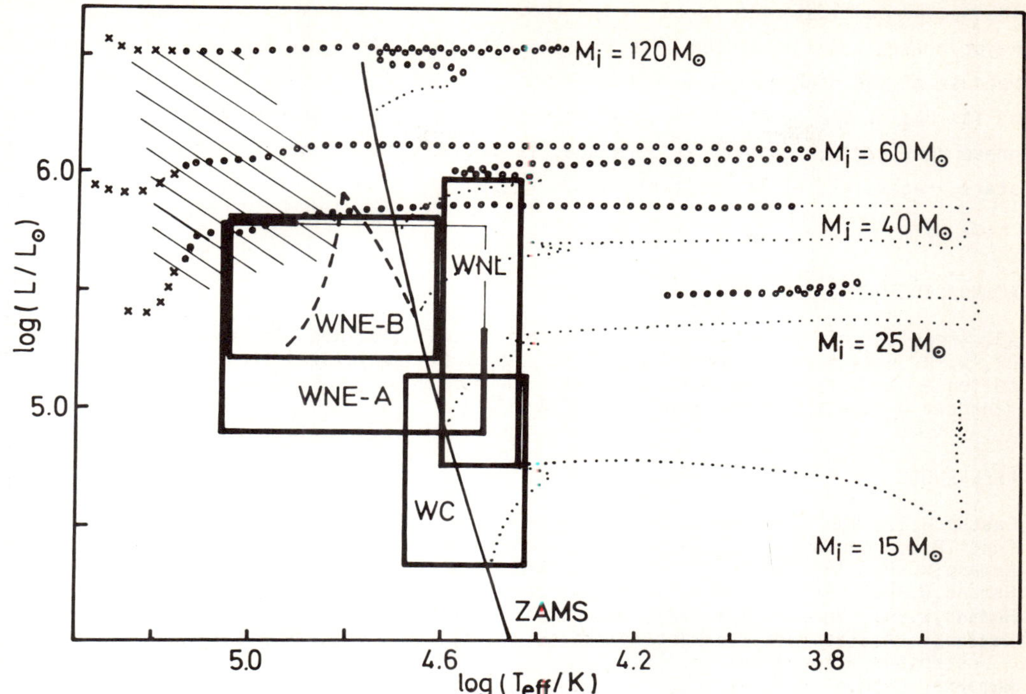


Figure 4. Location of the Wolf-Rayet stars in the HR-diagram superimposed on evolutionary tracks of massive stars computed with mass-loss and overshooting (Maeder and Meynet, 1987). During the phases marked by 'o' the C/N ratio at the surface is about 0.01, i.e. the star appears to be a WN star; in the phases marked by 'x' a WC stage would be visible. The hatched area in the upper left part of the diagram indicates where the Wolf-Rayet stars are likely to be observed, considering the evolutionary life-times

track in Fig. 4 is calculated with a rotationally induced turbulence that overcomes the μ-barrier. This track agrees well with the position of the WNE-B stars. The high rotation-velocity of WR138 (Schmutz et al., 1987b) points to rotationally induced mixing, at least for this particular star. A further hint that many Wolf-Rayet stars are not in the post-red-supergiant stage, comes from the inspection of the HR-diagram (Humphreys, 1978) of stellar associations containing Wolf-Rayet stars: Many Wolf-Rayet stars, especially the WNL, are located among luminous association members, which are in the hydrogen burning phase - a fact which can hardly believed to be accidental.

If indeed only a minority of the Wolf-Rayet stars were in the post-red-supergiant phase, the consequences would be twofold: **First, the estimate of the ratio of O-star to Wolf-Rayet life-times, $t_0/t_{WR}$, has to be revised, and second, all the indirect methods used to derive the initial masses of the Wolf-Rayet stars from the age of their surrounding would be not valid** (Conti et al., 1983; Schild and Maeder, 1984).

The second conclusion is based on the fact, that as soon as a star enters the Wolf-Rayet phase, its remaining life-time is extremely short (a few $10^5$ years), simply because of the high mass-loss rate. Hence, its total life-time is mainly determined by its age at the onset of the high mass-loss rate. If a star enters the Wolf-Rayet phase very early, or even at zero age, it cannot exist longer than the very massive stars, regardless of its initial mass.


ACKNOWLEDGEMENTS

All results of the "Kiel-research-group" are obtained in collaboration with Dr. W.-R. Hamann and U. Wessolowski. W.S. greatly appriciates the team-work with his collegues and thanks the DFG and the IAU for travel support. This research was financed by the DFG under grant Hu-39/25-1.


REFERENCES

Castor,J.I., van Blerkom,D.: 1970, Astrophys. J. **161**, 485
Conti,P.S., Garmany,C.D., deLoore,C., Vanbeveren,D.: 1983, Astropys. J. **274,** 302
Hamann,W.-R.: 1985, Astron. Astrophys. **145, ** 443
Hamann,W.-R.: 1986, Astron. Astrophys. **160,** 347
Hamann,W.-R., Schmutz,W.: 1987, Astron. Astrophys. **174,** 173
Hamann,W.-R., Schmutz,W., Wessolowski,U.: 1987, these proceedings and Astron.
   Astrophys. (in press)
Humphreys,R.M.: 1978, Astrophys. J. Suppl. **38,** 309
Hillier,D.J.: 1984, Astrophys. J. **280,** 744
Hillier,D.J.: 1986, in 'Luminous Stars and Associations in Galaxies', C.W.H.deLoore,
   A.J.Willis, P.Laskarides (eds.), IAU Symp. **116,** 261
Hillier,D.J.: 1987a, Astrophys. J. Suppl. **63,** 947
Hillier,D.J.: 1987b, Astrophys. J. Suppl. **63,** 965
Hillier,D.J.: 1987c, Astrophys. J. submitted
Hiltner,W.A., Schild,R.E.: 1966, Astrophys. J. **143,** 770
Maeder,A.: 1982, in 'Wolf-Rayet Stars: Observations, Physics, Evolution'
   C.W.H.deLoore, A.J.Willis (eds.), IAU Symp. **99,** 405
Maeder,A.: 1987, Astron. Astrophys. **178,** 159
Maeder,A., Meynet,G.: 1987, Astron. Astrophys. **182,** 243
Pauldrach,A., Puls,J, Hummer,D.G., Kudritzki,R.: 1985, Astron. Astrophys. **148,** L1
Schild,H., Maeder, A.: 1984, Astron. Astrophys. **136,** 237
Schmutz,W.: 1984, in 'Observational Tests of the Stellar Evolution Theory',
   A.Maeder, A.Renzini (eds.), IAU Symp. **105,** 269
Schmutz,W., Hamann,W.-R.: 1986, Astron. Astrophys. Letter **166,** L11
Schmutz,W., Hamann,W.-R., Wessolowski,U.: 1987a, in 'Circumstellar Matter',
   I.Appenzeller, C.Jordan (eds.), IAU Symp. **122,** 461
Schmutz,W., Hamann,W.-R., Wessolowski,U.: 1987b, these proceedings
Schmutz,W., Hamann,W.-R., Wessolowski,U.: 1987b,c, these proceedings and Astron.
   Astrophys. submitted
Smith,L.J., Willis,A.J.: 1982, Monthly Notices Roy. Astron. Soc. **201,** 451
Walborn,N.R.: 1974, Astrophys. J. **189,** 269
Wessolowski,U., Schmutz,W., Hamann,W.-R.: 1987, Astron. Astrophys. (in press)
Willis,A.J., Wilson,R.: 1978, Monthly Notices Roy. Astron. Soc. **182,** 897
Wright,A.E., Barlow,M.J.: 1975, Monthly Notices Roy. Astron. Soc. **170,** 41

# ANALYSIS OF 30 WOLF-RAYET STARS


W. Schmutz, W.-R. Hamann, U. Wessolowski

Institut für Theoretische Physik und Sternwarte der Universität Kiel

Olshausenstrasse, D-2300 Kiel 1, Federal Republic of Germany


Temperatures, mass-loss rates and luminosities of 30 galactic Wolf-Rayet stars (24 WN, 6 WC) are derived by fitting the observed equivalent widths of He I $\lambda 5876$ and He II $\lambda 5412$ and the absolute visual magnitude. A three-dimensional grid ($T_*$-$R_*$-$\dot{M}$) of model calculations provides the theoretical values.

The results are summarized in Figure 1. The different symbols denote the spectral classification of the individual stars, whereas four groups are distinguished: WNE-A, WNE-B, WNL and WC stars. The size of the symbols indicates the mass-loss rates. The uncertainties of the results are estimated to be 0.1 dex in $T_*$, 0.4 dex in $\dot{M}$ and 0.5 dex in L. For the 11 stars in common with the sample of Abbott et al. (1986) we find our mass-loss rates to be compatible with their radio flux if the correct ionization equilibrium in the radio emitting region is applied (Schmutz and Hamann, 1986). The model calculations show that for all but the WN2 and WN3 stars helium recombines to He$^+$ before the ions enter the radio-emitting region.

The results for the four groups are discussed briefly in the following:
- WNE-A: subtypes WN2 to WN6 with weak ("A") emission lines
  The stars of this group do not form a homogeneous set. Therefore, no typical parameters can be assigned to them, except that they have relatively small mass-loss rates. Only a lower limit of 70kK can be set to the temperatures of the WN2 and WN3 stars, because no He I lines are observed in their spectra. The other WNE-A stars have temperatures between 33 and 50kK. For the only binary of our sample, V444 Cyg, two solutions are given in Fig. 1 (brackets) for two different adopted absolute magnitudes, -4.0 or -4.8, respectively.
- WNE-B: subtypes WN4 to WN6 with strong ("B") emission lines
  This group is characterized by high mass-loss rates, and by luminosities of about $10^{5.5} L_\odot$. Their effective temperatures, referred to the core radius, are between 50kK and 90kK. But, owing to their high mass-loss rates, the continuum of the WNE-B stars emerges from regions in the wind. If the effective temperatures are referred to the radius of optical depth 2/3, they become 40kK or less for all stars of this group. The difference between core radius and continuum-emitting region is only important for the WNE-B stars.

- WNL: subtypes WN7 to WN9

  All stars of this group have almost the same temperature. Therefore a similar bolometric correction, BC, of about 3 mag applies for these stars. Because the temperatures are close to the threshold where the He II emissions break down, different spectral subtypes appear over a small range of temperatures.

- WC: subtypes WC5 to WC9

  The WC stars have lower luminosities than the WN stars. This is a consequence of their relatively low absolute visual magnitudes, together with their comparably low temperatures, which are resulting from the ratio of the strong He I lines (at 10830Å) to the rather weak He II emissions at 5412Å. The stellar temperatures of the WC stars range from 40kK for subtype WC5 to 30kK for WC9.

It is obvious that the derived parameters are not in agreement with the "standard" evolutionary calculations leading to Wolf-Rayet stars (Maeder and Meynet, 1987). The most extreme disagreement is found for the WC stars.

A detailed version of this paper is in preparation for "Astronomy and Astrophysics".

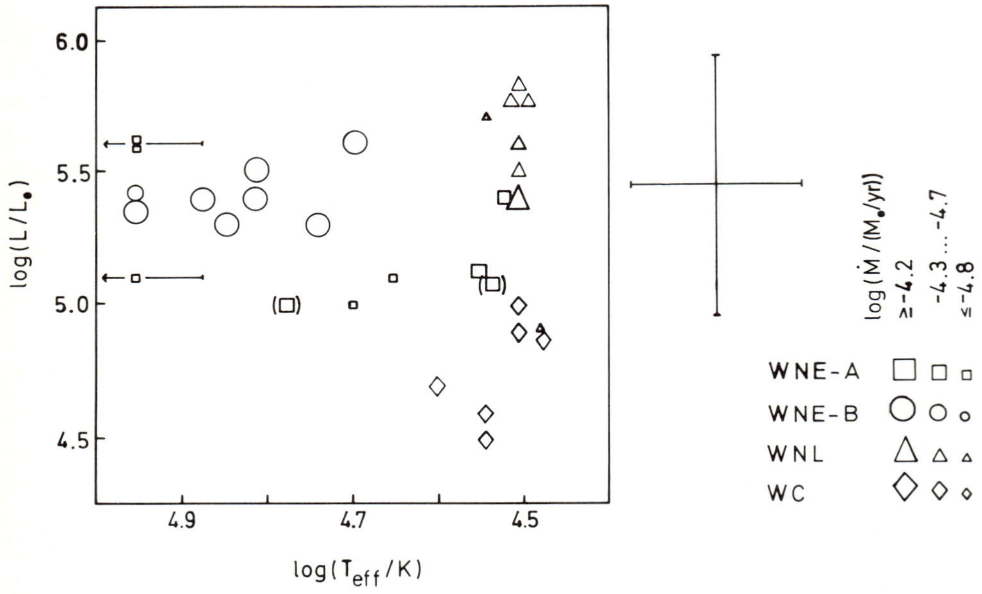



Figure 1.   The location in the HR-diagram of the 30 Wolf-Rayet stars as resulting from this work. The effective temperatures are referred to the "core" radii

REFERENCES


Abbott,D.C., Bieging,J.H., Churchwell,E., Torres,A.V.: 1986, Astrophys. J. **303**, 239
Schmutz,W., Hamann,W.-R.: 1986, Astron. Astrophys. Letter **166**, L11
Maeder,A., Meynet,G.: 1987, Astron. Astrophys. **182**, 243

# NLTE ANALYSIS OF THE WOLF-RAYET STAR HD193077 (WN5+abs)


W. Schmutz, W.-R. Hamann, U. Wessolowski
Institut für Theoretische Physik und Sternwarte der Universität Kiel
Olshausenstrasse, 2300 Kiel 1, Federal Republic of Germany


A model atmosphere code that accounts for the special physical conditions in Wolf-Rayet atmospheres (Hamann and Schmutz, 1986; Wessolowski et al., 1987) is used to analyse the spectrum of the Wolf-Rayet star HD193077 (WN5+abs). The stellar parameters are determined such that the profiles of the helium lines He I $\lambda\lambda$4471, 5876, He II $\lambda$5412, and the absolute visual magnitude are reproduced.

In order to estimate the systematic errors introduced by the model assumptions, we perform some test calculations. Instead of the velocity-law exponent of $\beta=1$, another fit is obtained with $\beta=0.5$ (Fig. 1). This fit yields an effective temperature of about 3000K higher than with $\beta=1$. In order to simulate the effect of the suspected hydrogen content, a further fit (Fig. 2) is made when one free electron per helium atom is added artificially. This has only marginal influence on the derived temperature (+100K). Thus, we conclude that our model assumptions may introduce a systematic error of the order of 5000K.

The best agreement with the line profiles (Fig. 2) is obtained with

$$T_{eff} / K = 33600 \pm 200 \quad \text{(for systematic errors see text)}$$
$$R_\ast / R_\odot = 15 \pm 3$$
$$\log(L / L_\odot) = 5.4 \pm 0.2$$
$$\dot{M} / (M_\odot/yr) = (2.5 \pm 1.5) \, 10^{-5} \text{ helium } + 1 \, 10^{-5} \text{ hydrogen (estimated)}$$
$$V_{rot} \sin i / (km/s) = 450 \pm 100$$

Special emphasis is put on the fit to the He I line at 4471Å, which appears as an unshifted absorption profile of quasi "photospheric" origin. A striking result is the very high rotational velocity which is indicated by its profile (Fig. 3).

The position of this star in the HR-diagram is among the bulk of luminous association members, which are in the hydrogen burning phase. This result, together with the star's high rotational velocity, favours the scenario of rotationally induced mixing (Maeder, 1987) for the formation of this Wolf-Rayet star.

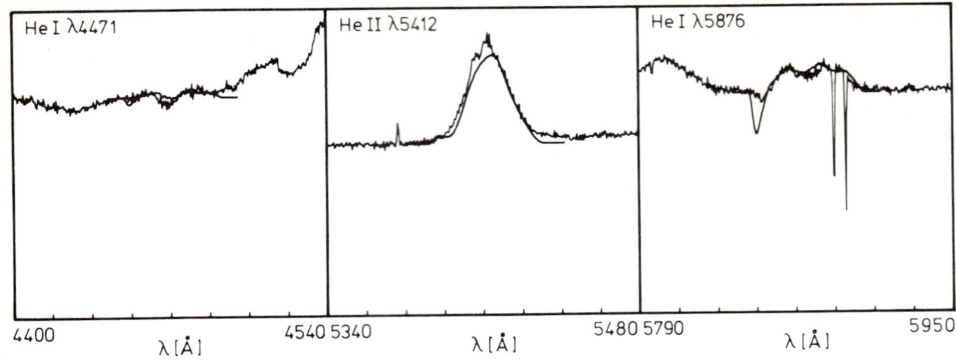


Figure 1. Theoretical line profiles of the final model with ß=0.5 superimposed on the observations. The profiles are rotationally broadend by 450 km/s

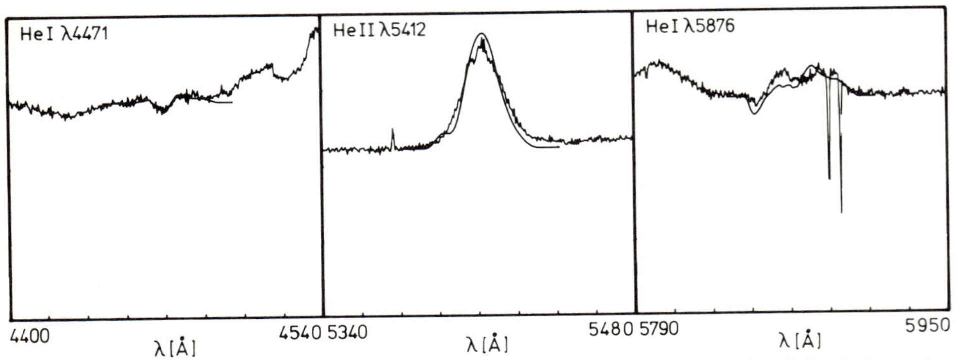


Figure 2. Same as Fig. 1, but for the final model obtained with ß=1 and one additional free electron per helium atom

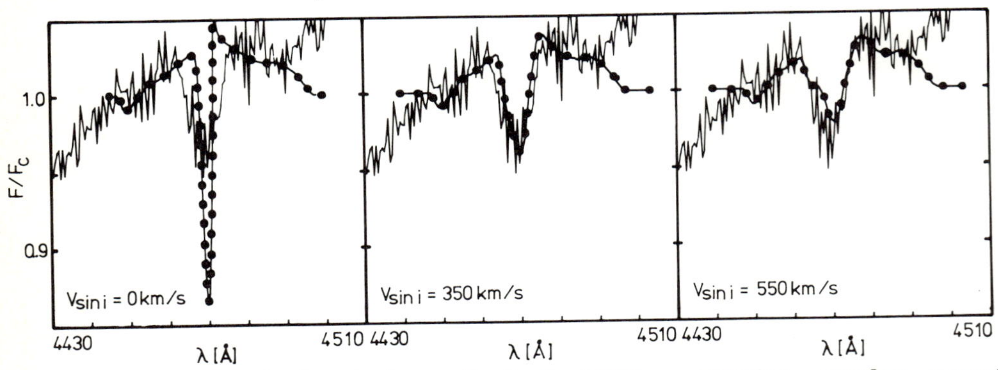


Figure 3. Rotational broadening of He I λ4471, assuming angular momentum conservation. The best profile fit is obtained with $V_{rot} \sin i$ = 450 km/s.


REFERENCES



Maeder,A.: 1987, Astron. Astrophys. **178**, 159

Hamann,W.-R., Schmutz,W.: 1986, Astron. Astrophys. **174**, 173

Wessolowski,U., Schmutz,W., Hamann,W.-R.: 1987, Astron. Astrophys. (in press)

**SPECTRAL ANALYSIS OF THE WOLF-RAYET STAR HD 50896**

W.-R. Hamann, W. Schmutz, U. Wessolowski
Institut für Theoretische Physik
und Sternwarte der Universität Kiel
Federal Republic of Germany


## 1. INTRODUCTION

The helium spectrum of the WN5 star HD 50896 (EZ Canis Majoris, WR6) is studied. Our aim is to establish a technique which allows the determination of the parameters of a Wolf-Rayet star from a systematic analysis of its spectral lines. Since the method of "iteration with approximate Lambda operators" became available for application to expanding atmospheres (Hamann, 1986, 1987), we are now able to compare observed spectra to realistic model calculations (Hamann and Schmutz, 1987; Wessolowski et al., 1987).


## 2. THE THEORETICAL MODELS

We consider the non-LTE spectral formation in a spherically expanding atmosphere. The velocity field $v(r)$ is specified in its supersonic part by the usual analytical law with the parameters $v_\infty$ (final velocity) and the exponent ß=1. The temperature structure is derived from the assumption of radiative equilibrium, but only approximately evaluated for the grey LTE case. The atmosphere is assumed to consist of pure helium. The model atom has a total of 28 energy levels, among these 17 levels of He I. The line radiation transfer is treated in the "comoving frame".


## 3. FIT OF THE OBSERVATIONS

Three important parameters enter the model calculations: the mass loss rate $\dot{M}$, the stellar radius $R_*$ and the temperature parameter $T_*$. The fit of the helium line profiles of HD 50896 requires a final wind velocity of 1700 km/s. Hence, we now calculate a small grid of models in the appropriate range of $R_*$ and $T_*$ and with the specially adapted $v_\infty$. The mass-loss rate is kept at log $(\dot{M}/(M_\odot/yr)) = -4.4$ as an arbitrary choice. The results are presented in the form of contour lines in the log $T_*$-log $R_*$-plane. Those of the contours which match the observed equivalent widths or peak intensities are extracted and yield a "fit diagram". We obtain a well-defined intersection region centered about $R_* = 2.6\ R_\odot$, $T_* = 60$ kK (hereafter quoted as "model B").

As the mass-loss rate is is considered as a free parameter, we now repeat the whole fit procedure for different values, namely log $(\dot{M}/(M_\odot/yr)) = -4.0$ (case "A") and $-4.7$ (case "C") (cf. Table 1). The resulting fit diagrams look qualitatively similar for all three mass-loss rates considered. There is no obvious preference in favour of a certain mass-loss rate from the internal consistency of the fits.

The obtained "fit point" parameters are now used for the calculation of "final models", and the resulting synthetic spectra are compared with the observation as a final check. We restrict the representation (Fig. 1) to the model C, since the corresponding profile fits for the other two mass-loss rates would appear very similar.

In total, we feel that the agreement between theoretical and observed line profiles is satisfactory and internally consistent. We take this as a confirmation of the basic physical input of our models, and as evidence that the results of the analysis are reliable.


145

**Table 1.** Stellar parameters for different adopted mass-loss rates

| Model | A | B | C |
|---|---|---|---|
| $\log\,(\dot{M}/M_\odot/yr))$ | -4.0 | -4.4 | -4.7 |
| $R_*/R_\odot$ | 4.7 | 2.6 | 1.5 |
| $T_*/kK$ | 57.5 | 60 | 63 |
| $\log\,(L/L_\odot)$ | 5.3 | 4.9 | 4.5 |

## 4. RESULTS AND DISCUSSION

The spectral analysis of HD 50896 performed in Sect. 3 resulted in a one-dimensional set of solutions for the three parameters M, $R_*$ and $T_*$, i.e. one of these parameters (e.g. the mass-loss rate) is still free (cf. Table 1). A comparison with Hillier (1987) reveals that his results are comparable with our model B. A restriction to an unique solution would be possible if the distance of this star could be determined.

An independent determination for the distance of HD 50896 does not exist. The "possible membership" (Lundström and Stenholm, 1984) to Collinder 121 (d = 0.91 kpc) can be ruled out from a study of the interstellar Na I lines in the spectrum of HD 50896 and of neighbouring stars (Schmutz and Howarth, in preparation). We conclude that HD 50896 may have any distance between 0.91 kpc and a few kiloparsecs, but there is some evidence for a value of about 2 kpc. The comparison between apparent and absolute magnitudes then leads to our model A, while the models B and C correspond to smaller distances.

The radio observations (Hogg, 1982) are in agreement with the results of our analysis, but do not help to restrict the one-dimensional variety of solutions of the analysis to an unique set of parameters.

The quotation of an "effective temperature" runs into conceptual difficulties in the case of extended atmospheres. The parameter $T_*$ ( 57.5 kK for model A) follows simply from the luminosity and the reference radius $R_*$. Rosseland optical depths of 1, 2/3 or 1/3 are reached at much larger radii: 2.5, 3.3 or 5.7 $R_*$, respectively (model A). The corresponding "effective temperatures", if related to these radii, are about 37, 32, or 24 kK.

The luminosity of HD 50896 was determined to $\log\,(L/L_\odot)$ = 5.3 (model A), or less in the case of a smaller distance (cf. Table 1). A comparison with recent evolutionary calculations by Maeder and Meynet (1987) indicates an actual mass of about 10 $M_\odot$ (case A) from the (extrapolated) mass-luminosity relation for WR stars. But unfortunately, stars of this luminosity do not become WR stars, according to these evolutionary models, since the corresponding tracks (e.g. for an initial mass of 20 $M_\odot$) never return to the blue side of the HR diagram.

A more detailed version of this paper will be published in "Astronomy and Astrophysics".


**REFERENCES**

Hamann, W.-R.: 1986, Astron. Astrophys. **160**, 347
Hamann, W.-R.: 1987, in "Numerical Radiation Transfer", W. Kalkofen (ed.), Cambridge University Press (in press)
Hamann, W.-R., Schmutz, W.: 1987, Astron. Astrophys. **174**, 173
Hillier, D.J.: 1987, Astrophys. J. Suppl. **63**, 965

Hogg, D.E.: 1982, in "Wolf-Rayet Stars: Observation, Physics, Evolution", IAU Symp. 99, C.W.H. deLoore and A.J. Willis (eds.), Reidel, Dordrecht, p. 221
Lundström, I., Stenholm, B.: 1984, Astron. Astrophys. Suppl. Ser. **58**, 163
Maeder, A., Meynet, G.: 1987, Astron. Astrophys. (preprint)
Wessolowski, U., Schmutz, W., Hamann, W.-R.: 1987, Astron. Astrophys. (in press)

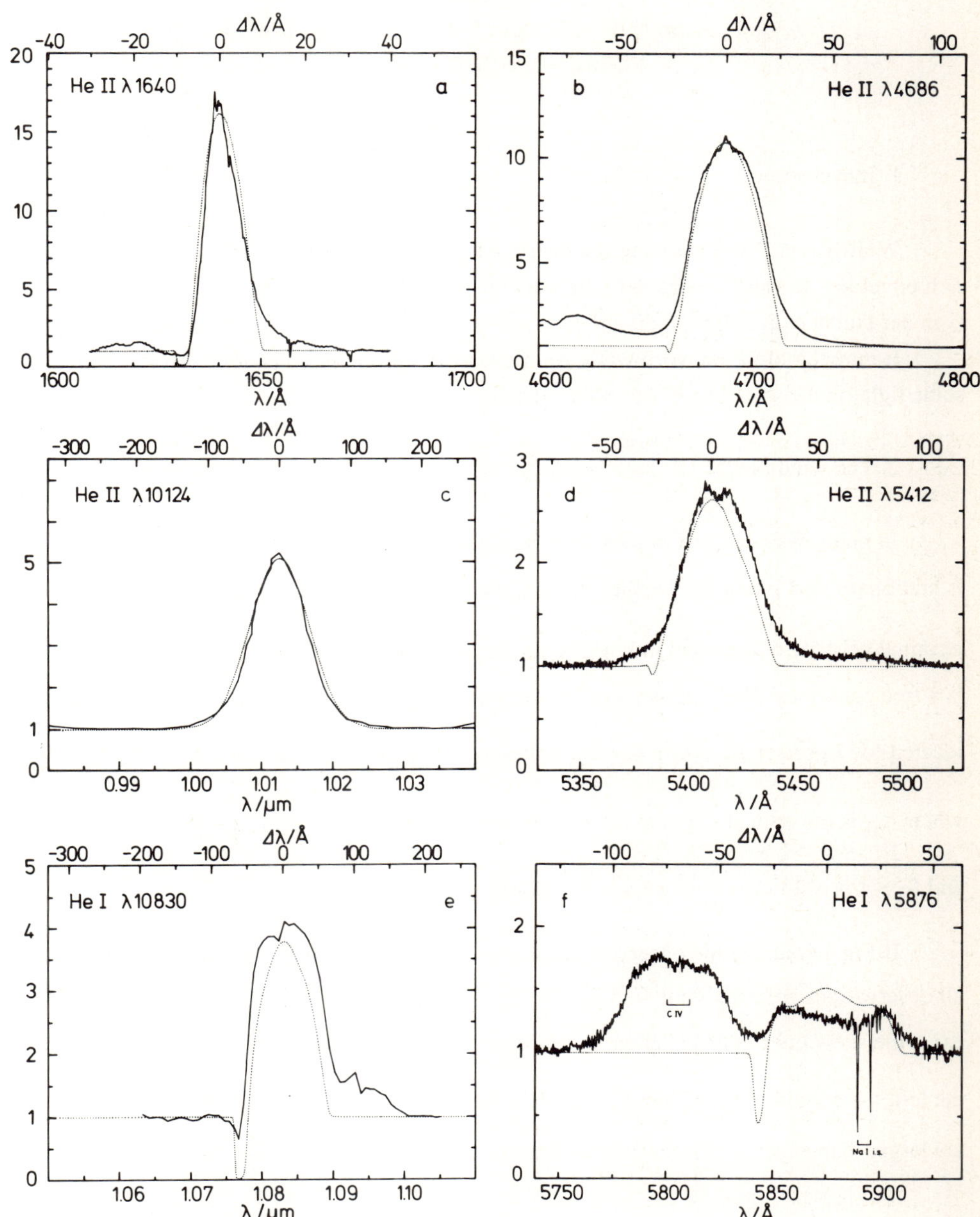



**Fig. 1.** Comparison of the observed line profiles to the theoretical profiles of model C (log $(\dot{M}/(M_\odot/yr))$ = -4.7, $R_* = 1.5\ R_\odot$, $T_* = 63$ kK)

# ATMOSPHERIC DIAGNOSTICS OF WOLF-RAYET STARS


C. Doom[*]

Astrophysical Insitute, Vrije Universiteit Brussel,

Pleinlaan 2, 1050 Brussel, Belgium


## 1. Introduction

Wolf-Rayet (WR) stars are the descendants of massive stars that have lost their hydrogen rich envelope. Recently more accurate data on WR stars have become available: mass-loss rates (van der Hucht et al. 1986), radii and luminosities (Underhill 1983, Nussbaumer et al. 1982).

It may therefore be worthwhile to investigate if combinations of observed parameters shed some light on the structure of the extended stellar wind of WR stars.

## 2. The wind model

In many WR stars the photosphere is situated in the stellar wind. We assume that the wind is stationary and isotropic. Further we assume a velocity law $v(r)=v_\infty(1-R_s/r)^\beta$ where $v_\infty$ is the terminal velocity of the wind in km/s, $R_s$ is the radius where the wind acceleration starts and $\beta > 0$ is a free parameter. We can then easily compute the level R in the wind where the photosphere is located (de Loore et al. 1982): R is the solution of the equation $6.27\ 10^{-9}\ \tau_{at}\ R\ v_\infty\ /\dot{M} = f_\beta(R_s/R)$

where $\tau_{at}$ is the optical depth at the photosphere (2/3 or 1), $\dot{M}$ (>0) is the mass loss rate in $M_\odot$/yr and $f_\beta > 1$ is a slowly varying function (Doom 1987).

If Fig.1 confront the observed parameters to the theory for $\beta = 2$. We can conclude that (1) all observations are consistent with the wind model; (2) WR stars with radii larger than $10\ R_\odot$ must exist: the skew lines in Fig.1 reflect the uncertainties on the radius; if we continue the lines down to the left, the condition $f_\beta > 1$ implies that $R > 10\ R_\odot$ for several WR stars and (3) the values for $R_s$ are larger than expected from stellar evolution theory ($R_s = 0.8 - 1.5\ R_\odot$, Doom et al. 1986).

[*] senior research assistant, NFWO Belgium

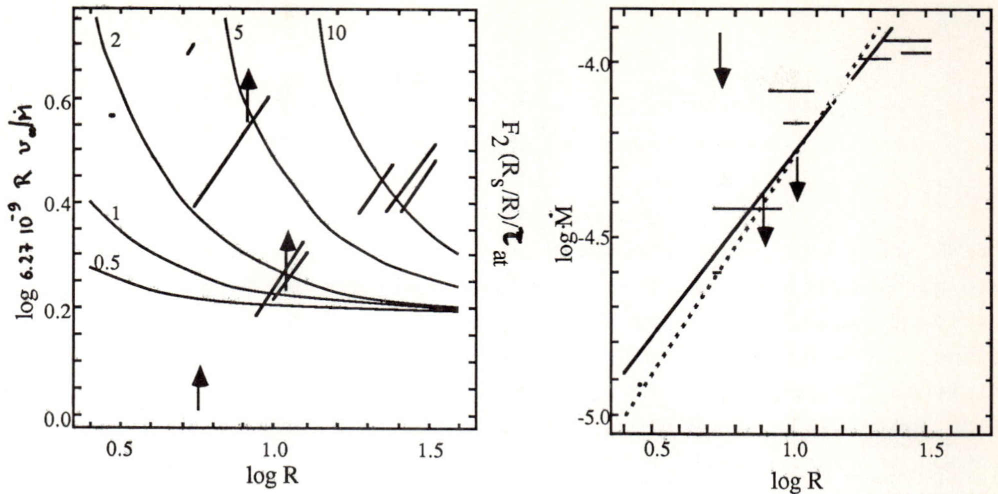


3. The mass-loss rates of WR stars.

Considering again the simple wind model we see that the quantity $R\,v_\infty/\dot{M}$ must be nearly constant since $\tau_{at}$ is fixed and $f_\beta$ is slowly varying. Both the observed radii and mass-loss rates vary by a factor of 10. If they were random we would expect variations of $v_\infty$ of at least a factor 20. Since only a factor of 4 is observed some positive correlation between R and $\dot{M}$ may exist.

The relation between R and $\dot{M}$ is given in Fig.2 (Doom 1988). One indeed finds a significant correlation with a slope of 1 to 1.2.

References

de Loore, C., Hellings, P., Lamers, H.J.G.L.M., 1982: in C.W.H. de Loore and A.J. Willis (eds.) *Wolf-Rayet stars, observtions, physics, evolution*, Proc IAU Symp. 99 (D. Reidel, Dordrecht), p53

Doom, C., 1987: *Rep. Prog. Phys.* (in press)

Doom, C., 1988: *Astron. Astrophys.* (in press)

Doom, C., De Greve, J.P., de Loore, C., 1986: *Astrophys. J.* **303**, 136

Nussbaumer, H., Schmutz, W., Smith, L.J., Willis, A.J., 1982: *Astron. Astrophys. Suppl.* **47**, 257

Schmutz, W., 1988: this volume

van der Hucht, K.A., Cassinelli, J.P., Williams, P.M., 1986: *Astron. Astrophys.* **168**, 111

NEW OBSERVATIONAL RESULTS OF Of/WN - TRANSITION TYPES

B. Wolf

Landessternwarte Königstuhl, Heidelberg

Of/WN stars have been introduced by Walborn (1977). These luminous stars
are particularly characterized by the appearance of both the characteristic Of emission lines of HeII and NIII and strong HeI and [NII] emissions. These objects have become particularly interesting due to the
discovery of their link to the S Dor variables (Stahl et al. 1983, Stahl
1986, Wolf 1987). They are supposed to represent key objects in the evolution of massive stars being the immediate progenitors of WN stars. We
observed the Of/WN transition type stars R127, R84 and S61 of the LMC
with high dispersion spectroscopy in the optical and satellite-UV range.
R127 is at present undergoing an S Dor-type outburst. In 1986 R127 became the visually brightest star of the LMC until it was surpassed by
SN1987A in Feb. 1987. The spectrum in the optical range during the extraordinary bright outburst phase is distinguished by numerous P Cygni
type lines of singly ionized metals. Since the beginning of the present
outburst the spectral type has gradually become later and the equivalent spectral type as derived from our latest CASPEC spectrum taken in
1986, August 22 was middle A. The IUE-spectrum of R127 during maximum
is characterized by copious lines of singly ionized metals formed in
the wind. A detailed disucssion of the optical and satellite UV high
dispersion spectrograms of R127 taken during its present extraordinary
bright outburst phase is forthcoming in Astronomy and Astrophysics (Wolf
et al. 1987a).

The Of/WN-characteristics at quiescence have been studied from the observations of R84 and S61. The high resolution and high S/N-spectra in
the optical range are particularly distinguished by strong emission
lines of H, HeI and [NII]. The UV spectra of both stars closely resemble those of late O-supergiants but all absorption lines are violet-
-shifted by about 250 km s$^{-1}$ (R84) and about 200 km s$^{-1}$ (S61). The absorption lines are stronger than in normal O-type stars. The UV-reso-
nance lines indicate low terminal wind velocities of $\approx$900 km s$^{-1}$ only.
Unlike to normal O-type stars the AlIII-resonance lines also show pronounced P Cygni profiles with an even lower edge velocity ($v_{edge} \approx$
400 km s$^{-1}$). The mass loss rates (6.10$^{-6}$ M$_\odot$yr$^{-1}$) are comparable to rates found in normal luminous hot stars. However, the wind appears to be

much more gradually accelerated similar to the wind of the galactic
supergiant P Cygni. It is suggested that the Of/WN transition type stars
are the hotter counterparts of the early B-type P Cygni stars. A detailed
paper on the Of/WN-transition type stars R84 and S61 is forthcoming in
Astronomy and Astrophysics (Wolf et al. 1987b).


References

Stahl, O., Wolf, B., Klare, G., Cassatella, A., Krautter, J., Persi, P.,
Ferarri-Toniolo, M.: 1983, Astron. Astrophys. 127, 49
Stahl, O.: 1986, Astron. Astrophys. 164, 321
Walborn, N. R.: 1977, Astrophys. J. 215, 53
Wolf, B.: 1987, IAU Symposium No. 122, pg. 409 (eds. I. Appenzeller,
C. Jordan)
Wolf, B., Stahl, O., Smolinski, J., Cassatella, A.: 1987a, Astron.
Astrophys. (submitted)
Wolf, B., Stahl, O., Seifert, W.: 1987b, Astron. Astrophys. (in press)

# THE GALACTIC DISTRIBUTION AND SUBTYPE EVOLUTION OF WOLF–RAYET STARS


B. Hidayat[1], A. G. Admiranto[1], K. R. Supelli[2] and K. A. van der Hucht[3]
[1] Observatorium Bosscha, ITB, Lembang, West Java, Indonesia
[2] Ministery of Science and Technology, BBPT, Jakarta, Indonesia
[3] SRON Space Research Laboratory, Utrecht, The Netherlands



**ABSTRACT.** The galactic distribution of Wolf-Rayet stars is redetermined. In the solar neighbourhood within d < 2.5 kpc the WN/WC number ratio is 0.54, indicating a larger influence of mass loss and convective overshooting than most present evolutionary models account for. The WR density gradient in the solar neighbourhood is constant. From a comparison of WR galactocentric distances and WR cluster and association membership, we find indications that WCE stars descend from WNL stars at R > 6.5 kpc, and that WCL stars descend from WNL stars at R < 8.5 kpc.


## 1. Observations used

Since the appearance of the 6-th WR Catalog (van der Hucht et al. 1981) the galactic distribution of WR stars has been studied by many authors (Hidayat et al. 1981, 1984; Meylan and Maeder, 1983; Conti et al. 1983). Recently the membership of 42 WR stars in open clusters and associations has been reassessed (Lundstrom and Stenholm, 1984) and improved photometry (Massey, 1984) and new classifications have become available for a considerable fraction of the 157 known galactic WR stars. This allows a redetermination of the intrinsic parameters $(b-v)_0$ and $M_v$ of WR stars and their photometric distances.

## 2. The WR distribution in the galactic plane

From the distribution of their heliocentric distances we estimate that the WR star cencus is practically complete out to $d \approx 3$ kpc. Considering a volume around the Sun with d < 2.5 kpc, we count 44 WR stars in the ratio $N_{WN} : N_{WC} : N_{WO} = 15 : 28 : 1$. If we accept that the WN and WC phases are consecutive phases in the evolution of massive stars, then this implies that in the solar neighbourhood the average WC phase lifetime is twice as long as the WN phase lifetime. More specifically, we count in the range $7.5 < R < 9.5$ kpc a WN to WC number ratio of 0.89, and in the range $6 < R < 7.5$ kpc a ratio of 0.35. The former ratio approaches the values of $t_{WN}/t_{WC} = 1.22$ to $1.34$ predicted by the recent evolutionary models of Maeder and Meynet (1987) for stars in the range of $M_i = 40-60\ M_\odot$ with mass loss and convective overshooting. As these authors state: *the ratio $t_{WN}/t_{WC}$ is very sensitive to mass loss and overshooting.* The value $N_{WN}/N_{WC} = 0.35$ found in the inner galactic region compares more favourably with the evolutionary calculations for stars in the range of $M_i = 50-60\ M_\odot$ by Prantzos et al. (1986), who allow for more overshooting in their models than Maeder and Meynet do, and calculate $t_{WN}/t_{WC} = 0.21$ to $0.34$. We confirm the strong decrease in $N_{WR}/kpc^2$ going from the inner to the outer galactic regions (a factor 9), but we observe no change in the WR density gradient, as suggested previously by Meylan and Maeder (1983) using the old data base of Hidayat et al. (1982). The WR density gradient is a factor 2.5 steeper than the O-type star density gradient. This could indicate that one or more of the WR formation channels (Maeder, 1982) are relatively more effective at smaller galactocentric distances. We also confirm that the binary channel is relatively more important at larger galactocentric distances. When observed in the volume with d < 3 kpc, the WR number density can be fitted by log $(N_{WR}/kpc^2) = 1.96 - 0.21\ R$, with $N_{WN}/N_{WC} = 0.47 + 0.05\ R$. If this density gradient is valid throughout the Galaxy, then the total number of galactic WR stars is about 2500 (~1500 WC stars), with half that number located within 3 kpc from the galactic center.

## 3. WR subtype distribution and evolution

The issue of WR subtype evolution has been addressed by various authors in the past few years. In general one agrees that WC stars descend from the WN sequence and that WO stars descend from the WC sequence, but on the question how WR stars evolve from one subtype to another different ideas have been proposed. In a very suggestive study Schild and Maeder (1984) argue on the basis of open cluster compositions for WN7->WC4-8 evolution in young clusters, WN5-6->WC9->WO evolution in older clusters, and WN3-5->SN evolution.

In order to judge whether the overall galactic WR distribution poses constraints on subtype evolution, we have plotted in Fig.1 the WR galactocentric distances R (based on $R_\odot \doteq 8.5$ kpc) versus WR subtype. This was performed earlier by Hidayat et al. (1982, Paper I), but with the improved data base we find as more pronounced trend in Fig.1 that the WC stars show a clear preference to have later subtypes at smaller distances from the center. WN2-5 stars are only present beyond $R \approx 6$ kpc, while WN6-8 stars are present at practically all observable galactocentric distances.

Of the evolutionary road system proposed by Schild and Maeder (1984) the WN7->WC4-8 trail is accessible in Fig.1, but there is an obstruction in their WN5-6->WC9->WO trail. Fig.1 allows the transition from WN6 to WC9, but definitely not the transition from WN5 to WC9, as was already visible in the results of Hidayat et al. (1982, Paper I). In addition, the transition from WC9 to WO appears also excluded: Fig.1 allows only WC4-7->WO evolution. In view of the alternative WC4-5p classification of the WO stars, a WC4-5->WO trail is clearly more probable.

Fig.1 shows that WN to WC subtype evolution is not just a function of $M_i$, but also of galactocentric distance, i.e. of gradients in galactic parameters like metallicity, which in turn may reflect itself in the mass loss rates of massive stars evolving to the WR phase. Of interest in this respect is that a similar distribution as found for galactic WC subtypes has been noted recently in M31 by Moffat and Shara (1987). Also of interest is the observation by Kunth and Schild (1986) of a positive correlation between metallicity and WR star formation rate in eleven emission line galaxies. The different WN3-5 and WC4-7 distributions could support the conclusion of Schild and Maeder (1984) that WNE stars by-pass the WC phase in their further evolution. This would imply that WCE stars can only descend from WNL stars at galactocentric distances beyond $R \approx 6.5$ kpc, and that WCL stars can only descend from WNL stars within $R \approx 8.5$ kpc.

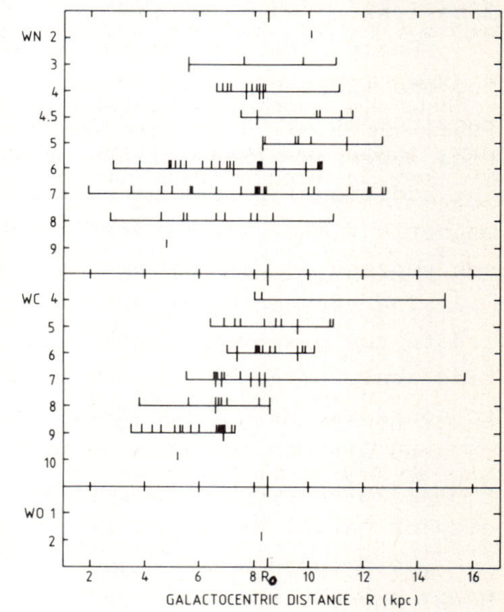



Figure 1. WR galactocentric distances.


## References

Conti, P.S., Garmany, C.D., de Loore, C., Vanbeveren, D: 1983, Astrophys. J. **274**, 302.

Hidayat, B., Supelli, K., van der Hucht, K.A.: 1982, In: C.W.H. de Loore, A.J. Willis (eds), Wolf-Rayet Stars: Observations, Physics, Evolution, Proc. IAU Symp. No. 99, p. 27.

Hidayat, B., Admiranto, A.G., van der Hucht, K.A.: 1984, In: B. Hidayat, Z. Kopal, J. Rahe (eds), Double Stars: Physical Properties and Generic Relations, Proc. IAU Coll. No. 80, Astrophys. Space Sci. **99**, 175.

van der Hucht, K.A., Conti, P.S., Lundström, I., Stenholm, B.: 1981, Space Sci. Rev. **28**, 227.

Kunth, D., Schild, H.: 1986, Astron. Astrophys. **169**, 71.

Lundström, I., Stenholm, B.: 1984c, Astron. Astrophys. Suppl. **58**, 163.

Maeder, A.: 1982, Astron. Astrophys. **105**, 149.

Maeder, A., Meynet, G.: 1987, Astron. Astrophys. **182**, 243.

Massey, P.: 1984, Astrophys. J. **281**, 789.

Meylan, G., Maeder, A.: 1983, Astron. Astrophys. **124**, 84.

Moffat, A.F.J., Shara, M.M.: 1987, Astrophys. J. **320**, 266.

Prantzos, N., Doom, C., Arnould, M., de Loore, C.: 1986, Astrophys. J. **304**, 695.

Schild, H., Maeder, A.: 1984, Astron. Astrophys. **136**, 237.

# A NEW DRIVING MECHANISM OF THE EPISODIC MASS-LOSS IN BE STARS


Hiroyasu ANDO

Tokyo Astronomical Observatory, University of Tokyo,
Mitaka, Tokyo 181.


Be stars are well known to be rapid rotator and to show intermittent emission-line activity. Such an activity is now interpreted as an abrupt mass-ejection and formation of a cool disk in the equatorial region and then its gradual disappearance on a time scale of several years to some decades. This intermittent mass-loss is called as episodic mass-loss.

Rapid rotation and mass-loss connection in Be stars was suggested for the first time by Struve(1931), which necessarily leads to the requirement of break-up velocity in Be stars. However, Vsin i statistics suggests almost all Be stars are well below the break-up velocity. The additional forces have been searched for so far; i.e., stellar wind, magnetic field, mass accretion, and so on. At the moment, none of them can succeed in explaining the episodic mass-loss in Be stars.

The very accurate spectroscopy by the solid state detectors have led to the discovery of the line profile variations in Be stars on a time scale of several hours to day. Such variations are considered due to nonradial pulsations(NRPs)(see Baade 1987), of which nature is correlated with Be emission activity. Mass-ejection driven by NRPs like radial pulsation was suspected by Willson(1986). But the quasi-periodicity of mass-loss cannot be explained naturally by this mechanism.

We propose a new driving mechanism of quasi-periodic mass-loss by NRPs. The observed NRPs being nonaxisymmetric modes, these modes can transport angular momentum from place to place. At the dissipative region near the surface, a prograde mode can accelerate the rotation, while a retrograde mode can decelerate the rotation (see Ando 1983). Observations show that retrograde modes are found when Vsin i of a Be star is larger than 170 km/s, and otherwise prograde modes. We presume this rule can be resulted from wave-rotation interaction even though a prograde and a retrograde mode are excited at a time.

In the presence of a prograde and a retrograde mode with equal characteristics, the rotation at the surface region is neither accelerated nor decelerated, and is in uniform rotation in a zero-th order. Is this state stable?  If the rotation profile is perturbed towards acceleration, intrinsic frequency (seen from rotating frame) of a pro-

grade mode decreases and that of a retrograde mode increases. When this
situation causes the increase of the accelerating effect of a prograde
mode and the reduction of the decelerating effect of a retrograde mode,
the rotation profile can be accelerated further more. Otherwise uniform
rotation is stable.

We will explain this physical meaning. If the inverse of intrinsic
frequency of a mode, that is, dynamical time scale is increased, the
corresponding region to the efficient radiation loss moves inwards. So
the temperature increases, and the opacity is decreased, if it is Kra-
mers' type opacity, which leads to the increase in radiation loss. According to Ando(1983), the increase in radiation loss leads to the increase both of the acceleration of a prograde mode and of the deceleration of a retrograde mode. Therefore, the rotation profile is unstable
in this case. When $H^-$ opacity is dominant, the situation is contrary and
the rotation profile is stable. So it is concluded that in the early
type stars the uniform rotation is unstable in the presence of a prograde and a retrograde mode, while in the late type stars it is stable.

In Be stars, the uniform rotation have been shown to be unstable
in the presence of a prograde and a retrograde mode. Ultimately their
rotation profile has been pointed out by Ando(1986) to oscillate quasi-
periodically around the uniform rotation (see Ando(1986) for the detailed explanation of the oscillatory behavior of the rotation profile).

From our calculations of wave-rotation interaction, we can summarize the two important aspects of this mechanism of episodic mass-loss.
1. Acceleration of rotation speed is, roughly speaking, wave pattern
   (phase) speed (20-100km/s), which is enough to enforce the equatorial rotation velocity to reach its break-up velocity.
2. this mechanism can explain the quasi-periodicity of episodic mass-
   loss quite naturally. Our calculations show its time scale is 1 to
   15 years, although it is a little shorter.

The observational test for a new mechanism may also be derived.
A. Quasi-periodic variation of Vsin i should be observed in accordance
   with episodic mass-loss.
B. There should be an anti-correlation of amplitude or of appearance
   of NRPs with Be emission activity.

References

Ando,H. 1983, Publ.Astron.Soc.Japan, <u>35</u>, 343.
Ando,H. 1986, Astron.Astrophys., <u>163</u>, 97.
Baade,D. 1987, in IAU Coll. No.92, Physics of Be stars, ed.A.Slettebak,
                Cambridge Univ. Press.
Struve,O. 1931, Astrophys.J., <u>73</u>, 94.
Willson,L.A. 1986, Publ.Astron.Soc.Pacific, <u>98</u>, 35.

# STELLAR WIND EQUATIONS IN A NEW STEADY STATE


Mariko Kato
Department of Astronomy, Keio University
Hiyoshi, Kouhoku-ku Yokohama   223   Japan



Abstract:
     A set of equations of stellar wind in a new steady state in spherically symmetry is presented. This equations are available also for the deep interior of stars, whereas the usual equations can be applied only to the surface region. The new equations have a variable mass flux which becomes zero at the inner boundary of the mass flow. The velocity also reduces zero, whereas it diverges at r=0 in the usual continuity equation $4\pi r^2 \rho v$=constant. In the surface region, the present equations approach the usual equations.


## 1. Introduction

In the usual steady state

$$\partial / \partial t |_r = 0 \qquad\qquad (1)$$

the equation of mass continuity becomes

$$4\pi r^2 \rho v = \text{const.} \qquad\qquad (2)$$

This equation has been widely applied to X-ray bursts and nova outbursts in which the acceleration occurs inside the photosphere (Kato 1983a,b, 1986). There is, however, a problem that the velocity could not be zero at the surface of the degenerate stars. Equation (2) gives the finite velocity at the surface of the degenerate stars, in spite of no matter actually flowing out from the interior of the degenerate stars.

Moreover if we apply equation (2) to the deep interior of stars (r=0), eather the velocity or the density should become infinitely large at r=0. Therefore we cannot get any normal stellar structures. This means that equation (2) is inadequate to the interior part of stars. This difficulty comes from the steady-state approximation (1). The mass flux must reduce zero at the center of the stars or the surface of the degenerate stars. The interior flow therefore should be described by another steady states, not by equation (1). Therefore we will present a new steady-state approximation and derive mass-loss equations which is available also to the deep interior of stars.

## 2. Equations in a new steady state.

We define the q-coordinate as

$$q = (M_r - M_c)/(M - M_c) , \qquad\qquad (3)$$

where $M_c$ and M are the mass within the inner boundary and the total mass of the star. We derive the equations in a new steady state

$\partial v/\partial t|_q = 0$, instead of usual steady state $\partial v/\partial t|_r = 0$. The equation of motion becomes

$$\frac{v}{Q}\frac{dv}{dr} + \frac{GM_r}{r^2} + \frac{1}{\rho}\frac{dP}{dr} = 0 \quad , \tag{4}$$

where the steady state factor Q is defined by

$$Q = 4\pi r^2 \rho v / |\dot{M}| q \quad . \tag{5}$$

The equation of motion (4) approaches the equation of hydrostatic balance near the inner boundary. The velocity also becomes smoothly zero; $v=0$, $dv/dr=0$ at $r=R_c$. With the assumption $\partial \rho/\partial t|_q = 0$ the equation of continuity becomes

$$\frac{1}{Q}\frac{d\ln\rho}{d\ln r} + \frac{d\ln v}{d\ln r} + 2 = 0 \quad . \tag{6}$$

This is integrated analytically as

$$4\pi r^2 \rho v = |\dot{M}|q - \frac{4\pi r^3 \rho |\dot{M}|}{3(M-M_c)}\left( 1 + \frac{C}{r^3} \right) \quad , \tag{7}$$

where C is the integral constant. Note that the mass flux $F = 4\pi r^2 \rho v = |\dot{M}|qQ$ is not constant but a function of r, whereas usual equation (2) gives a constant mass flux. The inner boundary conditions are that the mass flux F is zero at the inner boundary of the mass flow $R_c$.

$$\begin{aligned} C &= -R_c^3 \quad , \\ r &= R_c \qquad \text{at } q_c = 0. \end{aligned} \tag{8}$$

The inner boundary radius $R_c$ is the radius of the neutron star for X-ray bursts and is zero in the normal stars.

The equation of energy conservation defined by $\partial s/\partial t|_q = 0$ becomes

$$|\dot{M}|Tq\frac{ds}{dr} = 4\pi r^2 \rho_n - \frac{dL_r}{dr} \quad . \tag{9}$$

The equations of mass continuity and the energy transport are unchanged.

The present equations will be useful when the acceleration occurs deep interior or when the gravitational energy release is important in the energy conservation equation. In such cases we will have different structures from that of the usual solutions. Further investigation will be desired in many cases such as red giants or very massive stars.

references
Kato, M., 1983a, Publ. Astron. Soc. Japan, 35, 33.
Kato, M., 1983b, Publ. Astron. Soc. Japan, 35, 507.
Kato, M., 1986, Publ. Astron. Soc. Japan, 38, 29.

MOLECULAR SPECTROSCOPY AS DIAGNOSTICS OF OUTER ATMOSPHERE OF COOL LUMINOUS STARS:
QUASI-STATIC TURBULENT MOLECULAR FORMATION ZONE AND STELLAR MASS-LOSS


Takashi Tsuji
Tokyo Astronomical Observatory, The University of Tokyo
Mitaka, Tokyo,  181 JAPAN


## Abstract


The  origin of mass-loss in cool luminous stars is still obscure;  several known mechanisms such as thermally driven wind, radiation-driven wind(via dust),wave-driven wind etc all have serious difficulties,  if examined in the light of recent observations. At the same time,recent observations in the infrared and radio spectral domains revealed that outer envelope of red (super)giant stars has highly complicated spatial and  velocity structures,  while inner envelope may have new component that  had  not been  recognized before. For example,  recent high resolution infrared  spectroscopy revealed  a possible presence of a quasi-static turbulent molecular dissociation zone somewhere in the outer atmosphere. This new component may represent a transition zone between  the warm chromosphere and the huge expanding molecular envelope,  and may be a  cool component of chromospheric inhomogeneity or a moleclar condensation in a cool corona extended by turbulent pressure. Such a result can be regarded as observational evidence  in support of a recent theory of autocatalytic molecular formation by thermal instability due to molecular cooling.  Thus,  observation and theory consistently show  the presence  of  a new component  - quasi-static turbulent molecular formation zone -  in outer atmosphere  of cool luminous stars,  and a possibility of  a unified understanding of outer atmospheric structure and mass-loss,  in which turbulence  may play important role, can be proposed.



## I. Introduction

The importance of mass-loss in the late stages of stellar evolution, represented by the red giant phase, has long been known,  but the origin of the massive cool wind from  red giant stars is still not clear. This difficulty of understanding the origin of massive cool wind from red (super)giant stars should undoubtedly be related to our poor understanding of the physical structure of the outer atmosphere. This is evident if we remember the case of the Sun, in which  the solar wind has been understood as a necessary consequence  of  the outer atmospheric structure  consisting of  the photosphere,chromosphere,transition layer,and corona. Thus, more physical understanding of the outer atmospheric structure of red giant stars,  which may radically be different from  that  of  the Sun,  should be a prerequisite in understanding the  mechanism  of mass-loss in red giant stars.

In consideraing the problem noted above, one important characteristic of the cool wind from red giant stars is that the observed flow velocity is much smaller than the escape velocity at the stellar surface,  in marked contrast to that from hot luminous stars  in  which observed flow velocity exceeds the escape velocity  at  the  stellar surface. This  fact implies that the cool wind in red giant stars  cannot  originate from  the stellar surface but should originate from the outer layer at least  several stellar  radii above the stellar surface,  while stellar wind from hot luminous stars is originating from the stellar surface.  Thus, to clarify the origin of cool wind in red  giant stars,  it should be most important to know the physical structure of  the layer  located at several stellar radii above the stellar surface.  For this purpose, molecular spectroscopy,both in the infrared and radio spectral domains,should be most effective because of the relatively low temperature of such a region,while relatively hot region in outer atmosphere has been probed by atomic spectroscopy such as by IUE.

In this contribution, we first examine the known mass-loss mechanisms in the light of the recent observations(Sect.II),and then we review some recent infrared and radio observations that may be relevant to our understanding of the outer atmospheric structure and mass-loss phenomena(Sect.III). Based on these observations, new picture of the outer atmosphere of red (super)giant stars is proposed and its implications on circumstellar chemistry as well as on mass-loss phenomena are discussed(Sect.IV).


II. Difficulties of Mass-Loss Mechanisms in Red Giant Stars

As to the origin of mass-loss in cool luminous stars, several possibilities have been proposed, but none of them could provide satisfactory answer yet. As noted in Sect.I, the cool wind should originate not in the stellar surface but in outer layer at least several stellar radii above the stellar surface. We already know such a case in the thermally-driven wind from the hot corona of the Sun. However, the difficulty to apply the theory of thermally-driven wind to cool luminous stars has already been recognized at an early time( e.g., Weymann, 1963). More recent discovery of non-existence of hot transition region( and hence of corona) in these stars by IUE observations(Linsky,Haisch,1979) finally disclosed that the origin of mass-loss in cool luminous stars may be radically different from that in solar type stars.

Now,if hot corona plays little role in accelerating wind in cool luminous stars, how about a role of chromospheres? In fact,it was suggested that the flow velocity in chromospheres is nearly half of the terminal flow velocity( Reimers,1975), or that it already reaches the terminal flow velocity in the upper chromosphere and hence mass-loss out-flow already starts in chromospheres( Goldberg,1979). However, it must be remembered that the observed extension of the chromosphere, although much extended than in the Sun, is within a few stellar radii, as shown by H$\alpha$ images( e.g., White, Kreidl, Goldberg,1982; Hebden,Eckart,Hege,1986) or by radio continuum emissions(e.g., Wischnewski, Wendker,1981; Drake, Linsky, 1986). Then, observed flow velocity in the chromosphere is still much smaller than the local escape velocity in such an extended chromosphere. Thus, outflow in stellar chromospheres cannot be a direct origin of stellar mass-loss,unless there is further acceleration mechanism in the chromosphere. Despite extensive efforts,however,no such mechanism has been found(e.g.,Caster,1981). Thus, the material moving outward in chromospheres should eventually be decelerated. In fact, evidence of infall material is observed by FeII emission lines in $\alpha$ Ori (Boesgaard,Magnan,1975), although FeII emissions do not necessarily appear red shift-ed in red giant stars( Boesgaard, 1981). Also recent high resolution observations of several M-giant stars by IUE revealed that the velocity field in the chromosphere of red giant stars is highly compicated. For example, UV absorption lines gave large positive velocity while some UV emission lines showed negative velocity relative to the photospheric lines( Eaton,Johnson,1987). In this connection, it is interesting to note that a model which pictures the chromosphere as a complicated "fountain" of streaming material consisting of upflows and downflows has been suggested from observed line profiles of $\alpha$ Ori, after recognizing that Alfven wave driven model could not reproduce observed line profiles( Hartmann, Avrett, 1984).

Under such a situation, a mechanism of mass-loss that can be plausible in cool luminous stars may be radiation-driven model. Especially, it is known that radiation pressure on dust can effectively accelerate mass-loss outflow, once dust could be formed. The major problem in this case is where the dust could be formed. It is generally difficult to make dust near the photosphere because temperature may be too high for dust to be formed. Recently, a possibility to produce dust in the chromos-phere by the so-called condensation instability is proposed( Stencel,Carpenter,Hagen, 1986). Although this is an interesting idea that deresrves further attention( Sect. IV), it is not clear if dust could be formed near the star. In fact, recent speckle interferometry of many evolved cool stars in the near infrared has shown that there should be dust-free zone extending out to several stellar radii around the central star( Dyck et al.,1984; Ridgway et al,1986). Also, in the case of red supergiant star $\alpha$ Ori, direct infrared imaging revealed that the dust-free zone may be as large as 20 stellar radii( Bloemhof,Townes,Vanderwyck,1984). Thus, dust may be formed pretty far from the stellar surface, but such a dust formation is only possible if the gas

density  of the dust condensation point is high enough,  as has already been noted by
Kwok(1975).  One  possibility to have  high density region  in outer atmosphere is to
levitate the mater above the photosphere by stellar pulsation(e.g.,Jura,1986;Fedeyev,
1988),but this applies only to pulsating stars at best. For this reason,more consistent  understanding of the outer atmospheric structure should be needed before we  can
accept(or not accept) the dust-driven model of mass-loss in red (super)giant stars.

On  the  other  hand,  the velocity structure of the outer atmosphere  has  been
measured by radio interferometry:  For example,  Chapman and Cohen(1986) showed  that
the  velocity  extents of the maser emissions such as of SiO,  $H_2O$,  and OH  increase
systematically  with angular extents,  and they interpreted this fact as showing  the
continuous  increase of expansion velocity with radial distance.   They  pointed  out
that  the  driving force per unit mass must be increasing with the distance from  the
star to explain such a velocity structure.  However,  it is  difficult to provide the
required driving force by radiation pressure on dust in the inner envelope because of
the absence of dust itself as noted above.  Furthermore, masers of SiO and $H_2O$ do not
necessarily show  clear expansion pattern( see  Sect.III), and it is not sure  if the
inner envelope is really expanding. For carbon stars, however, carbon grains can condense at relatively high temperature near the photosphere,and a self-consistent model
of  dust-driven mass-loss can be constructed(e.g.,Sedlmayr,1988).   Even though,  the
solution of the equations of stellar wind may depend on the boundary condition, which
further depends on the inner structure of the circumstellar envelope,for which little
is known yet, as will be discussed in the next section.

Also, one classical question  is whether dust formation initiates  mass-loss  or
whether dust is formed as a result of mass-loss.  It is to be noted that  the  latter
process may be rather easy,  once mass-loss occurs by another mechanism. This problem
can  be examined on the basis of recent observations of CO radio emission  lines,  by
which  stellar  mass-loss rate has been determined with better accuracy than  by  any
other method for a large sample of red giant stars, and terminal flow velocities have
also been determined with high accuracy( e.g.,Knapp,Morris,1985). The result revealed
that  the  momentum  in the stellar wind and that in the  stellar  radiation  do  not
necessarily show good correlation(e.g.,Zuckerman,Dyck,1986). Also, a necessary condition for the winds to be accelerated by radiation pressure on dust( $\dot{M}v < L/c$) seems to
be not fulfilled for  considerable number of stars( Zuckerman,Dyck,1986; Knapp,1986).
Thus, the question addressed at the beginning of this paragraph still remains open.

Finally,  it is also evident that radiation pressure on dust cannot explain  all
the mass-loss phenomena in red giant stars,  since dust is not found in K and early M
giant stars,  which also show considerable mass-loss in general. In fact, recent IRAS
survey revealed infrared excesses that show the presence of dust only in stars  later
than M3-4III(Hacking  et al,1985).  Thus,  although dust-driven mass-loss  could  be
successful in  a  limited case  such as  cool carbon stars  with  massive wind( e.g.,
Sedlmayr,1988),it  could not provide unified understanding of mass-loss phenomena  in
red (super)giant stars in general.  Clearly,  more unified understanding of mass-loss
in cool luminous stars should be looked for.


III. Molecular Spectroscopy of the Outer Atmospheres of Red Giant Stars

A brief survey outlined in the previous section reveals that more information on
the  intermediate zone where mass-loss outflow starts should be needed.  As noted  in
Sect.I,such a region of outer atmosphere can best be probed by molecular spectroscopy
especially in the infrared and radio spectral regions.

First,one important result of the infrared molecular spectroscopy is a discovery
of stationary CO layer in Mira type variable star $\chi$ Cyg,  in which CO lines that stay
stationary have been  clearly separated from the photospheric CO  lines  that  show
cyclic Doppler-shifts  by the large amplitude pulsation of the photosphere  in  time
series spectra( Hinkle, Hall, Ridgway, 1982). It was suggested that the stationary CO
layer  may  be  located  at several stellar radii  above  the  stellar surface, since
excitation temperature of the stationary CO layer of $\chi$ Cyg was found to be 800K.

A possible presence of  a similar stationary layer in red supergiants has previously  been  suggested by Hall(1980).  The recognition of such a stationary layer  by
spectroscopic observations is generally difficult, because spectral lines originating
from  such a layer show little Doppler shift  against photosheric lines  in  non-Mira
stars.  More recently, however, a possible presence of a static layer has been demonstrated from a detailed analysis of CO overtone bands in several M-giant stars of the
spectral types  later than  M4III  as well as in  red supergiant stars such as $\alpha$ Ori
(Tsuji, 1986b, 1987a,b).  Although  CO lines  originating from such a static layer in
nornal  red (super)giants cannot be separated by Doppler shifts against  photospheric
lines, they can be recognized by the following evidences:

(1)  Quantitative spectroscopy applied to the CO first overtone  bands  in  high
resolution infrared  spectra showed reasonable success for M-giant stars  as cool  as
M7III, and  equivalent widths  of weak and medium strong CO  lines,  which  are  also
relatively high excitation lines,  can  quantitatively  be well understood  by  model
stellar  atmosphere with well defined parameters( Tsuji,1986a).  However,  strong  CO
lines,  which are also low excitation lines, show systematic excess  as compared with
expected ones based on the same model atmosphere( Fig.3 of Tsuji,1986b).

(2)  The low excitation lines show shifts and asymmetries  that indicate  excess
absorption in blue wing in some stars and in red wing in other stars( Figs.4 and 5 in
Tsuji, 1986b).

(3)  Radial velocities show differential time variations between  low  and  high
excitation lines.  For example,  low excitation lines remain almost stationary  while
radial velocities of high excitation lines show appreciable changes, possibly due  to
small amplitude pulsation of the photosphere, in the case of  $\alpha$ Her( Fig.1 in  Tsuji,
1987a).

While  the intensity anomaly alone(item 1  above) can be due to our  poor understanding  of atmospheric structure and/or of line formation in the upper  atmosphere,
peculiar line shift and line asymmetry(item 2) as well as differential time variation
note  above(item 3) suggest that at least a part of low excitation lines  should  be
originating  not  in  the photosphere but in an extra layer well separated  from  the
photosphere.  We tentatively refer  this separate extra layer  as CO absorption layer
(the "layer" does not necessarily imply  that it is well stratified but rather it can
also  be an inhomogeneous component in outer atmosphere,  and the "layer" is used  in
this extended meaning in what follows).

Now, the  problem is how could we separate the contribution by the CO absorption
layer from the stronger photospheric component.  For this purpose,  we  compared  the
observed  profiles  with  predicted  ones based on model atmospheres and found appreciable residual absorption for low excitation lines while no residual for high  excitation lines. Then, the contribution by the CO absorption layer has been separated by
subtracting the photospheric contribution from the observed profile,in low excitation
CO lines.  A detailed analysis on separated excess CO absorption gives the results as
follow( as for detail, see Tsuji,1987b):

(1)  Excess CO absorption can be seen not only in 2-0 band but also in 3-1 band,
and excitation temperatures of the CO absorption layer are between 1000 and 2000K.

(2)  Turbulent velocities in the CO absorption layer are larger  than 5 km/s  in
red giants and larger than 10km/s in red supergiants.

(3)  The CO column density of the absorption layer is as high as $10^{20}$/cm$^2$.

(4)  The movement of the CO absorption layer  relative to to the  photosphere is
very small, either positive or negative,  but it show expansion of a few km/s against
the central star for the cases of known stellar velocity( by thermal radio lines).

The physical parameters of the CO absorption layer in red (super)giant stars are
very similar to those of the stationary layer of Mira variable  star $\chi$ Cyg  mentioned
above(Hinkle,Hall,Ridgway,1982).  Thus,the presence of a separated stationary CO layer
may be not restricted to Mira variable stars, but rather it may be a basic characteristics of red giant and supergiant stars in general.

In this connection, it is to be remembered that $H_2O$ and SiO maser emissions also showed  the presence of a non-expanding turbulent layer within some 10 stellar  radii of late-type giant stars( e.g.,Reid,Moran,1981).  Maser emissions are powerful probes of the inner circumstellar envelope, especially if high resolution spatial interfero- metry could be applied.  For example, VLBI observations of SiO masers showed that the masers  occur  within a few stellar radii in outer atmosphere, but  do not  show  any systematic expanding patern(Lane,1984). Thus,SiO masers may originate in cloudlets of rather high density  in the  turbulent  region of  the inner  circumstellar  envelope (Alcock,Rose, 1986).  Also,  high  resolution observation of $H_2O$ maser emissions  for several late-type stars  by VLA revealed  that  $H_2O$ features  appear to be unresolved knots distributed over not more than 20 stellar radii( Johnston,Spencer,Bowers,1985). Thus, both thermal absorptions( CO infrared lines) and non-thermal emissions( $H_2O$ and SiO  masers)  consistently  show  the presence of  molecular formation zone, roughly between a few stellar radii and some 20 stellar radii.

In  contrast  to  a static  molecular  formation zone  in  the inner  part  of  the circumstellar  envelope,  expanding molecular envelopes are relatively easy to recog- nize.  However,  it is only recently that some details of velocity and spatial struc- tures  of the expanding envelope have been made clear. For example, it is interesting to  remember that presence of multiple velocity components in expanding CO  envelopes has been found  from the analysis of the fundamental bands of CO  in $\alpha$ Ori( Bernat et al.,1980)  as well as in normal red giants( Bernat,1981):  The expanding CO envelopes are charcterized by much lower excitation temperatures,  by smaller turbulent veloci- ties,and by smaller column densities, as compared with those of the static CO absorp- tion  layer  discussed  above.  It is natural to suppose  that  these  expanding CO components might have been outflowed from the static CO layer.

Further, expanding molecular envelope can be traced by the pure rotational tran- sitions  in  the millimeter wavelength region  out  to some  $10^5$ stellar radii( e.g., Zuckerman,1980; Olofsson,1985).  One interesting  feature of such  a huge  molecular envelope is that it may not necessarily be spherically symmetric( e.g.,Chapman,Cohen, 1986). Furthermore,asymmetric profiles suggesting bipolar structure has been observed in  M-type  giant such as OH231.8+4.2( Morris et al.,1987) as well as in carbon  star such as V Hya( Tsuji et al.,1987). Such a bipolar structure of the outer envelope may be related to binarity( Morris,1981).  Recently, infrared speckle interferometry also showed that the mass-loss geometry may be bipolar in nature,for several evolved giant stars( e.g.,Dyck et al.,1987; Cobb,Fix,1987). It is still not clear if bipolar struc- ture  of  the  outer envelope is a general feature of  mass-losing  stars,  and  more detailed  observations  on spatial structure of outer envelopes  of mass-losing stars should certainly be important.

IV.  A  New Component of Outer Atmosphere of Red Giant and Supergiant Stars:
     Quasi-Static Turbulent Molecular Dissociation Zone

A  survey  of  recent observations in the previous section revealed  that  there should be a new component in the outer atmosphere of red (super)giant stars;molecular formation zone,tentatively referred to as the CO absorption layer. This CO absorption layer  identified in Sect.III shows almost no expansion, in contrast to the expanding molecular envelope so far known, but has large turbulent motion. Thus, we may  assume a presence of  a quasi-static turbulent molecular dissociation zone  somewhere in the outer atmosphere.  One possibility is that  this zone may represent a transition zone between  the warm chromosphere and  the massive cool wind in  red (super)giant stars. Probably,  inhomogeneity may play important role in this transition region,  and this zone may be a cool component of the chromospheric inhomogeneity such as known for the Sun( Ayres,Testerman,Brault,1986) or suggested for $\alpha$ Boo( Heasley et al.,1978),  or a clumpy aggregation of  CO cloudlets  such as suggested for  SiO maser region( Alcock, Rose,1986). At present,however, a detailed structure of the inner part of the circum- stellar envelope is not clear at all,  and the recent observations have just revealed that it is far more complicated than has been supposed before.

In this connection,  one interesting suggestion is that  the thermal instability
due to extreme temperature sensitivity of molecular opacities induces molecular cooling and associated dynamical effect,in the outer atmosphere of red (super)giant stars
(Muchmore,Nuth,Stencel,1987). Also, a possibility that  molecules and  dusts could be
formed  in a quasi-static extended chromosphere by the so-called condensation  instability  has been suggested from the observation that chromospheres are not completely
quenched in the presence of dust( Stencel, Carpenter, Hagen, 1986). Such an idea that
the formation of molecules( or dusts) could be accelerated by the thermal instability
due to molecular( or dust) formation itself presents an interesting  possibility  for
understanding the large abundance of molecules(dusts) in the outer atmosphere of cool
luminous  stars.  Then,  possible  presence  of the quasi-static  turbulent  molecular
formation zone, revealed by our observations outlined in Sect.III, can be regarded as
observational  evidence for such a theory of molecular(dust) formation by  cooling(or
condensation) instability. Thus, although exact nature of the quasi-static  molecular
formation zone remains undefined until more detailed confrontation between theory and
observation  could  be  carried out,  such an idea may have a key to  understand  the
complicated structure of the inner part of the circumstellar envelope  of red (super)
giant  stars,  and some implications of the possible presence of such a  quasi-static
molecular  dissociation zone  in the outer atmospheres of red (super)giant stars  are
examined below:


(a) Structure of the Outer Atmosphere of Red Giant and Supergiant Stars

     Now,with the recognition of such a new component of the outer atmosphere in cool
luminous stars,  the division of cool stars  into two types  discovered by Linsky and
Haisch(1979)  can  be contrasted more clearly: While hot transition  layer  and  hot
corona  such  as  known in solar type stars do not exist in red  (super)giant  stars,
rather  cool transition layer  may exist instead  in red  (super)giant  stars.  This
transition  layer  in  noncoronal type stars can also be regarded as  a  cool  corona
extended  by  the turbulent pressure,  just as the hot corona in solar type stars  is
extended by the thermal pressure.  However,  it should be remembered that the concept
of "transition layer" or "corona" does not ncessarily imply a homogeneous  structure,
as is also known for the case of the Sun( for example,  coronal hole).  Probably, the
quasi-static  turbulent layer or cool corona may exist in all the red giant stars  of
the noncoronal type defined by Linsky and Haisch(1979), but it may be difficult to be
recognized in K and early M giants, simply because optical thickness of the CO  layer
is too small in these stars. In red supergiant stars,the presence of rather extensive
CO  absorption  layer  is already evident  in early M supergiants such as $\alpha$ Ori  and
$\mu$ Cep(Tsuji,1987b),and this fact implies that the quasi-static molecular layer can be
formed more easily in stars of the higher luminosity at the same spectral type.

     Now, if the presence of the quasi-static molecular formation zone is a fundamen-
tal property of red (super)giant stars,  how could we understand the origin of such a
component in outer atmosphere? One interesting possibility is that the molecular zone
in outer atmosphere could be formed by the cooling instability(Muchmore,Nuth,Stencel,
1987)  or  by the condensation  instability(Stencel,Carpenter,Hagen,1986),  as  noted
above.  Even  if  such a process could trigger the molecular formation in the  outer
atmosphere, however,there may still be two major problems: how matter can be supplied
to the outer static layer and how it can be supported in the non-expanding  envelope?
For  the  first  problem,  we have no answer at present  but we only notice  that the
surface  gravity in red (super) giant star is much smaller than that in the  Sun,  in
which matter can anyhow be transferred to the hot corona from the photosphere,despite
the large surface gravity. In this connection, it may be interesting to remember that
we  have already seen  upflow and downflow motions  in the chromosphere(Sect.II)  and
such  a  process may be transporting  the matter from the photosphere to the  quasi-
static  molecular formation zone.  For the second problem,  we may suggeste that  the
turbulence  may  be important,  since turbulent velocity in the CO  absorption  layer
turns out to be actually rather high(Sect.III). At present, however,origin of stellar
turbulence is not well understood,too.

(b) Circumstellar Chemistry

     Now,once the rather cool turbulent molecular formation zone or molecular conden-
sation is formed in the outer atmsphere of red giant stars, it may have further
important effects upon the chemistry of the outer atmosphere of these stars. Most
importantly, the rather high gas density due to a quasi-static character and to the
large scale height of the turbulent zone provides the necessary conditions for the
formation of various molecules. This fact may also be important for molecular
formation triggered by the cooling instability noted above. In fact, if matter is
always moving outward with velocity v, gas density at radius r falls as $v^{-1}r^{-2}$. Then
it may be difficlut to understand the presence of molecules in circumstellar envelope, since molecules formed in stellar photosphere should have been destroyed in
passing through the warm extended chromosphere( Hartmann,1983), and formation of new
molecules may be difficult at low density of the expanding envelope. In fact, the
presence of a static molecular formation layer had to be assumed at the base of
expanding envelope, either implicitly( Scalo,Slavsky, 1980) or explicitly( Clegg, van
IJzendoorn,Allamandole,1983),in discussing circumstellar chemistry in expanding outer
envelope.

     We now find observational evidence for the presence of such a molecular format-
ion zone in the quasi-static turbulent zone. From the CO column density of $10^{20}/cm^2$
(Sect.III) and the pressure scale height comparable to stellar radius of $10^{13}cm$( due
to large turbulence), the CO number density may be as high as $10^7/cm^3$ and this
implies that the hydrogen number density should be as high as $10^{11}/cm^3$. Thus, molecular formation in the quasi-static turbulent layer can follow chemical equilibrium
as a first approximation, and some examples of such equilibrium molecular formations
were given elsewhere( Tsuji,1986b). Also, it is to be noted that the clear contrast
between oxygen-rich and carbon-rich chemistries still prevails in observed circumstellar molecules( e.g.,Zuckermann,1980; Olofsson,1985),and this fact can be regarded
as an evidence that the chemical equilibrium is realized approximately for the
molecular formation in the inner part of the circumstellar envelope. Thus, prediction
based on a chemical equilibrium can be used as an initial condition for non-
equilibrium molecular formation in an expanding envelope.


(c)  A  Possibility of Turbulence-Driven Wind and Mass-Loss

     The situation is the same for dust formation as for molecular formation: the
turbulent molecular dissociation zone could provide density high enough and, at the
same time, temperature low enough for dust to be formed. Thus, we now find an observational clue as to the place where dusts could be formed, to initiate mass-loss outflow. In this connection,it is interesting to remember that the strongest correlation
of mass-loss rate with any of the observed properties was with the CaII K4
absorption width, and this fact may indicate that turbulence could contribute to
mass-outflow by increasing the scale height of the outer atmosphere and by raising
more material to the layer where the dust could be formed( Hagen, Stencel, Dickinson,
1983). However, it is not sure if the process of dust formation followed by mass
outflow is a steady process. It is also possible that the dust could not be formed
until the gas density of the transition layer exceeds certain critical value by the
transport of the matter from the photosphere,and hence mass-outflow may be sporadic.
Such a possibility is consistent with the presence of multiple expanding CO envelopes
around red (super)giant stars( Bernat,1981), as noted in Sect.III.

     Despite the observational identification of a possible site of dust formation,
however, the dust-driven wind could not be applied to stars without dust envelope, as
noted in Sect.II. Then, a more interesting possibility is a turbulence-driven wind,in
which the high turbulent pressure of the transition layer(or cool corona) pushes the
gas out of star,just as the high thermal pressure in corona does in solar-type stars.
In fact, if the turbulent zone is extended to about 10 stellar radii, the local
escape velocity there may already be small enough to be comparable with the
observed flow velocities. Thus, the Maxwellian tail of the turbulent motion in the
quasi-static molecular formation zone can directly lead to stellar mass-loss in all

the noncoronal stars. The mass-loss rate −CaII K4 absorption width correlation(Hagen, Stencel,Dickinson,1983) noted before  can also be regarded as supporting evidence for the  turbulence-driven wind.  If such a mechanism could work in cool luminous  stars, dust is not necessarily the prime mover of mass-loss,  but rather it is only a result of  mass-loss.  For example,  dust could most easily be formed  when  gas is  rapidly cooled  at the sonic point  during  the outflow.   Then dust will play only secondary role in accelerating the outflowed gas further.  The relative importance of  the  two major possibilities,  the turbulence-driven wind or the radiation-driven wind, should still be examined in detail.


V. Concluding Remarks

    Recent  developments of molecular spectroscopy in the infrared and  radio  wave-length  regions made it possible to probe the cool components of the huge outer enve-lope of red (super)giant stars,extending from the region near the stellar photosphere $(r \sim R_* \sim 10^{13} cm)$ out to the region where the stellar wind interacts with the ambient interstellar matter( $r \sim 10^5 R_* \sim 10^{18} cm$). Our present picture is that there should be a quasi-static  turbulent  molecular  formation zone,   in the inner part of  the  huge expanding circumstellar envelope,  possibly as an inhomogeneity in extended  chromos-phere  or  in cool corona.  Presence of such a molecular formation zone in the  outer atmosphere is consistent with the recent theory of molecular formation by the thermal instability due to  molecular cooling( Muchmore, Nuth, Stencel, 1987). Thus, with the quasi-static molecular formation zone,  observation and theory consistently show a new possibility  for understanding the physics and chemistry  in outer atmosphere of cool luminous  stars.  Also,  such a new component in the inner part of the  circumstellar envelope may have a key to understand the nature of huge expanding molecular envelope which  show  complicated velocity and spatial structures,  as revealed by the  recent infrared and radio observations.

    Further,  we  suggested a possibility that the turbulence-driven wind  from  the quasi-static intermediate zone could provide a unified understanding of stellar mass-loss in red (super)giant stars. However, this does not solve the problem yet, but the difficulty of  mass-loss mechanism is simply reduced to the problem  of the origin of stellar turbulence.  In this connection, it is to be remembered that one of the major unsolved problem in the theory of stellar atmospheres is how to understand the origin of stellar turbulence(e.g.,Gray,1978). Further, turbulence may already play an impor-tant  role in determining the physical structure of the photosphere of cool  luminous stars(de Jager,1980). We now suppose that the turbulence may play similarly important role in  the stellar outer atmosphere,  as it does in photospheres of cool  liminous stars.  Thus, fundamental problems in stellar photospheres as well as in outer atmos-pheres  could not be deemed resolved  until the origin of stellar turbulence is  well understood.


REFERENCES


Alcock,C.,Rose,R.R.:1986, Astrophys.J. 310, 838
Ayres,T.R.,Testerman,L.,Brault,J.W.:1986, Astrophys.J. 304,542
Bernat,A.P.: 1981, Astrophys.J. 246, 184
Bernat,A.P.,Hall,D.N.B.,Hinkle,K.H.,Ridgway,S.T.: 1979, Astrophys.J.Lett. 233,L135
Bloemhof,E.E.,Townes,C.H.,Vanderwyck,A.H.B.: 1984, Astrophy.J.Lett. 276,L21
Boesgaard,A.M.: 1981, Astrophys.J. 251,564
Boesgaard,A.M.,Magnan,C.: 1975, Astrophys.J. 198, 369
Caster,J.I.:1981, In Phys.Proc. in Red Giants eds I.Iben,Jr and A.Renzini,D.Reidel,
     Dordrecht, p.285
Chapman,J.M., Cohen,R.J.: 1986, M.N.R.A.S. 220,513
Clegg,R.E.S., van IJzendoorn,L.J., Allamandole,L.J.: 1983, M.N.R.A.S. 203,125
Cobb,M.L.,Fix,J.D.: 1987, Astrophys.J. 315, 325
de Jager,C.: 1980, The Brightest Stars, D.Reidel, Dordrecht
Dyck,H.M.,Zuckerman,B.,Leinert,Ch.,Beckwith,S.:1984,Astrophys.J. 287, 801
Dyck,H.M.,Zuckerman,B.,Howell,R.R.,Beckwith,S.:1987,Publ.Astron.Soc.Pacific 99, 99

Eaton,J.A.,Johnson,H.R.: 1987, Astrophys.J. in press

Fedeyev,Y.A.: 1988, this volume

Goldberg,L.: 1979, Quart.J.Roy.Astron.Soc. 20, 361

Gray,D.F.: 1978, Solar Phys. 59, 193

Hacking,P. et al.: 1985, Publ. Astron. Soc.Pacific 97,616

Hagen,W.,Stencel,R.E.,Dickinson,D.F.: 1983, Astrophys.J. 274,286

Hall,D.N.B.:1980, In Interstellar Molecules ed. B.H. Andrew,D.Reidel,Dordrecht,p.515

Hartmann,L.:1983, Highlights of Astronomy 6, 549

Hartmann,L.,Avrett,E.H.:1984, Astrophys.J. 284,238

Heasley,J.N.,Ridgway,S.T.,Carbon,D.F.,Milkey,R.W.,Hall,D.N.B.:1978, Astrophys.J.
    219,970

Hebden,J.C., Eckart,A., Hege,E.K.: 1987, Astrophys.J. in press

Hinkle,K.H., Hall,D.N.B., Ridgway,S.T.: 1982,Astrophys.J. 252, 697

Jura,M.:1986, Astrophys.J. 303, 327

Knapp,G.R.: 1986, Astrophys.J. 311, 731

Knapp,G.R., Morris,M.: 1985, Astrophys.J. 292, 640

Kwok,S.: 1975, Astrophys.J. 198, 583

Lane,A.P.: 1984, Proc. IAU Symp.No.110 "VLBI and Compact Radio Sources eds R.Fanti,
    K.Kellerman, and G.Setti, D.Reidel, Dordrecht, p.329

Linsky,J.L., Haisch,B.M.: 1979, Astrophys.J.Lett. 229,L27

Morris,M.: 1981, Astrophys.J. 249, 572

Morris,M.,Guilloteau,S.,Lucas,R.,Omont,A.:1987, Astropys.J. in press

Muchmore,D.,Nuth III.,J.A.,Stencel,R.E.:1987,Astrophys.J.Lett. 315,L141

Olofsson,H.: 1985, in Proc.ESO-IRAM-Onsala Workshop on "(Sub)Millimeter Astronomy"
    eds. P.A.Shaver and K.Kjär, ESO, München

Reid,M.J.,Moran,J.M.: 1981, Ann.Rev.Astron.Astrophys. 19,231

Reimers,D.: 1975, Mem. Soc. Roy. Sci. Liege, 6 Ser. 8, 369

Reimers,D.: 1977, Astron. Astrophys. 57,395

Ridgway,S.T., Joyce,R.R.,Connors,D.,Pipler,J.L.,Dainty,C.:1986,Astrophys.J. 302,662

Scalo,J.M.,Slavsky,D.B.: 1980, Astrophys. J. Lett.  239, L73

Sedlmayr,E.:1988, this volume

Stencel,R.,Carpenter,K.G.,Hagen,W.:1986, Astrophys.J. 308, 859

Tsuji, T.: 1986a, Astron. Astrophys. 156,8

Tsuji,T.: 1986b, Proc. IAU Symp.No.120 "Astrochemistry" ed. M.S.Vardya and
    S.P.Tarafdar, D.Reidel, Dordrecht, p.409

Tsuji,T.: 1987a, Proc.IAU Symp.No.122 "Circumstellar Matter" ed. I.Appenzeller and
    C.Jordan, D.Reidel,Dordrecht, p.377

Tsuji,T.: 1987b, Astron.Astrophys. submitted

Tsuji,T.,Unno,W.,Kaifu,N.,Izumiura,H.,Ukita,N.,Cho,S.,Koyama.K.:1987,Astrophy.J.Lett.
    submitted

Weymann,R.: 1963, Ann. Rev. Astron. Astrophys. 1,97

White,N.M., Kreidl,T.J., Goldberg,L.: 1982, Astrophys.J. 254, 670

Wischnewski,E.,Wendker,H.J.: 1981, Astron. Astrophys. 96,102

Zuckerman,B.: 1980, Ann. Rev. Astron. Astrophys. 18,263

Zuckerman,B.,Dyck,H.M.: 1986, Astrophys.J. 304, 394

## DUST DRIVEN WINDS IN LATE SUPERGIANTS


Erwin Sedlmayr                                Hans-Peter Gail
Carsten Dominik
Institut für Astronomie und Astrophysik       Institut für Theoretische Astrophysik
Technische Universität Berlin                 Universität Heidelberg
Hardenbergstraße 36, D-1000 Berlin 12         Im Neuenheimer Feld 561, 6900 Heidelberg 1



## 1. Introduction

The cool extended shells of giants and supergiants are well known to be places of
copious dust formation as indicated by the occurrence of pronounced extinction,
reddening, and polarization of the continuous star light and by the appearance of
particular absorption features, both manifesting the interaction of photons with
particles considerably larger than atoms or molecules.

The accepted explanation of these phenomena is the formation of circumstellar
dust, i.e. small solid particles with a typical size of the order of 0.1 micron.
However, the analysis of these effects yields only information on the interaction
of photons with certain functional groups within the clusters - e.g. Si-O, C-H, C-C
bending and Si-O, C-C, ... stretching vibrations - and thus allows no definite
determination of the "real" physical structure and chemical composition of the
grain particles. Therefore, observational conclusions concerning the properties of
circumstellar dust can provide only some "mean" information which allows no definitive conclusions regarding the true nature of the observed grains (geometrical
shape, crystalline structure, ...?).

An adequate answer to these questions must be based on the detailed study of the
processes of formation and growth of dust particles in these environments. However,
dust formation cannot be considered as an isolated problem because due to their
huge absorption cross sections even a small contamination of the atmospheres by
circumstellar dust may have a significant influence on the radiative transfer and
(via energy- and momentum-coupling) on the thermodynamic and hydrodynamic structure
of the dust forming shell.

Therefore, a consistent modelling of such objects requires the treatment of the
coupled nonlinear system comprising
- radiative transfer
- density, velocity, temperature structure of the shell
- chemistry and dust formation.

This has been pointed out by several authors in the past (e.g.
Wickramasinghe, 1972; Kwok, 1975; Salpeter, 1974; Deguchi, 1980; Jura, 1983), but

there was no consistent treatment of all relevant aspects of this important problem (especially for the complex of dust formation and growth) until the papers of Kozasa et al. (1984) and Gail and Sedlmayr (1984), who addressed the problem of dust driven winds for M-type stars and C-type stars, respectively.

Before presenting the self consistent theoretical formulation of the problem, let us have a look on the observational facts providing a general characterization of such objects (e.g. Knapp et al., 1982, 1986; Rowan-Robinson and Harris, 1983):
- typical luminosities $10^3...10^5$ $L_\odot$
-    "    stellar masses $0.7...2$ $M_\odot$
-    "    mass loss rates $10^{-7}...10^{-4}$ $M_\odot yr^{-1}$
-    "    wind velocities $10...30$ km $s^{-1}$
-    "    mass of the shells $0.1...0.6$ $M_\odot$
- wind phenomenon lasts for at least $10^4$ years, no abrupt ejection of material.


## 2. The basic equations

For giants and supergiants the relevant timescales which describe the evolution of the atmosphere generally are large compared to the characteristic timescales which control dust formation and growth and their thermodynamic and hydrodynamic feedback to the atmosphere. Therefore, we restrict our discussion to the stationary case. Further we assume a spherically symmetric outflow which beyond a certain distance from the star becomes cool enough for effective dust formation to occur. This assumption of spherical symmetry is not trivial in view of the nucleation process which is extremely sensitive to the temperature. Thus even small temperature fluctuations might have a large influence on the local nucleation rate and on the local atmospheric structure (cloud formation?).


### 2.1 The model equations

The hydrodynamical equations for the flow are given by the equation of mass conservation

$$\dot{M} = 4\pi r^2 \rho v \tag{1}$$

and the equation of momentum conservation

$$v \frac{\partial v}{\partial r} = - \frac{1}{\rho} \frac{\partial p}{\partial r} - \frac{GM_*}{r^2} (1 - \alpha) \tag{2}$$

with mass density $\rho$, hydrodynamical velocity $v$, thermodynamical pressure $p$ and radiative acceleration $\alpha$ (normated to gravitation). It is assumed that the total mass contained in the wind is negligibly small compared to the stellar mass $M_*$. Due to the assumption of stationarity, the mass loss rate $\dot{M}$ is a constant during the considered evolutionary state of the star.

By the equation of state

$$p = nkT \tag{3}$$

the gas pressure p is expressed as function of the total number density n and the gas kinetic temperature T. In the considered case

$$n = n_H + n_{H_2} + n_{He} \tag{4}$$

is a reliable approximation.

The temperature structure is determined by adopting radiative equilibrium in the shell. For optical thin winds (this condition is fulfilled for all objects discussed in the present paper) T(r) is given analytically according to Lucy (1976)

$$T^4(r) = \frac{1}{2} T_*^4 \frac{\kappa_J}{\kappa_B} \{ 1 - [ 1 - (\frac{R_*}{r})^2 ]^{1/2} + \frac{3}{2} \tau_L \} \tag{5}$$

with effective temperature $T_*$, stellar radius $R_*$ and the "diluted" optical depth

$$\tau_L = - \int_r^\infty dr' \, \kappa_H(r') \, (\frac{R_*}{r'})^2 \quad . \tag{6}$$

$\kappa_J$, $\kappa_B$ and $\kappa_H$ are the intensity-, the Planck- and the flux-mean of the monochromatic extinction coefficient $\kappa_\nu$, respectively. With $T_*$ and $R_*$ also the stellar luminosity

$$L_* = 4\pi R_*^2 \sigma T_*^4 \tag{7}$$

is given by the Stefan-Boltzmann law.

The external parameters of the problem are $M_*$, $T_*$, $L_*$ and $\dot{M}$. However, in a consistent treatment the optical depth $\tau_L$ is not only subject to the boundary condition

$$\tau_L(\infty) = 0 \tag{8}$$

but also to the additional condition

$$\tau_L(R_*) = \frac{2}{3} (2 \frac{\kappa_B}{\kappa_J} - 1) \tag{9}$$

in order to guarantee

$$T(R_*) = T_* \quad . \tag{10}$$

Hence, one of the above parameters is determined by an eigenvalue problem. For this reason, the problem is completely determined by specification of only **three** independent stellar parameters.


## 2.2 The equations for the dust


For describing dust formation and growth we adopt a frame of reference which is moving with the flow. Then the system of moment equations which describes these processes in a sufficient approximation reduces to

$$\frac{dk_o}{dr} = \frac{j_*^S}{v} \qquad \text{(cluster formation)} \qquad\qquad (11a)$$

and

$$\frac{dk_i}{dr} = \frac{i}{v_\tau} k_{i-1} \qquad i=1,2,3 \qquad \text{(grain growth)} \qquad (11b)$$

with $j_*^S = vr^2 J_*^S$ and $k_i = vr^2 K_i$ where $J_*^S$ is the stationary nucleation rate and

$$K_i = \int_1^\infty dN\; N^{i/3} f(t,r,N) \qquad\qquad (12)$$

the i-th moment of the local particle distribution function $f(t,r,N)$ describing the number density of clusters consisting of N monomers (Gail and Sedlmayr, 1987c).

The assumption of a stationary nucleation rate is justified by the fact that in situations not too far from equilibrium the relaxation time is small compared to those timescales which determine the hydrodynamical evolution of the shell. Yet this is not true for the growth process which always has to be treated as a time dependent problem (c.f. Gail and Sedlmayr, 1987c).

The treatment of grain formation and growth is of fundamental importance within the context of our problem. In the last decade, considerable progress has been achieved in this field based on the pioneering work of the Cornell group (Salpeter et al., Draine) and the Kyoto group (Hasegasa, Yamamoto, Kozasa, Seki).

All approaches are based either on the thermodynamical description of the gas-solid phase transition by classical nucleation theory or on a detailed discussion of the relevant chemical reactions leading finally to critical clusters (e.g. review by Gail, Sedlmayr, 1987d). We will refrain from a presentation of these various approaches but only list the basic molecules from which the primary condensates are likely to be formed:

   M-stars: SiO, MgS, Fe

   C-stars: $C_2H_2$.

All calculations presented in the following chapters are restricted to dust formation in C-stars and are performed under the assumption of chemical equilibrium for the molecular reactions among the various chemical elements. The element abundances have to be specified as additional external parameters. For dust driven winds the amount of condensed material determines the velocity field and in particular the terminal velocity of the wind. For this reason, the abundances of the dust forming elements and the terminal velocity of the wind are coupled very closely. For the case of M-stars similar calculations have been performed by Kozasa et al. (1984) considering formation of $MgSiO_3$-grains.

## 3. The structure of a dust driven wind

In order to demonstrate the typical results of a selfconsistent calculation, the detailed wind model of an arbitrarily selected star with the parameters $M_* = 1\ M_\odot$, $T_* = 2000$ K, $L_* = 3.3 \cdot 10^4\ L_\odot$ and a carbon/oxygen abundance ratio $\varepsilon_C/\varepsilon_O = 2$ (and cosmic abundances otherwise) is depicted. The gas extinction coefficients used for the radiative transfer calculations are adopted from Scholz and Tsuji (1984).

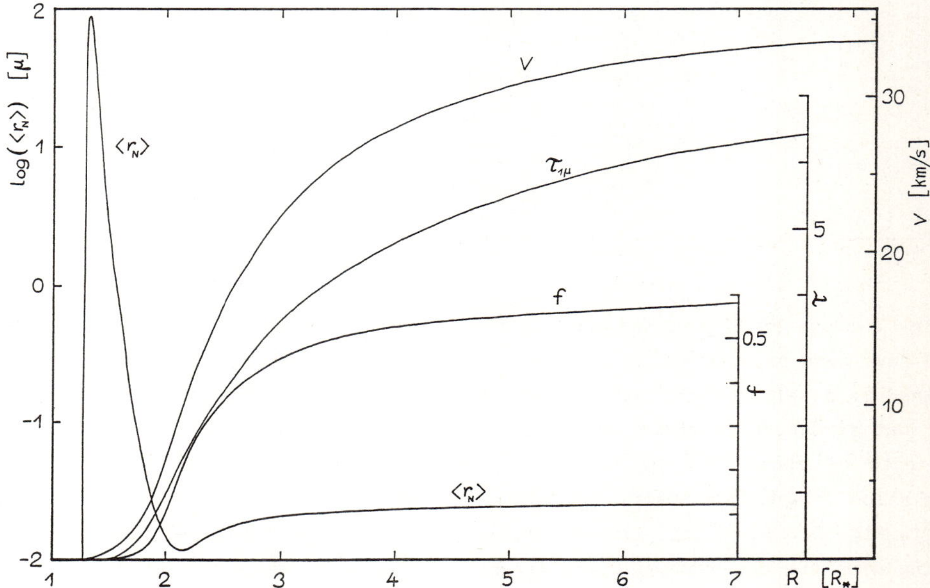

Fig. 1: Structure of a dust driven wind: mean grain radius $\langle r_N \rangle$, hydrodynamical velocity v, optical depth $\tau_{1\mu}$ of the dust shell at $\lambda = 1\mu$, degree of condensation f.

The eigenvalue problem is solved for $\dot{M}$ which results to be $3.5 \cdot 10^{-5}\ M_\odot yr^{-1}$. Fig. 1 shows the structure of the dust shell as it has been discussed in former papers (Gail, Sedlmayr, 1985, 1987a). We only want to point out two facts:
- hydrodynamics and dust formation are coupled so strongly that a consistent treatment is necessary,
- the mean grain radius is very small. Only very few big particles are formed in the early phase of the condensation process.


## 4. The influence of the model parameters

Our calculations allow to give an overview over the dust driven wind solutions in the case of C-stars and their dependence on the four parameters of the system of equations.

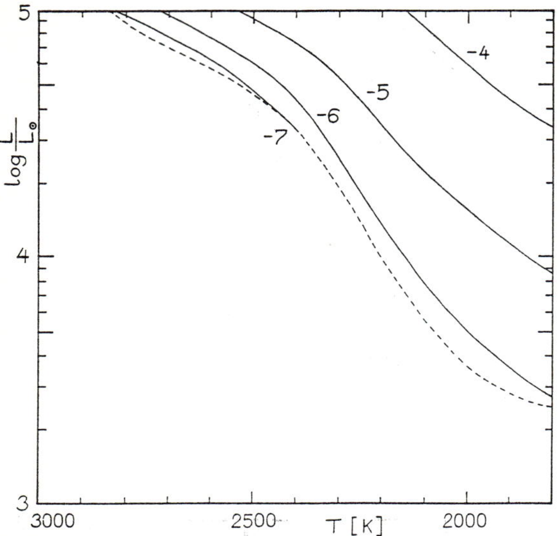


Fig. 2:
HR-diagram with lines of constant mass loss rate. The numbers indicate log $\dot{M}$ ($M_\odot yr^{1}$). Broken line: limit line for dust driven winds.

Fig. 2 shows an HR-diagram with contour lines of constant mass loss rates. The lines have been calculated for $M_* = 1\ M_\odot$ and $\varepsilon_C/\varepsilon_O = 2$. The dust driven winds are confined to a well defined area in the upper right corner of the HRD. Since effective dust formation requires sufficient high densities, there exists a minimum mass loss rate of about $10^{-7}\ M_\odot yr^{-1}$ (Gail, Sedlmayr, 1987b). Because of the obvious correlation between luminosity and mass loss, this defines a limit line (the broken line in Fig. 2). Below this line no dust driven winds are possible. This fact is in keeping with the observation that C-stars with extended envelopes always have mass loss rates above $10^{-7}\ M_\odot yr$ (Knapp and Morris, 1985).

As can be seen from Fig. 2, the mass loss rate increases strongly with increasing luminosity and decreasing temperature.

Similar HR-diagrams can be constructed for other quantities. The terminal velocity ranges from 10 km s$^{-1}$ to about 40 km s$^{-1}$. This again is in good agreement with observations by Knapp and Morris (1985) who yield velocities from 8 km s$^{-1}$ up to 30 km s$^{-1}$. v∞ increases with increasing luminosity.

The degree of condensation, i.e. the fraction of condensable material actually condensed into grains, is always greater than 10 % and increases with decreasing temperature. The mean value resulting for the dust to gas ratio by mass in the wind is $2.8 \cdot 10^{-3}$, which is in excellent agreement with the value given by Knapp (1985b).

The mean grain radius results to be rather small. In our calculations it ranges from 0.005 μ up to 0.05 μ which is significantly smaller than the size interval from 0.1 μ up to 10 μ which is usually adopted. The size of the grains increases with decreasing effective temperature of the star.

## 4.2 The influence of $M_*$

Variation of the stellar mass mainly affects the mass loss rate. For fixed $T_*$, $L_*$, $\varepsilon_C/\varepsilon_O$ the mass loss rate increases almost exponentially with decreasing mass. This effect, combined with the temperature and luminosity dependencies, strongly suggests that dust formation might be one possible driving mechanism for the so called **superwind** postulated for the AGB-PN transition.

Condensation degree, mean grain radius and the terminal velocity show only weak dependencies on the stellar mass.


## 4.3 The influence of $\varepsilon_C/\varepsilon_O$

Variations of the overabundance of carbon relative to oxygen mainly affects the terminal velocity. For fixed $T_*$, $L_*$, $M_*$ these two quantities are almost in linear correlation. This fact allows to determine $\varepsilon_C/\varepsilon_O$ by observing $v_\infty$. To fit the observed velocities ($v_\infty < 30$ km s$^{-1}$), $\varepsilon_C/\varepsilon_O$ has to be smaller than about 2.

An increase of $\varepsilon_C/\varepsilon_O$ also leads to a decreasing mean radius of the dust grains caused by the greater production rate of small clusters and the short growth time. The mass loss rate and the condensation degree show only very week dependencies on $\varepsilon_C/\varepsilon_O$.

The results of this chapter will be discussed in detail in a forthcoming paper.



## References

Deguchi, S.: 1980, Astrophys. J. <u>236</u>, 567
Gail, H.-P., Keller, R., Sedlmayr, E.: 1984, Astron. Astrophys. <u>133</u>, 320
Gail, H.-P., Sedlmayr, E.: 1987a, Astron. Astrophys. <u>171</u>, 197
Gail, H.-P., Sedlmayr, E.: 1987b, Astron. Astrophys. <u>177</u>, 186
Gail, H.-P., Sedlmayr, E.: 1987c, in preparation
Gail, H.-P., Sedlmayr, E.: 1987d, In "Physical Processes in Interstellar Clouds",
   eds. G.E. Morfil, M. Scholer, NATO ASI Series
Jura, M.: 1983, Astrophys. J. <u>267</u>, 647
Knapp, G.R., Phillips, T.G., Leighton, R.B., Lo, K.Y., Wannier, P.G.: 1982,
   Astrophys. J. <u>252</u>, 616
Knapp, G.R., Morris, M.: 1985a, Astrophys. J. <u>292</u>, 640
Knapp, G.R.: 1985b, Astrophys. J. <u>293</u>, 273
Knapp, G.R.: 1986, Astrophys. J. <u>311</u>, 731
Kozasa, T., Hasegawa, H., Seki, J.: 1984, Astrophys. Space Sci. <u>98</u>, 61
Kwok, S.: 1975, Astrophys. J. <u>198</u> 583
Lucy, L.B.: 1976, Astrophys. J. <u>198</u>, 583
Rowan-Robinson, M., Harris, S.: 1983, Mon. Not. R. astron. Soc. <u>202</u>, 797
Salpeter, E.E.: 1974a, Astrophys. J. <u>193</u>, 579
Scholz, M., Tsuji, T.: 1984, Astron. Astrophys. <u>130</u>, 11
Wickramasinghe, N.C.: 1972, in "Interstellar Matter", eds. N.C. Wickramasinghe,
   F.D. Kahn, P.G. Mezger, Geneva Observ., Sauverny

WINDS FROM RED GIANTS

Yu. A. Fadeyev
Astronomical Council of the USSR Academy
of Sciences, Pyatnitskaya 48 Moscow 109017 USSR


Abstract


Two principal mechanisms that may be responsible for mass loss from red giants
are considered: shock wave-driven winds and radiatively (dust)-driven winds.
Effect of the periodic shocks accompanying nonlinear oscillations of red giants
is most prominent in the outer layers of the stellar atmosphere where shocks are
able not only to expel gas but also increase gas density so that some molecular
components become supersaturated. In O-rich stars the most abundant condensible
species are silicon monoxide and iron, whereas in C-rich stars these are carbon,
silicon carbide and iron.


1. Introduction


For nearly 30 years since Deutch (1956) interpreted the displacement of narrow
absorption lines in the spectrum of $\alpha$ Her as a result of gas outflow, the red
giants are known to lose their mass due to cool stellar winds. These winds are
characterized by large mass fluxes and at the same time by relatively low
velocities. According to Knapp and Morris (1985) the rates of mass loss from red
giants are as high as a few$\times 10^{-4}$ $M_\odot$/yr, whereas the terminal wind velocities do not
exceed 20 km/s (Knapp 1985). At present four mechanisms are thought to drive
stellar winds from cool low-surface gravity stars. These are (1) thermally driven
winds; (2) Alfven wave driven winds; (3) shock-driven winds; and (4) radiatively
driven winds. However in the present paper I will discuss only two of these
mechanisms: shock-driven and radiatively driven winds. This is due to the two
following reasons. First, almost all red giants reveal the light variability
because of stellar pulsations. The nonlinear pulsations of low-surface gravity
stars are known to be accompanied by periodic shock waves that may expel gas from
the outer stellar atmosphere. Second, almost all red giants are surrounded by
circumstellar dust shells (Habing 1987). Therefore some role in mass loss may
belong to momentum transfer from dust grains accelerated by stellar radiation to
gas molecules. In order to estimate the contribution of such a wind to total mass
loss one should consider in detail the dust formation process. In recent years
some authors succeeded in application of the homogeneous nucleation theory to the
problem of mass loss from cool stars.

## 2. Shock-driven winds

The presence of shock waves in atmospheres of pulsating stars is due to the existence of a critical frequency $\sigma_{cr}$ for standing wave oscillations. For example, in the plane-parallel isothermal atmosphere oscillations exist in the form of standing wave if the dimensionless frequency

$$\sigma = \frac{2\pi}{P}\left(\frac{R_{ph}^{3}}{3G\,M}\right)^{1/2} \tag{1}$$

satisfies the following condition (Unno 1965):

$$\sigma^{2} < \sigma_{cr}^{2} = (1 + 3C^{2})/(12C) - 4/3 , \tag{2}$$

where P is the pulsation period, G is the gravitation constant, M is the mass of the star, $C = H_{p}/R_{ph}$ is the ratio of the pressure scale height to the radius of photosphere. In other words, the critical frequency derives the upper frequency limit for perfect reflection of the pulsation wave at the stellar photosphere. Substituting the appropriate values of mass, radius and period in (1) and (2) one can readily find that standing wave transforms into progressive running waves for $C > 0.01$. So, even in the framework of the linear theory the presence of running waves in the atmospheres of pulsating red giants seems to be inevitable since they are characterized by large values of C ranged from a few percent to of about ten percent.

There are two most important consequences of the presence of periodic shock waves in the stellar atmospheres:

1) Periodic compression waves of finite amplitude enhance the atmospheric scale height (Whitney 1956). In pulsating low-surface gravity stars the dynamic scale height may by an order of magnitude exceed the static scale height (Fadeyev 1984; Willson and Bowen 1986). The distension of the stellar atmosphere by periodic shocks is very crucial for dust formation since in the hydrodynamically equilibrium stellar atmosphere condensation is impossible due to extremely low gas density. The tight interrelation between the shocks and the existence of circumstellar dust has been recently corroborated by a correlation between the infrared excesses and pulsation periods of red giants (De Gioia-Eastwood et al. 1981; Jura 1986).

2) Periodic shocks are able to expel gas from the outer stellar atmosphere. Let us assume that motion of the gas shell of initial radius $R_{o}$ and initial velocity $v_{o}$ is governed by gravitation only. Then the time needed for the shell to return to its origin is (Willson and Hill 1979):

$$P_{grt} = 2R_{o}\zeta/v_{e} , \tag{3}$$

where

$$\zeta = \frac{\beta}{1-\beta^{2}} + \frac{1}{(1-\beta^{2})^{3/2}}\tan^{-1}\left(\frac{\beta^{2}}{1-\beta^{2}}\right)^{1/2} , \tag{4}$$

$\beta = v_{o}/v_{e}$, $v_{e}$ is the escape velocity. The radius of the shell averaged over the pulsation period increases with time if the shell is not able to return to the level of its origin before the passage of the next shock, that is for $P_{grt}>P$. For the sake of convenience, relation (3) can be rewritten in the terms of the pulsation constant Q and relative radius $x=R/R_{ph}$:

$$Q(GM_\odot/2R_\odot^3)^{1/2} = x_o^{3/2}\zeta \ . \tag{5}$$

For the fixed value of the initial velocity $v_o$ which is interpreted as the postshock velocity, relation (5) determines the radius of the layer $R_o$ above of which the gravitation return time $P_{grt}$ exceeds the pulsation period. The advantage of the model by Willson and Hill (1979) is that far away from the star the mass loss is characterized by a steady flow rather than shock events.

However the existing estimates of the shock-driven mass loss rates are still uncertain due to their strong dependence on the cooling rates. According to Wood (1979) the mass loss rate may be ranged from $10^{-12}$ $M_\odot/yr$ (isothermal shock) to $10^{-2}$ $M_\odot/yr$ (adiabatic shock). More certain estimates of mass loss rates might be derived from the models that take into account the interplay between the shock front propagating through the extended atmosphere, radiation field and atomic and molecular processes. Unfortunately no such a model has been constructed yet.

The most drastic effect on the losses of the thermal energy is due to dissociation of molecular hydrogen. According to Fox and Wood (1985) as much as a half of the thermal energy behind the shock front is absorbed due to dissociation of $H_2$ molecules. At the same time photodissociation of $H_2$ molecules in the precursor causes retardation of the collisional ionization in the relaxation zone, whereas the precursor structure is very sensitive to the radiation flowing from the wake (Gillet and Lafon 1983; 1984). So, the self-consistent model of the radiative shock is urgently needed.

## 3. Dust-driven winds

Calculations of molecular equilibrium for gas with solar-like composition show that various molecular compounds condense in solid state when gas temperature drops below of 1000K-1500K (see e.g. Wood 1963; Lord 1965; Gilman 1969; Lattimer et al. 1978). However we should be interested only in a few condensible species that are characterized by both highest evaporation temperature and highest abundance. As was shown by Gail and Sedlmayr (1986) the most important condensible species are formed out a few most abundant elements that can be divided into the three following groups: (1) hydrogen; (2) carbon, nitrogen, oxygen; (3) magnesium, silicon, sulphur, iron. Within each of these groups the elements are characterized by nearly comparable abundance. Other elements, such as natrium, aluminium, calcium etc., can be neglected due to their lower abundances.

A sequence of condensates formed in the stellar atmosphere can be calculated provided that gas density and gas temperature T are known as a function of the radius R. Let us assume that gas outflow is described by a spherically-symmetric steady-state continuity equation:

$$\dot{M} = 4\pi R^2 \rho v \,, \tag{6}$$

and gas temperature is $T = T_{eff} W^{1/4}$, where $\dot{M}$ is the mass loss rate, $T_{eff}$ is the stellar effective temperature, $W$ is the radiation dilution factor. Then for a given gas velocity $v$ we can estimate the radius $R_{sat}$ of a layer where the supersaturation ratio of a certain condensible species is $S=1$. This layer presents an inner boundary of the condensation region for a given molecular component since the condition of supersaturation $S>1$ fulfills at radii $R > R_{sat}$.

Table 1 and Table 2 give the radial distance $x$ expressed in units of $R_{ph}$ and gas temperature at the saturation level of the most abundant condensible species in O-rich and C-rich stars with $M_{bol}=-5$ mag, respectively. It is assumed that gas velocity is equal to the local escape velocity, that is approximately is in the range from 10 km/s to 20 km/s. Fortunately both the saturation radius $R_{sat}$ and gas temperature $T_{sat}$ only slightly depend on the expansion velocity. For example, the change in the ratio $v/v_e$ from 0.5 to 2.0 corresponds to the change in $R_{sat}$ and $T_{sat}$ less than 5%.

Table 1. Properties of saturation levels at [C]/[O]=0.5

| $T_{eff}$ | $\log \dot{M}$ | MgO | | Fe | | MgS | | SiO | |
|---|---|---|---|---|---|---|---|---|---|
| | | x | T | x | T | x | T | x | T |
| 3000 | -3 | 3.26 | 1182 | 4.04 | 1059 | 5.46 | 910 | 6.38 | 841 |
| | -5 | 3.87 | 1083 | 5.01 | 950 | 6.49 | 834 | 7.78 | 761 |
| | -7 | 4.53 | 1000 | 6.07 | 862 | 7.71 | 765 | 9.31 | 696 |
| 3500 | -3 | 4.41 | 1182 | 5.48 | 1060 | 7.42 | 910 | 8.66 | 841 |
| | -5 | 5.24 | 1083 | 6.80 | 950 | 8.82 | 834 | 10.58 | 761 |
| | -7 | 6.15 | 1000 | 8.25 | 862 | 10.48 | 765 | 12.66 | 696 |

Table 2. Properties of saturation levels at [C]/[O]=2.0

| $T_{eff}$ | $\log \dot{M}$ | C | | SiC | | Fe | |
|---|---|---|---|---|---|---|---|
| | | x | T | x | T | x | T |
| 3000 | -3 | 1.53 | 1771 | 2.15 | 1469 | 4.04 | 1059 |
| | -5 | 1.73 | 1653 | 2.55 | 1343 | 5.01 | 950 |
| | -7 | 1.99 | 1530 | 3.00 | 1235 | 6.07 | 862 |
| 3500 | -3 | 2.02 | 1772 | 2.88 | 1470 | 5.48 | 1060 |
| | -5 | 2.30 | 1654 | 3.43 | 1343 | 6.80 | 950 |
| | -7 | 2.66 | 2530 | 4.05 | 1235 | 8.25 | 863 |

In O-rich stars almost all atoms of carbon are tied in molecules of carbon monoxide CO so that the condensible species are mainly refractory oxides and atomic iron. The existence of circumstellar silicate grains is compelling due to the Si-O stretching spectral feature observed in either emission or absorption at 9.7 $\mu$m, and the weaker O-Si-O bending feature observed always in emission at 18.8 $\mu$m. The existence of MgS dust particles in O-rich stars is questionable because this material may be rapidly destroyed via reactions with OH and $H_2O$ molecules (Nuth et al. 1985).

In C-rich stars almost all atoms of oxygen are blocked in CO molecules, whereas the large fraction of carbon is tied in molecules $C_2H$ and $C_2H_2$. The rest of carbon exists in the form of free atoms and molecules of silicon carbide SiC. Both carbon and silicon carbide are the most abundant condensible species in C-rich stars. The presence of SiC circumstellar grains is compelling due to the 11.3 $\mu$m spectral feature identified firstly by Treffers and Cohen (1974).

Dust formation is an irreversible, non-equilibrium process so that in order to derive the final radii and number densities of dust grains one should numerically integrate the differential equations describing the mass exchange between gas phase and solid state. It is such an approach that was used by Fix(1969), Woodrow and Auman(1982), Fadeyev (1983; 1987), Gail et al. (1984) Gail and Sedlmayr (1985; 1986; 1987) for calculation of dust formation in circumstellar environments. All these studies are based on the classical theory of homogeneous nucleation (see e.g. Feder et al. 1966; Abraham 1974 for reviews) and the assumption of spherical dust grains. Results of these calculations can be summarized in a following way.

When the supersaturation ratio S becomes greater than unit, the small liquid droplets (i.e. molecular clusters) commence to appear. Almost all the droplets are immediately destroyed due to evaporation and only small fraction of the droplets (critical clusters) with radii greater than a critical radius $r_*$ have a chance to survive and grow by accretion of vapor molecules (monomers) onto their surface. It is assumed that macroscopic thermodynamics is applied to the critical clusters that are considered as liquid droplets containing the large number of monomers, that is $n_* \gg 1$. The number of the critical clusters formed per unit time per unit volume is the nucleation rate J so that the number density of dust grains is $N_d = \int J dt$. Expressions for calculation of the nucleation rate and other quantities can be found in the review paper by Draine (1981).

At the beginning of phase transition the nucleation rate quickly increases with time until growth of dust grains perceptibly depletes vapor. After that both the supersaturation ratio and the nucleation rate drop and phase transition is due to growth of dust grains. The existence of the maximum of the nucleation rate is the result of the competition between decreasing gas temperature and increasing mean collision time of monomers $t_{col}$. The interplay between these two effects accompanying expansion of gas can be described by a dimensionless parameter $\Lambda = t_{sat}/t_{col}$ where $t_{sat}$ is supersaturation ratio e-folding time. Clearly, phase transition negligibly departs from thermal equilibrium if $\Lambda \gg 1$. In this case the problem of dust formation is simplified since it can be reduced to the solution of algebraic equations (Yamamoto and Hasegawa 1977; Draine and Salpeter 1977). Unfortunately the condition $\Lambda \gg 1$ usually does not fulfill in circumstellar flows so that these approaches should be used with caution.

According to Fadeyev and Henning (1987) the final radii of SiO grains formed in O-rich stars with $M_{bol} = -5$ mag and $2500 < T_{eff} < 3000$K relate to mass flux as

$$\log r = \begin{Bmatrix} 4.38 \\ 4.45 \end{Bmatrix} + \begin{Bmatrix} 0.78 \\ 0.70 \end{Bmatrix} \log \dot{M} , \qquad\qquad (7)$$

where $r$ and $\dot{M}$ are expressed in angstroms and $M_\odot$. The fitting coefficients were estimated for the surface tension energy $\sigma=500$ erg/cm$^2$ and 750 erg/cm$^2$, respectively.

The studies of carbon dust formation in C-rich stars have been recently done by Gail et al. (1984) and Gail and Sedlmayr (1985; 1987). They found that the final radii of solid carbon particles are in the range from 130A to 670A for mass loss rate ranged from $10^{-6}$ $M_\odot$/yr to $2 \times 10^{-4}$ $M_\odot$/yr. The existence of large dust grains is due to the fact that grain growth proceeds not only by accretion of free carbon atoms and $C_2$, $C_3$ molecules via the following exchange reactions:

$$C_n + C_2H_2 = C_{n+2} + H_2 , \qquad\qquad (8)$$

$$C_n + C_2H \quad\quad = C_{n+2} + H , \qquad\qquad (9)$$

where n is the number of carbon atoms in the dust grain.

The strict solution for the problem of the resistance to the motion of a small sphere moving through gas has been obtained by Baines et al. (1965). They considered both specular and diffuse reflection of the molecules at the surface of the sphere mass of which is large in comparison with the mean mass of gas molecules and the radius to be small compared with the mean free path of gas molecules. All these assumptions are applicable for circumstellar outflows. Fadeyev and Henning (1987) used these solutions for calculation of momentum transfer from silicate dust grains to gas molecules in cool O-rich red giants with $M_{bol}=-5$ mag and found that at the radial distances less than 10 $R_{ph}$ the drift velocity of dust grains does not exceed 0.1 km/s. Gail and Sedlmayr (1986) studied dust formation in gas outflows from O-rich red giants with $L=10^4$ $L_\odot$, $T_{eff}=2500K$ and $M=10^{-5}$ $M_\odot$/yr. They found that at radial distances less than $32R_{ph}$ the gas expansion velocity does not exceed 5 km/s. This is due to the small final radii of dust grains that are ranged from 10A to 20A. In C-rich stars the terminal velocities are ranged from 8 km/s to 30 km/s for the mass loss rates ranged from $10^{-6}$ $M_\odot$/yr to $2 \times 10^{-4}$ M /yr, respectively.


References


Abraham F.F. 1974, Homogeneous Nucleation. Academic Press, New York.

Baines M.J., Williams I.P., Asebiomo A.S. 1965, Mon.Not.R.astr.Soc., **130**, 63.

De Gioia-Eastwood K., Hackwell J.A., Grasdalen G.L., Gehrz R.D. 1981, Astrophys.J.(Letters), **245**, L75.

Deutch A.J. 1956, Astrophys.J., **123**, 210.

Draine B.T. 1981, in Physical Processes in Red Giants, ed. I.Iben and A.Renzini, D.Reidel, Dordrecht, p. 317.

Draine B.T., Salpeter E.E. 1977, J.Chem.Phys., **67**, 2230.

Fadeyev Yu.A. 1983, Astrophys.Space Sci., **95**, 357.

Fadeyev Yu.A. 1984, Astrophys.Space Sci., **100**, 329.

Fadeyev Yu.A. 1987, in Circumstellar Matter, ed. I.Appenzeller and C.Jordan, D.Reidel, Dordrecht, p. 515.

Fadeyev Yu.A., Henning T. 1987, in preparation.

Feder J., Russel K.C., Lothe J., Pound G.M. 1966, Adv.Phys., **15**, 11.

Fix J.D. 1969, Mon.Not.R.astr.Soc., **146**, 37.

Fox M.W., Wood P.R. 1985, Astrophys.J., **279**, 455.

Gail H.-P., Keller R., Sedlmayr E. 1984, Astron.Astrophys., **133**, 320.

Gail H.-P., Sedlmayr E. 1985, Astron.Astrophys., **148**, 183.

Gail H.-P., Sedlmayr E. 1986, Astron.Astrophys., **166**, 225.

Gail H.-P., Sedlmayr E. 1987, Astron.Astrophys., **171**, 197.

Gillet D., Lafon J.-P.J. 1983, Astron.Astrophys., **128**, 53.

Gillet D., Lafon J.-P.J. 1984, Astron.Astrophys., **139**, 401.

Gilman R.C. 1969, Astrophys.J.(Letters), **155**, L185.

Habing H.J. 1987, in Circumstellar Matter, ed. I.Appenzeller and C.Jordan, D.Reidel, Dordrecht, p. 197.

Jura M. 1986, Irish Astron.J., **17**, 322.

Knapp G.R. 1985, Astrophys.J., **293**, 273.

Knapp G.R., Morris M. 1985, Astrophys.J., **292**, 640.

Lattimer J.M., Schramm D.N., Grossman L. 1978, Astrophys.J., **219**, 230.

Lord H.C. 1965, Icarus, **4**, 279.

Nuth J.A., Donn B. 1982, Astrophys.J.(Letters), **257**, L103.

Nuth J.A., Donn B. 1982, J.Chem.Phys., **77**, 2639.

Nuth J.A., Donn B. 1983, J.Chem.Phys., **78**, 1618.

Nuth J.A., Moseley S.H., Silverberg R.F., Goebel J.H., Moore W.J. 1985, Astrophys.J.(Letters), **290**, L41.

Treffers R., Cohen M. 1974, Astrophys.., **188**, 545.

Unno W. 1965, Publ.Astron.Soc.Japan, **17**, 205.

Whitney C. 1956, Ann.d'Astrophys., **19**, 34.

Willson L.A., Bowen G.H. 1985, Irish Astron.J., **17**, 249.

Willson L.A., Hill S.J. 1979, Astrophys.J., **228**, 854.

Wood J.A. 1963, Icarus, 2, 152.

Wood P.R. 1979, Astrophys.J., **227**, 220.

Wooddrow J.E.J., Auman J.R. 1982, Astrophys.J., **256**, 247.

Yamamoto T., Hasegawa H. 1977, Prog.Theor.Phys., **58**, 816.

# DISCOVERY OF A REFLECTION DUST ENVELOPE AROUND IRC+10216


M. Tamura[1]

T. Hasegawa[2], N. Ukita[2]

I. Gatley[3], I. S. McLean[3], M. G. Burton[3]

J. T. Rayner[4], and M. J. McCaughrean[4]

[1]Department of Physics, Kyoto University, Japan.
[2]Nobeyama Radio Observatory, Tokyo Astronomical Observatory, University of Tokyo, Japan.
[3]United Kingdom Infrared Telescope Unit, Hawaii.
[4]Department of Astronomy, University of Edinburgh, U.K..


Infared polarimetric and photometric mapping observations at K(2.2 $\mu$m) and H(1.65 $\mu$m) have revealed an extended dust envelope around the late-type star IRC+10216.  The observations were made on the 3.8-m United Kingdom Infrared Telescope on Mauna Kea, Hawaii, in 1985 December and 1987 January and February.  The polarization observations were made by emplying the Kyoto polarimeter (Sato et al. 1987).  Great care was taken to check the contamination by stray light in the telescope and instruments as the source on peak was extremely bright (K $\sim$ 0 mag).  From the observations of normal stars, we found that the polarized intensity (degree of polarization times the intensity) was a good measure of the envelope, free from contamination by stray light, although the intensity and the degree of polarization suffered from the contamination separately.

Figure 1 shows the polarization vectors at K superposed on the surface brightness map and Figure 2 shows the radial distribution of the polarized intensity.  The infrared envelope is roughly circular symmetric with radial extent greater than 1 arcmin ( $\sim$ 0.09 pc), comparable to the size of the molecular gas envelope (Bieging, Chapman, and Welch 1984, Wilson, Schwartz, and Epstein 1973, Wannier <u>et al.</u> 1979, Wootten <u>et al.</u> 1982).  The polarization vectors show a clear centro-symmetric pattern, indicating that the infrared nebulosity is due to scattering by dust grains in the envelope.  The radial distribution of dust grains derived from Figure 2 is most likely to obey an inverse-square law, suggesting a steady mass loss between r ~15" and 60".

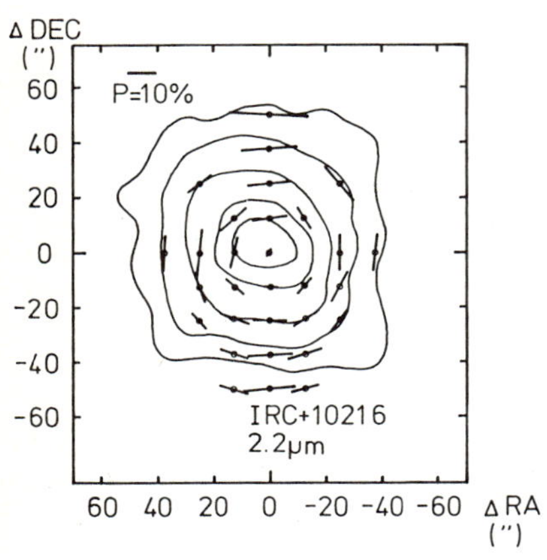

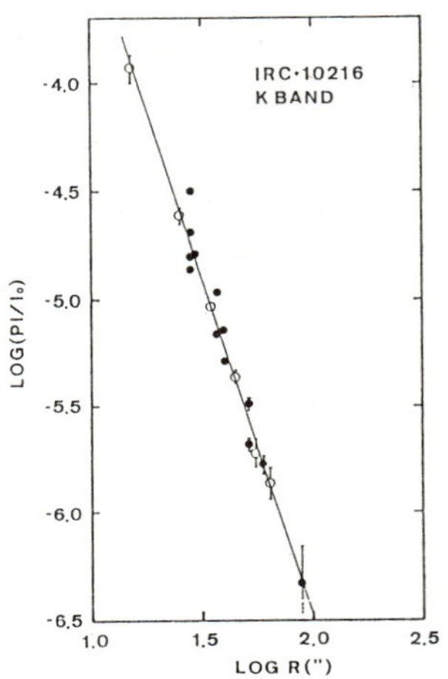

Figure 1: Surface brightness and polarization map of IRC+10216 in the K band, with the beam size of 20" and 12", respectively. Filled and open circles in the middle of the polarization vectors indicate positions of polarimetry made in 1987 January and February, respectively.

Figure 2: Logarithmic radial distribution of the polarized intensity of the envelope of IRC+10216 in the K band. Open and closed circles are data taken with different photometers on different days. A least-square-fitted line is shown by a thin line.

REFERENCES


Bieging, J.H., Chapman, B., and Welch, W.J.
      1984, Ap. J., **285**, 656.
Sato, S. <u>et al.</u> 1987, M.N.R.A.S., in press.
Wannier, P.G., Leighton, R.B., Knapp, G.R., Redman, R.O.,
      Phillips, T.G., and Huggins, P.J. 1979, Ap. J., **230**, 149.
Wilson, W.J., Schwartz, P.R., and Epstein, E.E.
      1973, Ap. J., **183**, 871.
Wootten, A., Lichten, S.M., Sahai, R., and Wannier, P.G.
      1982, Ap. J., **257**, 151.

# Distance Determination of Mass Losing Carbon Stars from CO and HCN Radio Observations


Wasaburo Unno[1], Takashi Tsuji[2], Kou-ichi Koyama[1], Hideyuki Izumiura[3], Nobuharu Ukita[2], and Norio Kaifu[2]

[1] Kinki University, Higashi-Osaka, Osaka 577.  [2] Tokyo Astronomical Observatory, Mitaka, Tokyo 181

[3] Department of Astronomy, University of Tokyo, Tokyo 113



**Abstract**   A method of distance determination of evolved carbon stars is proposed and applied to six stars for which the terminal velocity of expansion and the angular size are obtained from the CO and HCN radio observations made at the Nobeyama Radio Observatory.  The method assumes the radiation driven wind in spherical geometry.  Within those small samples, however, two types of mass losing stars that are systematically different in the antenna temperature ratio $T_A(CO)/T_A(HCN) \gtrsim 1$, in the velocity ratio $v(CO)/v(HCN) \lesssim 1$, and in the angular size ratio $\theta(CO)/\theta(HCN) \gtrsim 1$ seem to be present, suggesting nonspherical geometry of the outflow.


## §1. Distance Determination of Evolved Carbon Stars

In astrophysics, little can be understood without the knowledge of the distance.  The determination of mass-loss rate from evolved carbon stars depends also on the distance determination.  For red giant stars especially for pulsating variables, the distance determination may be fairly reliable.  However, about a half of the mass-losing carbon stars that are mainly IRC and CRL objects are infrared objects, for which the usual distance determination is not feasible except through the kinematic or similar rough assumptions, e.g., $L = 10^4 L_\odot$ .  The principle for distance determination is based on the radiation driven stellar wind solution for which the velocity $v$ is given by,

$$v^2 = \frac{\eta}{2\pi} \frac{\kappa L}{c r_c}, \qquad (r \gg r_c), \tag{1}$$

where $\eta$ denotes the efficiency factor taking care of the change in the dust opacity $\kappa$( at $r \geq r_c$), $L$ the luminosity, and $r_c$ the sonic point radius assumed to be equal to the dust formation radius.  Observations give $v$, $F_{bol}$ ( $= L/4\pi d^2$), and the angular size $\theta$ ( the source radius $r_t = \theta d$), where $d$ denotes the distance.  We have approximately $(r_t/r_c)^3 = (T_d/T_t)^4$ from the radiation transfer, where $T_d(\sim 10^3 \text{ K})$ and $T_t(\sim 10 \text{ K})$ denote the dust formation temperature and the molecular shell temperature at $r_t$.  Then, we obtain

$$d = d_0 \frac{\theta v^2}{F} \quad \text{where} \quad d_0 = \frac{1}{2\eta} \frac{c}{\kappa} \left(\frac{T_t}{T_d}\right)^{4/3}. \tag{2}$$

Fig. 1 shows just one example of the five point observation (each separated by 15" from the neighboring point) of HCN (J = 1–0). The terminal velocity $v$ is read directly from such tracings. The angular size $\theta$ is obtained from the deconvolution of the Gaussian fit of the observed peak intensities with the antenna beam profile assumed to be also Gaussian. Two $\theta$ values are obtained from the N-S and E-W comparisons and they are averaged. The discrepancy between this and the true $\theta$ corresponding to the edge may be absorbed in the efficiency factor $\eta$. Table 1 shows the result for $d_0 = 4.0 \times 10^7 \mathrm{g}\ \mathrm{cm}^{-1}\mathrm{s}^{-1}$ The value of $d$ is apparently too large for IRC+00365, and for CRL1922. Futher improvement of the method is now in progress.

Table 1.

| Name | $m_{bol}$ | line | $T_A$(K) | $v(\frac{km}{s})$ | $\theta_{NS}$" | $\theta_{EW}$" | $\theta$" | $d$(kpc) | Type |
|---|---|---|---|---|---|---|---|---|---|
| IRC+00365 | +5.31 | CO | 0.20 | 35.2 | — | — | 12.7 | 48 | I |
| | | HCN | 0.55 | 37.3 | 8.9 | 7.2 | 9.0 | 38 | |
| IRC+10216 | +0.51 | CO | 4.53 | 16.6 | 19.9 | 22.6 | 21.3 | 0.22 | I |
| | | HCN | 9.00 | 15.4 | 24.4 | 22.6 | 23.5 | 0.16 | |
| IRC+10401 | +4.51 | CO | 0.11 | 29.4 | — | 13.5 | 13.5 | 0.71 | I |
| | | HCN | 0.27 | 28.5 | — | 8.1 | 8.1 | 0.90 | |
| IRC+40540 | +4.39 | CO | 0.72 | 14.2 | 18.8 | 17.6 | 18.2 | 0.48 | II |
| | | HCN | 0.75 | 18.5 | 6.7 | 7.6 | 7.2 | 0.32 | |
| IRC+50096 | +4.24 | CO | 0.42 | 16.1 | 17.2 | 11.8 | 14.6 | 4.5 | II |
| | | HCN | 0.34 | 17.2 | 20.3 | 3.4 | 13.7 | 4.8 | |
| CRL 1922 | +5.95 | CO | 0.47 | 18.7 | 17.2 | 13.5 | 15.7 | 31 | II |
| | | HCN | 0.33 | 20.4 | 4.2 | 4.2 | 4.2 | 9.6 | |

The antenna temperature $T_A$ listed above is for the 45 m radio telescope having beam size $17 \pm 1$", aperture efficiency $26 \pm 3\%$ and beam efficiency $45 \pm 5\%$ at 115 GHz and $24 \pm 1$", $37 \pm 5$ %, and $68 \pm 5$ % at 86 GHz.

**§2. Are There Two Types of Mass Losing Carbon Stars?** Three stars IRC+00365, IRC+10216, and IRC+10216 belonging to Type I given in the last column of Table 1 show $T_A(\mathrm{CO})/T_A(\mathrm{HCN}) \leq 1$, $v(\mathrm{CO})/v(\mathrm{HCN}) \geq 1$ but $\theta(\mathrm{CO})/\theta(\mathrm{HCN}) \leq 1$, and the other three stars IRC+40540, IRC+50096, and CRL 1922 belonging to Type II show the opposite tendency. Such strange tendency (cf. Nguyen-Q-Rieu et al., 1987) is difficult to understand if the spherically symmetric geometry should be assumed. Further study is required.

**Reference**

Nguyen-Q-Rieu, N. Epchtein, Truong-Bach and M. Cohen, Astron.Astrophys. **180**,117,1987

# A Sensitive Line Search in Circumstellar Envelopes

Nguyen-Q-Rieu [1] , S. Deguchi, H. Izumiura
N. Kaifu, M. Ohishi, H. Suzuki, and N. Ukita

Nobeyama Radio Observatory, Minamimaki, Minamisaku,
Nagano 384-13, JAPAN

[1] Present Address, Observatoire de Paris-Meudon

__Abstract__    A molecular line search in the range between 85 and 89 GHz has been performed in the circumstellar envelopes of 11 evolved stars. Emissions of $^{29}SiO$ J=2-1, $^{28}SiO$ J=2-1, HCN J=1-0, $H^{13}CN$ J=1-0, $HC_5 N$ J=33-32, $HCO^+$ J=1-0 transitions and other transitions of $C_2 H$, $C_4 H$, and $C_3 N$ have been observed in 11 stars. We have detected the ground state $^{29}SiO$ J=2-1 maser in several stars. We have also detected HCN emission in VY CMa. A narrow $H^{13}CN$ spike feature near the central velocity has been found in the spectrum of CRL 2688.

__Introduction__    An increasing number of molecular transitions have been detected  in the circumstellar envelopes of  evolved objects  through recent sensitive  searches  with large  radio telescope. Some progress has been made toward the understanding of  chemical processes  at work  in the circumstellar  medium, based on observational results and no-LTE chemical calculations. Using the  VLA  and  Hat Creek  interferometers,  Nguyen-Q-Rieu et al. (1987) and  Bieging and Nguyen-Q-Rieu (1988)  found that the $NH_3$  and HCN shells of CRL 2688  are troid  while $HC_7 N$  is distributed in a spheroidal halo. The molecule HCN which was be-lieved to exist  only in carbon-rich atmosphere  has been found in oxygen-rich envelopes (Deguchi and Goldsmith 1985).   Cyano-polyyne  and  hydrocarbon have been detected   in carbon stars (Saito et al. 1987). Above considerations suggest that the enve-lope of cool stars is rich in physical phenomena  and prompt us to perform a search for molecular emission in stellar envelopes.

<u>Observations and Results</u>          Observations were made in March
and May 1987 using the 45m radio telescope at Nobeyama. We
search for molecular transitions in the range between 85.00 and
89.25 GHz in sample of 11 envelopes known to be rich in carbon
and oxygen. An acousto-optical spectrometer (AOS) with 8 arrays
and 2048 channels each was used. We have detected many lines in
a sample of 11 stars. The HCN molecules have been discoverd in
VY CMa, an oxygen-rich star, in which carbon was supposed to be
bound as CO and HCN was not expected. The abundance of HCN in
the envelope of VY CMa is calculated to be $6 \times 10^{-9}$ per H2. We
have found a spike feature in the $H^{13}CN$ spectrum in CRL 2688.
This source, a bipolar reflection nebula known as Egg Nebula,
has been observed at infrared and radio frequencies. The origin
of the spike feature at 29 Km s$^{-1}$ in the HCN spectrum is not
clear and no counterpart appears in the $H^{12}CN$ spectrum. Plausible is that it is an anomalous excitation.

References


Bieging, J. H., and Nguyen-Q-Rieu, 1988, Ap. J. (in press)
Deguchi, S., and Goldsmith, P. F., Nature **317**, 336
Nguyen-Q-Rieu et al., 1986, A. Ap. **165**, 204
Saito, S. et al., 1987, Publ. A. S. Japan **39**, 193


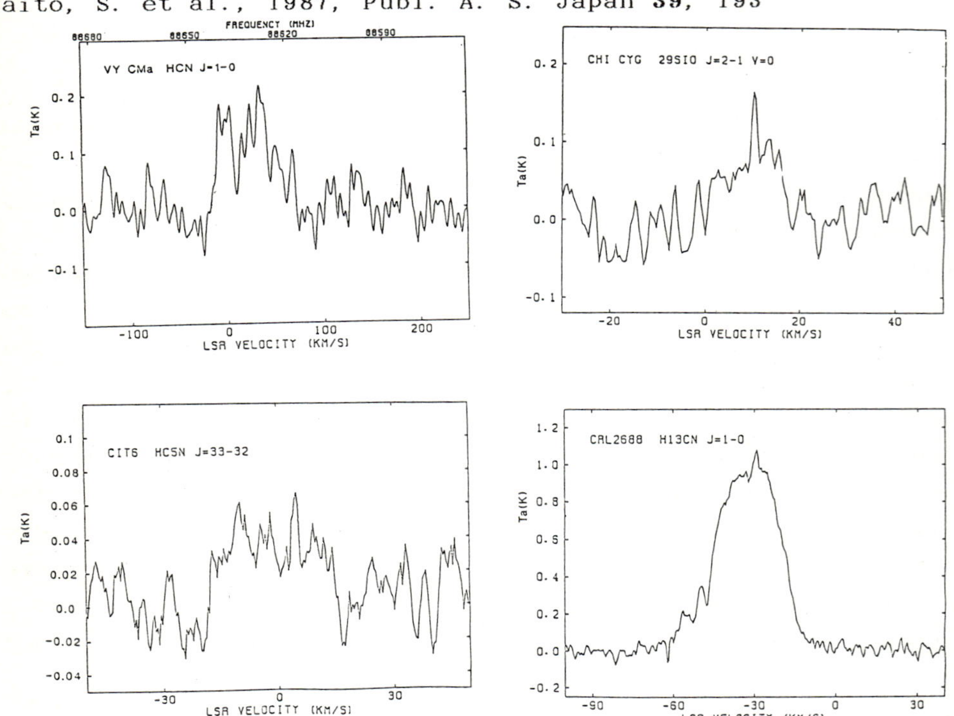

186

PHOTOSPHERES OF MIRA VARIABLES

M. S. Bessell[1], J. M. Brett[1], M. Scholz[2], Y. Takeda[2],
R. Wehrse[2], P. R. Wood[1]
[1]Mount Stromlo and Siding Spring Observatories, Institute of Advanced
Studies, Australian National University
[2]Institut für Theoretische Astrophysik, Universität Heidelberg, F.R.G.

Mira variables form an important subgroup of the red giant stars which
are typical representatives of stars showing burnt material at their
surfaces. Since the photospheres of Miras are not in hydrostatic equilibrium but are characterized by spherically very extended density stratifications, their properties and emitted spectra differ substantially
from those of non-Miras, and any attempts to analyse Mira spectra by
means of conventional techniques must fail. Non-hydrostatic models are
needed for analysis work.

We computed a small set of exploratory M type (solar abundances) Mira
model photospheres whose density distributions were taken from an improved modification of a pulsation model proposed by Wood (1979). Besides the fundamental model parameters, i.e. mass, luminosity and radius
(or effective temperature), the parameters of the density stratification
(e.g. the heights of the density discontinuities at the shock front positions, or the effective gravity acting between two successive shock
fronts) must be treated as free parameters within certain limits, owing
to insufficient knowledge of the velocity stratification entering the
pulsation model. Typical density and temperature stratifications are
shown in Fig.1 of Scholz (1987). Both CO lines, measuring the outflow
and infall velocities of matter between the shock fronts, and selected
molecular band features (colors) were found to react sensitively to ad-
justments of the model parameters. As an example, Fig.1 shows the differences between CO line profiles computed from the near-maximum model
of Fig.1 of Scholz (1987) (Fig.1a) and those computed from a model in
which the effective gravity between the shock fronts was increased by a
factor of 1.5 (Fig.1b). The same velocity stratification was adopted in
both cases. In addition to spectral features,the wavelength dependence
of monochromatic radii proves to be a powerful tool of diagnostics of
Mira photospheres (Scholz and Takeda 1987; Bessell et al. 1987).

Our exploratory models also predict remarkable differences between the
spectra of Miras and non-Miras, in good agreement with observations. In

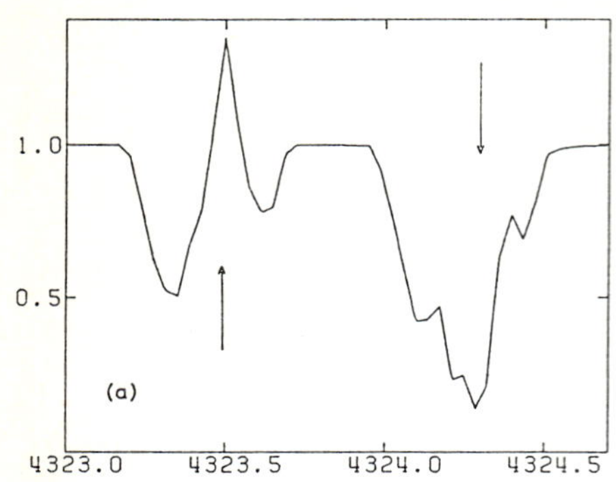
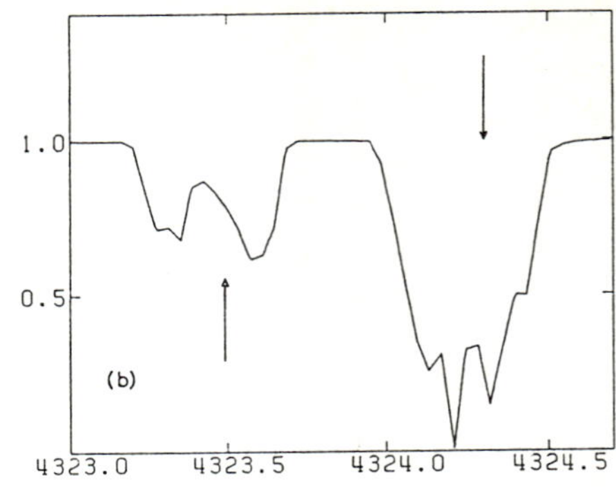


Fig.1  Profiles of two sample absorption lines of the first overtone CO band emerging from two slightly different Mira model photospheres (see text). The line at 4323.5 $cm^{-1}$ has an excitation potential of 1.55 eV and lg gf = −4.92. The line at 4324.3 $cm^{-1}$ has 0.10 eV and −5.69. The unshifted absorption profiles whose centers are indicated by arrows are Voigt profiles with zero microturbulence.

particular, the strengths of TiO bands used for spectral type classification may deviate substantially, and $H_2O$ bands are systematically stronger in Mira spectra. Consequently, Miras and non-Miras which have identical fundamental parameters may have drastically different colors, and color indices which are to measure effective temperatures of both Miras and non-Miras must be selected scrupulously. For instance, the computed J-K versus K-L two color diagram in which J-K approximately measures the temperature and K-L is strongly affected by $H_2O$ absorption in the L filter, shows a clear separation between Mira and non-Mira models in accordance with observations. Most Miras are cooler than non-Miras and, in contrast to non-Miras, their infrared blackbody temperatures are good approximations to the effective temperatures. (See Bessell et al. 1987 for a detailed discussion.)

REFERENCES

Bessell, M. S., Brett, J. M., Scholz, M., Wood, P. R. 1987, to be published
Scholz, M. 1987, IAU Symp. No. 122, p. 385
Scholz, M., Takeda, Y. 1987, Astron. Astrophys., in press
Wood, P. R. 1979, Astrophys. J. 227, 220

THE $\underline{V}$-STOKES PARAMETER AS A MANIFESTATION OF ENVELOPE ACTIVITY
FOR COOL, BRIGHT, EVOLVED STARS

B. D. Holenstein and R. H. Koch
Department of Astronomy and Astrophysics
University of Pennsylvania

R. J. Pfeiffer
Department of Physics
Trenton State College

At Pennsylvania's Flower and Cook Observatory, instrumentation has
been developed to measure simultaneously the four Stokes parameters of
the filtered radiation field from a celestial source.  The instrumental
$\underline{Q}/\underline{U}/\underline{V}$-parameters have been found to be very small and well-behaved.
Thus far, the program has concentrated on cool bright giants and
supergiants and on hot, evolving close binaries.  A single season's
investigation of Alp Ori has already been reported (Holenstein 1987)
and the present paper is a summary of current results for the cool,
evolved program stars.

For  Psi$^1$ Aur, V CVn, 6 Gem, 72 Leo and 119 Tau no $\underline{V}$-signal at the
level of $3\sigma$ has been detected from data from the 1986-1987 season.  At
the level of 0.0n%, unambiguous and variable $\underline{V}$-signals have been
detected for VV Cep, Mu Cep, Alp Her, Alp Ori, Bet Peg, and Alp Sco.
For these, we make a number of recognitions: (1) for no star has there
been an enduring correlation among $\underline{Q}$ or $\underline{U}$ and $\underline{V}$, (2) for all stars as a
group the absolute value of $\underline{V}$ is not correlated with the value of $\underline{p}$,
(3) no star has remained at $\underline{V}$ = 0.00% for an extended (>50 days)
interval although an interstellar component of $\underline{V}$ could be present in
the data, (4) for red through blue, polarity changes of $\underline{V}$ have been
common and both senses of change appear equally common, and (5) time
scales for significant changes of $\underline{V}$ range from about 2 days to about
100 days.  Information from the $\underline{p}$- and $\underline{V}$-polarization spectra for the
same six stars may be summarized: (6) for a given star the linear
spectrum is variable changing from monotonic to non-monotonic and back
again and with a continuous range of gradient, (7) for a given star the
$\underline{V}$-spectrum is variable, and (8) the most conspicuous detail of the $\underline{V}$-
spectrum is the negative ultraviolet circular polarization for five of
the stars.

Items (1) and (2) may be interpreted to indicate that the linear
and circular signals arise in different locales or that conversion
efficiencies vary with time because the scatterers themselves vary with

time.  Item (3) signifies that the concentrations of scatterers may
diffuse to an insignificant density but that some other mechanism
concentrates them again quickly or that other scattering centers form
quickly.  It may also be that an asymmetric distribution of secondary
scatterers becomes symmetrical for a brief time.  It also follows that
all linear surveys have underestimated the polarization of cool
supergiants by the neglect of the V-component.  Item (4) may indicate
that more than one type of birefringent scatterer exists in a stellar
envelope or that there are twisted alignments of them which can change
with time.  The time scale range in Item (5) is shorter than the
fundamental pulsational time scales for these stars, but light time
effects in the binary stars and mass motions in the M-star envelopes
are permitted.  Item (6) is familiar from older linear surveys and has
been ascribed to a time-variable assortment of scatterers as modelled
by Shawl (1975).  It may be considered that Item (7) is a necessary
consequence of Items (1), (2), and (6).  Item (8) may be a consequence
of observational selection from a small number of stars or may reveal
something of the birefringence of the scattering medium.

The most evident understanding of most of these results would rest
upon the explanation developed by Angel and Martin (1973) for the near-
IR V-signal seen from a few, very cool objects: multiple scattering in
an asymmetric, dusty envelope.  For the five stars other than VV Cep,
the electric vector rotates significantly.  Therefore, by the criterion
of Angel and Martin, elongated grain alignment is unnecessary to cause
their circular signals.

IRAS data indicate that for most of the stars there does not
appear to be an emission peak near $10\mu m$.  On the other hand, merged,
low resolution IUE spectra clearly show the $0.22\mu m$ dip for VV Cep and
Alp Sco and show it weakly for Mu Cep and Alp Ori.  The spectra for VV
Cep and Alp Sco have been fitted successfully using Seaton's (1979)
reddening law, when the hot companions are taken into account.
Therefore, it appears their circumstellar dust envelopes do not inflect
the $0.22\mu m$ feature.

Support from NAG 468 is gratefully acknowledged.

Reference List


Angel, J. R. P., and Martin, P. G. 1973, Ap. J. <u>180</u>, L39.
Holenstein, B. D. 1987, Bull. A. A. S. <u>19</u>, 754.
Seaton, M. J. 1979, M. N. R. A. S. <u>187</u>, 73P.
Shawl, S. J. 1975, A. J. <u>80</u>, 595.

# THE MIXING LENGTH RATIO, EDDY DIFFUSIVITY, AND ACOUSTIC WAVES


Kwing L. Chan
Applied Research Corp.
Landover, Maryland, U.S.A.

and

Sabatino Sofia
Center fo Solar and Space Research, Yale University
New Haven, Connecticut, U.S.A.


Many processes in the convection zone of a star affect the
evolution and the atmospheric diagnostics.  Here, a progress report is
given on our numerical study of some convection related phenomena.
The numerical results are obtained by solving the Navier Stokes
equations in a three dimensional rectangular domain.  The units are
chosen such that the initial temperature, pressure, density, and the
depth of the domain are all normalized to 1.

## A. The mixing length ratio

The mixing length theory relates the convective (enthalpy) flux
$F_C$ to the envelope structure quite well (Chan and Sofia 1987).  For
efficient convection that occurs in deep convective regions, the
numerical results are compatible with a mixing length ratio of 2.1.
The mixing length theory fails to address the significance of the flux
of kinetic energy $F_{KE}$ (see Figure 1).  $F_{KE}$ is negative and has a
magnitude comparable to the total flux.  These results are
qualitatively similar to those of two dimensional computations
(Hurlburt et. al. 1984).

## B. Diffusive action of the convective turbulence

The convective turbulence tend to dissipate large scale shears
(wave length $\lambda$ > pressure scale height H ).  The rate of dissipation
is approximately proportional to $\lambda^{-2}$, and the effective kinematic
viscosity turns out to be about $\frac{1}{4}$ rms($V_z$) H ($\pm50\%$) where rms($V_z$) is
the root-mean-square vertical velocity.  The convective turbulence
also diffuses large scale temperature perturbations.  The effective
diffusivity is approximately $\frac{1}{3}$ rms($V_z$) H.  The temperature
perturbation excites acoustic oscillations right away (see next
section).

## C. Acoustic waves in the convection zone

Finite-amplitude oscillations co-exist with the convective
turbulence.  The frequencies of the modes are in agreement with the
acoustic frequencies obtained by eigenvalue analysis (see Figure 2).


This research is partially supported by the NSF (AST-8504399).

**REFERENCES**

Chan, K. L. and Sofia, S. 1987, SCIENCE, **235**, 465.
Hurlburt, N. E., Toomre, J., and Massaguer, J. M. 1984, Ap. J., **282**, 557.

Figure 1.

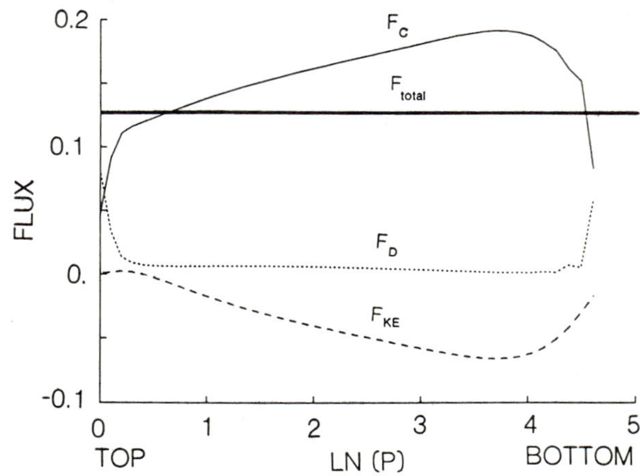


An example of the distributions of fluxes versus depth, ln(pressure). The diffusive flux (dotted curve) is very small except near boundaries. The flux of kinetic energy (dashed curve) is downward and its magnitude reaches about 50% of the total flux (thick solid curve). This makes the enthalpy flux (thin solid curve) nonuniform.


Figure 2.

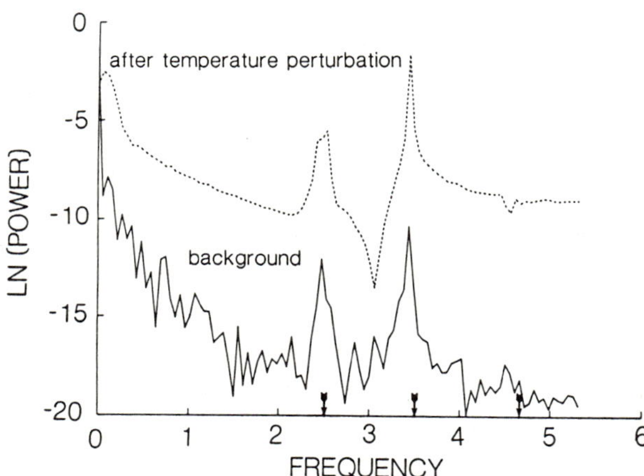


Frequency spectra showing peaks at acoustic frequencies predicted analytically (arrows). The dotted curve is for the case with temperature perturbation, and the solid curve is for the background state.

INTERMITTENT TRANSITION TO IRREGULAR PULSATION
WITH MASS LOSS IN HYDRODYNAMIC MODELS

Toshiki Aikawa

Faculty of Engineering, Tohoku-Gakuin University,
Tagajo 985, Japan

## Introduction

Hydrodynamic simulations of non-linear pulsation for less-massive
cooler supergiants have been perfoemed by several authors (Tuchman,
Sack and Barkat, 1979; Fadeyev and Tutukov, 1981; Fadeyev, 1982, 1984;
Nakata, 1987; Buchler et al., 1987).  The outburst of large amplitude
pulsation at times is one of common features of these models, and
renders mass-loss from the atmosphere of pulsating stars by generating
strong shock waves.

## Models and results

To find out routes of the transition from limit cycles to this
type of irregular pulsation, we performed hydrodynamic simulations for
a series of models the luminosity $\log(L/L_{\odot}) = 3.505$, and $Te = 5300$ K with
a narrow range of the mass, $1.4\ M_{\odot} \leq M \leq 1.5\ M_{\odot}$ by using the hydrodynamic
code, TGRID (Simon and Aikawa, 1986).

With decreasing the mass, we confirm a transition from limit
cycles to the irregular oscillations.  The models which show the
irregularity keep stationary oscillations with the amplitude of the
corresponding limit cycle of massive models for a while (the laminar
phase), but gradually get kinetic energies of pulsation and move
towards oscillations with much larger amplitudes, and then finally
dissipate the kinetic energies by generation of strong shock waves,
causing an outburst of irregular oscillations.

## Analysis

The nature of the transition is finally specified by examining
the dissipation of the kinetic energies in stable models, as the
pulsation starts with larger amplitudes than their limiting amplitudes.
We find that the rates of the dissipation are so small that the
pulsation with these amplitudes might be marginally stable.

Furthermore, the oscillation starting with even larger amplitudes
gets the kinetic energies until it reaches a limit where the oscillation induces strong shock waves and dissipates its kinetic energies
(see Fig. 1).  Thus we conclude that the model which has the stable
limit cycle near the transition has another unstable fixed point with
a larger amplitude.  The transition therefore is induced by the

disappearance of these two fixed points, as the mass, the control
parameter in our case, is varied, and exactly fits the intermittency
of Pomeau and Manneville (1981) proposed as an universal route to
chaos in dissipative systems.

It is suggested that the outbursts of the irregular oscillations
with mass-loss often observed in hydrodynamic models of less-massive
supergiants may be in consequence of this intermittent transition.

<u>References</u>


Buchler, J.R., Goupil, M.-J. and Kavács, G.: 1987, preprint.
Fadeyev, Y.A.: 1982, Astrophys. and Space Sci., 86, 143.
Fadeyev, Y.A.: 1984, Astrophys. and Space Sci., 100, 329.
Fadeyev, Y.A. and Tutukov, A.: 1981, Mon. Not. R. Astr. Soc., 195, 811.
Tuchman, Y., Sack, N. and Barkat, Z.: 1979, Astrophys. J., 234, 217.
Nakata, M.: 1987, Astrophys. and Space Sci., 132, 337.
Pomeau, Y. and Manneville, P.: 1981, Commun. Math. Phys., 74, 189.
Simon, N.R. and Aikawa, T.: 1986, Astrophys. J., 304, 249.


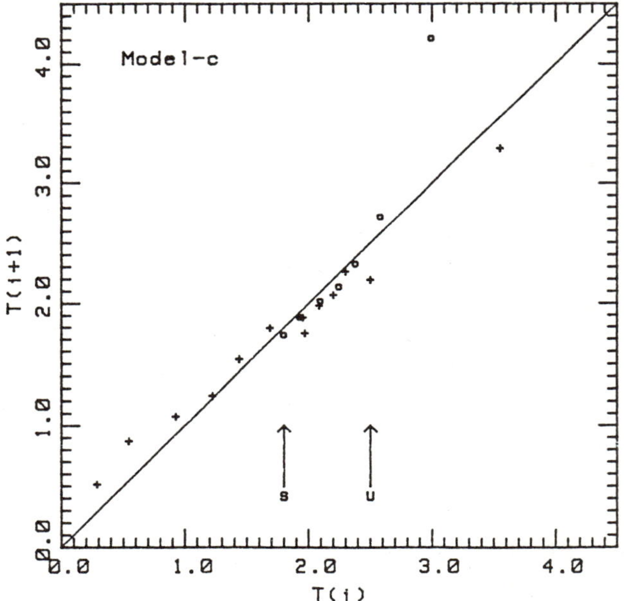


Fig.1 - Dissipation structure of a stable model. The resulting kinetic
energies T(i+1) after one cycle against the input kinetic energies T(i)
are plotted. If T(i+1) < T(i), the input oscillation is damped, and if
T(i+1) > T(i), the oscillation is excited. The model has a stable
limit cycle indicated by an arrow with "s". The rates of dissipation
for oscillation with larger amplitudes than the limit cycle are so
small that the oscillations might be marginally stable. Furthermore,
for even larger amplitudes, the model has another unstable fixed point
indicated by an arrow with "u". These are characteristics of stable
models near the transition.

# A Model Oscillator of Irregular Stellar Variability

Mine Takeuti
Astronomical Institute, Tohoku University, Sendai 980, Japan
and
Yasuo Tanaka
Faculty of Education, Ibaraki University, Mito 310, Japan

Stellar pulsation is one of the condidates for strong mass loss from red giant stars.  Recent investigations have shown sporadic outbursts of the pulsation can eject a considerable mass from the stars.  Such a sporadic increase of the amplitudes seems to have a connection with the irregularity found in the pulsations.  Recently the iregular propreties in stellar pulsation are investigated in the relation to nonlinear dynamics(see for instance Perdang, 1985).  Unfortunately no single model oscillator of a star of which the equilibrium state is dynamically stable had been found.  In the present paper, we shall discuss a simple oscillator which shows period-doubling and chaotic, that is, irregular oscillations.

We consider a model oscillator as

$$dx/dt = y, \quad dy/dt = Ax+z+my,$$
$$dz/dt = -Bdx/dt-pz-qy+syz \ , \tag{1}$$

where x, y and z are the displacement from an equilibrium state, the velocity and additional forces defined by the thierd equation, respectively.  The coefficients, A, B, m, p, q and s are constants. When  m, p, q and s equal zero, the system of the equations shows an adiabatic oscillation for  $B > A$.  We restrict ourselves for the case of dynamical stability as  $B-A+q-mp > 0$.

When the system is pulsationally unstable and in secular stabitlity, we can have the period two limit cycles and succeeding period-doubling for the change of p.  The values of the parameters are, for instance, as follows:
 i)    A=-0.5, B=0.5, m=0.5, q=0.5, s=1, p=3,
 ii)   A=4, B=5, m=-0.5, q=-0.7, s=5, p=-0.06,

where the values of p are given at the period two limit cycles.  For
the case of i) where it is convenient to compare with the Rossler
equation (1966), we have the period-doubling as the increase of p.

For ii), the system of equations is rewritten in the equivalent
form as follows:

$$dx/dt = y, \quad dy/dt = Ax+z,$$
$$dz/dt = -By+ax+bz+(n_0+n_1z+n_2y)y, \tag{2}$$

which corresponds to the equations derived from the one-zone model for
stellar pulsation (Baker, 1966).  We find that the limit cycles of
equation (2) corresponding to that of equation (1) appear and chaos
occurs again for the change of b.

The detailed results will be published in Tanaka and Takeuti
(1987).  The results show that the system of equations with nonlinear
nonadiabatic terms behaves as a simple oscillator which produces
irregular and chaotic oscillations.


References


N.H. Baker, 1966, in "Stellar Evolution", ed. R.F. Stein and A.G.F.
     Cameron, (Plenum Press), 333.
J.Perdang, 1985, in "Chaos in Astrophysics", ed. J.R.Buchler, J.M.
     Perdang, and E.A. Spiegel, (D. Reidel Publ.), 11.
O.E.Rossler, 1976, Phys. Letter 57A,397.
Y. Tanaka and M. Takeuti, 1987, in preparation.

# The Structure of a Stationary Atmosphere with a Heat Source or Sink

Kyoji Nariai

*Tokyo Astronomical Observatory, The University of Tokyo,
Mitaka, Tokyo 181, Japan*

There are several methods for solving a structure of a stellar atmosphere in radiative equilibrium. Among them, the matrix method (Nariai and Shigeyama 1984, Nariai and Ito 1985) uses only the energy equation at the nodes; therefore, it is easy to extend this method to an atmosphere that is not in radiative equilibrium. The matrix method gives direct solutions for the cases of a gray model or a picket-fence model, and it is so effective that the calculation converges in 3 or 4 iterations in the case of a non-gray atmosphere with scattering even if the starting model is isothermal. In order to simplify the problem, in this paper, however, we will limit ourselves to the case of a gray atmosphere with pure absorption.

A plane-parallel stellar atmosphere is a semi-infinite medium. In the numerical calculation, we divide it into $n$ finite elements and 1 semi-infinite element. Let us define a node as a point between two elements. Node 0 is defined as the boundary between the surface element and the vacuum. In total, we have $n+1$ nodes. The distribution of any physical quantity is represented by a vector of $n+1$ dimensions with its values at the $n+1$ nodes as elements. The mean intensity of radiation $J$ is written in the ordinary expression as

$$J(\tau) = \frac{1}{2}\int_0^\infty B(t)K(|t-\tau|)dt \tag{1}$$

Let the value of $B$ at node $i$ be represented by $b_i$. Then, we can integrate equation (1) analytically if we assume an appropriate interpolation formula within each element. Linear interpolation was adopted in Nariai and Shigeyama(1984) while third order polynomials were used in Nariai and Ito(1985). Results can be expressed in a form

$$\mathbf{j}=\mathbf{Ab}+\mathbf{y}, \tag{2}$$

where the constant vector $\mathbf{y}$ is mainly determined by the gradient term in the semi-infinite element and its element is expressed approximately as

$$y_i = 0.375 \, F \, K_2(\tau_n - \tau_i), \tag{3}$$

where $F$ is flux.

The equation of radiative equilibrium is written as b=j. When there is a heat source or sink, we add a term q to the right side of the equation so that

$$\mathbf{b} = \mathbf{j} + \mathbf{q}. \tag{4}$$

We can write the solution of the energy equation as

$$\mathbf{b} = (\mathbf{U} - \mathbf{A})^{-1}\mathbf{y} + (\mathbf{U} - \mathbf{A})^{-1}\mathbf{q}. \tag{5}$$

The first term is the solution of the equation of radiative equilibrium. The second term represents the effect of the perturbation by the source or sink layer in the atmosphere. When q represents a source term, the second term is positive at the surface, has a maximum near the center of the source region, and approaches a finite value with zero gradient at the bottom of the atmosphere $(\tau = \tau_n)$.

Nariai and Murata (1987) solved the structure of an atmosphere in a binary system by treating the radiation from the other component explicitly in the radiative equilibrium equation. However, it is possible to treat the same problem as an example with a source term which is generated by the decay of the direct radiation from the other component.

An atmosphere that has quasi-steady horizontal flow is another interesting example. As the flow exchanges energy with the surrounding medium through radiation, entropy along the flow line is not constant, therefore, q is finite. An atmosphere with a sink layer gives a very low temperature in the line-forming region. Therefore, analysis of such an atmosphere by means of normal atmospheric model may lead to false abundance values. The sunspot penumbra may be interpreted with the present model. If the assumption of the existence of a penumbra type atmosphere in some part of the atmosphere of Ap stars is justified, abundances for Ap stars will have to be revised completely. The existence of strong magnetic field in Ap stars may justify such an assumption.

References


Nariai,K. and Shigeyama,T. 1984, Publ.Astron.Soc.Japan,36,593.
Nariai,K. and Ito,M. 1985, Publ.Astron.Soc.Japan,37,553.
Nariai,K. and Murata,Y. 1987, Publ.Astron.Soc.Japan,39,163.

EVOLUTION DIAGNOSTICS IN INTERACTING BINARIES

Jorge Sahade
FCAG, UNLP, C.C. 677, 1900 La Plata
IAR, C.C. 5, 1894 Villa Elisa
Member of the Carrera del Investigador Científico, CONICET, Argentina

Through the analysis and interpretation of observational material, particularly on the part of Otto Struve and his collaborators, the structure of an interacting binary is depicted (cf. Sahade and Wood 1978) as formed by a) the two stellar components; b) a gaseous stream from the less massive and more evolved component of the system towards the companion; c) a circumstellar gaseous envelope —designated as "ring" or "disk" depending on the density of the material; d) a circumbinary gaseous envelope that surrounds the whole system and is normally in expansion, as suggested by the conventional, ground-based observations.

The evolution of interacting binaries involves the effect of matter outflow and transfer in at least a stage of rapid mass loss and a stage of slow mass loss. As a result, the evolution of the components appreciably departs from the evolution of single stars and produces very bizarre objects which find no counterparts among non-binaries. Both the observational and the computational results suggest that the amount of mass involved in the process of mass outflow must be a large percentage of the total mass of the evolving component. It seems, therefore, reasonable to expect to find evidence, in evolved systems, for processed material from the interior of the mass-losing component.

1. With such an idea in mind, attempts were made in the middle 60's, to use Strömgren's metallic index m' to prove metal abundances in the atmosphere of the mass-losing member of the pair. The results obtained by several investigators were interpreted as indicating underabundance of metals, a conclusion that was questioned by Baldwin (1976), by Hall (1969) and by Koch (1972) on several grounds. More recently, Parthasarathy _et al_. (1979) pointed out that "the photometric index used as an indicator of metal abundances measured the strength of the G-band at 4300 A. The weakness of this index ... may indicate that CH, which is a mayor contributor to the G band, is weak in secondaries. This would suggest an underabundance of C ... consistent with the effect of CNO processing that, as a result of mass loss, the secondaries are expected to evidence".

2. Almost a quarter of a century before the photometric attempts we have just mentioned were carried out, Greenstein (1940) had called attention to the fact that in the spectrum of the 137.94-day period binary $\cup$ Sagittarii (B2 by the high excitation lines; early A by the metallic lines; fainter companion smaller and of larger surface brightness) H is very weak and He is strong.

Ten years later another similar case, that of KS Persei = HD 30353, a non-eclipsing 360.47-day period binary (about A5, by Ca II-H and K; F5, by ionized metals), was brought to the limelight by Bidelman (1950), who stated that the object "appears to be a lower-temperature analogue of υ Sagittarii".

And, again ten years later, Boyarchuk (1960) announced that the spectrum of the famous, peculiar interacting binary β Lyrae (B8 II; P= 12.9 days) suggested H-deficiency.

To these objects, which are always mentioned whenever the question of abundance anomalies in interacting binaries comes up, LSS 4300, which, according to Schönberner and Drilling (1984), should be the hot counterpart of υ Sgr and of KS Per, may perhaps be added.

Table 1 lists the abundance determinations in β Lyr.

Table 1

ABUNDANCE DETERMINATIONS IN β LYRAE

| Author | | $He/H$ | $C/N$ | $O/N$ | Method |
|---|---|---|---|---|---|
| Boyarchuk | 1960 | ∿25 | | | curve of growth |
| Struve, Zebergs | 1961 | | | | comparison of E.W.'s |
| Hack, Job | 1965 | 1-2.25 | | | E.W. comp. α Cyg (A2 Iae) |
| Leushin et al. | 1977 | 1.55 | | | |
| Leushin et al. | 1979 | 1.5 | 0.6 | 20 | comp. with theoretical profiles |
| | | | 0.03 | 0.04 | |
| Leushin, Snezhko | 1980 | 1.5 | 0.25 | | with the use of |
| Bahýl | 1982 | 0.125 | | | Klinglesmith's |
| Balachandran et al. | 1986 | 1.5 | ≨0.011 | 0.025 | (1971) models |
| | | | ± .022 | ±.050 | |

Note: normal Balmer discontinuity

In the case of υ Sgr Hack and Pasinetti (1963) derived, from a comparison with γ Pegasi, a B2 IV (Hoffleit and Jaschek 1982), that He is 40 times more abundant than H. υ Sgr does not appear to show a Balmer discontinuity.

As for KS Per, which also does not display a Balmer discontinuity Danziger et al. (1967) and Wallerstein et al. (1967) compared fluxes and profiles with those suggested by Böhm-Vitense's H-poor model atmosphere calculations and derived a value of the ratio of H over He of the order of $10^{-4}$, and a high abundance of N relative to C and to O.

2.1 In regard to the three binaries that are considered as H-deficient objects, I would make the following comments:

2.1.1 In the first place, β Lyr's brighter component is a B8 II object, whose effective temperature is of the order of 11.500ºK (Böhm-Vitense 1981), and, although there is no doubt that its mass is smaller than that of the companion, by no means it is certain how large or how small the actual value is.

In the second place, the spectrum of β Lyr is a very peculiar one, characterized not only by photospheric absorption lines but also by emissions that are strong in H, and are also present in He, and by other features, which arise partly in the gaseous stream from the B8 II component and partly in the outer edges of the opaque disk that surrounds and hides the so far unobservable companion. As it has been shown by Batten and Sahade (1973), in the case of Hα , and by Aylin et al. (1987), in the case of the resonance lines of C IV, Si IV and N V in the IUE ultraviolet range, the line profiles can be interpreted in terms of a superposition of two profiles, one of them a broad, relatively faint emission that shifts back and forth throughout the orbital cycle. The behavior of the radial velocities from this feature suggests that it shares roughly the expected orbital motion of the companion to the B8 II component and, therefore, that it may arise in the optically thick disk that surrounds it.

In view of these considerations, is it at all reasonable to derive abundance anomalies from such a complex spectrum by using, as comparison, stars that appear to be normal and models that are valid for certain sets of physical and abundance parameters with no consideration of complicated and strongly emitting gaseous structure as one finds in  β Lyr?

As a consequence, I feel skeptical in regard to the conclusions thus far derived in regard to H-deficiency in the case of β  Lyr.

2.1.2 I am somewhat skeptical also in relation to the similar conclusion in the case of υ Sgr, because here again there is a gaseous structure in the systems which gives rise, at some phases, to strong Hα  emission and weak diluted He I, the latter having been discovered by Sahade and Albano (1970) nearly twenty years ago.

2.1.3 As for KS Per, which also displays Hα  in emission and weak H in absorption, at the moment I have nothing to comment upon, perhaps because I know less about this system. However, it might be appropriate to point out that Schönberner and Drilling (1984) have suggested that the degree of H-deficiency that is attributed to objects like KS Per imply that the same component is in its second episode of mass loss, that is, that the system is undergoing Delgrado and Thomas (1981) case BB of binary star evolution.

3. Let us now deal with the cases that do not display H-deficiency but have been announced as showing evidence of CNO-processed material in the system.

3.1 Rather recently, Parthasarathy *et al.* (1979, 1983) carried out a study of the secondaries of U Cephei (G8 III-IV) and U Sagittae (G3 III-IV) on spectra with a resolution of 1.9 A with 0.46 A per diode. The approach used was that of a spectrum synthesis, the comparison star for U Cep being $\kappa$ Geminorum (G8 III) and, for U Sge, $\delta$ Coronae Borealis (G3.5 III-IV) and the result was that, for both cases, $[Fe/H] = 0.0 \pm 0.3$.

The same authors (Parthasarathy *et al.* 1983) undertook also to examining the C and N abundances in the secondaries of the same two Algol systems, with application of the spetrum synthesis method. The result obtained suggests that in the two cases the late-type mass-losing components are carbon-deficient, $[C/Fe] \sim -0.5$, and nitrogen-rich, $[N/Fe] \sim +0.5$.

3.2 As it is well known and has been pointed out in a recent review paper (Sahade 1986), all the ultraviolet spectra of close binary systems – except Algols – are characterized by the presence of high temperature resonance doublets of C IV, Si IV and N V, normally in emission, the ions being listed in order of decreasing intensity.

In the Algols observed outside of eclipse, however, the resonance doublet line appear in absorption and, contrary to what we have described for the rest of the close binary systems, C IV is the weakest of the ions. This fact prompted Peters and Polidan (1984) to advance the idea that in the atmosphere of the more massive components in Algol systems, C is underabundant and N overabundant, in confirmation of Parathasarathy *et al*'s (1983) results.

It was interesting that a few Algols, namely, U Cep (Plavec 1983), RW Tauri (Plavec and Dobias 1983) and TT Hydrae, U Sge and UX Monocerotis (Plavec *et al.* 1984) were observed with the IUE satellite at principal eclipse. And on these images the resonance doublets of C IV, Si IV and N V are displayed in emission and, as far as intensities and sequence of intensities go, they behave like in the rest of the close binary systems, that is, C IV is the strongest feature of the three.

These observations, which are reminiscent of the behavior of H, particularly of Hα , in Algol systems in the photographic spectra (cf. for instance, Sahade and Wood 1978), and the emission intensity measurements made by Plavec (1983) at partial and total phases during the principal eclipse of U Cep, led McCluskey and Sahade (1987) and Sahade (1986) to suggest that the behavior of C IV outside of eclipse must arise from the filling in of the absorption by "incipient" emission from the circumstellar envelope. This suggestion is in the process of being checked with a scrutiny of IUE archival data.

3.3. An interesting peculiar interacting system which is considered to evidence the result of CNO-processing in the interior of the evolving component, is V453 Scorpii=HD 163181 which was found by Walborn (1972) to display strong N lines in its spectrum which was classified as BN 0.5 Iae. A few years later, Hutchings (1975) compared the spectrum of V453 Sco with those of κ Cassiopeiae (B1 Iae), χ² Orionis (B2 Iae) and HD 190603 (B1.5 Iae) and reached the conclusion that there is practically no difference in Hγ , that O II is weaker by a factor of around 2 and that N II is stronger by a factor larger or equal than 2.

In regard to this object, I would point out that its spectrum varies with phase, shows the effect of a veiling at some phase interval and that the circumstellar envelope around the fainter, larger mass component is quite opaque, at least it is not completely transparent to the stellar radiation.

3.4 We have discussed in a condensed way the question related to the objects and evidence that are normally considered when talking about abundance anomalies found in interacting binaries, and offered relevant comments. It seems that the evidence for CNO-processed material is probably there but the evidence for H-deficiency is at least subject to doubt. And the doubts are even greater when one realizes that in interacting binaries, well advanced in their evolution, one does not find the sort of abundance anomalies that we would seem to find in less evolutionary advanced systems.

4. For the sake of completeness I would like to finish this review paper by referring to the working hypothesis that I suggested (Sahade 1986, 1987a) to try to make sense out of the peculiar interacting binaries. If the working hypothesis was proven to be right, then it would imply that we have a sequence for the evolution of the gaseous structure that characterizes interacting binaries.

Since I have earlier given the rationale for the working hypothesis, in order not to take up more space than alloted to invited papers I refer the reader to the previous papers. I would only add the different stages that are consistent with the working hypothesis (see also Sahade 1987b):

a) stage of rapid mass loss which is preceded by the presence of variable, optically thick plasma:                                 The R Arae stage;


b) stage at which systems are embedded in gaseous matter, not totally opaque, that dominates the spectrum:                          The W. Serpentis stage ;


c) stage at which systems are embedded in thick plasma and the spectrum shows emissions arising in the circumstellar and in the circumbinary envelopes:                                                     The GG Carinae stage;


d) stage at which one component, the now more massive one – after mass – ratio reversal – is surrounded by an opaque disk:
                                                     The β Lyrae stage;


e) stage at which the opaque disk has become semitransparent:
                                                     The V453 Scorpii stage;


f) the envelope around the more massive component becomes thinner as time goes by and we have then U Cephei and then the typical Algols:
                                                     The Algol stage.


In order to check this suggestion we would have to being able to determine mass loss rates and abundances in the typical objects.


I would like to express my most cordial thanks to the SOC and the LOC of the Colloquium for the invitation and for having provided the means that enabled me to attend.


REFERENCES


Aydin, C., Brandi, E., Engin, S., Ferrer, O.E., Hack, M., Sahade, J., Solivella, G. and Yilmaz, N. 1987, Astron. Astrophys. in press.
Bahýl, V. 1982, in Be Stars, IAU Symp. Nº 98, eds. M. Jaschek and H.G. Groth (Dordrecht: Reidel), p. 205.
Balachandran, S., Lambert, D.L., Tomkin, J. and Parthasarathy, M. 1986, Mon. Not. R. Astron. Soc. 219, 479.
Batten, A.H. and Sahade, J. 1973, Pub. Astron. Soc. Pacific 85, 599.
Bidelman, W.P. 1950, Astrophys. J. 111, 333.
Böhm-Vitense, E. 1981, Ann. Rev. Astron. Astrphys. 19, 295.
Boyarchuk, A.A. 1960, Soviet Astron. A.J. 3, 748.
Danziger, I.J., Wallerstein, G. and Böhm-Vitense, E. 1967, Astrophys. J. 150, 239.
Delgrado, A.J. and Thomas, H.C. 1981, Astron. Astrophys. 96, 142.

Greenstein, J.L. 1940, _Astrophys. J._ 91, 438.

Hack, M. and Job, F. 1965, _Zs. f. Astrophys._ 62, 203.

Hack, M. and Passinetti, L. 1963, _Contr. Oss. Astron. Milano-Merate_, № 215.

Hall, D.S. 1969, _Bull. American Astron. Soc._ 1, 345.

Hoffleit, D. and Jaschek, C. 1982, _The Bright Star Catalogue_ (Yale U. Obs.)

Hutchings, J.B. 1975, _Pub. Astron. Soc. Pacific_ 81, 245.

Klinglesmith, D.A. 1971, _Hydrogen Line Blanketed Model Stellar Atmospheres_, NASA SP-3065.

Koch, R.H. 1972, _Pub. Astron. Soc. Pacific_ 84, 5.

Leushin, V.V. and Snezhko, L.I. 1980, _Soviet Astron. J. Lett._ 6, 94.

Leushin, V.V., Nevskii, M. Yu. and Snezhko, L.I. 1979, _Bull. Spec. Astrophys. Obs. N. Caucasus_ 11, 34.

Leushin, V.V., Nevski, M. Yu. and Snezhko, L.I. and Sokolov, V.V. 1977, _Bull. Spec. Astrophys. Obs. N. Caucasus_ 9, 1.

McCluskey, G.E., Jr. and Sahade, J. 1987, in _Exploring the Universe with the IUE Satellite_, ed. Y. Kondo (Dordrecht: Reidel), p. 427.

Parthasarathy, M., Lambert, D.L. and Tomkin, T. 1983, _Mon. Not. R. Astron. Soc._ 203, 1063.

Peters, G.J. and Polidan, R.S. 1984, _Astrophys. J._ 283, 745.

Plavec, M.J. 1983, _Astrophys. J._ 272, 296.

Plavec, M.J. and Dobias, J.J. 1983, _Astrophys. J._ 272, 296.

Plavec, M.J., Dobias, J.J., Etzel, P.B. and Weiland, J.L. 1984, in _Future of UV Astronomy Based on Six Years of IUE Research_, NASA CP-2349, eds. J.M. Mead, R.D. Chapman and Y. Kondo, p. 420.

Sahade, J. 1986, in _New Insights in Astronomy_, ESA SP-263, p. 267.

Sahade, J. 1987a, _Comments on Astrophys._ 12, 13.

Sahade, J. 1987b, _J. Space Astron. Research_, in press.

Sahade, J. and Albano, J. 1970, _Astrophys. J._ 162, 905.

Sahade, J. and Wood, F.B. 1978, _Interacting Binary Stars_ (Pergamon Press), p. 40.

Schönberner, D. and Drilling, J.S. 1984, _Astrophys. J._ 276, 229.

Struve, O. and Zebergs, V. 1961, _Astrophys. J._ 134, 161.

Walborn, N.R. 1972, _Astrophys. J. Letters_ 176, L119.

Wallerstein, G., Greene, T.G. and Tomley, L.J. 1967, _Astrophys. J._ 150, 245.

ATMOSPHERIC EVIDENCE OF EVOLUTIONARY PROCESSES IN INTERACTING BINARIES


Yoji Kondo
Goddard Space Flight Center
Greenbelt, MD 20771, U.S.A.


The preceding invited paper by Sahade (1987) gave a balanced overall picture for the subject matter.  I shall attempt to complement his talk in discussing two classes of atmospheric diagnostics of the evolutionary processes in interacting binary systems.  These are:  (I) Abnormal abundances of elements in the atmosphere resulting from the nuclear processes in the stellar interior.  (II) Mass flow as a consequence of the evolution of either of the components and the Algol type binaries.


(I)  Abundance Anomalies:  There are several classes of objects with abundance anomalies, such as Am stars.  At least some of these objects are known close binaries.  In particular, most Am stars appear to be close binary stars.  However, the physical causes of the metallic overabundances in those stars are still unclear.  We shall in the following examine Algol type binaries for abundance anomalies.

The best place to look for abundance anomalies is the atmosphere of the evolved late-type component.  However, the star is typically fainter than the early type companion and is difficult to observe spectroscopically at high resolution for abundance analysis.  An optimum time to observe its spectrum is during a primary (total) eclipse when the spectrum of only that star is visible. Regrettably, the total eclipse usually does not last long enough to obtain a good exposure at high resolution.  Spectroscopic observations with large telescopes using imaging device may provide valuable data on the relative abundances of N, C and O.  A good candidate is S Cancri, whose late-type component has an estimated mass of only 0.18 solar mass while its unevolved hotter companion has an estimated mass of some 2.3 solar masses.  The late-type star's atmosphere should reveal nuclear-processed material and overabundance of heavier elements.  S Cnc may provide a rare opportunity to peel down the outlayer of an evolved star to see what it looks like inside.

Using Digicon spectra obtained during the total phase of eclipse, Parthasarathy, Lambert and Tomkin (1983) concluded that the late-type components in U Cephei and U Sagittae had an underabundance of C and an overabundance of N, which they construed as the result of the conversion of C into N.

It is perhaps time that we utilized large telescopes and modern detector technology to study the abundance anomalies in close binaries during total primary eclipse, in a vigorous manner similar to that used, say, in extragalactic research. An up-to-date ephemeris would also be essential in such efforts.

(II) _Mass_ _Flow_: Binaries that are the products of internal evolutionary processes also include Novae and X-ray binaries that are thought to contain compact objects. Also, there is strong evidence that many, if not all, symbiotic stars are binaries.

The very existence of the Algol-type binaries, in which the less massive late type component is the more evolved of the two stars, is strong evidence of the past mass loss from the late-type star as a result of its evolution. Presumably, the present late-type star was originally the more massive (component A) of the two and thus evolved off the main-sequence before its initially less massive companion (component B) did so.

Recent observations of Algol type binaries observed in the ultraviolet, in particular with the IUE satellite, show that some form of mass flow is occurring in virtually all of them (McCluskey and Sahade 1987).

Not many binaries are known to be in a stage where component A has evolved and expanded to fill its critical equipotential Roche surface and is in an active phase of mass transfer. However, several binaries observed in the ultraviolet with OAO-A2, Copernicus, ANS and IUE exhibit observational evidence that they are currently in such a phase; the examples include beta Lyrae (Hack et al. 1975), U Cephei (during its active phase), R Arae and HD 207739 (Kondo, McCluskey and Parsons 1985).

Observational evidence for the dynamic mass-flow phase includes (A) light-curves where the secondary eclipse becomes deeper at shorter wavelengths (Kondo et al. 1985), (B) the non-monotonic variation of the spectral energy distribution which is pronounced in the ultraviolet (Kondo et al. 1985), and (C) continued presence, both inside and outside the eclipses, of emission features observed in beta Lyrae (Hack et al. 1977). Phenomena (A) and (B) have been attributed to the presence of variable, optically-thick, extrastellar plasma.

In the mid-1960's a popular view was that the mass transfer occurred wholly conservatively between the two stars in a binary and that the process would be repeated back and forth, although there existed no theoretically compelling reason or viable observational evidence in support of such a speculation. With the availability of ultraviolet spectra from space, it has become clear that the mass flow is not conservative; however, a fraction of the outflowing matter from component A is often accreted to its companion, giving rise to spectral features (e.g., C IV and N V) that are too hot for component B (Kondo, McCluskey and Stencel 1979).

During the active mass outflow episode in 1986 June, U Cephei was observed
with the IUE (McCluskey, Kondo and Olson 1987) and from ground.  The unusual
nature of the mass flow is quite evident:  (1) partical covering, (2) secular and
phase-dependent variation, and (3) maximum velocity of some 800 km/s, well in
excess of the escape velocity from the binary.



## References

Hack, M., Hutchings, J. B., Kondo, Y., McCluskey, G. E., Plavec, M. and Polidan,
R. S. 1975, _Ap.J._, _198_, 453.

Hack, M., Hutchings, J. B., Kondo, Y. and McCluskey, G. E. 1977, _Ap.J. Suppl._,
_34_, 565.

Kondo, Y., McCluskey, G. E. and Stencel, R. E. 1979, _Ap.J._, _233_, 906.

Kondo, Y., McCluskey, G. E. and Parsons, S. B. 1985, _Ap.J._, _295_, 580.

McCluskey, G. E., Kondo, Y. and Olson, E. C. 1987, _Ap.J._, submitted.

Parthasarathy, M., Lambert, D. L. and Tomkin, J. 1983, _Mon.Not.R.Astr.Soc._, _203_,
1063.

Sahade, J. 1987, invited review, _Proc. IAU Colloq. No. 108_.

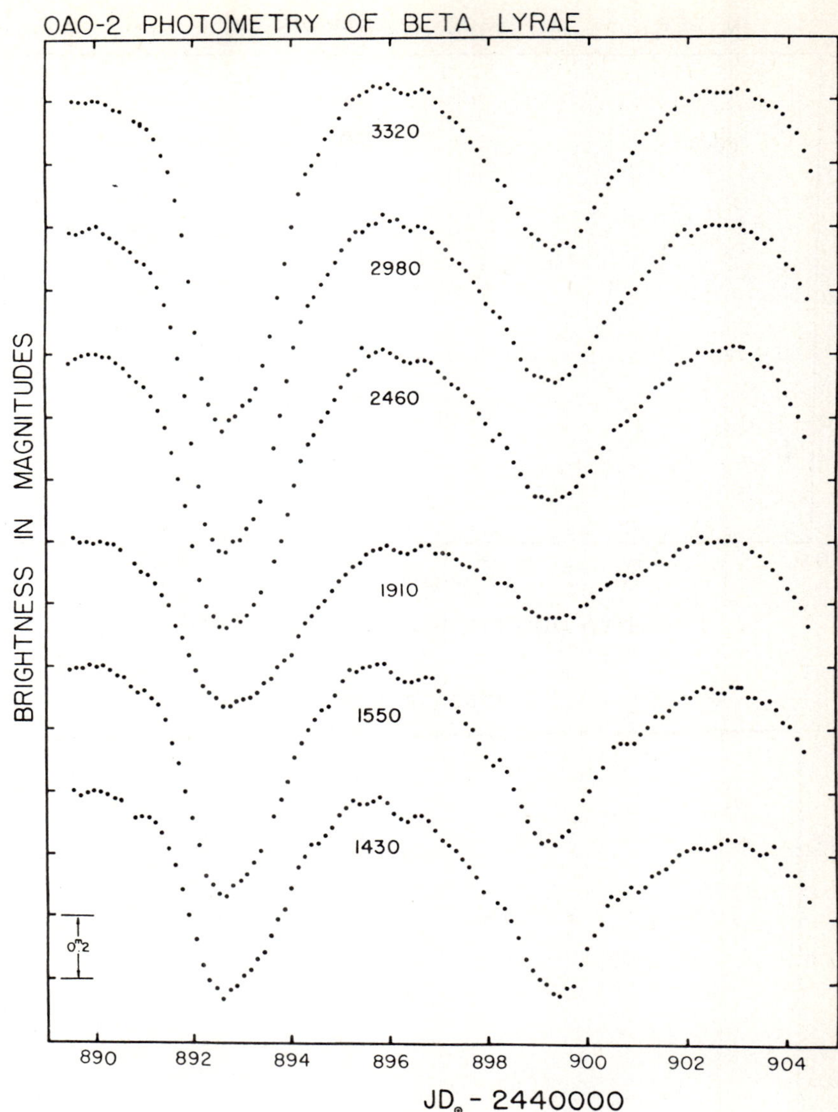


Fig. 1.  Ultraviolet light curves of beta Lyrae observed with the OAO-2 (Kondo, McCluskey and Eaton 1976, Ap. Space Sci., 41, 121).  Note the deepening of the secondary minimum in the far-ultraviolet.

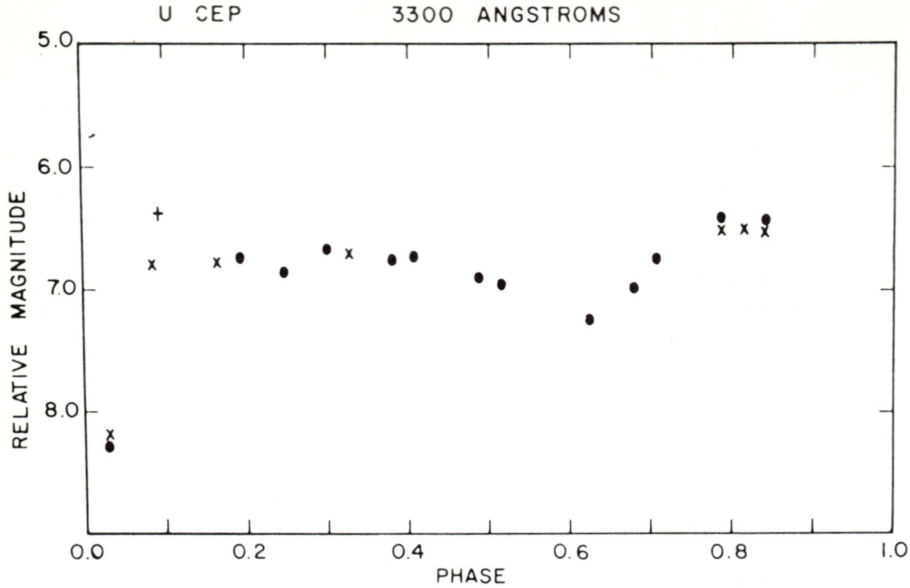


The *ANS* observations of U Cep 3300 Å.

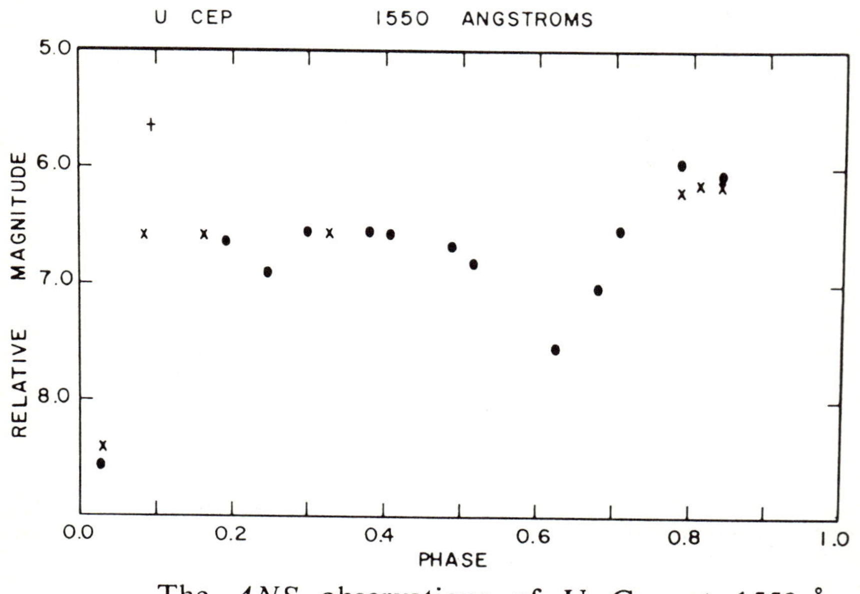


The *ANS* observations of U Cep at 1550 Å.

Fig. 2a & 2b.  Ultraviolet light curves of U Cephei observed with the ANS.  Note the deepening of the displaced secondary minimum in the far-ultraviolet.

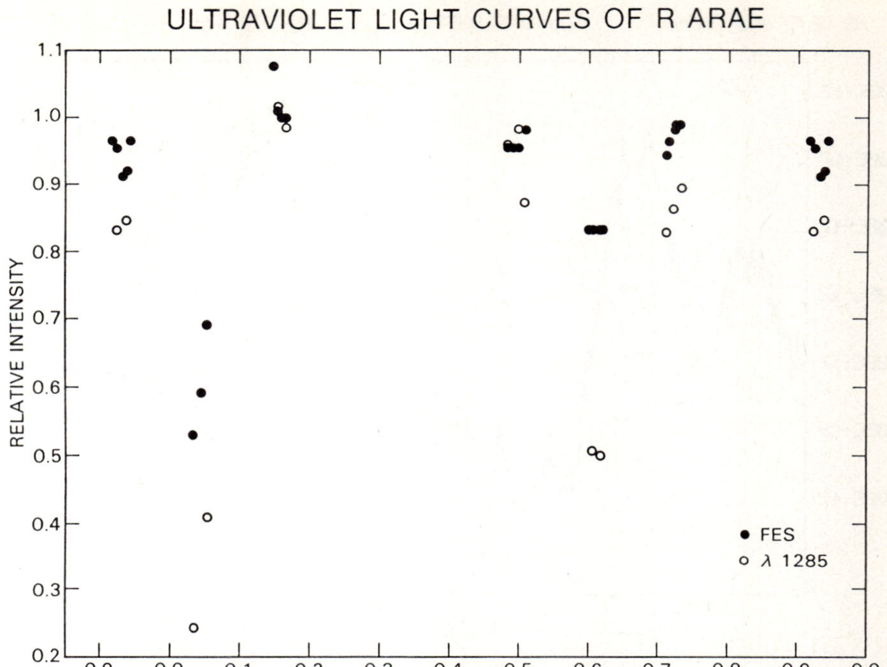


Fig. 3.  Ultraviolet light curves of R Arae observed with the IUE.  FES stands for the IUE Fine Error Sensor, whose color is approximately Blue.

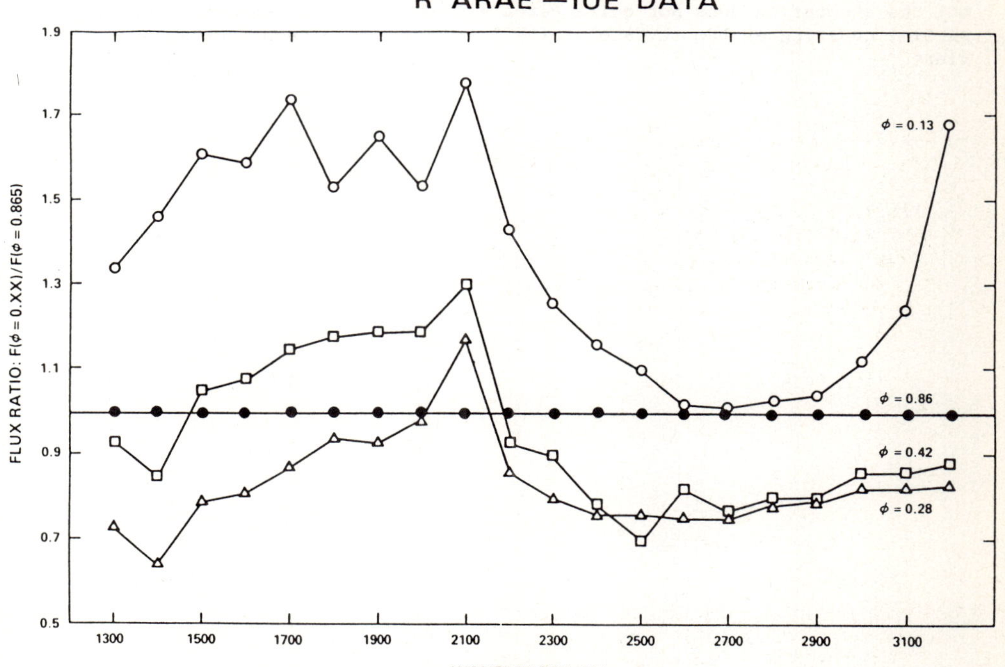


Fig. 4.  Spectral engergy variations of R Arae observed with the IUE; flux at phase 0.86 is normalized to unity.

211

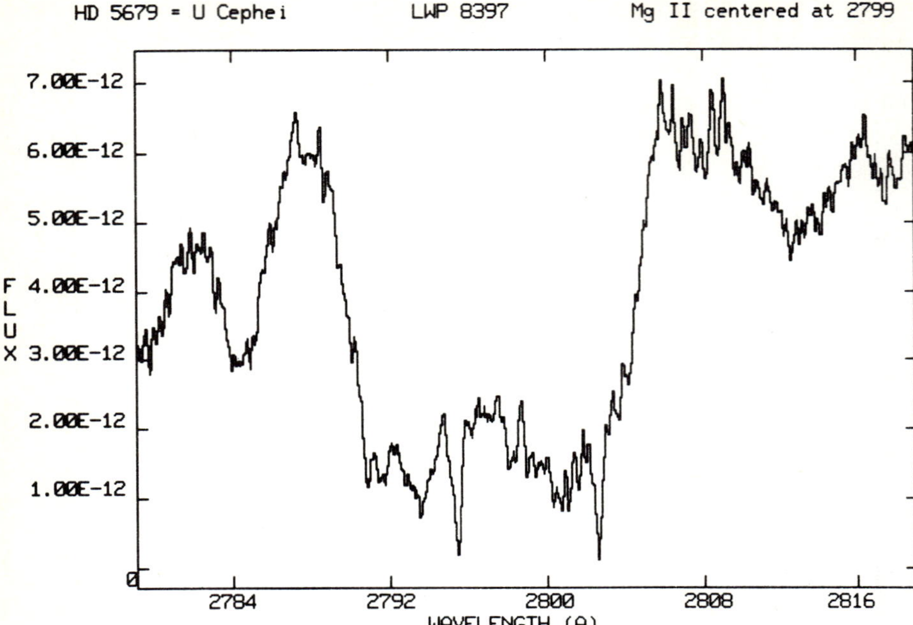


Fig. 5.  High resolution spectra of U Cephei in the Mg II resonance doublet region at 2975 and 2802 A, at phase 0.58, during its active mass flow episode in 1986 (McCluskey et al. 1987).  The maximum-velocity toward the observer is about 800 km/s.  The flat bottom of the broad absorption feature indicates saturation but the absorption does not quite reach zero-flux level, suggesting only a partial covering of the surface of the B star by the plasma flowing out of the G giant.

W URSAE MAJORIS STAR MODELS:  OBSERVATIONAL CONSTRAINTS

Albert P. Linnell
Department of Physics and Astronomy
Michigan State University
East Lansing, MI  48824-1116 U.S.A.

W Ursae Majoris stars can be understood as contact binary stars with a common envelope (Lucy 1968).  They subdivide into two types:  The A-type are earlier in spectral class than about F5, are believed to have radiative envelopes, and associate primary (deeper) eclipse minimum with transit eclipse.  The W-type have spectral classes later than F5, are believed to have convective envelopes, and associate primary minimum with occultation eclipse.  Controversy has surrounded the explanation of W-type light curves.

Four distinct models have been introduced to describe the envelopes or photospheres of W UMa stars.  (1) The Rucinski hot secondary model directly explains W-type light curves on a postulational basis.  Since 70%-90% of the emitted radiation from the secondary (less massive) component is believed to reach the secondary via circulation currents from the primary, there is an apparent thermodynamic mystery why the secondary should be hotter.  (2)  The Lucy Thermal Relaxation Oscillation (TRO) model argues that the secondary component is perpetually out of thermal equilibrium and that the components are in contact only during part of a given TRO cycle.  During contact the photosphere is supposed to be barotropic.  In this case primary minimum always associates with transit eclipse, in disagreement with observation for W-type systems.  (3) The Shu et al. thermal discontinuity (DSC) model also argues for a barotropic photosphere but differs from Lucy on the gravity brightening exponent.  The changes are insufficient to produce W-type light curves.  (4) Webbink (1977), and, separately, Nariai (1976), argue for a baroclinic envelope.  If the baroclinicity extends to the photosphere there is a possibility that W-type light curves could be explained.  In particular, the Webbink scenario produces a hot secondary.

On the other hand, an ingenious proposal by Mullan (1975) apparently rescues the Lucy model. The Mullan proposal populates the primary component photosphere with starspots. These reduce the average surface brightness, thereby reduce the light loss at transit eclipse, and produce W-type light curves. This proposal has been received favorably by many specialists in the binary star field.

The study of binary star color curves, in addition to light curves, provides helpful observational discrimination among competing models. Color curves produce good temperature diagnostics. A comparison of theoretical color curves with W UMa observational data appears in a recent paper (Linnell 1987), together with a discussion of the models described earlier. If one adopts a physically reasonable spot temperature contrast from the adjacent photosphere, it is possible to calculate the fractional coverage of the primary component necessary to produce the observed W-type light curve in the $\underline{V}$ band. The resulting $\underline{B}$-$\underline{V}$ color curve then differs only slightly from the theoretical color curve for a barotropic photosphere. This color curve disagrees with observation. On the other hand, a hot secondary model has a corresponding $\underline{B}$-$\underline{V}$ color curve in reasonable accordance with observation. Other objections to the starspot model are in the paper cited (Linnell 1987). The best accordance of all the visible wavelength data is with a hot secondary model.

An apparent difficulty for the hot secondary model is the uv data obtained by Eaton, Wu and Rucinski (1980) for W UMa, using the ANS satellite. If the hot secondary model is correct, the difference in eclipse depths increases in the uv. Rather than increasingly W-type, the observations at 2200 A show a marginally A-type light curve. It should be noted that the starspot model is of no help here. Since starspots are cool, the emergent flux difference between spot and photosphere increases in the uv. Then the reduction in average surface brightness in the uv, for the primary component, cannot be less than in $\underline{V}$. A W-type light curve in $\underline{V}$, produced by the Mullan starspot model, will not become A-type in the far uv. A possible explanation of the uv observations is a uv excess on the primary component, produced by inferred flare activity (Linnell 1987).


REFERENCES

Eaton, J.A., Wu, C.-C., and Rucinski, S.M.  1980, Ap. J., 239, 919.
Linnell, A.P.  1987, Ap. J., 316, 389.
Lucy, L.B.  1968, Ap. J., 151, 1123.
Mullan, D.J.  1975, Ap. J., 198, 563.
Nariai, K.  1976, Publ. Astron. Soc. Japan, 28, 587.
Webbink, R.F.  1977, Ap. J., 215, 851.

# Mass and Energy Transfer
# in Semi-Detached Binary Systems

Wasaburo Unno, Masayoshi Kiguchi
*Research Institute of Science and Technology,*
*Kinki University, Kowakae, Higashi-Osaka-shi, Osaka 577, Japan*
and
Masatoshi Kitamura
*Tokyo Astronomical Observatory, Mitaka-shi, Tokyo 181, Japan*

**1. Introduction**   Recently, Kitamura and Nakamura[1] have found that the anomalous gravity darkening occurs in semi-detached binary systems. The exponent of gravity darkening for the secondary components, which is defined by $\alpha_c = \frac{d \log F}{d \log g}$ where $F$ is the radiative flux and $g$ is the gravitational accerelation, is significantly greater than the unity as shown in Table 1. We interpret this in terms of the energy transport by the mass outflow from the secondary component filling the Roche-lobe.

Table 1.

Empirical $\alpha_c$–values and physical quantities of five semi-detached binary systems.

| Star | $T_c(°K)$ | $M_c/M_\odot$ | $M_c/M_1$ | $\overline{R}_c/R_\odot$ | $\log \overline{g}_c$ | $\alpha_c$ |
|---|---|---|---|---|---|---|
| u Her | 12000 | 2.9 | 0.39 | 4.4 | 3.61 | 8.47±0.51 |
| z Vul | 8500 | 2.3 | 0.43 | 4.5 | 3.48 | 9.73±0.12 |
| LT Her | 5200 | 0.50 | 0.20 | 1.6 | 3.73 | 5.86±0.08 |
| RZ Cas | 5000 | 0.63 | 0.36 | 1.8 | 3.73 | 2.25±0.15 |
| VV UMa | 5300 | 0.46 | 0.22 | 1.2 | 3.96 | 3.78±0.12 |

From the analysis based on this interpretation, we show that 1) the mass out-flow carries energy if $\nabla < \nabla_{ad}$ and 2) the anomalous gravity darkening requires that the flow must originate from deep interior. The mass loss rate can be estimated from the anomalous gravity darkening. Thus, the anomalous gravity darkening can be used not only to estimate the mass transfer rate but also to probe the interior structure and the evolution of the secondary star of the semi-detached binary. This model links Morton's instability[2] of the secondary with the accretion disk of the primary.

**2. Basic Equations**   We make following assumptions that 1) the mass loss occurs steadily and 2) the flow is polytropic with index $n$ within the secondary. In these assumptions, the energy conservation is given by

$$\nabla \cdot \{\mathbf{F} + (W + \phi + \frac{v^2}{2})\rho \mathbf{v}\} = 0 \tag{1}$$

and the kinetic energy conservation is given by

$$\nabla \cdot \{[\frac{v^2}{2} + \phi + (n+1)\frac{\Re T}{\mu}]\rho \mathbf{v}\} = 0, \tag{2}$$

where $F$ denotes the radiative flux, $W$ the enthalpy and $\phi$ the gravitational potential.

For the simplicity of analysis, we make 1-dimensional approximation. Then, from these equations, the radiative flux $F$ is given by

$$F(\xi, \zeta) = \frac{A_b}{A(\zeta)}F_b - (n - \frac{1}{\gamma - 1})(1 - \frac{T}{T_b})\frac{q(\xi)}{A(\zeta)}\frac{\Re T_b}{\mu}, \tag{3}$$

215

where $\xi$ and $\zeta$ designate the coordinate to specify the flow tube and the coordinate along the tube, respectively, $A(\zeta)$ is the cross section along the $\xi$-tube, $q$ is the mass flux through the tube $q(\xi) = \rho v A(\zeta)$ which is constant in each flux tube, and the subscript $_b$ means the deep interior where radiative flux $F_b$ is constant. The polytropic index n is to be determined by the radiative transfer, but is tentatively assumed here as $n = 3$ in the radiative region.

Now, we assume an isothermal siphon flow outside the photosphere which passes trough a sonic point designated by subscript s. Then the velocity $v_R$ of the mass flow at the surface of the secondary is determined from

$$\frac{v^2}{c_s^2}\exp\left(-\frac{v^2 - c_s^2}{c_s^2}\right) = \left(\frac{A_s}{A}\right)^2 \exp\left(-\frac{2}{c_s^2}(\phi_s - \phi)\right). \tag{4}$$

The mass flux $q$ is given by $q(\xi) = \rho_R v_R A(\zeta_R)$, where $\rho_R$ is the density at the photosphere which is determined by the usual atmosphere theory with effective gravity modified by the flow as

$$\rho_R = \frac{1}{\kappa \frac{\Re}{\mu} T_e} \frac{\kappa_T + 1 + n(\kappa_\rho + 1)}{n + 1}(1 + \chi)g_{Teff}, \tag{5}$$

using the opacity $\kappa = \kappa_0 \rho^{\kappa_\rho} T^{\kappa_T}$. The modification factor $(1 + \chi)$ is given by

$$\chi = \frac{\frac{v^2}{2} - \frac{v_b^2}{2}}{\phi - \phi_b}. \tag{6}$$

Finally, we obtain for the effective temperature at the surface of secondary

$$\sigma T_e^4 = \sigma T_{e,pol}^4 - K\frac{g_{Teff}}{\kappa}\frac{T_b}{T_e}v_R \tag{7}$$

where

$$K = \left(\frac{1}{\nabla} - \frac{1}{\nabla_{ad}}\right)\left(1 - \frac{T_e}{T_b}\right)\left(1 + \frac{\kappa_T + n\kappa_\rho}{n + 1}\right)(1 + \chi). \tag{8}$$

**3. Result** We specify the flow outside the surface of secondary by two parameters $\lambda$ and $l$ defined by

$$z = \lambda r\sqrt{\frac{2}{2l+1}(1 - \cos^{2l+1}\theta)},$$

where $z$ is the distance from the $L_1$ Lagrange point along the ridge of the potential, $r$ is the radius of the secondary and $\theta$ is the angle between lines which direct the $L_1$ point and a surface point from the center, and we specify the internal model by the temperature $T_b$ at the point where $\nabla = \nabla_{ad}$.

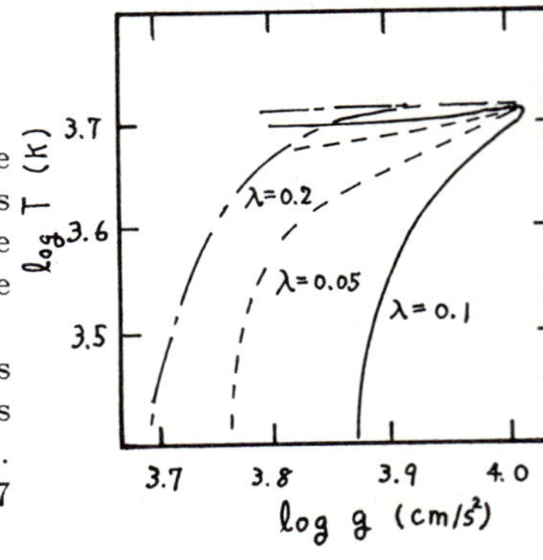

a) LT Her.   We show in Fig. 1 the effective temperature versus surface gravity for various $\lambda$. The other parameters are fixed as $T_b = 8 \times 10^5$K and $l = 1$. The $\lambda = 0.1$ model is the best fitted. The mass loss rate is 0.41 for $\lambda = 0.2$, 1.9 for $\lambda = 0.1$ and 1.7 for $\lambda = 0.05$ in units of $10^{-6} M_\odot/y$.

Fig. 1



b) Z Vul.   As the effective temperature of this star is relatively high, the surface density becomes low due to the large opacity. This reduces the mass flow, so that it becomes necessary to make $T_b$ high ($T_b > 4 \times 10^7$K) in order to give rize to a large gravity darkening.

## References


[1] M. Kitamura and Y. Nakamura, Ann. Tokyo Astron. Obs. 2nd Series **21**, 387(1987).
[2] D. C. Morton, Astrophy. J. **132**, 146(1960).

EMPIRICAL DETERMINATION OF THE GRAVITY-DARKENING EXPONENT
FOR THE SECONDARY COMPONENTS FILLING THE ROCHE LOBE IN
SEMI-DETACHED CLOSE BINARY SYSTEMS


Masatoshi  Kitamura

Tokyo Astronomical Observatory, Mitaka,
Tokyo 181, Japan

and

Yasuhisa  Nakamura

Komaba Senior High School, Meguro-ku,

Tokyo 151, Japan


The ordinary semi-detached close binary system consists of a main-sequence
primary and subgiant (or giant) secondary component where the latter fills the
Roche lobe.    From a quantitative analysis of the observed ellipticity effect,
Kitamura and Nakamura (1986) have deduced empirical values of the exponent of
gravity-darkening for distorted main-sequence stars in detached systems and
found that the empirical values of the exponent for these stars with early-
type spectra are close to the unity, indicating that the subsurface layers of
early-main sequence stars in close binaries are actually in radiative equilibrium.    The exponent of gravity-darkening can be defined by $H \propto g^{\alpha}$ with
H as the bolometric surface brightness and g as the local gravity on the stellar
surface.

Thus, by assigning   $\alpha_1 = 1$ to the early-type main-sequence primaries in
semi-detached systems, similar practical analysis  as done in the previous work
of Kitamura and Nakamura (1986) has been carried out to determine empirical
values of the exponent for the secondaries of those systems which fill the
critical Roche lobe.

The result of the analysis indicates that the $\alpha_2$-values deduced for the
secondary components of nine well-understood semi-detached close binary systems
are significantly greater than the unity, as shown in Table 1.    Such greater
values of the exponent for the secondaries of semi-detached systems could not
be reconciled by any adjustment of the elements used as the input physical
parameters within the extent of reduction errors (including uncertainties of
the  adopted temperature scales and limb-darkening coefficients).    Also, such
an excess of $\alpha_2$-values for the secondaries is evidently in a direction opposite
to that expected from convection for stellar atmospheres.    In this connection,
it may be noted that Budding and Kopal (1970) previously discussed the degree

of gravity-darkening of the secondary component of Algol based on an analysis of
the infrared light curve of its secondary minimum and reached the same conclusion
of large gravity-darkening.    They have deduced $T_\lambda/T_0$ = 0.26 $\pm$ 0.06 for the
secondary of Algol, which corresponds to $\alpha_2$ = 3.8 $\pm$ 0.6 in our notation.

Table 1.   Empirical $\alpha$-values determined for the
secondary components of nine well-understood
semi-detached close binary systems.

| Star | $T_2$ | $\log g_2$ | $\alpha_2$ |
|---|---|---|---|
| V Pup | 26600°K | 3.95 | 5.44 $\pm$ 0.16 |
| TT Aur | 17300 | 3.90 | 3.84 $\pm$ 0.22 |
| u Her | 12000 | 3.61 | 8.47 $\pm$ 0.51 |
| Z Vul | 8500 | 3.48 | 9.73 $\pm$ 0.12 |
| LT Her | 5200 | 3.73 | 5.86 $\pm$ 0.08 |
| RZ Cas | 5000 | 3.73 | 2.25 $\pm$ 0.15 |
| VV UMa | 5300 | 3.96 | 3.78 $\pm$ 0.12 |
| V356 Sgr | 10300 | 2.83 | 3.83 $\pm$ 0.19 |
| GT Cep | 10000 | 3.05 | 3.00 $\pm$ 0.43 |

At present, we do not know what is the real origin of such excessive gravity-
darkening for the secondary components of semi-detached systems.    However, one
possibility to explain such observational evidence may be found in the effect
of mass loss from the secondaries (e.g., Unno, Kiguchi and Kitamura, in this
Proceedings).

Full detail of the present analysis has been published quite recently elswhere
(1987).

References


Budding, E. and Kopal, Z. 1970, Astrophys. Space Sci., 9, 343.
Kitamura, M. and Nakamura, Y. 1986, Ann. Tokyo Astr. Obs., 21, 229.
Kitamura, M. and Nakamura, Y. 1987, Ann. Tokyo Astr. Obs., 21, 387.
Unno, W., Kiguchi, M., and Kitamura, M. (in this Proccedings).

CaII  H  AND  K  EMISSION  IN  THE  SECONDARY  COMPONENT  OF  U  CEPHEI

Akira Okazaki
Tsuda College, Kodaira, Tokyo 187, Japan

Yasuhisa Nakamura
Komaba Senior High School, Meguro-ku, Tokyo 153, Japan

Jun-Ichi Katahira
Science Education Institute of Sakai, Sakai 591, Japan

U Cephei (V = 6.8 — 9.0, P = 2.493 d) is an eclipsing binary consisting of a B7V primary and a G8III-IV secondary component.  This binary is one of the semi-detached Algol systems showing soft X-ray emission which is probably associated with a hot corona surrounding the secondary component (White and Marshall 1983).

We made spectroscopic observations of U Cep with the coudé image-tube spectrograph of the 1.9-m telescope at Okayama Astrophysical Observatory on October 14, 1986.  We obtained four spectrograms with a dispersion of 16 Å mm⁻¹ covering $\lambda\lambda 3700-4300$ Å during the primary eclipse.  The first two exposures were made in a total eclipse, while the last two were slightly after the third contact.  The CaII H and K emission lines appear clearly in all the spectrograms. Figure 1 represents an intensity tracing of one of these spectrograms.

These emission lines have a half width of $\sim 2$ Å, which is consistent with the rotational broadening of the secondary component ($V_{rot} \sim 95$ km-s).  In the four spectrograms, no radial velocity difference is found between the CaII H and K emission lines and the secondary component's absorption lines.  Thus, these CaII H and K emission lines are considered due to a chromospheric activity of the secondary component.

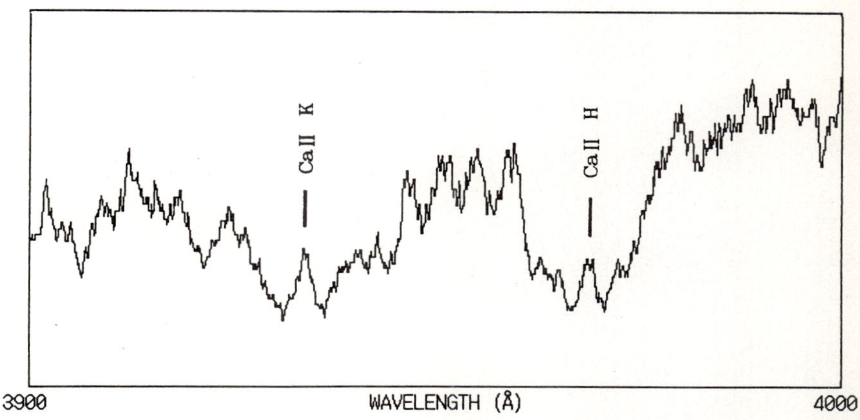


Figure 1: An intensity tracing of the spectrogram of U Cep at phase $\sim 0.99$ around CaII H and K lines (exposure time $\sim 24$ min.).

According to the procedures given by Linsky et al.(1979), we obtained the corrected surface fluxes $F'(K_1) \sim 5(+6)$ and $F'(H_1) \sim 4(+6)$ [erg cm$^{-2}$ s$^{-1}$] for the secondary component of U Cep.

A full account of this work will be published elsewhere.

This work was supported in part by the Grant-in-Aid of the Ministry of Education, Science, and Culture of Japan (62540193).


REFERENCES

Linsky, J.L., Worden, S.P., McClintock, W., and Robertson, R.M.: 1979, Astrophys. J. Suppl., <u>41</u>, 47.
White, N.E., and Marshall, F.E.: 1983, Astrophys. J. Letter, <u>268</u>, L117.

# CARBON ABUNDANCE IN MASS-EXCHANGING BINARIES


**H. CUGIER**

Institut Astronomiczny

Universitety Wroclawskiego

ul. M. Kopernika 11

51-622 Wroclaw - Poland

**J.P. DE GREVE**

Astrophysical Institute

V.U.B.

Pleinlaan 2

B-1050 Brussels - Belgium


The evolution of the carbon abundance at the surface of both components of a mass-exchanging (Algol-type) binary is examined (fig. 1). Distinction is made between case B and case AB (fig. 2) of mass transfer, in view of the different timescales involved. In the mass accreting component thermohaline mixing is adopted when matter with decreasing hydrogen abundance is deposited on the surface.

It is shown that at the surface of the loser a very low C-abundance is present, while at the surface of the gainer different regimes occur. On the average the expected C-abundance on the gainer is clearly lower than the observed solar value, but far above the value at the surface of the loser. The variation in time during the mass-exchange process is compared to the values, derived from observation of several Algol-type systems.

Theoretical models were calculated by Packet (1987) and the authors. The evolutionary codes are discussed by De Greve et al. (1985) and Prantzos et al. (1986).

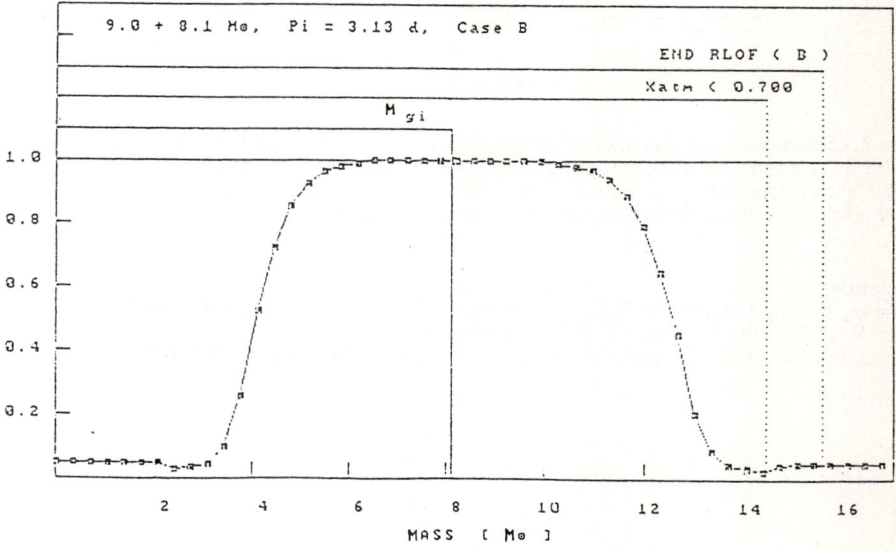


Figure 1. Carbon abundance as a function of mass for both components of a close binary system at the onset of mass transfer. The region from Mr=0 to Mr=Mgi=8.1 Mo corresponds to the originally less massive component (gainer), whereas the carbon distribution of the loser is plotted from 8.1 Mo (surface) to 17.1 Mo (center). The first occurrence of hydrogen depleted layers (Xat<0.7) and the end of the Roche Lobe Overflow are indicated.

The observed carbon depletion of mass-losing components in Algol-type systems may be explained as the result of the conversion of C to N, while these stars were on the main sequence. Furthermore a largescale mixing of matter in mass accreting components is suggested. Indeed, the present computations show that, especially for case AB of mass transfer, large but also small underabundances of carbon may be observed at the surface of the gainer, depending on the initial parameters of the system and on the mass fraction removed from the mass-losing star.

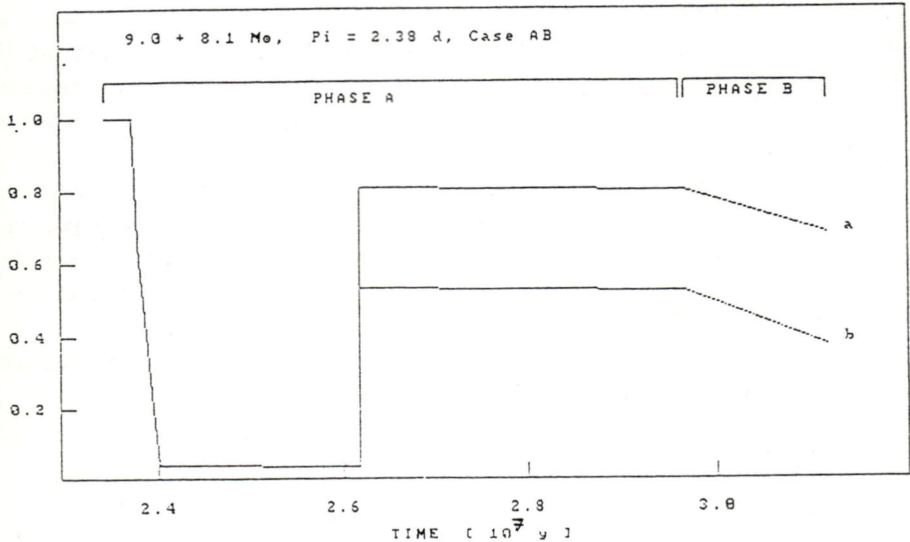


Figure 2. Carbon surface abundance of the gainer in the system 9 M☉ + 8.1 M☉, Pi=2.38 d, as a function of time, during a case AB of mass transfer. The line marked 'a' was obtained under the assumption that no significant conversion of C to N takes place above the convective core, during mass transfer.

## References


De Greve, J.P., de Landtsheer, A.C., Packet, W.:  1985, Astron. Astrophys. 142, 367.
Packet, W.: 1987, Ph. D. thesis, V.U.B. Brussels (preprint)
Prantzos, N., Doom, C. Arnould, M., de Loore, C.: 1986, Astrophys. J. 304, 695.

PHOTOMETRIC AND POLARIMETRIC OBSERVATIONS OF
THE RV TAURI STAR AR PUPPIS

A.V.Raveendran and N.Kameswara Rao
Indian Institute of Astrophysics
Bangalore 560 034, India

RV Tauri stars are variable yellow supergiants and they display anomalous excesses in infrared radiation indicating the presence of cool and extended dust thermospheres, presumably formed from the matter ejected by them (Gehrz 1972). AR Pup, recognised as a member of RV Tauri stars, shows strong CH and CN bands in the optical spectrum and is considered to be carbon-rich (Lloyd Evans 1974). During a programme of broad band polarimetric observations of carbon variables, AR Pup was found to exhibit a large and highly time-dependent linear polarisation (Raveendran & Kameswara Rao 1987).

During the period from 1987 January 20 to April 6, AR Pup was observed on 33 nights through standard B and V filters with the 34-cm reflector of the Vainu Bappu Observatory, Kavalur. We also obtained UBVRI polarimetry of AR Pup with the PRL - Polarimeter (Deshpande et al. 1985) attached to the 102-cm reflector on six nights during the same period. Present observations clearly indicate that the large time dependent variations in polarisation, both in the amount and wavelength dependence, observed at earlier epochs by us were not isolated events but were segments of a continuous variation which is cyclic and related to the light curve.

In Fig.1, we have plotted the results of BV photometry and the linear polarisation in V band against the corresponding Julian days of observation. It is evident from Fig.1 that polarisation increased during the ascending branch of the light curve, a case qualitatively very similar to what is observed in the long period variable o cet (cf. Shawl 1975). The rapid increase in polarisation is probably related to the outward  passage of an atmospheric shock. We find that larger changes in polarisation and position angle occur in ultraviolet than in red. The nature of wavelength dependence of polarisation and rotation of position angle with wavelength seen in AR Pup indicate that the main mechanism which produce the observed polarisation is grain scattering. Since the position angle of polarisation is not constant in time, there is no fixed axis of symmetry for the scattering material. This would suggest that a major portion of polarisation in AR Pup arises from more localised transient regions and not from the extended infrared emitting circumstellar envelope.

The photometric data indicates a 75 ± 1 day period for the light variation.

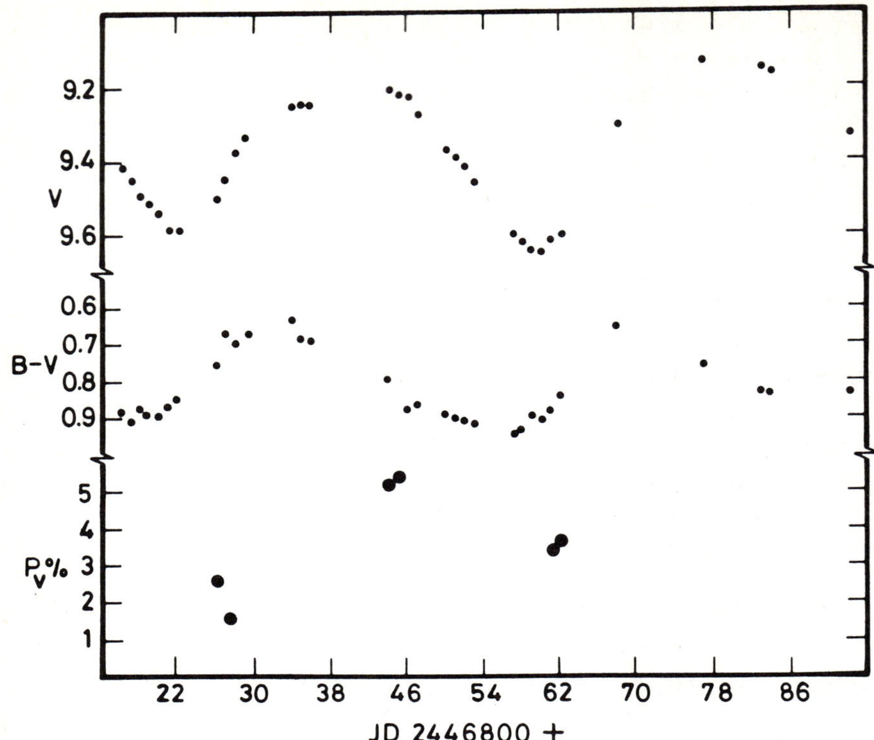

Fig. 1.  Plots of V, (B-V) and linear polarisation in V ($P_v$%)
against the Julian days of observation.


A detailed version of the paper will be published elsewhere.


## References


Deshpande, M.R., Joshi, U.C., Kulshrestha, A.K., Banshidhar, Vadher, N.M.,
Mazumdar, H.S., Pradhan, S.N. & Shah, C.R., 1985. Bull. Astr. Soc. India,
13, 157.

Gehrz, R.D., 1972. Astrophys. J., 178, 715.

Lloyd Evans, T., 1974. Mon. Not. R. Astr. Soc. 167, 17p.

Raveendran, A.V. & Kameswara Rao, N., 1987. Astron. Astrophys. (accepted).

Shawl, S.J., 1975. Astron. J. 80, 602.

# III. Chemical and Dynamical Structures of Exploding Stars

*Chair: G.S. Kutter, R.P. Kirshner*
*S. Hayakawa, and M. Dopita*

CLASSICAL NOVAE - BEFORE AND AFTER OUTBURST

Mario Livio

Dept. of Astronomy, Univ. of Illinois,
DARC Observatoire de Paris, Meudon
and Dept. of Physics, Technion, Haifa.


I <u>Introduction</u>


Classical nova (CN) and dwarf nova (DN) systems have the same binary components (a low-mass main sequence star and a white dwarf) and the same orbital periods. An important question that therefore arises is : are these systems really different ? (and if so, what is the fundamental difference ?) or, are these the same systems, metamorphosing from one class to the other ?

The first thing to note in this respect is that the white dwarfs in DN systems are believed to accrete continuously (both at quiescence and during eruptions). At the same time, both analytic (e.g. Fujimoto 1982) and numerical calculations show, that when sufficient mass accumulates on the white dwarf, a thermonuclear runaway (TNR) is obtained and a nova outburst ensues (see e.g. reviews by Gallagher and Starrfield 1978, Truran 1982). It is thus only natural, to ask the question, <u>is the fact that we have not seen a DN undergo a CN outburst (in about 50 years of almost complete coverage) consistent with observations of DN systems</u> ? In an attempt to answer this question, we have calculated the probability for a nova outburst <u>not</u> to occur (in 50 years) in 86 DN systems (for which at least some of the orbital parameters are known).
The data and assumptions used were :
1) White dwarf masses were taken from Ritter's Catalogue of Cataclysmic Binaries (1987, and references therein). When the mass was not known, we took the average of known masses above or below the period gap, depending on the system's orbital period.
2) Accretion rates were taken from Patterson (1984) and Verbunt and Wade (1984). When the accretion rate was not known, we used Patterson's accretion rate-orbital period relation.
3) We used the fact that a TNR occurs when the pressure at the base of the accreted envelope reaches a critical value, $P_{crit} \simeq 2 \times 10^{19}$

dyne cm$^{-2}$ (Truran and Livio 1986, Livio 1987) to calculate the recurrence timescale of CN outbursts for each system.
4) We used a mass-radius relation for white dwarfs. Based on (1)-(4) we
found that the probability for a nova not to occur in 50 years in these
systems is 0.78-0.84, with the range resulting from differences in mass
radius relations. We therefore find, that the fact that a nova outburst
has not been observed in these systems is not surprising and does not
imply that DN and CN systems are different. Incidentally, the systems
found most likely to undergo a CN outburst in the near future were :
RU Peg, RX And, SS Cyg and possibly V Sge (V1017 Sgr may also erupt for
other reasons, see Webbink et al. 1987).

We have thus established that (in principle at least)DN can metamor-
phose into CN. The remaining question is therefore, do CN become DN du-
'ring some phases of their evolution ? Before attempting to answer this
question, we shall examine some problems with the mass accretion rates
deduced from observations of CN systems.

II $\underline{\text{Problems with the Accretion Rates in CN systems}}$.

Patterson (1984) and Warner (1987) attempted to deduce the mass accretion rates, $\dot{M}$, in CN systems. The values which they obtained present
some problems,which we shall now briefly discuss.
a) $\underline{\dot{M} \text{ deduced from observations is too high to produce strong nova outbursts}}$.
Both numerical and analytical calculations have shown that in order to
obtain strong TNRS on the surface of $1M_\odot$ white dwarfs, the accretion rate must be lower than $\sim 10^{-9}M_\odot$/yr (Kutter and Sparks 1980, Prialnik et al.
1982, Fujimoto 1982). For higher accretion rates, strong compressional
heating leads to ignition under only mildly degenerate conditions, thus
producing weak outbursts. Yet, observations seem to imply $\dot{M} \gtrsim 10^{-8}M_\odot$/yr.
$\underline{\text{Possible solutions}}$
The deduced values of $\dot{M}$ are higher than the real ones or the calculations of TNR development have to be modified, or
$\underline{\dot{M} \text{ changes as a function of time}}$, assuming $\underline{\text{lower values during at least a part}}$
$\underline{\text{of the interoutburst period}}$.
b) $\underline{\text{A number of CN systems were found to exhibit occasionally DN eruptions}}$.
The class of such systems includes V446 Her, Q Cyg, V3830 Sgr, Nova Vul
(1979), WY Sge, GK Per, BV Cen and V1017 Sgr (Livio 1987 and references
therein). The problem with this observation lies in the fact that in the
disk instability model for DN eruptions (Meyer and Meyer-Hofmeister 1983,
Faulkner, Lin and Papaloizou 1983, Cannizzo and Wheeler 1984, Mineshige

and Osaki 1983), for mass accretion rates above a certain critical value, $\dot{M}_{crit}$, no eruptions are expected to occur. This is a consequence of the fact that for high accretion rates the disk lies on the hot, stable branch of the effective temperature -surface density curve. All CN systems (except for the very long period ones, e.g. GK Per), were found to have accretion rates above $\dot{M}_{crit}$ (Warner 1987) and thus, are not expected to undergo DN eruptions.

Possible solutions

The predictions of the disk instability model (or the model itself) could be uncertain, or
$\dot{M}$ changes as a function of time, assuming lower values during at least a part of the interoutburst period.

c) Observations of the oldest recovered novae imply lower accretion rates.
Observations of CK Vul (1670) found it in a state of very low $\dot{M}$ ($\dot{M} < 10^{-11.5} M_\odot$/yr, Shara, Moffat and Webbink 1985). In addition, WY Sge (1783) is fainter by at least a factor 10 than normal CN (Shara et al. 1984). The nova RR Pic (1925) was observed to decrease in brightness by ~0.25 mag in 1975 (Warner 1986).

Possible explanations

These old novae are unusual, or
$\dot{M}$ changes as a function of time, assuming lower values during at least a part of the interoutburst period.

d) The spread in the values of $\dot{M}$ at a given orbital period.
The most fashionable theories of cataclysmic variables evolution assume that mass transfer above the period gap is driven by magnetic braking (e.g. Lamb and Melia 1987, Hameury et al. 1987). Available magnetic braking laws give accretion rates that are dependent on the orbital period but are quite insensitive to other parameters of the binary system (Verbunt and Zwaan 1981, Mestel and Spruit 1987). In particular, the spread in the values of $\dot{M}$ observed at a given orbital period (Warner 1987), is significantly wider than predicted by the theory.

Possible solutions

The predictions of magnetic braking models (or the model itself) could be uncertain, or $\dot{M}$ changes as a function of time so that the spread is introduced by the time variability of $\dot{M}$.


III The "Cyclic Evolution" Scenario.


Points (a)-(d) in section II clearly suggest the possibility that the accretion rates in CN systems are reduced to lower values (than the ones deduced presently from observations) for at least a part of the

time between outbursts. We would like to note that it is not impossible
that in fact all of the other proposed solutions (or different solutions)
should apply, since many uncertainties are associated with all of the models involved. However, the variable $\dot{M}$ offers a somewhat simpler solution.

One scenario in the context of which $\dot{M}$ changes between outbursts is
the "hibernation" scenario (Shara et al. 1986, Livio and Shara 1987). In
this scenario, it has been suggested that $\dot{M}$ decreases a few hundred years
following the nova outburst (by a factor 10-100) and then returns slowly to its initial value. The mechanism responsible for the reduction
in $\dot{M}$ in the original model, was the increase in the binary separation,
resulting from mass loss during the outburst. Such an increase has indeed
been observed in BT Mon (Schaefer and Patterson 1983). The model assumed
that the separation increase causes the secondary to underfill its Roche
lobe, thus reducing the mass transfer. The model further suggested that
the system stays bright for 50-200 years following the outburst, due to
the presence of the hot white dwarf, which both produces reprocessed
light and induces mass transfer from the secondary by irradiation (Livio
and Shara 1987, Kovetz et al. 1987). Mass transfer was assumed to increase slowly (after the reduction) through the action of magnetic braking.

We note, however, the following possible difficulty with the original "hibernation" scenario. If the secondary's atmosphere is isothermal
(due to irradiation by the WD), then it can be expected that the mass
transfer rate will be reduced by a factor 10-100 depending on $\Delta a/H$, where
$\Delta a$ is the change in the separation and H is the (constant) scaleheight
(Livio and Shara 1987). If, however, the secondary's atmosphere is convective, then $\dot{M} \sim (\Delta R)^3$, where $\Delta R$ is the distance by which the Roche
lobe is overfilled. In such a case, an increase in the separation by
$\Delta a/a \sim 10^{-4}$ will result in a decrease in $\dot{M}$ by at most a factor 2 (Edwards
and Pringle 1987). Thus, it is not clear at all whether an increased
separation can produce a significant decrease in $\dot{M}$.

We would like to propose here a modified "cyclic evolution" scenario,
which does not require any special mechanism for the reduction in $\dot{M}$. This
is based on the idea that <u>many CN systems are brighter both after and before the outburst</u>. The fact that post outburst systems stay (for a few tens
to ~ 200 years) brighter than their real quiescent values is more or less
established. It is sufficient to note that a number of systems, such as
V1229 Aql, IV Cep, HR Del, FH Ser, V1500 Cyg and CP Pup did not return

to their pre-outburst magnitude (Robinson 1975, Warner 1985). The idea
that CN systems experience a pre-outburst brightening (for perhaps a
few tens of years before the outburst) is suggested first of all by
observations of V533 Her, LV Vul, Nova Vul (1979), CP Lac and GK Per
(Robinson 1975 and references therein). The question is of course, what
causes this pre-outburst brightening. We suggest that this could result
from the increase in the $\underline{bolometric}$ luminosity of the white dwarf, due
to nuclear burning prior to the TNR. The radiation is reprocessed (in
the accretion disk and the secondary) and perhaps also induces an increased mass transfer. Recent calculations have shown that the increase in
$L_{BOL}$ prior to the TNR can be quite gradual, especially if the white dwarf
is relatively hot and not too massive (Livio, Shankar and Truran, unpublished). Similar results were obtained in the quasi-static calculations
of Iben (1982).

The picture of cyclic evolution that therefore emerges is one, in
which the mass transfer rate in many CN systems, is the same as in DN
with the same orbital period throughout most of the time between outbursts. The system then undergoes DN eruptions. A few tens of years
prior to the outburst, the bolometric luminosity of the white dwarf
starts increasing, producing a brightening of the system (the accretion
disk is stabilized then, by heating or increased mass transfer). Following the outburst the system stays bright due to the presence of the
hot white dwarf and then returns gradually to its quiescent, low accretion rate phase. The period of reduced $\dot{M}$ ensures that a strong TNR will
ensue (Livio, Shankar and Truran 1988).

## Acknowledgements


I would like to thank Jim Pringle, Anurag Shankar, Mike Shara, Jim
Truran, Romek Tylenda and Ron Webbink for very useful discussion.

I thank the LOC and in particular Ken Nomoto for support. I am grateful to J.P. Lasota and D.A.R.C. for their hospitality at the Observatoire de Paris at Meudon. This research has been supported in part by
US NSF grant AST 86-11500 at the University of Illinois.


## References


Cannizzo, J.K. and Wheeler, J.C., 1984, $\underline{Ap.J.Suppl.}$, **55**, 367.
Edwards, D.A. and Pringle, J.E., 1987, preprint.
Faulkner, J., Lin, D.N.C. and Papaloizou, J.C.B., 1983, $\underline{MNRAS}$, **205**, 359.
Fujimoto, M. 1982, $\underline{Ap.J.}$, **257**, 767.
Gallagher, J.S. and Starrfield, S.G. 1978, $\underline{Ann.Rev.Astr.Ap.}$, **16**, 171.
Hameury, J.M., King, A.R., Lasota, J.P. and Ritter, H. 1987, $\underline{MNRAS}$,
    submitted.

Iben, I., Jr., 1982, Ap.J., **259**, 244.
Kovetz, A., Prialnik, D. and Shara, M.M. 1987, preprint.
Kutter, G.S. and Sparks, W.M. 1980, Ap.J., **239**, 988.
Lamb, D.Q. and Melia, F. 1987 in The Physics of Accretion onto Compact Ob-
    jects, eds. K.O. Mason, M.G. Watson and N.E. White (Berlin : Springer
    Verlag) p. 113.
Livio, M. 1987, Comments on Astrophys., **12**, 87.
Livio, M. and Shara, M.M. 1987, Ap.J., **319**, 819.
Livio, M., Shankar, A. and Truran, J.W. 1988, Ap.J., in press.
Mestel, L. and Spruit, H.C. 1987, MNRAS, **226**, 57.
Meyer, F. and Meyer-Hofmeister,E. 1983, Astron.Ap., **121**, 29.
Mineshige, S. and Osaki, Y. 1983, Pub.Astr.Soc.Japan, **35**, 377.
Patterson, J. 1984, Ap.J.Suppl., **54**, 443.
Prialnik, D., Livio, M., Shaviv, G. and Kovetz, A. 1982, Ap.J., **257**,312.
Ritter, H. 1987, Astron.Ap.Suppl., in press.
Robinson, E.L. 1975, A.J., **80**, 515.
Schaefer, B.E. and Patterson, J. 1983, Ap.J., **268**, 710.
Shara, M.M., Moffat, A.F.J., McGraw, J.T., Dearborn, D.S., Bond, H.E.,
    Kemper, E. and Lamontague, R. 1984, Ap.J., **282**, 763.
Shara, M.M., Moffat, A.F.J. and Webbink, R.F. 1985, Ap.J., **294**, 286.
Shara, M.M., Livio, M., Moffat, A.F.J. and Orio, M. 1986, Ap.J., **311**,163.
Truran, J.W. 1982 in Essays in Nuclear Astrophysics, eds. C.A. Barnes,
    D.D. Clayton, and D.N. Schramm (Cambridge : Cambridge University
    Press), p. 467.
Truran, J.W. and Livio, M. 1986, Ap.J., **308**, 721.
Verbunt, F. and Zwaan, C. 1981, Astron.Ap., **100**, L7.
Verbunt, F. and Wade, R.A. 1984, Astron.Ap.Suppl., **57**, 193.
Warner, B. 1985, in ESA SP-236, Recent Results on Cataclysmic Variables
    p.1.
Warner, B. 1986, MNRAS, **219**, 751.
Warner, B. 1987, MNRAS, **227**, 23.
Webbink, R.F., Livio, M., Truran, J.W. and Orio, M. 1987, Ap.J.,**314**,653.

SPECTRAL PECULIARITIES IN NOVA VULPECULAE 1984 N° 2

Yvette ANDRILLAT, Laboratoire d'Astronomie, Université des Sciences et Techniques du Languedoc, Montpellier, France.
Léo HOUZIAUX, Département d'Astrophysique, Université de Mons, Mons, Belgium.

This slow nova has been observed spectroscopically at Observatoire de Haute-Provence in two wavelengths regions : 330-460 nm (plate factors : 4 and 8 nm), and 750-1100 nm (plate factors : 23 and 26 nm/mm).
Time coverage includes the following stages : close to maximum of light, early nebular phase and advanced nebular stage.

1. Days 8 to 12 after maximum.

Spectra are similar to those of classical novae : strong continuum, wide emission bands with double structure, flanked by a moderate strength absorption component at -1400 km/s.The following ions are identified : H I, O I, C I, N I, Mg I and II, Ca I and II, Fe II, Ti II, Si II and Sr II.
Magnesium lines are remarkably strong : Mg I at 880.6 nm is detected for the first time in a nova spectrum while the Mg II doublet at 787.7 and 789.6 nm is quite conspicuous. Measurements are a lot easier than in the blue region crowded with blends. The strength of the Magnesium features is confirmed by IUE observations.

2. Days 273 to 296 after maximum.

At this early nebular phase the continuum has dropped and neutral or weakly ionized elements are no more observed, with the exception of O I 844.6 nm (probably excited by Hydrogen Lyman β coïncidence). Lines belonging to the following spectra are present : He I (1083 nm is very strong), He II, N III. Forbidden lines belong to [O III], [O II], [Ne III], [S II], [Ne V]. [Fe III] at 376 nm, and [N I] at 1039.5 and 1040.4 nm are also seen. Lines are broad (1450 km/s) and structureless. Although the spectrograph sensitivity is very low around 330 nm, the [Ne V] lines are saturated indicating an abnormally high strength for these features.

3. Days 491, 565 to 578, 628 and 641, 893 and 922 after maximum.

During this extended period, we observe a stengthening of the nebular lines due to [O II], [O III], [Ne III] and [Ne V]. Many changes occur in the near infrared. On day 496, the visual brightness variation ΔMv is still as low as 5.4 magnitudes, but we note an important weakening of the Hydrogen Paschen lines compared to day 273, with a very similar ΔMv of 5.2. The same occurs for [N I]

232

and  O  I  lines.    On the other hand, He I (1083.0 nm) and He II (1012.4 nm) remain
strong while $[\text{S III}]$ lines (906.9 and 953.2 nm) increased, as well as  a  line  at
991.8  nm of the highly ionized ion S VIII.  The presence of a "coronal" type plasma
is confirmed by the $[\text{Fe VII}]$ (at 376.0) and $[\text{Fe XI}]$ lines at 789.1 nm.

We attempted to evaluate energy radiated per unit time in the  $[\text{Ne III}]$  and
$[\text{Ne V}]$ lines  on  day  577  in comparing the nebular line intensities to 109 Vir
observed under the same conditions.  The resulting fluxes F are shown in  the  table
below with an estimated error of the order of 35% :

| Ion | Line (nm) | F (ergs cm$^{-2}$ s$^{-1}$) | $L_{line}/L_s$ |
|---|---|---|---|
| Ne$^{2+}$ | 386.9 | $1.0 \ 10^{-10}$ | 29.1 |
| Ne$^{2+}$ | 396.7 | $2.2 \ 10^{-10}$ | 61.3 |
| Ne$^{4+}$ | 334.6 | $8.6 \ 10^{-11}$ | 24.1 |
| Ne$^{4+}$ | 342.6 | $3.9 \ 10^{-11}$ | 10.4 |

In  the last column of the table, we estimated the power radiated in each line,
in units of solar luminosity, referring to a distance of 3 kpc, following  Gehrz  et
al. (Astrophys.  J.  _298_, L 47, 1985).  These figures may be compared to the 80 $L_s$
radiated in the 12.8 m $\mu$ $[\text{Ne II}]$ line on day 140, according to these authors.

The observed  intensity  ratio  F(334.6)/F(342.6)  =  0.365,  compared  to  its
theoretical  value  for an optically thin case (0.475) does not indicate any optical
depth effect in the $[\text{Ne V}]$ lines, especially if we realize that the $[\text{Ne III}]$ $^1$S
$-^1$D  line at 334.2 nm may also contribute to the strength of 334.6 nm.
The same conclusion can be also drawn from the F(386.9)/F(396.7) ratio, for which we
observe a 3.3 value instead of 2.3, noting however that neighbouring Hydrogen  lines
somewhat vitiate these measurements.
He  I,  He  II, $[\text{O III}]$  and  $[\text{S VIII}]$ lines continue to increase on day 629.
Another noticeable feature is the well marked P Cygni profile of  C  IV  (770.9  and
772.6  nm).   The  absorption  component is displaced by -3000 km/s.  It indicates a
strong stellar wind leading to a high mass loss rate.  Such a velocity  has  already
been  found  for the C IV ions in the resonance doublet of C IV at 154.8 - 155 nm in
Nova HR Del.

On day 893, the visual brightness decrease since maximum is about 7  magnitudes
and  the  strongest features in the near infrared are the $[\text{S III}]$ lines with reach
about 10 times the local continuum.  $[\text{S VIII}]$ has dropped dramatically, but  other
coronal  lines  develop, namely $[\text{Fe XI}]$ , $[\text{Ni XV}]$ and the nebular lines $[\text{A III}]$
at 713.5 nm and $[\text{Cl IV}]$ at 804.6 nm.
An unidentified emission around 825.2 nm arises around day 500. The full paper will
be published elsewhere.

THE CHEMICAL COMPOSITION OF THE WHITE DWARFS IN CATACLYSMIC
VARIABLE SYSTEMS WHICH PRODUCE NOVAE

Warren M. Sparks
Los Alamos National Laboratory

Sumner G. Starrfield
Arizona State University and Los Alamos National Laboratory

James W. Truran
University of Illinois

G. Siegfried Kutter
Lab for Astronomy and Solar Physics, NASA/GSFC

Recently a number of studies have been published on the nuclear
abundance of nova ejecta, as summarized by Truran and Livio (1986).
H is always underabundant (compared to solar) and He is overabundant
except for the cases where the heavier elements are far overabundant.
The abundances of C, N, and O range from nearly solar to highly overabundant.  A few novae are very rich in Ne and Mg as well as O, which
has led to the discovery that these novae occur on O/Ne/Mg white dwarfs
(Williams, et al., 1985).  We will assume that the abundances are an
accurate and consistently determined set of data for our purposes.  The
nova ejecta is a combination of original white dwarf material, remnant
material, remaining on the white dwarf from the previous outburst, and
accreted material, all of which has undergone thermonuclear processing
during the outburst.  The question we address here is "Can we untangle
the observational abundances to determine the contributions of each
source?"  A positive answer would allow us to tell whether the white
dwarf's mass is increasing or decreasing and thus have implications on
the accreting white dwarf model for a SNI.

We will make the following assumptions: 1) the accreted material is
of a solar composition; 2) the original white dwarf material is either
He or C/O or O/Ne/Mg; 3) during an outburst the material is well mixed
such that the ejected material and the remnant material initially have
the same abundance; 4) the H and C of any remnant material are burned
to He and N before next accretion period; 5) all of the excess N comes
from proton capture unto C and little O is created or destroyed; 6) He
depletion is negligible; 7) the ejected material is composed of white
dwarf material, remnant material and accreted material (hereafter referred to as the components of the ejecta); and 8) two sequential out-

bursts have similar composition.  Assumptions 3, 4, 5 and 6 are based
on nova calculations.

Table I gives the component fractions and the white dwarf compositions for various novae.  In all cases, a remnant component fraction
of zero is possible.  Naturally the remnant component is ultimately
composed of white dwarf and accreted material and the first values given
represent their relative contributions.  Table I is divided into C/O
and O/Ne/Mg which represents the types of white dwarfs that the outburst
has occurred on.  The four C/O novae show a surprisingly consistant
~50%C and 50%O for the white dwarf composition, and the white dwarf
component of the ejecta ranges from 10-50%.  For the two O/Ne/Mg novae,
the white dwarf component varies from 40 to 85%.  The very high abundance of S and the relative low abundance of Mg and Si for the white
dwarf of V1370 Aql are very difficult to understand from a stellar
evolution point of view.  A nova with nearly solar heavy element abundance like RR Pic, HR Del and USco could have erupted on a He envelope
white dwarf that is decreasing in mass or on any white dwarf that is
increasing in mass.

Table I

|  | C/O Novae | | | | O/Ne/Mg Novae | |
|  | T Aur | V1500 Cyg | V1668 Cyg | DQ Her | V693 CrA | V1370 Aql |
|---|---|---|---|---|---|---|
| WD | .12 − .08 | .26 − .24 | .30 − .25 | .53 | .39 − .24 | .86 − .41 |
| Rem | .0 − .31 | .0 − .11 | .0 − .15 | .0 | .0 − .39 | .0 − .53 |
| Acc | .88 − .61 | .74 − .65 | .70 − .60 | .47 | .61 − .37 | .14 − .06 |

| | | T Aur | V1500 Cyg | V1668 Cyg | DQ Her | | V693 CrA | V1370 Aql |
|---|---|---|---|---|---|---|---|---|
| WD | C | .57 | .51 | .56 | .45 | O | .55 | .24 |
|  | O | .43 | .49 | .44 | .55 | Ne | .43 | .53 |
|  |  |  |  |  |  | Mg | .02 | .01 |
|  |  |  |  |  |  | S | − | .22 |

In summary, those novae which show large metal overabundances are
decreasing in mass while those that show an overabundance of He can be
either increasing or decreasing.  Obviously, only the latter that are
increasing in mass can eventually produce a SNI.  U Sco seems to be
the best prospect (Starrfield, Sparks, and Truran 1985).

BIBLIOGRAPHY:
Starrfield, S., Sparks, W. M., and Truran, J. W. 1985, Ap.J., 291, 136.
Truran, J. W., and Livio, M. 1986, Ap.J., 308, 721.
Williams, R. E., Sparks, W. M., Starrfield, S., Ney, E. P., Truran, J. W.
and Wyckoff, S., 1985, M.N.R.A.S., 212, 753.

NUMERICAL MODELLING OF THE CLASSICAL NOVA OUTBURST


G. Siegfried Kutter[*]
Lab. for Astronomy and Solar Physics, NASA/Goddard Space Flight Center

and

Warren M. Sparks
Applied Theoretical Physics Division, Los Alamos National Laboratory



Abstract:  We describe a mechanism that promises to explain how classical nova outbursts take place on white dwarfs of 1 $M_\odot$ or less and for accretion rates of 4 x
$10^{-10}$ $M_\odot$ $yr^{-1}$ or greater.


Observations suggest that the average mass of white dwarfs in classical nova systems
is 0.8 $M_\odot$ and that the typical mass transfer rate in such systems is $10^{-9}$ - $10^{-8}$ $M_\odot$
$yr^{-1}$ (Patterson 1984).  However, model calculations that take the accretion process
into account have been unsuccessful in reproducing the classical nova outburst for
these parameters.  The reason is insufficient confinement of the nuclear runaway zone
and, hence, insufficient release of energy (Sparks and Kutter 1987).  To overcome this
problem, white dwarf masses equal to or greater than 1.25 $M_\odot$, low accretion rates, or
hybernation scenarios have been invoked (Kutter and Sparks 1980, Livio and Shara 1987,
Prialnik and Kovetz 1984, Starrfield et al. 1985).

We describe here a mechanism that promises to produce nova outbursts for the observational data listed above, though to date we have explored it only for these parameters:

> white dwarf         1 $M_\odot$, consisting of $^{12}C$ and $^{16}O$, $10^{-3}$ $L_\odot$;
> accreting matter    4.23 x $10^{-10}$ $M_\odot$ $yr^{-1}$, solar mix with $^{12}C$ and $^{16}O$ each enhanced
>                      to 10% by mass, arriving on white dwarf with Keplerian angular
>                      momentum.

The computational method is described in Kutter and Sparks (1987) and Sparks and Kutter
(1987).

The shear created on the white dwarf's surface by the rotational velocities of the
accreting H-rich matter produces an accretion belt of mixed stellar and accreted matter.  After 7,080 years, the surface of the belt rotates as rapidly as the newly arriving matter and shear mixing ceases.  Accretion continues to age 256,400 years, when
nuclear burning exceeds $10^6$ erg/(g·sec) at a fractional mass from the star's surface
of Q = 2.570 x $10^{-4}$ and the nuclear runaway gets under way.  Up to this point the

---


[*] National Research Council -- NASA/GSFC Research Associate, on leave from The Ever-
green State College, Olympia, WA

star's evolution resembles the models described in Sparks and Kutter (1987); and, if
the evolution calculations are continued, 7.6 years later the nuclear runaway peaks
with $\varepsilon_{nuc}$ = 2.26 x $10^{15}$ erg/(g·sec). No nova results.

According to our experiences, a nova outburst will occur on a 1 $M_\odot$ white dwarf and for
accretion parameters shown above only if the nuclear reactions run away at greater
depth than in the above calculations and/or if the white dwarf's envelope rids itself
of some of its angular momentum during the early phase of the runaway.  The greater
depth increases the gravitational confinement of the burning region; and the loss of
angular momentum increases the mechanical confinement, due to reduction of the centrifugal forces.

Is there a mechanism capable of moving, during the early phase of the nuclear runaway,
the nuclear runaway zone to greater depth and of ridding the envelope of angular momentum?  We suggest that it is the convective instability, which in the course of the runaway grows until it reaches the star's surface.  Convection mixes matter in the envelope's outer rapidly rotating layers with matter in the inner less rapidly rotating
layers and thereby transports angular momentum to the nuclear runaway zone.  The resulting shear between that zone and the adjacent layer interior to it creates a Kelvin-
Helmholtz instability, which in turn leads to mixing of matter.  We think that this
mixing moves both the zone of peak nuclear burning and angular momentum inward.  It
also mixes H-rich matter with layers rich in carbon and oxygen.

To test whether the Kelvin-Helmholtz instability, induced by convection, is capable of
producing nova outbursts, we ran our model through a nuclear runaway, artificially
moving the zone of peak nuclear burning inward to $Q$ = 3.072 x $10^{-4}$ and removing 10% of
the envelope's angular momentum.  Furthermore, we enhanced the $^{12}$C and $^{16}$O content of
the accreting matter (see above) to simulate the mixing of H-rich and original white
dwarf matter.  The nuclear reaction rates rise to 2.52 x $10^{17}$ erg/(g·sec) and the entire envelope above the runaway zone is ejected with speeds of 190 to 1,600 km $sec^{-1}$.
The mechanism seems to work because rotational kinetic energy of the accreting matter
is not immediately released, but only when it is needed to increase the burning region's confinement during the early phase of the nuclear runaway.  Details of the
results will be presented in an Ap. J. paper now in preparation.


BIBLIOGRAPHY:



Kutter, G. S., and Sparks, W. M. 1980, Ap. J., 239, 988.
_________. 1987, Ap. J., 321, 386.
Livio, M., and Shara, M. M. 1987, Ap. J., 319, 819.
Patterson, J. 1984, Ap. J. Suppl., 54, 443.
Prialnik, D., and Kovetz, A. 1984, Ap. J., 281, 367.
Sparks, W. M., and Kutter, G. S. 1987, Ap. J., 321, 394.
Starrfield, S., Sparks, W. M., and Truran, J. W. 1985, Ap. J., 291, 136.

# SUPEROUTBURSTS AND SUPERHUMPS OF SU UMA STARS


Yoji OSAKI and Masahito HIROSE

Department of Astronomy, Faculty of Science, University of Tokyo,

Bunkyo-ku, Tokyo 113


## 1. INTRODUCTION

SU UMa stars are one of subclasses of dwarf novae. Dwarf novae are semi-detached close binary systems in which a Roche-lobe filling red dwarf secondary loses matter and the white dwarf primary accretes it through the accretion disk. The main characteristics of SU UMa subclass is that they show two kinds of outbursts: normal outbursts and superoutbursts. In addition to the more frequent narrow outbursts of normal dwarf nova, SU UMa stars exhibit "superoutbursts", in which stars reach about 1 magnitude brighter and stay longer than in normal outburst. Careful photometric studies during superoutburst have almost always revealed the "superhumps": periodic humps in light curves with a period very close to the orbital period of the system. However, the most curious of all is that this superhump period is not exactly equal to the orbital period, but it is always longer by a few percent than the orbital period.

## 2. MODEL FOR SUPEROUTBURSTS AND SUPERHUMPS

One of the present authors proposed a model for superoutbursts and superhumps of SU UMa stars [1]. In this model, the normal outbursts of SU UMa stars are thought to be essentially the same as the ordinary outbursts of dwarf nova and they are supposed to be caused by the disk instability (see e.g., Smak [2]). Superoutbursts are explained in the following way. The heating of the secondary's atmosphere by strong far UV and soft X-ray radiation due to accretion and the resulting increase in mass-overflow rate from the secondary may lead to a positive feed back instability between accretion and mass overflow in a certain circumstance. This "irradiation-induced mass-overflow instability" is the suggested mechanism of "superoutburst" of SU UMa stars.

The "superhump" phenomenon may be explained by this model in the following way. It is hypothesized that a slowly precessing eccentric accretion disk develops during superoutbursts. (The eccentric disk was first proposed by Vogt [3] in a slightly different context with the present model). Irradiation on the secondary's atmosphere varies periodically because of varying shadow cast by the eccentric disk. Since mass overflow rate from the secondary is strongly influenced by irradiation effect during superoutburst, the mass transfer rate itself will also vary periodically. The superhump phenomenon may be caused by this periodic variation in the mass transfer rate. The period of this mass transfer variation (and therefore the period of superhump) is determined by the repetition period of the same secondary and eccentric disk configuration (i.e., the synodic period). The latter period is not exactly equal to the orbital period, but it is slightly longer than that, because the apsidal line of the eccentric disk is not fixed in space but it advances slowly in the same direction as the orbital motion of the secondary. This is the reason why the superhump period is slightly longer than the orbital period of the binary.


## 3. PRECESSING ECCENTRIC DISK AND LIGHT CURVE SIMULATION

If a precessing eccentric disk ever exists, its effects should show up in light curves for systems with high orbital inclination. That is, the light from accretion disk may vary with the precessing period because of the variable aspect of non-axisymmetric disk from the observer. Furthermore, if the system is an eclipsing binary, asymmetric eclipse light curves may result. Such effects have in fact been observed in an eclipsing SU UMa star called OY Car. Schoembs [4] has found that the brightness level of OY Car outside of hump and eclipse was modulated with a long period of about 3 days, which is the beat period between the orbital and superhump periods. Krzeminski and Vogt [5] have also found that OY Car during supermaximum exhibited asymmetric eclipse light curves and the sense of asymmetry oscillated with the beat period. These observations may most naturally be explained by the precessing eccentric disk.

In order to confirm this and to find the configuration of the eccentric disk, we have tried to simulate these observations by adopting a simple eccentric disk model and by calculating light curves. It is found that the basic feature of the two phenomena (i.e., brightness variation outside of eclipse and asymmetric eclipse light curves) can be reproduced with appropriate parameters describing the precessing eccentric disk and the secondary. However, it has turned out that the orientation of the apsidal line of the eccentric disk has to be chosen almost in the opposite direction for the two cases. In order to explain the timing of asymmetric eclipse observed by Krzeminski and Vogt [5], the supermaximum must occur when the apsidal line of the eccentric disk is directed to the secondary (consistent with Vogt's model). On the other hand, in order to explain the phase of the brightness modulation outside of eclipse observed by Schoembs [4], the superhump maximum must occur when the apsidal line is directed away from the secondary (consistent with Osaki's model). So far this kind of observations has been limited to the two observations mentioned above. More observations are clearly needed either to confirm or reject the existence of the eccentric disk during superoutbursts.


## References


1. Osaki, Y., 1985, *Astron. Astrophys.*, **144**, 369.
2. Smak, J., 1984, *Publ. Astron. Soc. Pacific.*, **96**, 5.
3. Vogt, N., 1982, *Ap. J.*, **252**, 653.
4. Schoembs, R., 1986, *Astron. Astrophys.*, **158**, 233.
5. Krzeminski, W. and Vogt, N., 1985, *Astron. Astrophys.*, **144**, 124.

Accretion-Disk-Instability Model for Outbursts of FU Orionis

Hitoshi HANAMI
Department of Physics, Hokkaido University
060 Sapporo, JAPAN

**1. Introduction;**    The  FU Orionis objects show us the abrupt brightening by ~5
mag.   The  two  best studied examples, FU Ori and V1057Cyg, brightened and have
remained  very luminous for years and began to fade gradually (cf. Herbig 1977).
On  the other hand, a large number of young stellar objects have been discovered
with  energetic molecular bipolar outflows ( Lada 1985, for a review ). On high-
resolution  radio  observations, the disks have been detected around the central
infrared  sources  (cf.  Kaifu  et al. 1984, Hasegawa et al. 1984).  Most of the
disks  seem  to  be  perpendicular  to  the bipolar outflows.  These observations
suggests  that  the  disk is strongly related with the energetics phenomena like
bipolar  outflow  in star forming regions. (cf.Okuda and Ikeuchi 1986, Hanami and
Sakashita 1986, Pudritz 1985, Shibata and Uchida 1985)

**2. Models  for  Self-Gravitationally  Unstable  Accretion Disk;**    Hartmann and
Kenyon  (1985,1987)  have proposed the possibilities that the outburst of FU Ori
is  similar to those proposed for dwarf nova (e.g. Osaki 1974, Hoshi 1979, Meyer
and  Meyer-Hofmeister  1981,  Smak 1982, Cannizzo and Wheeler 1984, and Mineshige
and  Osaki  1983).  Their  analysis  is  essentially  based  on  the  thermal
instabilities in accretion disk for standard $\alpha$-model (Shakura and Sunyaev 1973).
On  the  other hand, recently, Lin and Pringle (1987) propose a new prescription
for  treating  the  transfer of angular momentum within a gaseous differentially
rotating  disk  to  self-gravitational  instability  in  terms  of  an effective
kinematic viscosity.  Then, we consider <u>the unsteady accretion disk model for FU
Orionis phenomena with hybrid prescription of self-gravitational instability and
the viscosity of standard $\alpha$-disk.</u>  This  instability  may  induce viscosity
increase and modulate mass transfer throughout the disk.

**3. Hysteresis  Feature  in  $(\nu\Sigma,\Sigma)$  Plane;**    Following Lin and Pringle (1987), we
now  consider this hybrid prescription.  Firstly, according to Toomre (1964), The
basic  principles  of the self-gravitational instability in a thin rotating disk
can  be  elucidated.   If the size $\lambda$ of the disturbances is greater than $\lambda_c \sim G\Sigma\Omega^{-2}$
($\Sigma$; surface density, $\Omega$;angular velocity), the perturbation are stabilized by the
shear.  Over  the size of region $\lambda_c$, angular momentum with this turbulent eddies
is  transferred  in time scale $\Omega^{-1}$.  Then, we can write  $\nu_s \sim \lambda_c^2\Omega = (G\Sigma)^2/\Omega^3 \sim$
$Q^{-2}H^2\Omega$, where  $Q \sim C_s\Omega/G\Sigma$  ($C_s$; sound velocity).  This viscosity prescription
would  be  valid  for  $\lambda_j < \lambda_c$, which equivalent to $Q\lesssim 1$, the criterion by Toomre
(1964).   This  criterion  give  the relation $\Sigma > C_s\Omega/G = \Sigma_{cl}$.  This prescription
for  effective  viscosity  is  equivalent  to  the standard $\alpha$ model by Shakura and
Sunyaev (1973) with replacing $\alpha$ by $Q^{-2}$, but $Q^{-2} > 1$.
On  the other hand, when the self-gravity does not dominate in the accretion
disk,  the prescription for the effective viscosity should become the standard $\alpha$-
disk prescription in which typical turbulent eddy scale $\lambda$ is nearly the hight of
the  disk.   According to standard $\alpha$-disk model, we get for viscosity, $\nu_\alpha = \alpha C_s H$
$=\alpha C_s^2/\Omega$.  We  obtain  the  criterion  of  the  validity for the standard $\alpha$-disk

prescription $\Sigma < \Omega^2 r/G = \Sigma_{c2}$, from that the self-gravity of disk must be small compared with the centrifugal force. Then, we adopt the $(\nu\Sigma, \Sigma)$ relatioship $\nu\Sigma = G\Sigma^3/\Omega^3$, ($=\alpha c_s^2\Sigma/\Omega$) in state 1,(2) when $\Sigma > \Sigma_{c1}$, ($\Sigma < \Sigma_{c2}$). The hysteresis feature of the relationship in $(\nu\Sigma, \Sigma)$ plane is sketched in a figure below.

## 3. Applications to Protostellar Disk and Outburst of FU Orionis

We consider the criterion for the validity of giving occasion to the recurrence outburst of FU Orionis in our model. As shown in Figure, there are three condition which must be satisfied for our model. (A) <u>the viscosity due to the prescription of self-gravitational instabilities is higher than that of standard $\alpha$-disk prescription,</u> $\nu_\alpha < \nu_s$. (B) in the phase plane of $(\nu\Sigma, \Sigma)$, <u>two different prescription branches coexist for the same value of $\Sigma$,</u> $\Sigma_{c1} < \Sigma_{c2}$. (C) $\underline{\Sigma}$ <u>must increases in low-viscosity state,</u>

$\dot{M}_{acc} > \nu_\alpha \Sigma$. From (A) and (B),

$$2.2\ M_\odot\ \alpha\left(\frac{C_s}{0.34\text{km/s}}\right)^{-2}\left(\frac{\Sigma}{10^2 g\cdot cm^{-2}}\right)^2\left(\frac{R_d}{2\cdot10^{15}\ cm}\right)^3 > M_c$$

$$> 0.039\ M_\odot\ \left(\frac{C_s}{0.34\text{km/s}}\right)^2\left(\frac{R_d}{2\cdot10^{15}\ cm}\right),$$

where $R_d$ and $M_c$ are the radius of a disk and the mass of central star. This mass range constraint naturally explains the reason why FU Orionis phenomenon occurs for only low mass star ($\sim 1M_\odot$) in the early phase of the stellar evolution. From (C),

$$\dot{M}_{acc} > 1.7\times10^{-5}\ M_\odot\ yr^{-1}\ \alpha^{6/5}\left(\frac{\Omega}{10^{-10}s^{-1}}\right)^{-4/5}.$$

The accretion flow with this instability will produce strong accretion shocks, turbulence, and heating of the outer layer of the star. Matter will also be ejected, and this could account for much of the mass and energy from young stars. The luminous FU Ori flare-ups may be important energy source for driving molecular out flows around low mass young stars.

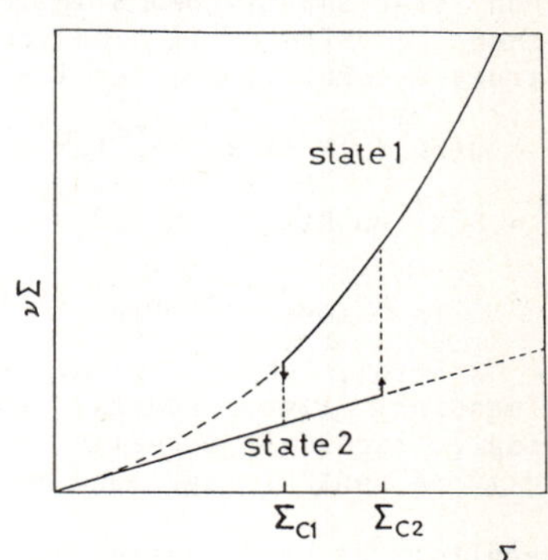


## References


Cannizzo, J. K. and Wheeler, J. C., 1984, Ap. J. Suppl., 55, 367.
Hanami, H., 1987, "NATO ASI. Galactic and Extragalactic Star Formation".
Hanami, H. and Sakashita, S., 1986, "IAU Symposium No. 115 on Star Forming Regions", edited by Peimbert and Jyugaku.
Hartman, L. and Kenyon, S. J., 1985, Ap. J., 299, 462.
Hartman, L. and Kenyon, S. J., 1987, Ap. J., 312, 243.
Hasegawa, T., et al., 1984, Ap. J., 283, 117.
Herbig, G.H., 1977, Ap.J., 214, 747.
Hoshi, R., 1979, Prog. Theor, Physics, 61, 1307.
Kaifu, N., et al. 1984, A. Ap., 134, 7.
Lada, C. J., 1985, Ann. Rev. Astrophys. 23, 267.
Lin, D. N. C. and Pringle, J. E., 1987, M.N.R.a.S., 225, 607.
Meyer, Y. and Meyer-Hofmeister, E., 1981, A. Ap., 104, L10.
Mineshige, S. and Osaki, Y., 1983, Publ. A, Soc, Japan, 35, 377.
Okuda, T. and Ikeuchi, S., 1986, Publ. A, Soc, Japan, 38, 199.
Osaki, Y., 1974, Publ. A, Soc, Japan, 26, 429.
Pudritz, R. E., 1985, Ap. J., 293, 216.
Shibata, K. and Uchida, Y., 1985, Publ. A, Soc, Japan, 37, 31.
Smak, J., 1982, Acta. Astron., 32, 199.

# Nonlinear Oscillation of the Magnetosphere around Neutron Stars


Hitoshi HANAMI
Department of Physics, Hokkaido University
Sapporo 060, JAPAN.



**SUMMARY;**  We investigate the unsteady motion of mass reservoir formed by the accretion onto the magnetosphere around rotating neutron stars. The unsteady motion of the reservoir induces secondary accretion to neutron star by R-T instability. The nonperiodic or quasiperiodic phenomena of X-ray bursters seems to be related to this property of mass reservoir on the magnetosphere. We classify the typical dynamical state of the reservoir into three types with the parameters which are accretion rate $\dot{M}_{acc}$ and angular velocity of neutron star $\Omega_s$. They are nonsequential oscillation, sequential periodic (quasi-periodic) oscillation, and chaotic oscillation states.


1. **Physical Model;**  We propose a symplified model for the non-linear-like phenomena in X-ray sources, considering the properties of Rayleigh-Taylor instability on the magnetopause which is formed by the accreting matter to the neutron star, and then obtain the motion of this magnetopause.
   We assume that the stellar magnetic field is dipolar ($\propto \tilde{\omega}^{-3}$), and has axial symmetry everywhere. We use cylindrical coordinates ($\tilde{\omega}$, $\Phi$, $z$) centered on the neutron star and aligned with the stellar rotation axis. This configuration is sketched in Figure 1. We obtain the nondimensionalized equations which construct a complete set for the dynamics of reservoir ring, as following,

$$\Sigma\frac{dV}{dt} = Q(Bp^2-\alpha)x +1/2\cdot(-x^{-2}+2H^2x^{-3}) -(Vx^{(d+1)}+\dot{\Sigma}_{acc})^2\cdot x^{-(d+1)}\leftarrow(1), \quad \frac{dx}{dt} = V \leftarrow(2),$$

$$\Sigma\frac{dH}{dt} = 2Qx^2\cdot Bp\cdot Bt +\dot{\Sigma}_{in}(x/2)^{1/2}\leftarrow(3), \quad \frac{dBt}{dt} = \frac{Bp}{L}(H/x -\Omega x) -\dot{\Sigma}_{loss}/\Sigma\cdot Bt \leftarrow(4),$$

$$\frac{d\Sigma}{dt} = \dot{\Sigma}_{in} -\dot{\Sigma}_{loss}\leftarrow(5), \quad \dot{\Sigma}_{in}= \theta(Vx^{(d+1)} +\dot{\Sigma}_{acc}) \leftarrow(6), \quad \dot{\Sigma}_{loss}= \Sigma\cdot(\theta(\frac{dV}{dt}/x))^{1/2}\leftarrow(7),$$

$Bp = x^{-3}\leftarrow(8)$, where t, x, V, H, Bp, Bt, $\Sigma$, $\dot{\Sigma}_{acc}$, $\Omega$, and $\dot{\Sigma}_{loss}$ are the nondimensinal time, radius, radial velocity, specific angular momentum, mass, poloidal, toroidal magnetic field of the reservoir, accretion rate, angular velocity of central star, and mass loss rate from reservoir.

Figure 1

3. **Results;**  We have started our calculation giving the initial condition fixed on $\Sigma$ = 1.0, V=0.0, x=1.0, H = 0.1, Bt=0.0, and the various values of the control parameters $\dot{\Sigma}_{acc}$ and $\Omega$. We get three types solutions, as shown below.

|  | $\Omega$ | $\dot{\Sigma}_{acc}$ | dynamics |
|---|---|---|---|
| case (a) | 1.0 | 3.00 | Periodic |
| case (b) | -1.0 | 2.50 | Expansion |
| case (c) | -1.0 | 2.53 | Chaotic |
| case (d) | -1.0 | 3.00 | Periodic |
| case (e) | 0.0 | 1.00 | Periodic |
| case (f) | -1.0 | 2.80 | Periodic |

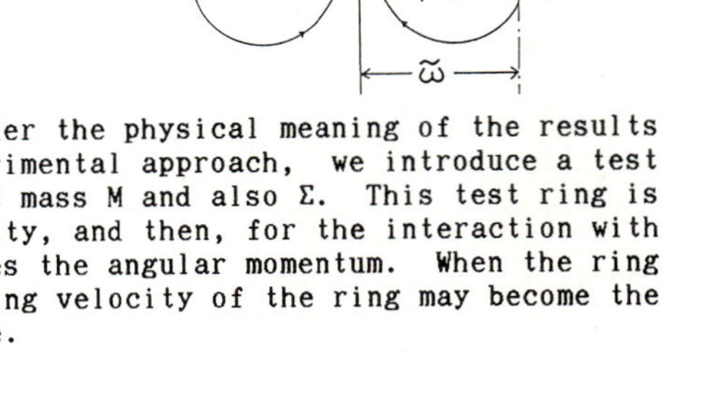

4. **Physical Meanings;**  We will consider the physical meaning of the results shown in above section. Now, for experimental approach, we introduce a test circular ring which does not change its mass M and also $\Sigma$. This test ring is initially rotating with Keplerian velocity, and then, for the interaction with stellar magnetic field, the ring losses the angular momentum. When the ring falls into the magnetosphere, the rotating velocity of the ring may become the same velocity as that of the magnetosphere.

These situations, which represent the difference between the direct and the retrograde rotation of the star against the disk, shows schematically in figure 3 (a) and (b).

Direct Rotation;        A (Stable) ——————— Periodic Oscillation
                                    ↑ Perturvation
                        B (Stable)↓
Retrograd Rotation;        +Mass Change → Chaotic Oscillation
                        D (Stable)↑        (Shift of Stable Point)

For retrograde case, the shift of two stable equilibrium points may randumly mix the oscillation modes related two points by changing of the mass. Then, the chaotic oscillation of this system may be generated.

**5. Discussion;** We introduce typical time scale and burst energy of this system.

$$td = \tilde{\omega}_0 \left( \frac{2GM_s}{\tilde{\omega}_0} \right)^{-1/2} = 0.2 \left( \frac{\tilde{\omega}_0}{10^9 \text{ cm}} \right)^{3/2} \left( \frac{M_s}{1M_\odot} \right)^{-1/2} \text{(sec)},$$

$$E = 2.58 \times 10^{38} \cdot \left( \frac{M_s}{1M_\odot} \right)^{1/2} \left( \frac{\tilde{\omega}_s}{10^6 \text{cm}} \right)^{3/2} \left( \frac{\dot{M}_{acc}}{10^{18} \text{g/s}} \right) \left( \frac{\tilde{\omega}_0}{10^9 \text{cm}} \right)^{3/2} \Delta\tau \quad \text{(erg)}.$$

where $\Delta\tau$ is nondimensionalized burst interval time.

$$B = 2.98 \times 10^{13} \cdot \left( \frac{M_s}{1M_\odot} \right)^{1/4} \left( \frac{\tilde{\omega}_s}{10^6 \text{cm}} \right)^{-3} \left( \frac{\dot{M}_{acc}}{10^{18} \text{g/s}} \right) \left( \frac{\tilde{\omega}_0}{10^9 \text{cm}} \right)^{7/2} \Delta\tau^{1/2} \quad \text{(gauss)}.$$

From X-ray observations ( van der Klis et al. 1985), the typical frequency of quasi-periodic oscillations $\omega$ = 20 ~ 40 Hz. The typical period of oscillation cases is 2td for case (a) and 5td for case (b). If the typical radius of this magnetosphere is 1000 km, typical frequency of our model is a good agreement with the observational data. This value of $\tilde{\omega}$ requires surface magnetic field log(B(G))=10. On the other hand, for Rapid Burster, the time to following burst spreads from 10s to 1000 s. These time scale requires $\tilde{\omega}$ > r$_\odot$cm, and log(B(G))>15 for explaining with our model. Furthermore, the retrograde stellar rotation against that of the disk is important for chaotic behavior of Rapid Burster for our model.

**References**
van der Klis et al., 1985, Nature, 317, 681.
Hanami, H. 1987, M.N.R.A.S., submitted.

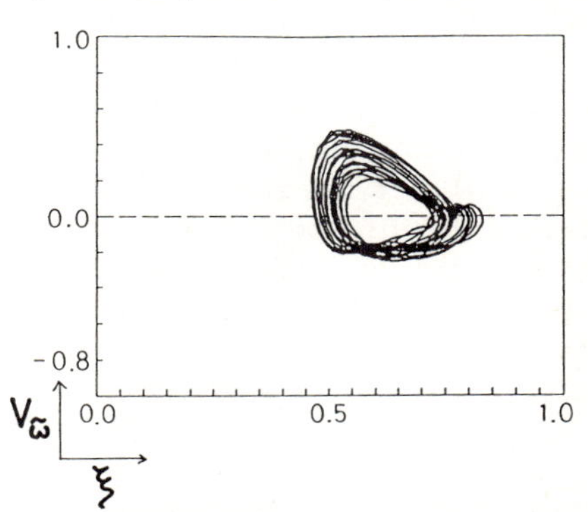


Figure 2       ;Orbits in phase space (velocity, radius) for cases (c).
Figure 3 (a);Schematic sketch of the rotation velocity as a function of the radius of test circulating ring along experimental orbit. (b);Each absolute value of the acceleration on test circular ring by magnetic pressure and effective gravity which includes the centrifugal force of rotation motion. The subscripts ↑ and ↓ of g represent the values for direct and retorograd rotation cases. Open circles A, B, C, and D represent dynamical equilibrium points.

## MASS EJECTION DURING HELIUM SHELL FLASHES FROM A MASSIVE WHITE DWARF

Hideyuki Saio[1], Mariko Kato[2], and Izumi Hachisu[3]

[1]Department of Astronomy, University of Tokyo
[2]Department of Astronomy, Keio University, Yokohama
[3]Department of Physics and Astronomy, Louisiana State University
   and Department of Aeronautical Engineering, Kyoto University

**Abstract:** We have simulated the helium shell flashes on an 1.3 $M_\odot$ white dwarf and estimated the amount of mass loss. Our results may suggest a serious difficulty for the theories of the formation of Type I supernovae and of the accretion-induced formation of neutron stars  because a significant amount of envelope mass is ejected during a helium shell flash.

Nova explosion which is caused by thermal flash of the hydrogen burning shell in an accreting white dwarf ejects a large fraction of the mass above the hydrogen burning shell (e.g., Kato 1983, Kato and Hachisu 1988). This poses a serious difficulty for the theories of the formation of Type I supernovae and of the accretion induced formation of a neutron star, because for these theories to work mass must be accumulated efficiently on the white dwarf. One way to avoid this difficulty may be to assume that the accreting matter consists mostly of helium (no hydrogen). Although nova-like explosions due to helium shell flashes may occur, the shell flashes  and hence mass ejection are expected  to be weaker because the energy liberated per unit mass  in helium burning is ten times smaller than that in hydrogen burning.  Nearly pure helium accretion onto a white dwarf is expected to occur in a double white dwarf system or a white dwarf-helium star pair. Before constructing senarios for the formations of neutron stars and Type I supernovae, we need to know how much mass is ejected during the helium shell flash as a function of the accretion rate.

First, we have examined whether optically thick wind occurs or not. Using the same method as  Kato and Hachisu's (1988) for nova outburst, we have found that the wind occurs in a massive white dwarf ( $M \geq 1.23 M_\odot$) if the radius of the envelope extends sufficiently. For a $1.3 M_\odot$ white dwarf, the wind begins when the envelope expands up to about $5 R_\odot$ and  ceases when the radius decreases to 2.3 $R_\odot$ after passing the most extended stage. The envelope mass is $3.8 \times 10^{-5} M_\odot$ at the cessation of the wind mass loss. In order to know for what accretion rate the helium shell flash is strong enough to cause mass loss, we have calculated time dependent models for helium accreting white dwarfs of 1.3 $M_\odot$ by using a Henyey type code. The calculations were started with the steady state model for an assumed accretion rate (Kawai, Saio, and Nomoto 1988). It is assumed that when the radius of a model exceeds 1 $R_\odot$ during helium shell flash, mass loss is started at an arbitrarily chosen rate of $2 \times 10^{-5}$ $M_\odot \mathrm{yr}^{-1}$. The mass loss is assumed to continue until the radius decreases down to $0.02 R_\odot$. (If the mass loss is terminated when the radius is considerably larger than the above value the radius begins to increase again.) The calculations were continued until the ratio of the lost mass to the accreted mass became constant from cycle to cycle.

The accretion rates considered are $2 \times 10^{-6}$, $1 \times 10^{-6}$, $6.6 \times 10^{-7}$, $3 \times 10^{-7}$, and $1 \times 10^{-7}$ $M_\odot \mathrm{yr}^{-1}$. The envelope mass at the ignition of helium shell flash and the strength are larger for lower accretion rates. After passing the phase of the maximum nuclear luminosity, the radiative luminosity at the photosphere and hence the radius increase. About half of the helium in the envelope is consumed before the mass loss starts. When the envelope mass decreases to $\sim 3.8 \times 10^{-5}$ $M_\odot$ by losing mass, the radius decreases enough for the accretion to start again.

Figure 1 shows the ratio of the lost mass to the accreted mass during one cycle of the helium shell flash as a funciton of accretion rate. The helium shell flash is strong enough to cause wind type mass ejection if the accretion rate is smaller than $1.7 \times 10^{-6}$ $M_\odot \mathrm{yr}^{-1}$. For smaller accretion rates, a larger fraction of the accreted mass is ejected from the system. For example, 20% of the accreted mass is ejected if $M = 1 \times 10^{-6} M_\odot \mathrm{yr}^{-1}$, while 90% of the accreted mass is ejected if $M = 1 \times 10^{-7} M_\odot \mathrm{yr}^{-1}$. The wind type mass loss affects the evolution of the binary systems because total mass and the angular momentum are lost from the system. These effects are more serious for slower mass accretion onto the white dwarf. The existing scenarios about the formations of a Type I supernova and a neutron star through accretion-induced collapse should be revised, because in these scenarios the mass of the white dwarf must be increased up to about 1.4 $M_\odot$ by mass transfer from the companion. One example of the binary evolutions taking into account the systemic mass loss due to helium shell flashes is demonstrated by Hachisu, Miyaji, and Saio (1987).

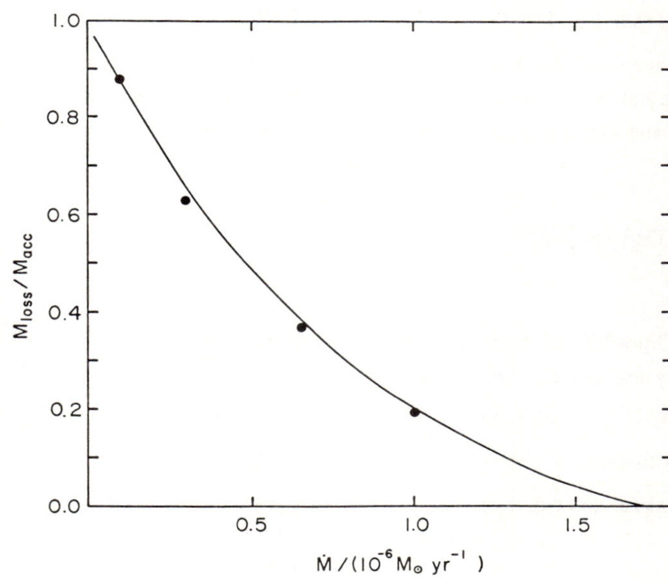

Figure 1: The ratio of the lost mass to the accreted mass for each cycle of helium shell flashes as a function of the helium accretion rate.

## References


Hachisu, I., Miyaji, S., and Saio, H. 1987, submitted to Ap. J..
Kato, M. 1983, Pub. Astr. Soc. Japan, **35**, 507.
Kato, M. and Hachisu, I. 1988, submitted to Ap. J..
Kawai, Y., Saio, H., and Nomoto, K. 1987, Ap. J. in press.

THE EVOLUTION OF THE PROGENITOR OF SN 1987A

C. de Loore[*] and C. Doom[**]
Astrophysical Institute
Vrije Universiteit Brussel, Pleinlaan 2, 1050 Brussel Belgium
* also University of Antwerp, RUCA, Belgium
** senior research assistant NFWO, Belgium

## 1. INTRODUCTION

It is now almost certain that the progenitor of SN 1987 is the B3 supergiant Sk -69°202 (e.g. Walborn et al. 1988). This provides us with valuable constraints on the SN precursor: According to West et al.(1987) the star has $M_v$=-6.8. Using the calibrations of Humphreys and McElroy (1984) we find that the precursor had $T_{eff}$ = 13000 - 16000K, $-M_{bol}$= 7.7 - 8.1 (log $L/L_{\odot}$ = 5.04) and a radius of 48 - 60 $R_{\odot}$.

The SN precursor was not a red supergiant, but a rather blue supergiant, which also explains the peculiar form of the light curve (Woosley, 1988; Schaeffer et al. 1987). However, standard evolutionary calculations would expect such stars to explode as red supergiants.

## 2.CONSTRAINTS ON THE PROGENITOR

If we compare the HRD position of the progenitor to evolutionary tracks (keeping in mind that the luminosity is nearly constant after helium ignition) we see that the progenitor had a mass between 15 and 20 $M_{\odot}$ (Fig.1). Depending on the inclusion of overshooting in the model computations, the mass at explosion may vary from 12.5 $M_{\odot}$ (overshooting) to 17.5 $M_{\odot}$ (Schwarzschild). Table 1 summarizes some data on both models. They are compatible with the findings of Woosley (1988).

Table 1: parameters of the progenitor of SN1987A

|  | Schwarzschild[1] | | Overshooting[2] | |
| --- | --- | --- | --- | --- |
| Initial Mass | 17.9 | 20.0 $M_\odot$ | 16.2 | 18.6 $M_\odot$ |
| Helium core mass | 5.4 | 6.5 | 7.3 | 9.0 |
| Mass at end of H-burning | 17.3 | 19.1 | 12.9 | 14.3 |
| Mass at explosion | 16.8 | 18.1 | 12.3 | 12.9 |

[1] Maeder and Meynet, 1987
[2] Pylyser et al., 1985


## 3. WHY WAS THE PROGENITOR BLUE?

In order to determine what effects cause a star to be a blue supergiant instead of a red supergiant we calculated a number of envelopes of stars, varying the hydrogen content and the metallicity. We chose a star of 12 $M_\odot$, of which the outer 6 $M_\odot$ are modeled. The luminosity was fixed at $10^5 L_\odot$.

The results are shown in Fig. 2. We immediately see that envelopes with larger hydrogen content and/or large metallicity produce red supergiants. For a given metallicity, the stellar radius decrteases rapidly if the hydrogen content drops below a certain value (e.g. Z = 0.03, below X = 0.6 - 0.5). Moreover this threshold value is larger for smaller metallicity.

When a star has evolved into a red supergiant its atmospheric hydrogen content decreases because the mass loss exposes helium rich layers at the surface. It is well known (e.g. Maeder, 1981) that galactic red supergiants return to the blue when the hydrogen content drops below 0.5. Since for smaller metallicity this threshold is larger, less mass loss in the red supergiant stage is needed to produce a blue supergiant. If the metallicity is low enough, it may be that no mass loss is required.

In the model by Maeder (1988), the red supergiant is peeled off until the atmospheric hydrogen abundance is low enough to make the star blue. The models of Maeder evolve along a line of constant Z in Fig.2 The same is true for the models of Wood (1988).

On the other hand the models of Arnett (1987) and Truran et al. (1987) exploit the low metallicity. Z is so low that red supergiants are never produced.

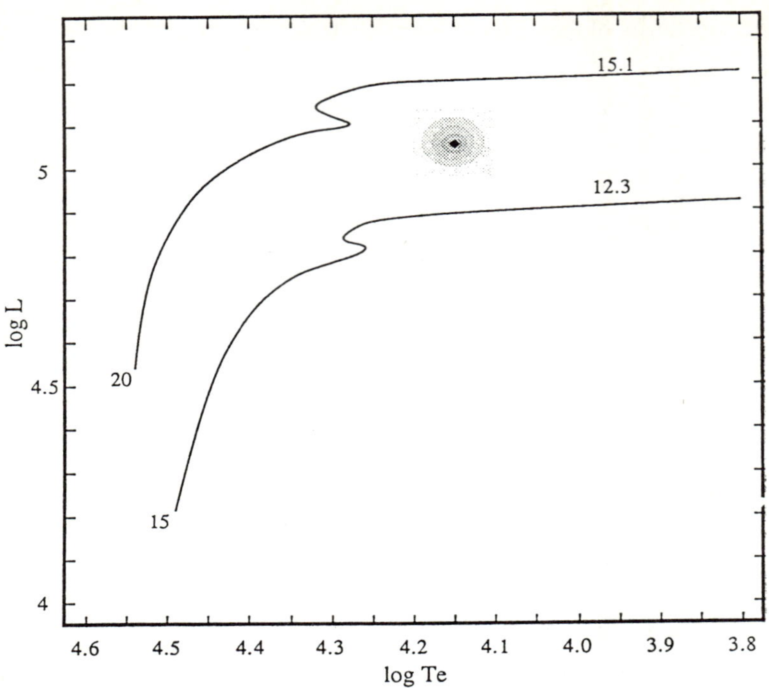



Fig. 1: The position of the progenitor of SN1987A, compared to the evolutionary tracks of a 15 $M_\odot$ and a 20 $M_\odot$ star.

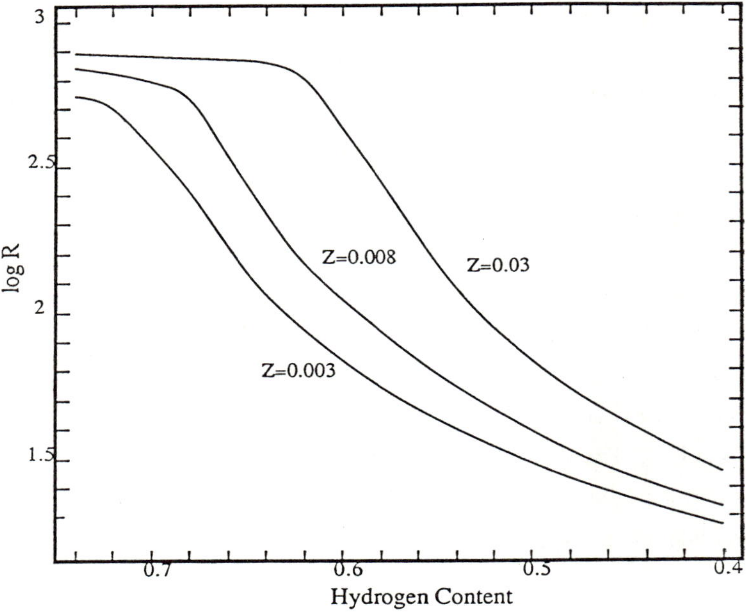



Fig. 2: The radius of supergiant envelopes calculated with varying hydrogen content X and metallicity Z.

# 4. A REVIEW OF SOME CALCULATED MODELS

## a) MAEDER (1988)

This model starts from a 20 $M_\odot$ star, adopting a moderate amount of overshooting ($d_{OV}$ = 0.3 $H_p$) and a mass loss rate of 3/5 of the values of de Jager et al. (1987) for the Galaxy. The evolution is followed up to the end of carbon burning.

During hydrogen burning the star loses about 0.5 $M_\odot$. It then evolves to the red supergiant region, loses a large part of its hydrogen envelope and returns to the blue. The final masses range between 8.6 and 10 $M_\odot$, depending on the (huge) mass loss rate as a red supergiant.

There is a large sensitivity of the final radius on the mass loss rate: if the mass removed is not large enough, the star remains a red supergiant, if too much mass is removed the star "shoots over" to the left and becomes a hot Wolf-Rayet like-star. The reason for this is the presence of a composition gradient at the surface at the time of explosion. This graqdient is left behind by the retreating convective core during hydrogen burning. Removing a little bit more matter from the star produces a much lower atmospheric hydrogen content at the time of explosion, which, according to Figure 1 causes the large dependence of the radius at explosion on the mass lost during the lifetime of the star.

## b) WOOD AND FAULKNER (1988)

This model starts from a 17.5 $M_\odot$ star, without overshooting, and evolves after extensive mass loss in the red supergiant stage into a 5.4 $M_\odot$ blue supergiant at the time of explosion. The HRD position is less dependent on the mass loss rate: because no overshooting is adopted, the composition gradient, left behind by the convective hydrogen burning core, is leveled out by an intermediate convective zone. Lateron, mass is removed until this plateau is exposed. Since there is no composition gradient near the surface at the timne of explosion, a little more or less mass loss does not change the atmospheric hydrogen abundance and the radius.


### c) ARNETT (1987) AND TRURAN ET AL.(1987).

The basdic idea of these models is that 15 - 20 $M_\odot$ models with a low metallicity do not become red supergiants, but remain blue supergiants until the explosion (cf. discussion on Fig.1). Taken at face value these models are in direct contradiction with the HRD of the Magellanic Clouds ( Humphreys and McElroy (1984): in both the LMC and the SMC red supergiants are observed in the mass range 15 - 20 $M_O$. The models of Arnett (1987) and Truran et al. (1987) must therefore represent exceptional stars in the LMC: a very low metallicity and no mass loss at all. Although the presence of such stars in the LMC is not excluded (there may be a large spread in metallicity), these factors make the model less plausible.


## 4. CONCLUSIONS

The evolution of the progenitor of SN1987A is by far not clear. There are two "classes" of models: without mass loss and with a large mass loss. The first models rely on the fact that low metallicity induces bluewards evolution. But the Woosley (1988) model only has limited convection - a matter of debate - while the other nodels of this class do not produce red supergiants - difficult to reconmcile with the observed characteristics of the HRD.

The second class of models relies on mass loss to peel off the star until helium rich layers are exposed and the star turns from a red- into a blue supergiant. All computed models need a huge mass loss rate durinmg the red supergiant phase - much larger than suggested by the observations. A way out of this would be the inclusion of a large amount of overshooting: only 0.6 $M_\odot$

of matter has to be removed to expose the appropriate layers. But in this case the mass loss rate has to be timed in a very delicate way, because of the very steep composition gradient at te surface at the time of explosion.

We may conclude that all models for the progenitor evolution of SN1987A require some special conditions: either a huge mass loss or no mass loss at all or an exceptionally low metallicity. If one of these models represents the reaslity of nature, SN1987A was indeed an exceptional event.

## REFERENCES

Arnett, W.D., 1987, *Astrophys. J.*, **319**, 136

Humphreys, R.M., McElroy, D.B.:1984, *Astrophys. J.* **284**, 565

de Jager, C., Nieuwenhuizen, H., van der Hucht, K.A.:  *Astron. Astrophys.* ( in press)

Maeder, A.: 1981, *Astron. Astrophys.*  99, 97

Maeder, A.:1988, *this volume*

Maeder, A., Meynet, G.: 1987, *Astron. Astrophys.* **182**, 243

Pylyser, E., Doom,C., de Loore, C.: 1985, *Astron. Astrophys.* **148**, 379

Schaeffer, R., Cassé,M., Mochkovitch, R., Cahen, S.:1987, *Astron. Astrophys.* **184**, L1

Truran, J.W., Hoflich, P., Weiss, A., Meyer, F.: 1987, *Messenger, ESO* **47**, 26

Walborn, N.R., Lasker, B.M., Laidler, V.G., Chu,Y.H.: 1988, *this volume*

West, R.M., Lauberts, A., Jørgensen, H.E., Schuster, H.E.: 1987, *Astron. Astrophys.* **177**, L1

Wood, P.R., Faulkner, D.J.: 1988, *this volume*

Woosley,S.E.: 1988, *this volume*

SN 1987A: ULTRAVIOLET OBSERVATIONS AND MASS LOSS

Robert P. Kirshner

Harvard-Smithsonian Center for Astrophysics

60 Garden St., Cambridge, MA 02138 USA

<u>ABSTRACT</u>

Mass loss from the B3 Ia progenitor star for SN 1987A is revealed by the recent emergence of narrow ultraviolet emission lines. The emitting gas is nitrogen-rich, has low velocity, and may be located a light-year from the supernova. This gives every sign of having been ejected from the SK -69 202 progenitor when it was a red supergiant, prior to its brief and ultimately violent life as a blue supergiant. Changes in the hydrogen line profiles during the early evolution provide a way to estimate the density distribution in the supernova atmosphere, and the mass of hydrogen it contains. A preliminary estimate is that the power-law index of density in the envelope goes as $v^{-11}$ and the mass that lies above a velocity of 6,000 km s$^{-1}$ is between 1 and 6 solar masses.

<u>ULTRAVIOLET OBSERVATIONS</u>

Promptly following the discovery of SN 1987A by Shelton at the University of Toronto's observatory at Las Campanas, a target of opportunity program to observe the supernova was put into action. The IUE was used in both its high and low resolution modes to measure the spectrum from 1200Å to 3200Å. Initial reports based on IUE data can be found in Kirshner et al. (1987) and Dupree et al (1987) and in European work (Panagia et al 1987, Cassatella et al 1987, Fransson et al 1987).

The ultraviolet flux was declining before the first optical maximum was reached on 27 February (day 58 of 1987), and the short wavelength bands declined much more rapidly than the long wavelength bands, due to rapid cooling of the photosphere from a high initial temperature. The decline in the shortest wavelength band is truly precipitous, descending a factor of 1000 in 3 days. The non-zero level reached is the result of two hot stars located near SK -69 202 in the aperture for the short wavelength, but is due to the supernova itself in the long wavelength bands.

These earliest UV observations can be combined with optical data to derive a bolometric light curve (Blanco et al. 1987). Although the color temperature was only of order 15,000 K at the time of our first data, it must have been far higher on February 23.  If we take the moment of the core collapse as Feb 23.316  to agree with the Kamiokande (Koshiba 1987) and IMB (Svoboda 1987) results, then McNaught's observation at 6 mag on Feb 23.443 probably refers to an epoch when the temperature was of order 100 000 K with a photospheric radius of order $10^{13}$ cm.  The ionizing output of the supernova cannot be reliably estimated from the ultraviolet light curves because so much of the ionizing flux that results when the shock hits the surface of the star came before the first ultraviolet observations.  A reasonable estimate might be $10^{47}$ erg available to ionize and heat the circumstellar gas observed later.


<u>SPECTROSCOPY</u>


The detailed spectral evolution includes complex changes over the first few days. In the early spectra, a broad absorption and emission stretching from 2500A to 2800A is likely to be a P-Cygni line of Mg II, with the blue wing of the absorption minimum extending beyond 30,000 km s$^{-1}$ blueshift.  It seems likely that the ordinary photospheric abundance of iron peak elements is capable of producing the strong UV line blaneting observed after the first few days (Branch 1987, Lucy 1987).
The successful models for the ultraviolet spectra of previous SN II consider the emission from a substantial circumstellar layer (Fransson et al. 1984), such as might be expected around a red supergiant with a slow dense wind.  The absence of these features makes it unlikely that SK -69 202 had such a wind immediately before the explosion, although it does not rule out a wind at an earlier stage of evolution.  The picture in which SN 1987a has a low density circumstellar envelope is consistent with the weak, brief radio emission (Turtle et al 1987) and the absence of early X-ray flux (Makino 1987) as described by Chevalier and Fransson (1987).

The persistence of UV flux in the short wavelength range is the result of stars which are present in the aperture when the satellite is pointed at the supernova. With accurate position measurements and image synthesis of the SK -69 202 field in hand (Walborn et al 1987, West et al 1987), it is possible to conduct a careful deconvolution of the IUE data and to establish that the two stars present are stars 2 and 3.  The 12 mag B3 I supergiant is not the source of the UV light.  It has disappeared and it may be identified as the progenitor of SN 1987a (Sonneborn, Altner, and Kirshner 1987, Gilmozzi et al 1987).

<u>CIRCUMSTELLAR EMISSION</u>

Starting in 1987 May, narrow emission lines began to appear in the short wavelength IUE spectra (Wamsteker et al 1987, Kirshner et al 1987). The lines are narrow ($<1500$ km s$^{-1}$) and near zero velocity. It is hard to see how they could emerge from the fast moving debris, especially as the opacity of that material is very high. A more likely picture is that they arise from circumstellar material surrounding the supernova. This idea is strengthened by the line identifications. Strong lines of N III, N IV, and N V are present, and the corresponding carbon and oxygen lines are weak. This suggests that the material is nitrogen-rich, as might be expected for the envelope of a star which has undergone extensive CNO hydrogen burning. While the details depend on a proper understanding of the origins of the emission, the ratio C/N implied by the observations must be down from normal values by a factor of 30 or more. A plausible picture would be that the B3 I star which exploded spent some time as a red supergiant, and that the material which we see in the short wavelength UV results from mass loss during that interlude. The investigations of mass loss in massive stars by Maeder (1987 and in this volume) are especially relevant here. The high nitrogen abundance observed requires very substantial mass loss, down to the zone where CNO cycle burning of hydrogen prevailed. It is interesting to note that a very similar UV spectrum is seen in the nitrogen-rich material ejected from Eta Carina (Davidson et al 1986), which is another star that is presumably on the path to becoming a supernova.

If we are seeing the results of a wind, the material would be at a distance that corresponds to the red giant wind velocity multiplied by the time the star spent as a blue supergiant. A reasonable estimate would be 10 km s$^{-1}$ for 30000 years, or a distance of 1 light year. The ionizing flux from the supernova outburst would ionize and heat some of this circumstellar gas. The emission that we see would come either from the excitation of these ions, or possibly from recombination, at a temperature of order 50 000 K, as described by Lundquist and Fransson (1987). If this picture is correct, then the flux we see should grow with time as the light travel effects allow us to see more of the circumstellar shell. The available radius should grow as $t^{1/2}$; its area, and the flux should grow as t. In fact, the increasing flux of N III] 1750A matches that rate, although the data are still quite sketchy. The flux should continue to increase until the age is comparable to the light travel time across the shell.

For a reasonable guess of 1 LY scale, we can expect the flux from this shell to continue to increase for the next several months. If it grows strong enough to make a high-dispersion exposure practical, very stringent limits on the radial velocity of the emitting material will be measured. One amusing consequence of the

circumstellar shell will be a violent interaction when the fast moving debris from
the supernova collides with it.  Since the debris has a velocity in excess of 0.1c,
this renaissance of the supernova can be expected in 10 years.  The collision will
result in a high-temperature shock with copious X-ray emission as discussed by Masai
et al in this meeting and possibly a new burst of radio emission.  If these new
sources do appear, the same light travel time effect that shapes the growth of the
UV lines may dominate their temporal evolution.

The presence of a circumstellar shell is very important in establishing the
evolutionary history of the SK -69 202 star.  Theory alone does not provide a robust
guide to understanding the travel through the H-R diagram for massive stars,
including this one.  In particular, this shell seems like very good evidence that
the SK -69 202 star was once losing mass, presumably in a slow dense wind like that
seen around red supergiants.

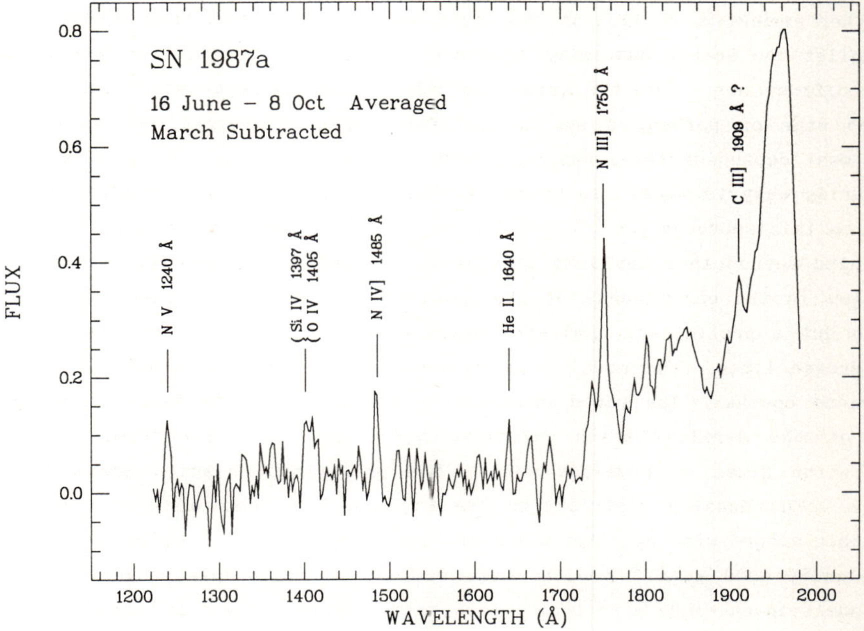


Figure 1.  Averaged short wavelength spectrum, formed by subtracting the March 87
spectrum, which contains only flux from the two background stars, from the
observations.  Note the strong emission lines of Nitrogen.

<u>ENVELOPE MASS</u>

The luminosity of SK -69 202 is consistent with a helium core mass of about 6 solar masses, and this corresponds to a main sequence mass of order 15 solar masses. An important question that will affect the future behavior of the supernova is how much of the 9 solar masses not in the core was present on the surface of the star when it exploded. If mass loss was large, then the mass on the star could have been small, although we know that the surface layers were hydrogen-rich from the optical spectra.

One way to find the envelope mass would be to estimate the circumstellar mass, and subtract. This is unlikely to lead to a satisfactory estimate because the ionizing flux from the supernova is not so large that we can be sure it has ionized all the surrounding material. This approach is also vulnerable to large errors from clumping effects, and in any event requires an accurate knowledge of the density which we are unlikely to possess.

Another avenue is to look at the light curves. The large bump extending through May implies the energy deposited by radioactivity or other sources was trapped and had to diffuse out. But the details of the energy sources (which could include a neutron star or perhaps a substantial recombination energy) and of the opacity (which must depend on the composition and ionization of the stellar mantle) are not necessarily easy to model, so we may not have an accurate way to find the hydrogen mass from this approach.

A third method that has some promise is to examine the hydrogen lines and their evolution in the early weeks of the supernova. Eastman and Kirshner (1987) have carried out a preliminary analysis, based on a simple theory for the formation of the hydrogen lines. The model fits the shapes of the blue edge of the Balmer lines at various epochs. The rapid recession in velocity of the Balmer lines allows a model of the density versus velocity to be derived. Integrating that density distribution gives an envelope mass. The preliminary results are shown in the figures. The density distribution has a power law index of $v^{-11}$ , which is in reasonable accord with expectations from hydrodynamic models, and a total mass above the velocity 6000 km s$^{-1}$ of 1.8 solar masses. A reasonable range for the errors would admit an envelope mass from 1 to 6 solar masses. The chief uncertainty in the method is that it requires an estimate of the level populations in hydrogen. It is certainly the case that the population in n=2 is small compared to the ground state population. The population in n=2 is closely tied to the observations, but the desired quantity is the total amount of hydrogen, which depends on doing this atomic model correctly. Eastman is creating a detailed model atmosphere that will help sharpen this estimate. If we take it at face value, we would have a plausible picture in which the star would have had a 6 solar mass core, 1 to 6 solar masses in the envelope, and 8 to 3 solar masses in the circumstellar shell.  If this picture is correct, then mass loss from a late type star is an essential part of the SN 1987A story.

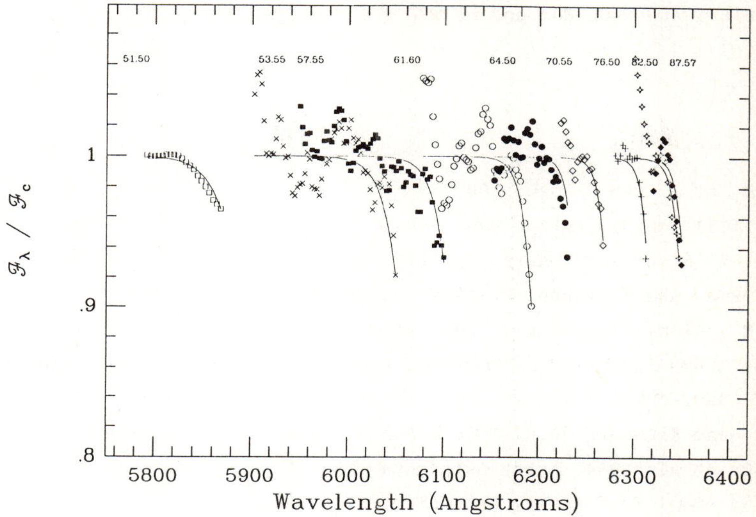

Figure 2.  Observed H alpha line profiles from CTIO data fit by a simple model.

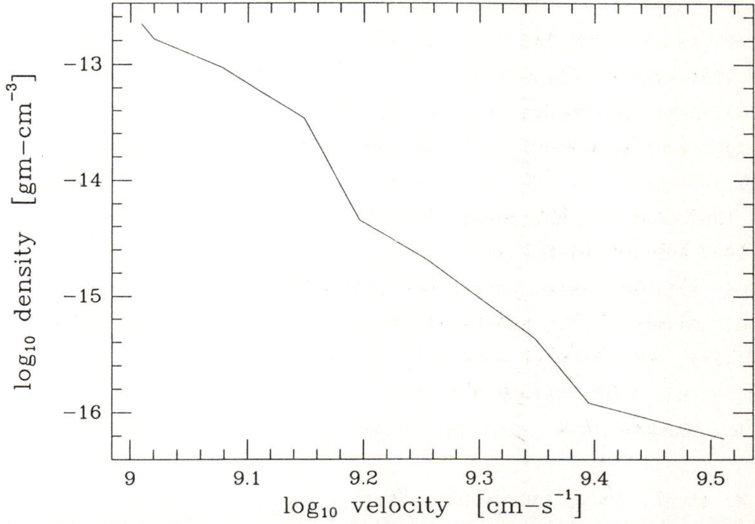

Figure 3.  Density distribution required to fit the observed line profile evolution. The density falls as $v^{-11}$, and the integrated mass above 6000 km s$^{-1}$ is about 2 solar masses.

I am very grateful for the outstanding cooperation of the IUE Observatroy staff, and for the tolerance of the many IUE observers whose work was dislocated by the observations of SN 1987A. The special contributions of George Sonneborn, George Nassiopoulos, Eric Schlegel, and Ronald Eastman were essential to the work reported here. Work on supernovae at the Harvard College Observatory is supported by NASA grants NAG5-841 and NAG5-645 and by NSF grant AST85-16537.

## REFERENCES

Blanco, V.M. et al. Ap.J. **320**, 589.

Branch, D. 1987, Ap.J. (Lett.) **320**, L121.

Cassatella, A. et al 1987, Astr. Ap. **177**, L29.

Chevalier, R.A. and Fransson, C. 1987, Nature **328**,44.

Davidson, K., et al. 1986, Ap.J. **305**, 867.

Dupree, A.K., Kirshner, R.P., Nassiopoulos, G.E., Raymond, J.C., and Sonneborn, G. 1987, Ap.J. **320**, 597.

Eastman, R. and Kirshner, R.P. 1988 B.A.A.S. Vancouver late papers.

Fransson, C. et al. 1984, Astr. Astrophys. **132**, 1.

Fransson, C. et al 1987, Astr. Astrophys. **177**, L33.

Gilmozzi, R. et al. 1987, Nature **328**, 318.

Kirshner, R.P., Sonneborn, G., Crenshaw, D.M., and Nassiopoulos, G.E. 1987, Ap.J. **320**, 602.

Kirshner, R.P. et al. 1987 IAU Circular 4435.

Koshiba, M. 1987, IAU Circular 4338.

Lucy, L. 1987, Astr. Astrophys. **182**, L31.

Lundquist, P. and Fransson, C. 1987 in "ESO Workshop on the SN 1987A" (I.J. Danziger, ed.)

Maeder, A. 1987, Astron. Astrophys. **173**, 287.

Makino, F. 1987 IAU Circular 4336.

Panagia, N. et al 1987, Astr. Astrophys. **177**, L25.

Sonneborn, G., Altner, B.A., and Kirshner, R.P. 1987, Ap.J. (Lett.) in press.

Svoboda, R. 1987, IAU Circular 4340.

Turtle, A.J. et al. 1987, Nature **327**, 38.

Walborn, N.R., Lasker, B.M., Laidler, V.M., and Chu, Y-H 1987, Ap.J. (Lett.) **321**, L41.

Wamsteker, W. et al. 1987, IAU Circular 4410.

West, R.M., Lauberts, A., Jorgensen, H.E., and Schuster, H-E, 1987, Astr.Astrophys. **177**, L1.

SPECTROSCOPIC AND PHOTOMETRIC OBSERVATIONS OF SN1987a OBTAINED AT SAAO


R.M. Catchpole
South African Astronomical Observatory
P.O. Box 9, Observatory 7935, Cape, South Africa


Introduction

The dedication of at least 24 local and visiting observers has enabled the SAAO to
accumulate a large body of spectroscopic and photometric observations of SN1987a.
Broad band photometry has been obtained on every possible photometric night while
spectra at 7A(FWHM), including the first spectrum in the world, were obtained every
night for the first 14 nights and thereafter on a weekly basis.   These data, which
have been more fully presented elsewhere (Menzies et al. 1987, Catchpole et al. 1987)
are briefly discussed here for the time interval 24 February until 31 August 1987.
Throughout this paper we adopt a distance modulus of 18.5 and a reddening of Av=0.6
for SN1987a.


Spectroscopy

All spectra were obtained with the Grating Spectrograph attached to the Cassegrain
focus of the 1.9-m telescope at Sutherland.   These spectra, covering the wavelength
range 3400 to 7600 A, have been first converted to relative fluxes using spectrophotometric standards and then to absolute fluxes by comparison with simultaneous
broad band V photometry.   Representative spectra are shown in Fig. 1 where the time
in days since the Kamiokande-II neutrino event is shown at the top right of each
spectrum.   The first spectrum, obtained on day 1.6 shows a very blue continuum with
broad blue-shifted hydrogen absorption lines and He I 5876A.   Over the next 5 days
the velocity of H$\alpha$ decreased from 18500 km s$^{-1}$ at the rate of 789 km s$^{-1}$ day$^{-1}$.   By
day 8 SN1987a had completed its initial rapid fading, the photospheric temperature
had declined from an initial 14000K to 6000K, H$\alpha$ had developed a P Cygni profile and
many other absorption lines had appeared.   This is well illustrated in the remaining
spectra which also show the sudden rapid decline in brightness shortward of 4200A
undoubtedly due to line blanketing.   The strength of H$\alpha$ emission steadily increases
with time as does [Ca II] at 7300A which first appeared on about day 70.

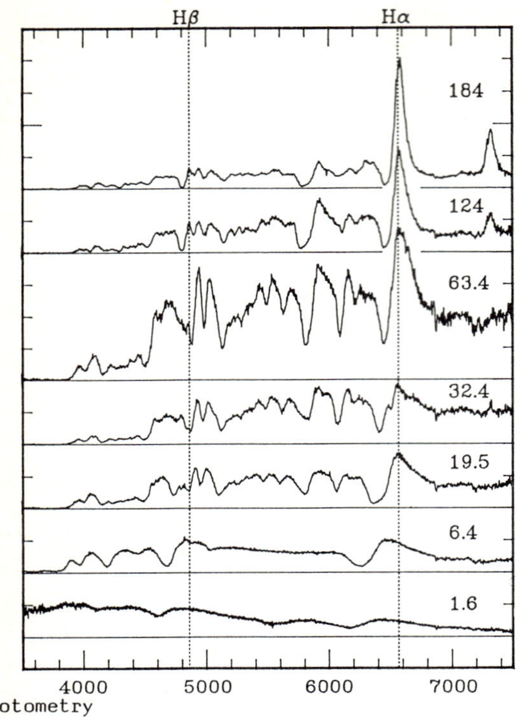


Figure 1.    Spectra of SN1987a on a relative flux scale.    The numbers indicate days since the Kamiokande-II neutrino event.

Photometry

UBV(RI)$_c$JHKL photometry has been obtained on 109 of the 184 days discussed in this paper.    Details of the observations and the reduction are given in Menzies et al. 1987 and Catchpole et al. 1987.    However it is important to draw attention to one aspect of the reduction method.    The observations obtained on the telescope natural system are transformed to the Johnson UBV and Cousins RI photometric systems using colour equations defined for normal stars.    It is important to realise that although this method will produce a self-consistant body of data it will be systematically different from that produced at another observatory, unless the two natural systems are identical.    This happens because the supernova spectrum differs considerably from those of the stars which define the colour transformations.    Care should therefore be taken when combining data from different observatories, especially when looking for subtle changes of slope and claims that one observatory is closer to the Johnson and Cousins systems than another should be treated with great caution.    This problem is not so serious when working with narrow band photometric systems where data should be published on the natural system along with the wavelength sensitivity curves.

One of the most useful data that can be obtained from the photometry for
comparison with theory is the variation of the bolometric flux with time.   SN1987a
provides an ideal opportunity to make this comparison because we know the distance to
the LMC with an uncertainty of about 10% based on independent distance indicators,
the reddening is well constrained by observations of the progenitor and we have broad
band photometry covering the region in which after day 3 more than 95% of the flux is
emitted.   We have used two methods to determine the bolometric flux.   The first
method fits a blackbody, which has the advantage of providing the parameters,
temperature and radius while the second method is to integrate under a spline curve
which by definition passes through all the points.   The end points of the spline
curve are taken at $F_\nu=0$ at $\nu=0$ and $F_\nu=0$ at the frequency at which a line through
$F_U$ and $F_B$ intersects the frequency axis.   Examples of blackbody and spline fits
to the data are shown in Fig. 2 for three representative days.   Day 8 is the time of
the first bolometric minimum, day 86 is at bolometric maximum while day 163 occurs
during the radioactive decline and is a day on which the M magnitude was also
measured.   The variation of the bolometric flux with time given by the two methods
is shown in Fig. 3 for two values of the reddening.   In Fig. 3 the data points have
been suppressed for the sake of clarity and the curves are made up of chords joining
individual data points.

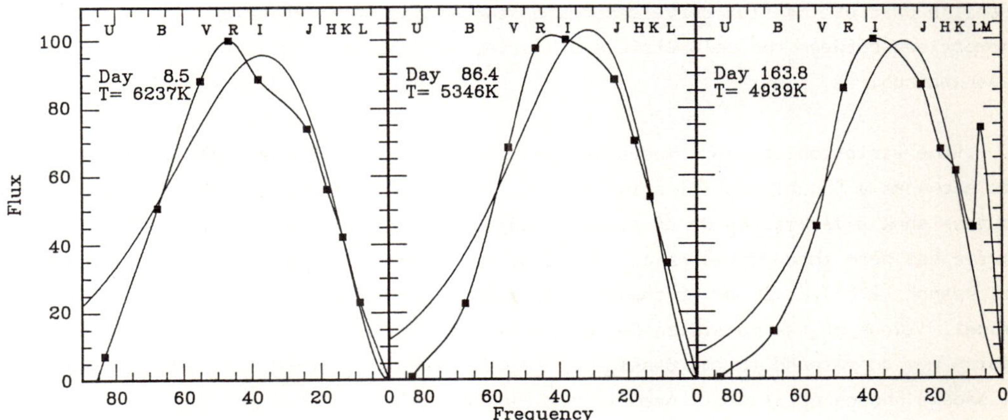


Figure 2.   Normalized broad-band fluxes, corrected for $A_V=0.6$ and joined by spline
curves, are shown for three representative days.   The best fitting blackbody is also
shown.   An M measurement made on day 163 is illustrated but was not included in the
determination of the bolometric flux curves shown in Fig. 3.

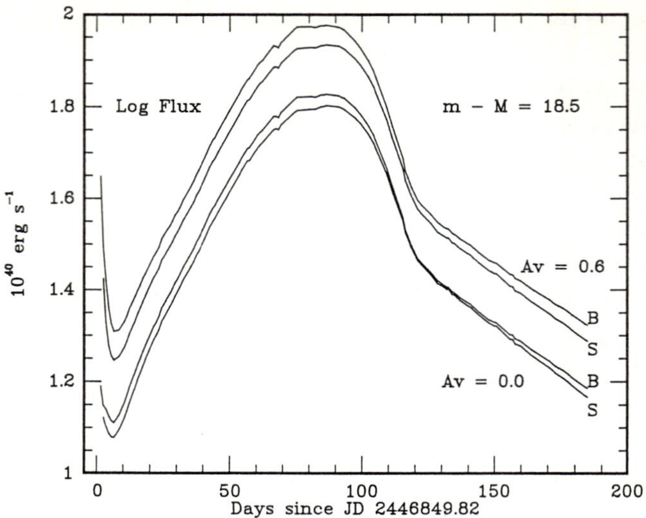


Figure 3.   The logarithm of the bolometric flux given by blackbody (B) and spline (S) fitting is shown as a function of time for 2 values of $A_v$.

The maximum difference between the blackbody and the spline method is always less than 12%.   However it is important to note that the slope on the radioactive decline part of the flux curve depends on the method of integration.   Direct comparison between the bolometric flux curves and models is given elsewhere in these proceedings.

The variation of the temperature and radius deduced from the blackbody fitting is shown as a function of time in Fig. 4.   Both the radius and the temperature curves show a marked change of slope at day 5.   The slope of the angular radius curve has been interpreted as an indicator of the density gradient in the atmosphere by Branch (1987).   On day 5 the density gradient changes from $R^{-11.7}$ to $R^{-5}$ on his model.   None of the models so far presented for the explosion predicts this change, which may correspond to the passage of the photosphere through the hydrogen shell. A second change of slope is seen in the temperature curve at about day 32.

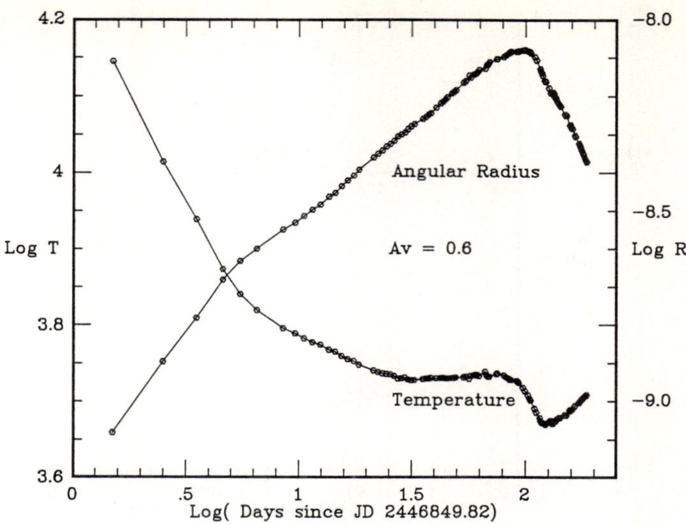


Figure 4.   Logarithm of radius and temperature given by blackbody fitting are shown as a function of log time.


Infrared M band (4.8 micron) Photometry

We have obtained a number of 4.8 micron observations which are combined with those kindly obtained by J Albinson and R Maddison and are illustrated in Fig. 5.   Until day 60 (L-M) remained constant while M increased in lock step with the increase in brightness of SN1987a.   As the bolometric flux of the supernova has declined the M flux has remained constant while (L-M) has increased so that SN1987a now shows a significant excess at M.   The effect of the M excess on the bolometric flux is at present small.   On day 163, which is illustrated in Fig. 2, the effect of including M in the spline integration is to increase the flux by 4% while the effect on the blackbody integration is to increase the flux by only 0.4%.   There is however a correspondingly greater change in the derived temperature and radius.
At present we are making observations in an attempt to distinguish between the possibilities of a light echo, a dust excess or possible CO emission.

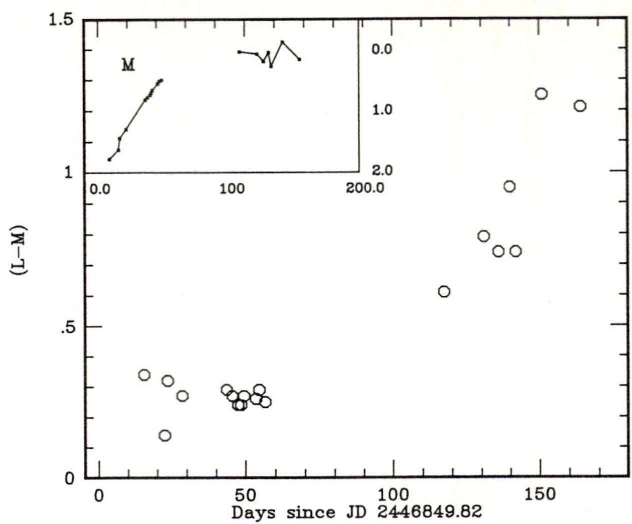

Figure 5.    Variation of (L-M) and (inset) M as a function of time.


References

Branch, D. 1987.    Astrophys. J. Lett., **320,**  123.
Catchpole, R.M., et al. 1987.    Mon. Not. Roy. astr. Soc. accepted.
Menzies, J.W., et al., 1987.    Mon. Not. Roy. astr. Soc. **227,**  39p.

SN 1987A: AN AUSTRALIAN VIEW.

Michael A. Dopita

Mt. Stromlo and Siding Spring Observatories

The Australian National University

## 1. Introduction.

The explosion of SN 1987A in the LMC during the early hours of February 23 1987, has presented southern observers opportunity, unique in our lifetime, to gain a new insight into the supernova phenomenon. Fortunately, SN1987A has risen to the challenge, and has shown us that many of our comfortable pre-conceptions about Type II events will have to be revised in fascinating and sometimes unexpected ways. This paper is an attempt to describe what has been learnt so far wth an emphasis on the research being carried out in Australia. All references are dated 1987 unless otherwise specified.

## 2. The Nature and Evolution of the Precursor Star.

There is now no doubt that SN 1987A is positionally coincident with the B3 I star, Sk-69 202. This has been established by astrometry with a positional accuracy of 0.1 arc sec, or less (White and Malin a,b; West *et al.* ). This star had the following parameters ( Rousseau *et al.* 1978):

$$V= 12.24 \quad (B - V) = 0.04 \quad (U - B) = -0.65.$$

The reddening is somewat uncertain. From a measurement of the Balmer Decrement of the surrounding HII region, Danziger *et al.* find $A_V = 0.66 \pm 0.2$. Wampler *et al.* use the observed absolute strength of the $\lambda 5780$Å diffuse interstellar band (Vladilo) to estimate $A_V = 0.45$. On the other hand, the diffuse interstellar band at $\lambda 6613$Å shows the LMC feature to be only about 75% as strong as the Galactic component (Vidal-Majar *et al.* ). Since the LMC line of sight Galactic reddening has been fairly accurately measured at E(B-V)= 0.034 (McNamara and Feltz, 1980), we can estimate an $A_V$ of order 0.22 on this basis. We adopt a value of $A_V = 0.44$, E(B - V) = 0.14, being the average of these estimates. Wood and Faulkner estimate the following parameters of Sk -69 202:

$$M_{bol} = -7.71 \quad \log (L/L_\odot) = 4.98 \quad \log T_{eff} = 4.11$$

These parameters define an entirely unremarkable blue supergiant. Conventional wisdom had it that Supernovae of Type II occur either in red supergiants, or perhaps, in the Wolf-Rayet phase of evolution. The central problem for the evolutionary models is therefore, how can the moment of core collapse be contrived to occur in a blue supergiant star? There have already been many attempts made to address this question, and from these it is apparent that the main sequence mass must have been in the range 15-20 $M_\odot$ . These models teach us that the end-point of evolution is remarkably sensitive to the assumptions and approximations made. The major parameters that determine the outcome are, in no particular order, the opacity, the treatment of convection and the treatment of mass loss.

In general, the decrease in opacity obtained with the lower abundances characteristic of the LMC tends to help to confine the evolutionary tracks to the blue side of the H-R Diagram. However, most models are computed with the an abundance set taken as solar divided by, say, four. In practice, the LMC abundance distribution is not this simple.Dopita (1986) and Russell, Bessell and Dopita have shown that, in the LMC, the underabundance of various elements with respect to solar is dependent upon their atomic number. For example, Cand N are depleted by about 0.8 dex, O and Ne by about 0.5 dex, Ca by anbout 0.3 dex and the heavy elements from Ti through Fe to Ba by about 0.2 dex. This pattern is similar to that produced in models of deflagration supernovae, and may indicate that these have been relatively more important in enriching the interstellar medium in the LMC.

An important constraint in the evolutionary models is that they should correctly describe the observed ratio of red to blue supergiants in the LMC (Wood and Faulkner; Maeder; Miyaji and Saio, all this conference). Many models without mass loss are unsuccessful in this, since they fail to evolve to the red supergiant phase at all (Arnett, Hillebrandt $et\ al.$ ). However, this problem is code dependent, and others do evolve to the red and then return (Woosley $et\ al.$ ; Wood and Faulkner; Woosley, this conference).

Models involving mass loss are more successful in reproducing the observed ratio of blue to red supergiants. Those which include convective overshooting (Maeder, this conference) have a somewhat higher core mass, and a larger residual hydrogen envelope than those which do not. The Wood and Faulkner model for a 17.5 $M_\odot$ precursor star is in many respects a fully self-consistent model. It contains the correct opacity for the observed LMC abundances, and uses Waldron (1985) mass-loss rates which correctly reproduce the blue to red supergiant ratio.The supernova occurs in these models when the helium core mass is 5.2 $M_\odot$ and only 0.2-0.6 $M_\odot$ of hydrogen is left on the star. Hydrogen shell burning has been extinguished before the supernova explosion.

Theoretical estimates of the residual mass of the hydrogen envelope at the time of the explosion thus range from about 0.3 $M_\odot$ all the way up to about 10 $M_\odot$. However, some observational data tends to support the lower values. The strength of the nitrogen lines in the precursor (Walborn), and the appearance of narrow NV, NIV] and NIII] lines in the UV some weeks after the explosion (Kirshner) all suggest that CN processed material was not only abundant in the surface layers, but indeed, had been ejected in a previous red-giant phase. The appearance of X-ray emission at early epochs and the early appearance of lines of s-process elements in the spectra (Williams) (He-burnt material at the photosphere) also tends to support lower values for the total mass of the pre-supernova star unless substantial post-SN mixing has occured. However, the energy arguments presented at this conference by Woosley; Wheeler Harkness and Barkat; and Nomoto Shigeyama, and Hashimoto would suggest high residual hydrogen mass, since in the low-mass scenario the hydrogen layers would be ejected at too high a velocity.

## 3. The Light and Colour Evolution

Monitoring at U, B, V, R and I has been carried out on the 76 cm telescope at MSO by Flynn and Meatheringham, and by a larger group of observers, using the 40 cm and 60 cm telescopes at SSO (Achilleos, Dawe, Rawlings, Mc Naught and Shobbrock). This data (partly reported in Dopita *et al.*) is in no sense as accurate or complete as that collected in South Africa (Menzies *et al.* ; Catchpole *et. al.* ) or at Cerro Tololo (Gregory *et al.* ; Hamuy *et al.* ).In the following analysis, therefore, the Australian data has been combined with the above material.

The compact initial state of the supernova ensured that the initial colour evolution is about five times faster than a normal Type II. At five to seven days the (B - V) colour evolution accelerated as line blanketing, initially from higher members of the Balmer series, but later from metal lines, mainly from the iron group, becomes important. The climb towards maximum was accomplished at an almost constant temperature as measured by the (V - I) colour, but a further decline in temperature was seen in the post-maximum phase as the luminosity collapsed towards the radioactive tail. An increse in (U - B) occurred at 40 - 50 days, but this does not correspond to any change in photospheric temperature. This is presumably the result of a composition change working its way out to the photosphere.

The reduction of these colours and the V magnitude to physical quantities is of necessity rather approximate and has been described in detail elsewhere (Dopita *et al.* ). Suffice it to say that the Bolometric luminosities are obtained on the assumption that the apparent distance modulus of the supernova is 18.8, corresponding to a visual absorption $A_V$ = 0.44, and using the Bolometric Corrections of Carney (1980). The derived Bolometric luminosity is given in Figure 1.

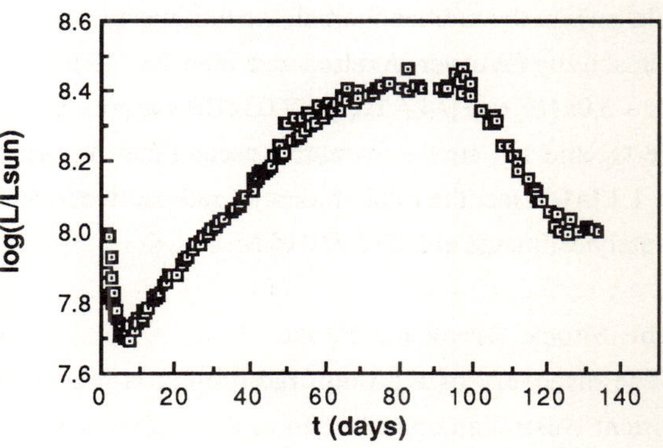


Fig 1: The evolution of the absolute luminosity of SN 1987A. Note the initial "flash", the slow climb to maximum, and the rapid decine to the radioactive powered tail. The maximum corresponds to the "plateau" phase of of normal Type II, provided that Ho ~ 100 km/s/Mpc.

If we can assume that the shocked SN ejecta expands homologously, and that locally, the density structure can be represented by a power law with index $\alpha$, then, for a atmosphere with constant ionisation fraction, it follows that $R \propto [t / t_0]^{(\alpha-3)/(\alpha-1)}$. Therefore, the evolution of the radius with time follows a power law with an index which is directly related to the index of the density gradient at the photosphere. An analysis of the photometrically defined temperature and radius shows at least four such power-law segments (fig. 2). The initial expansion is effectively ballistic, with a slope corresponding to an $\alpha \geq 11$. In this phase adiabatic cooling is reducing the

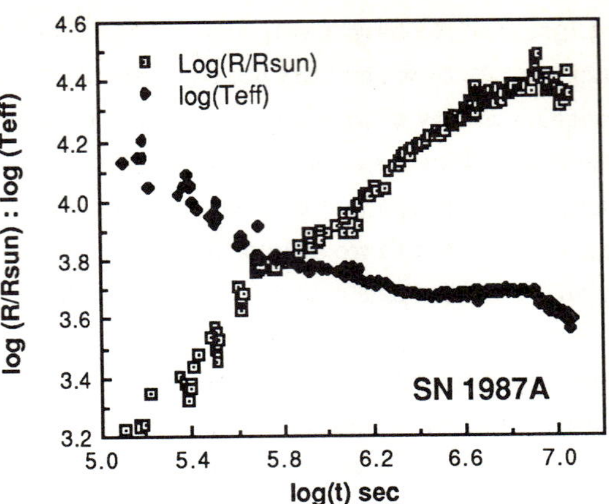



Fig 2: The Logarithmic variation of temperature and radius. Note the discontinuous change of slope at log(t) = 5.7. This corresponds with the onset of hydrogen recombination in the photosphere (see text).

effective temperature. Eventually, the temperature falls to the point at which a recombination wave starts to propagate inwards (log(t) = 5.7). Since in the early phase, electron scattering is the dominant continuous opacity source, the photosphere is locked to the recombination front, and follows it as it sweeps inward. However, the recombination front moves in so rapidly that it eventually leaves the photosphere behind. The $\alpha$ determined by the radius-time relationship should then be an accurate measure of the true $\alpha$, or about 11 (6.2 < log(t) < 6.7).

The decline from the peak is relatively rapid, since the opacity of the core is determined by electron scattering. As the heavy elements recombine, this declines precipitately and the diffusion timescale rapidly becomes shorter than the dynamical timescale (see, for example, Shaeffer, Cassé and Cahen). In the radioactive tail, the luminosity can be set equal to the rate of energy generation in radioactivity (Weaver, Axelrod and Woosley, 1980):

$$L = 3.9 \times 10^9 \exp\left[- t / \tau_{Ni}\right] + 7.03 \times 10^9 \left\{\exp\left[- t / \tau_{Co}\right] - \exp\left[- t / \tau_{Ni}\right]\right\} \; erg.g^{-1}.s^{-1}.$$

where $\tau_{Ni}$ and $\tau_{Co}$ are the respective decay times on radioacive Nickel and Cobalt. Since $L = 10^8 L$ at $t = 1.12 \times 10^7$ sec, the total amount of radioactive nickel produced in the explosion can be fairly accurately estimated at 0.087±0.015 M .

## 4. The Shock Breakout Phase

The discovery of a prompt radio burst is related to the epoch of shock break-out was an important Australian contribution to the studies of SN1987A. The burst lasted for only about a week, (Turtle *et al.* ), and can be understood in terms of free-free absorption of synchrotron emission (Storey and Manchester). The location of this radio emission was probably in the shocked stellar wind region of the star, with the free-free absorption arising from within this layer. The thickness of the emitting layer was estimated at only 4.8% of the radius, which would be consistent with a power law density distribution of matter with an index of 11.8. Since the first week, the radio emission continued to fade (despite some other reports to the contrary), becoming effectively unobservable at 843MHz after about 50 days.

The epoch of shock breakout from the photosphere certainly resulted in a "flash" of UV photons, although probably not a very intense one. From the early data we find;

$$\log (T_{eff}) = (7.47\pm0.23) - (0.64\pm0.04)\log(t)$$

Shock breakout occured at about $\log(t) = 3.6$, and at this time the above regression gives a photospheric temperature, $T_{eff} \sim 150000K$. This should be compared with the estimate derived on the assumption that, at the time of shock breakout, the shock is driven by radiation pressure, which gives $T_{eff} \sim 230000K$. The UV flash can therefore be characterised by a temperature of order $10^5$ K, a luminosity of $2x10^8 L_O$ and a duration of 2-4 hr. These parameters are consistent with the absence of any detectable effect on the earth's ionisosphere (Edwards). Dopita *et al.* proposed that this flash would ionise the precursor stellar wind (see also Chevalier and Fransson). Fluorescent ionisation of a dense blob of gas lost to the star in the red giant phase may give an explanation of the "mystery spot" discovered by speckle interferometry (Karosova *et al.* ; Marcher, Meikle and Morgan) , and is certainly the cause of the narrow emission lines which have developed in the UV (Kirshner, this conference).

## 5. Spectral Monitoring of the Supernova

Observations have been obtained in the 3100-7500Å with the Anglo-Australian telescope, generally at fairly low resolution, although a few higher resolution observations exist. These have been reduced by Raylee Stathakis, and a data base is being built up. The FORS spectra which cover the range 5500-10000Å are particulary useful for monitoring the development of the Ca II triplet feature, and the relative intensity of the. A more extensive spectral library has been accumulated using the MSSSO 1.8m telescope at Coudé with its 32 inch camera and the Photon Counting Array as a detector. This data has a typical spectral coverage of 3200-7400Å , a resolution of 40km/s, and a signal to noise of up to 100 per resolution element. Of complete spectra, the data collected at MSSSO has the highest resolution, but is difficult to reduce to absolute flux.

The very earliest spectra were nearly featureless, but there was a very rapid development of strong Balmer lines with very broad P-Cygni profiles. Initially, the maximum expansion velocity in the absorbing material was as high as 30000km/s, with the absorption maximum in H$\alpha$ at about 17000km/s. This was almost twice as large as in "normal" Type II supernova events, and is certainly associated with the compact nature of the precursor star. Like Danziger *et al.*, we found that the radial velocity of the hydrogen absorption features rapidly decreased, initially by about 800km/s each day. This corresponds to the fastest-moving material expanding, and becoming optically thin. The development of a CaII P-Cygni feature occured very early, but by about March 3 as the photospheric temperature falls below 6000K, and the recombination wave developed, many absorption features corresponding to FeII, NaI and Mg I lines appeared and deepened. The appearance and strength reached by the s-process element lines of Ba and Sc are particulary interesting, and may show real enhancements of these elements, which are produced in the He-burning layers (Williams). However, non LTE effects are certainly important in the formation of the hydrogen lines after only a couple of days, and so abundance estimates should be treated with caution.

Throughout the second half of March and through April, the rate of spectral evolution slowed considerably. Line blanketing below 4400Å became almost total, and the depth and width of the absorption lines continued to decrease slowly. The Hβ line showed a particulary interesting behaviour, almost dissapearing by the end of April, before returning as a prominent and almost saturated absorption line in late June. This may have been the effect of veiling by the FeII features according to Chugaj (prive communication). An alternative explanation is that hard X-rays started to heat the layers above the photosphere leading to an increase in the excitation temperature of the hydrogen. The re-emergence of the Hβ lines corresponds to the onset of the nebular phase, with Hα and CaII developing in emission and with the appearance of [CaII] emission near 7400Å. The ratio of the CaII to [CaII] emission will provide a very useful density diagnostic. The OI 7774Å feature also appeared in absorption for the first time at this epoch, suggesting that oxygen-rich material may now be reaching the photosphere.

Spectral monitoring has also been carried out in the IR, between 1.0-1.4μm, 1.45-1.85μm, 1.9-2.5μm and 2.9-4.1μm (by Mc Gregor, Hyland and Ashley at MSSSO and by Allen at the AAO) and at 8-13μm ( by Aitken and collaborators using the AAO). The spectrum has evolved from a relatively featureless continuum with a few broad lines of hydrogen with P-Cygni profiles, to a rich emission-line spectrum with features reminiscent of a Nova. The continuum distribution in the 8-13μm band is consistent with the opacity being dominated by the free-free contribution. A particulary interesting discovery was the development of the 1st overtone band of CO in emission after about 120 days (see figure 3). The lines are broad, and the emission is almost certainly produced by collisional excitation near the photosphere.

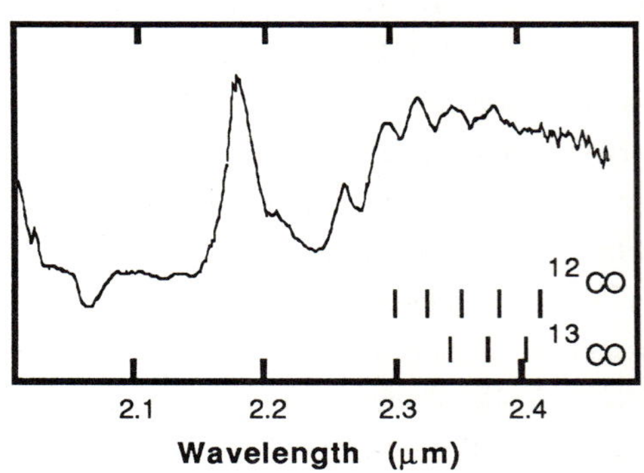



Fig 3: IR Spectrum taken on Sept 1 by P. Mc Gregor, showing prominent CO emission. The bright isolated emission feature is the hydrogen Brackett - Gamma line.

Spectropolarimetric monitoring is being carried out on the Anglo-Australian telescope by Cropper *et al.* (1987). The initial continuum polarisation was about 0.8%, but this subsequently decreased. However, the polarisation in the lines, particulary in the absorption component of Hα has increased sharply. Since the polarisation is determined by the interstellar dust, the shape of the supernova fireball and the scattering processes in the photosphere, these results are difficult to interpret. However, they can provide us with very useful modelling constraints.

## 6. Interstellar Absorption Line Studies.

The rich interstellar absorption spectrum of SN1987A has already been well described by Vidal-Majar *et al.* . Using the Parkes radio telescope Wayte (in prep) has shown that the clouds at 64, 125 and 167 km.s$^{-1}$ can be identified with very faint HI features. If these clouds in fact fill the 15 arc min.beam, then the HI column densities are $1.1 \times 10^{18}$ cm$^{-2}$, $5.6 \times 10^{18}$ cm$^{-2}$ and $0.6 \times 10^{18}$ cm$^{-2}$, respectively. The total Galactic and LMC column densities in the direction of SN 1987A are $5.1 \times 10^{20}$ cm$^{-2}$ and $2.6 \times 10^{21}$ cm$^{-2}$, respectively. Using a special set-up at the Coudé focus of the AAT, Pettini and Gillingham (1988) have been able to measure the hyperfine splitting for a number of clouds along the line of sight at a resolution of about $10^6$. However, perhaps the most remarkable result has been the discovery of [FeX] in absorption in the LMC (Pettini *et al.*.1987; D'Odorico *et al.* 1987). This extends from 205 to 380 km.s$^{-1}$, and has an equivalent width of 16.4mÅ, implying a column density N(FeX) = $2.1 \times 10^{17}$ cm$^{-2}$. If this is global to the LMC it would require an ionised hydrogen column density of order $10^{22}$ cm$^2$. It seems more likely that SN1987A is sitting in a local  bubble of hot gas, possibly produced by the precursor star.


## References

Arnett, W.D. 1987a *Astrophys. J.*, **319**, 136.

__________. 1987b *Astrophys. J.*, (in press).

Blanco, V.M., Gregory, B., Hamuy, M., Heathcote, S.R., Phillips, M.M., Suntzeff, N.B., Terndrup, D.M., Walker, A.R. , Williams, R.E., Pastoriza, M.G. and Storchi-Bergmann, T. 1987 *Astrophys. J.*,  **320**, 589.

Carney, B.W. 1980 *Publ. Ast. Soc. Pacific,* **92**, 56.

Catchpole, R.M., Menzies, J.W., Monk, A.S., Wargau, W.F., Pollacco, D., Carter, B.S., Whitelock, P.A., Marang, F., Laney, C.D., Balona, L.A., Feast, M.W., Lloyd Evans, T.H.H.,  Sekiguchi, K.,  Laing, J.D., Kilkenny, D.M., Spencer Jones, J., Roberts, G., Cousins, A.W.J., van Vuuren, G. and Winkler, H., 1987 *Mon. Not. Roy. Ast. Soc.,* (in press).

Chevalier, R.E. and Fransson, C. 1987 *Nature,* **328**, 44.

Cristiani, S., Babel, J., Barwig, H., Clausen, J.V., Gouiffes, C., Günter, T., Helt, B.E. Heynderickx, D., Loyola, P., Magnusson, P., Monderen, P., Rabattu, X., Sauvageout, J.L., Schoembs, R., Schwarz, H. and Steenman, F. 1987 *Astron. Astrophys (Lett),* **177**, L5.

Cropper, M., Bailey, J., McCowage, J., Cannon, R.D., Couch, W.J., Walsh, J.R., Straede, J.O. and Freeman, F 1987 *Mon. Not. Roy. Ast. Soc.* , (in press).

Danziger, I.J., Fosbury, R.A.E., Alloin, D., Cristiani, S., Dachs, J., Gouiffes, C., Jarvis, B. and Sahu, K.C. 1987 *Astron. Astrophys (Lett),* **177**, L13.

D'Odorico, S., Molaro, P., Pettini, M., Stathakis, R. and Vladilo, G. 1987 in Proceedings of ESO Workshop *"SN 1987A"*, ESO:Garching.

Dopita, M.A., 1986 IAU Symposium #115 *"Star-Forming Regions"*, ed M.Peimbert and J. Jugaku, Dordrecht:Reidel, p501.

Dopita, M.A., Achilleos, N., Dawe, J.A., Flynn, C. and Meatheringham, S.J.,
   1987 *Proc. Ast. Soc. of Australia,* **7** (in press).

Dopita, M.A., Meatheringham, S.J., Nulsen, P. and Wood, P. 1987 *Astrophys. J. (Lett.),* **322**

Edwards, P.J. 1987 *Proc. Ast. Soc. Australia,* **7**, (in press).

Filippenko, A.V. and Sargent, W.L.W. 1985 *Nature,* **316**, 407.
__________________________________. 1986 *Astron. J.*, **91**, 691.

Hamuy, M., Suntzeff, N.B., Gonzalez, R. and Martin, G., 1987 *Astrophys. J.* (in press).

Hillebrandt ,W. , Höflich, P. , Truran, J.W., and Weiss, A. 1987 *Nature,* **327**, 597.

Karosvska, M., Nisenson, P., Noyes, R. and Papaliolios, C. 1987 IAU Circular # 4382.

Kirshner, R.P., and Kwan, J., 1974 *Astrophys. J.,* **193**, 27.

Kirshner, R.P. 1987 (this conference).

Nomoto, K. and Shigeyama, T. 1987 (this conference).

Rousseau, J., Martin, N., Prévot, L., Rebeirot, A.R., and Brunet, J.P. 1978
   *Astron. Astrophys. Suppl. Ser.,* **31**, 243.

Russell, S.C., Bessell, M.S. and Dopita, M.A., 1987 NATO Advanced Study Institute on
   *"Star Formation"*, Calgary, Canada.

Schaeffer, R., Cassé, M. and Cahen, S. 1987 *Astrophys. J. (Lett.)* , **316**, L31.

Shigeyama, T., Nomoto, K., Hashimoto, M., and Sugimoto, D. 1987 (preprint).

Storey, M.C. and Manchester, R.N. 1987, Paper presented at 21st. A.G.M. of the Ast.Soc. of
   Australia, Canberra, May 12-15,*Proceedings Ast. Soc. of Australia,* (in press).

McNamara, D.H.and Feltz, K.A. Jr. 1980 *Pub. Astron. Soc. Pacific* , **92**, 587.

Maeder, A. 1987 (this conference).

Marcher, S.J., Meikle, W.P.S. and Morgan, B.L. 1987 IAU Circular # 4391.

Menzies, J.W., Catchpole, R.M., van Vuuren, G., Winkler, H., Laney, C.D., Whitelock, P.A.,
   Cousins, A.W.J., Carter, B.S., Marang, F., Lloyd Evans, T.H.H., Roberts, G.,
   Kilkenny, D., Spencer Jones, J., Sekiguchi, K.,  Fairall, A.P. and Wolstencroft, R.
   1987 *Mon. Not. Roy. Ast. Soc.,* (in press).

Miyaji, S. and Saio, H. 1987 (this conference).

Nomoto, K. , Shigeyama, T., and Hashimoto, T. 1987, Proceedings of ESO Workshop on
   *"SN 1987A"*, ed. J. Danziger (ESO:Garching)

Pettini, M. and Gillingham, P.R. 1987 (in prep).

Pettini, M., Stathakis, R., D'Odorico, S., Molaro, P. and Vladilo, G. 1988
   *Astrophys. J.* (in press)

Rousseau , J., Martin, N., Prévot, L., Rebeirot, A.R. and Brunet, J.P. 1978
   *Astron. Astrophys. Suppl. Ser.,* **31**, 243.

Schaeffer, R., Cassé, M. and Cahen, S. 1987 *Astrophys. J.,* **316**, L31.

Storey, M.C. and Manchester, R.N. 1987 *Nature,* (in press).

Turtle, A.J., Campbell-Wilson, D., Bunton, J.D., Jauncey, D.L., Kesteven, M.J.,
   Manchester, R.N., Norris, R.P., Storey, M.C. and Reynolds, J.F. 1987 *Nature* , **327**, 38.

Vidal-Madjar, A., Andreani, P., Cristiani, S., Ferlet, R., Lanz, T. and Vladilo, G. 1987
    *Astron. Astrophys (Lett)*, **177**, L#.

Vladilo, G. 1987 *The Messanger (E.S.O)*, **47**, 29.

Walborn, N.R. (this conference).

Waldron, W.L. 1985 *"The Origin of Nonradiative Heating/Momentum in Hot Stars"*,
    eds. A.B. Underhill and A.G. Michelitsianos, NASA Conf. Publ. CP-2358, p95.

Wampler, E.J., Truran, J.W., Lucy, L.B., Höflich, P. and Hillebrandt, W. 1987
    *Ast. Astrophys.*, **182**, L51.

Weaver, T.A., Axelrod, T.S. and Woosley, S.E. 1980, Proceedings of Texas Workshop on
    *"Type I Supernovae"*, ed. J.C. Wheeler (Austin: U.Texas), p113.

West, R.M., Lauberts, A., Jørgensen, H.E. and Schuster, H.-E 1987
    *Astron. Astrophys (Lett)*, **182**, L51.

Wheeler, J.C., Harkness, R.P. and Barkat, Z. 1987 (this conference).

White, G.L. and Malin, D.F. 1987a IAU Circular #4330.

______________________________. 1987b *Nature*, **327**, 36.

Williams, R.E.1987 *Astrphys. J. (Lett.)*, **320**, L117.

Wood, P.R., and Faulkner, D.J. 1987 (this conference).

Woosley, S.E., Pinto, P.A., Martin, P.G. and Wheeler, T.A., 1987 *Astrophys. J.*, **318**, 664.

Woosley, S.E. 1987 (this conference).

ENRICHMENT OF s-PROCESS ELEMENTS IN THE PROGENITOR OF SN 1987A

by

R.E. Williams

Cerro Tololo Inter-American Observatory
National Optical Astronomy Observatories
Casilla 603, La Serena, CHILE

Abstract:  Existing calculations of s-process nucleosynthesis in massive stars such
as Sk -69° 202 show enhancements in the He-burning shells which, when mixed to the
surface, lead to Sc, Sr, and Ba enrichments of factors of 2 to 15.  Abundances
derived from the supernova absorption lines using a simple scattering model yield
enhancements of this magnitude.  Thus, the observed and predicted s-process
abundances are roughly in accord, although the Ba/Sr ratio derived for the supernova
may be higher than the s-process can account for.

The early development of the optical spectrum of SN 1987A has been presented by
Danziger et al. (1987), Blanco et al. (1987), and Menzies et al. (1987).  After an
initial high-excitation phase which lasted less than a week and featured strong
absorption from the Balmer lines and helium, the excitation and color temperature
dropped.  Numerous lines rapidly appeared during this subsequent phase of low-
excitation, after which the spectrum underwent little change for four months until
the low-excitation lines gradually weakened and the spectrum began to develop
stronger emission lines.

An interpretation of the absorption spectrum in the low-excitation phase has
been made by Williams (1987), based on line identifications of scandium, strontium,
and barium---elements which are selectively enriched from nucleosynthesis by the
slow capture of neutrons (Clayton 1968).  Since Sr and Ba are enhanced much more by
the s-process than by the r-process, these elements must have been present in the
progenitor star.  Furthermore, since they appeared in the spectrum within two weeks
after the outburst, they were present in the outer part of the ejected envelope.
They were either formed with the progenitor from the interstellar medium, or they
were created in it as an adjunct to helium-burning and mixed to the surface.  In the
latter event, some enrichment of the elements would be expected, although
calculations of the evolution of massive stars have indicated that no products from
the helium burning core should be convected to the surface (Lamb et al. 1977).

The extent to which the s-process may have operated in the supernova progenitor
is an important source of information concerning its structure and evolutionary
history, as well as the origin of s-process elements in the galaxy.  Standard
procedures of spectral analysis can be used to determine the relative element
abundances from the strengths of the spectral lines.  A rigorous analysis would
involve constructing a curve of growth from unblended and unsaturated lines.  In
SN 1987A this is not possible since each of the ions has at best only one or two
usable lines.  However, each of the ions Sc II, Fe II, Sr II, and Ba II does have

one line of moderate strength, not strongly saturated, that can be used to derive
relative abundances.  Fortunately, the lines all have similar absorption depths and
so uncertainties caused by saturation partly cancel out.

Ion abundances can be obtained either from the equivalent widths $W_\lambda$ of the
absorption lines, or from their residual intensity at line center, $r_{\lambda_0}$.  The two
parameters are basically equivalent in that they are directly related to each other,
although the use of $W_\lambda$ is preferable because it is directly proportional to
abundance in a model-independent way when the lines are unsaturated, e.g., finite
spectral resolution increases the observed residual line intensity, but not the
equivalent width.  In the supernova the lines are all easily resolved, and the
residual intensity at the bottom of the lines is more readily measured than $W_\lambda$,
hence we use $r_{\lambda_0}$ to obtain the abundances.

Many of the absorption lines in SN 1987A are quite deep, and are therefore
produced largely by scattering rather than true absorption.  This is consistent with
the large deviations from LTE that must exist in the rapidly expanding, tenuous
envelope.  In this situation the line formation can be approximately represented by
the Schuster-Schwarzschild model of a reversing layer superposed on top of a
continuum, in which case the residual intensity of a line $r_{\lambda_0}$ is related to the
optical depth at line center $\tau_0$ by the relation (Mihalas 1969)

$$r_{\lambda_0} = \frac{1}{1 + \tau_0} . \tag{1}$$

The optical depth,

$$\tau_0 = N_\ell \, \frac{\sqrt{\pi} e^2}{m_e c^2} \, \frac{f_\ell \lambda_\ell}{V_0}, \tag{2}$$

where $N_\ell$ is the column density of absorbers in the lower level of the transition,
and $f_\ell$ and $\lambda_\ell$ are the oscillator strength and wavelength.  The Doppler velocities $V_0$
of the lines are all roughly the same because they are dominated by the large-
scale velocity structure of the envelope.  The abundance of the ion $N_i$ producing
each absorption line can be related to $N_\ell$ by the Boltzmann equation,

$$N_\ell = N_i \, \frac{g_\ell}{Z_i} \, e^{-X_\ell / kT_{exc}}, \tag{3}$$

where $Z_i \approx g_1$ is the partition function of the ion, and $X_\ell$ is the excitation
potential.  Thus, the ion abundance

$$N_i = \frac{m_e c^2}{\sqrt{\pi} e^2} \, \frac{g_1}{g_\ell} \, \frac{V_0}{f_\ell \lambda_\ell} \, e^{X_\ell / kT_{exc}} \, \left( \frac{1}{r_{\lambda_0}} - 1 \right) . \tag{4}$$

We have selected the most appropriate lines suitable for use for the ions observed
in SN 1987A, and they are listed in Table 1 together with the relevant atomic data
for the transitions and the observed absorption depths.  The f-values have been
taken from the sources referenced.

The optical Sr II lines are all strongly blended and saturated, so we have used

the IR line at 1.03$\mu$ from a scan obtained on 9 May 1987.  The nearest date for which
a good signal-to-noise optical scan was obtained is 14 May, and so all of the
optical data pertain to this date, although neither the IR nor the optical spectrum
was changing appreciably near this time.  The largest uncertainties in
the determination of the abundances derive from the poorly defined continuum flux
levels and the unknown excitation temperature.  It is not likely that the continuum
flux is uncertain by more than 40 percent, and it should be in error by similar
amounts for all the optical lines.  In addition, although the excitation
temperature is not well known, the lines being considered all have similar
excitation potentials and so the relative Boltzmann factors between the various
lines are not sensitive to temperature.

The spectral line identifications for SN 1987A indicate that the heavy elements
are predominantly singly ionized.  Limits to the amount of neutral material can be
obtained from the absence of the neutral lines Fe I $\lambda$5269 and Ba I $\lambda$5535, both of
which should be quite strong for any substantial fraction of neutral species because
of their low excitation potentials and high f-values.  The limits to the absorption
strengths of these lines listed in Table 1 require the amount of neutral iron and
barium to be Fe I/Fe II $\leq$ 0.004 and Ba I/Ba II $\leq$ 0.01, thus the heavy elements may
be considered to be singly ionized.  The possibility does exist that a non-
negligible fraction of doubly ionized species could be present, yet undetected,
since none of those ions have low-excitation transitions in the optical or IR,
although we believe this to be unlikely.

TABLE   1

ABSORPTION LINE PROPERTIES OF SN 1987A
(MAY 1987)

| Line | $r_{\lambda_0}$ | log $g_\ell f_\ell$ | $\chi_\ell$(eV) | $g_1$ | Ref. |
|---|---|---|---|---|---|
| H I   $\lambda$4861 | 0.37 | −0.02 | 10.19 | 2 | a |
| Sc II $\lambda$5527 | 0.73 | +0.13 | 1.76 | 15 | b |
| Fe I   $\lambda$5269 | >0.5 | −1.32 | 0.86 | 25 | c |
| Fe II $\lambda$5018 | 0.42 | −1.40 | 2.88 | 30 | d |
| Sr II $\lambda$1.033$\mu$ | 0.48 | −0.25 | 1.83 | 2 | e |
| Ba I   $\lambda$5535 | >0.75 | +0.20 | 0.00 | 1 | b |
| Ba II $\lambda$6142 | 0.39 | −0.08 | 0.70 | 2 | b |

(a) Wiese, Smith, and Glennon (1966);  (b) Reader, Corliss, Wiese,
and Martin (1980);  (c) Fuhr, Martin, Wiese, and Younger (1981);
(d) Phillips (1979);  (e) Lindgard and Nielsen (1977).

Using the line strengths and parameters given in Table 1, and the solar
abundances for Sc, Fe, Sr, and Ba listed by Cameron (1982), we have computed the
relative abundances of these elements with respect to their solar values as a
function of excitation temperature from eqn. (4), and the results are shown in
Figure 1.  The corresponding color temperature for the supernova at this time has

been computed by Hamuy _et al._ (1987) from optical and IR photometry to be $T_c$ = 5,600 K. By analogy with supergiant atmospheres, the excitation temperature should be less, in the vicinity of $T_{exc}$ ≈ 5,000 K, since it is coupled to the lower kinetic temperature higher in the envelope than the region where the continuum is formed.

It should be emphasized that solar abundance ratios are used here only as a convenient reference point. The LMC is known to have a total heavy element abundance that is approximately two to three times less than solar (van Genderen, van Driel, and Greidanus 1986; Dufour 1984). The abundances of Sc, Sr, and Ba in the LMC are not known because of the difficulty in detecting lines of these elements in objects. They are probably not solar; however, unless the history of nucleosynthesis in the Large Cloud is completely different from that in our Galaxy, the relative abundances of the s-process elements with respect to each other and to Fe should not differ greatly from those of the sun.

The Fe II lines have the highest excitation potentials of the heavy elements we are considering, therefore a lower assumed temperature leads to a higher derived abundance for Fe relative to Sc, Sr, and Ba, as is seen in Figure 1. For temperatures exceeding $T_{exc}$ = 4,000 K, all of the s-process elements are enriched relative to Fe. At lower temperatures, the Sc and Sr are increasingly enhanced relative to Ba. The uncertainties in the logarithmic abundances relative to each other are of the order of ±0.3, independent of the uncertainties associated with knowing $T_{exc}$. For $T_{exc}$ = 5,000 K, the s-process elements have enrichments of factors of 3 to 8 with respect to iron.

In mid-May on the dates of the present observations the highest Balmer lines were visible in the spectrum, and their existence helps immeasurably in the determination of $T_{exc}$. Because of the much higher excitation potential of the Balmer lines ($\chi$ = 10.2 eV) in comparison with the other lines under consideration, the strengths of the hydrogen lines relative to those of Fe II are very sensitive to temperature for $T_{exc} \leq 10^4$ K. There is no a priori reason to suspect that the Balmer lines are formed under conditions much different from those of the heavy elements since their strengths are comparable, although their formation is selectively favored in higher temperature regions of the envelope. Since Hα is more saturated than Hβ, we have used the strength of Hβ to determine the relative abundance of H with respect to Fe from eqn. (4) in the same manner as the other elements, and that result is also plotted in Figure 1. The curve represents a lower limit to [H/Fe] since some hydrogen could be ionized. An abundance of [H/Fe] ≈ 0.4 in the progenitor is expected on the basis of the lower metallicity of the LMC. This value occurs for a temperature $T_{exc}$ = 5,100 K, which is essentially the same $T_{exc}$ inferred from the color temperature. Thus, we take as the derived abundances for the supernova envelope the following values based on the observed absorption line strengths [Sc/Fe] = 0.90, [Sr/Fe] = 0.75, and [Ba/Fe] = 0.55.

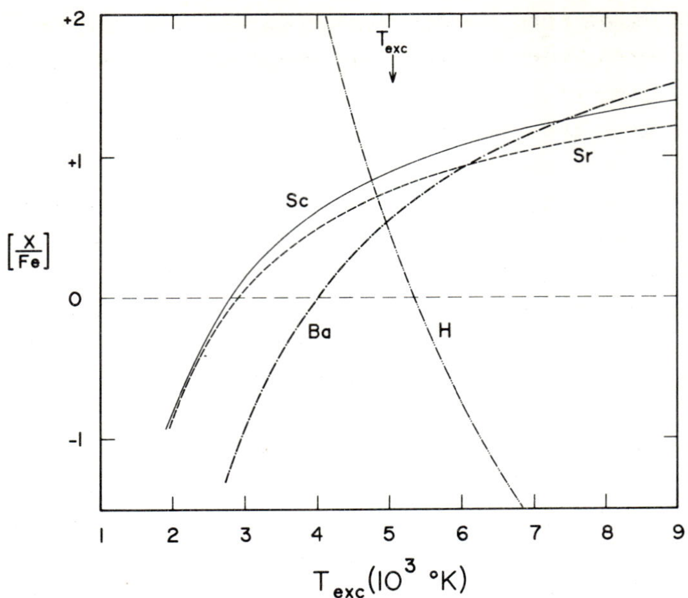


FIG. 1-- The logarithmic abundances of H, Sc, Sr, and Ba with respect to Fe in SN 1987A relative to their solar system values, as a function of the envelope excitation temperature $T_{exc}$. The abundances are derived from the absorption lines in May 1987, and the vertical arrow shows the appropriate temperature at that time.

The calculations confirm what the spectral scans suggest, i.e., the unusual strengths of absorption from Sc, Sr, and Ba are due in part to enhancements of these elements. The enrichments are not very large, partly a result of dilution caused by mixing of the processed gas from the He-burning zone with the outer hydrogen envelope of the star. One of the predictions of individual s-process episodes in stars is a greater enrichment of Sc and Sr compared to that of Ba. The calculations of Lamb et al. (1977) and Prantzos, Arnould, and Arcoragi (1987) have shown that the enhancement factors for Sr are roughly 6 times that for Ba from the s-process in stars with initial masses in the range $15 < M < 60\ M_\odot$. Repeated episodes in which a larger fraction of the star is subjected to He-shell burning with multiple pulses over a long time-scale can eventually lead to an equilibrium situation which gives similar enhancement factors for Sr and Ba, as occurs in Ba II stars, however this situation is not expected to pertain to massive stars such as Sk -69° 202.

Models of the supernova progenitor have indicated that SN 1987A possessed a helium-core mass of $M_{He} \approx 6\ M_\odot$, with a total mass at the time of outburst in the range of 10-20 $M_\odot$, and a carbon-core mass of 4 $M_\odot$ (Woosley et al. 1987). If the s-process enhancements produced in the carbon core are approximately those computed by Lamb et al. (1977) for this mass regime, and if these are subsequently mixed via

convection or the outburst itself with the outer hydrogen envelope of the star, the
resulting abundance ratios would, assuming an initial solar distribution, be given
by

$$\left[ \frac{X_A}{X_B} \right] = \log \left[ \frac{E_A \, M_C + M_T}{E_B \, M_C + M_T} \right] , \tag{5}$$

where $E_A$ and $E_B$ are the enrichment factors of elements $X_A$ and $X_B$, and $M_C$ and $M_T$ are
the C-core and total mass of the progenitor. For an initial mass of around 18 $M_\odot$,
interpolation of the results of Lamb <u>et al</u>. gives $E_{Sc} = 20$, $E_{Sr} = 25$, and $E_{Ba} = 3.5$.
We have used these over-production factors to compute the expected s-process
enhancements in SN 1987A on the basis of complete core/envelope mixing. Since the
helium-core (and therefore carbon-core) mass is fairly well fixed by the supernova
models, with $M_C \sim 4 \, M_\odot$ (Woosley <u>et al</u>. 1987), the primary variable in the
calculations is the uncertain envelope mass. We have assumed several values of $M_T$
for the calculations, and have computed the resulting abundance ratios of Sc, Sr,
and Ba with respect to Fe, as given in Table 2. For homogeneous mixing in the
progenitor, which is indicated from the absence of appreciable spectral evolution
during the first four months following outburst, the calculations demonstrate that
enhancements of the order of factors of 2 to 20 should be present in the supernova.
This is roughly what the observations of the absorption line strengths yield.

TABLE  2

EXPECTED s-PROCESS ENRICHMENTS IN THE PROGENITOR

| Elements | $M_T = 7 \, M_\odot$ | $M_C = 4 \, M_\odot$<br>$M_T = 10 \, M_\odot$ | $M_T = 15 \, M_\odot$ |
|---|---|---|---|
| [Sc/Fe] | 1.1 | 0.9 | 0.8 |
| [Sr/Fe] | 1.2 | 1.0 | 0.9 |
| [Ba/Fe] | 0.5 | 0.4 | 0.3 |

The extent of the s-process enrichments in SN 1987A could enable limits to be
placed on the mass of the hydrogen envelope if the uncertainties in the observed
abundances can be improved upon. The over-all enrichments deduced from the present
scattering-atmosphere analysis are consistent with total progenitor masses in the
range of 7 - 15 $M_\odot$. Also, it is worth noting that the enhancement of Ba with
respect to Sr and Fe is observed to be larger than it is predicted to be from the
s-process calculations. The differences are near the limits of the uncertainty of
the analysis, so they may not be significant. On the other hand, they may signify
that the initial abundance distribution of the elements was not solar, or that the
r-process also operated to modify the distribution.

The present calculations show that the Sc, Sr, and Ba abundances are higher
with respect to Fe than found in the sun. Since these elements are secondary
nucleosynthesis products, synthesized from CNO and the iron-peak nuclei by the

s-process, we expect from the lower metallicity of the LMC that their normal
abundances with respect to Fe in the Cloud are lower than the solar values. Their
enhancements in SN 1987A may therefore be taken as evidence for the s-process in the
progenitor, as is indeed expected to occur in massive stars. The extent of the
observed enrichments is in rough agreement with that computed from the models of the
progenitor if mixing occurs, however the Ba/Sr ratio in the supernova does appear to
be higher than that expected. This could be due to the fact that the Ba/Sr
abundance is generally higher in the LMC than in our Galaxy, having been produced in
the ISM of the two galaxies by different classes of objects.

As an adjunct to the s-process enrichments, carbon should also be enhanced
since the s-process occurs in the helium burning zone. At low temperatures no
absorption lines of the lower ionization stages of carbon are expected in the
optical. However, when the spectrum evolves to an emission spectrum, emission from
C II $\lambda 4267$ (or C IV $\lambda 1549$ and C III] $\lambda 1909$ in the UV) should be detected to test
this expectation. Further refinements to the abundance analysis and the s-process
calculations applied to the progenitor should ultimately allow more detailed
statements to be made concerning the nature of the progenitor and its evolution.

The author wishes to thank D. Dearborn and I. Iben for informative discussions,
A. McWilliam for assistance with the atomic data, and acknowledges the stimulating
atmosphere of the Aspen Center for Physics where part of this work was done. Most
important, great credit must be given to the entire CTIO staff who have gone to such
lengths to secure and reduce the photometric and spectroscopic data each clear
night, making it readily available to every interested scientist. Their tremendous
attitude and spirit of inquiry is a pleasure to be associated with.

REFERENCES


Blanco, V.M., et al., 1987, Ap. J., **320**, 589.
Cameron, A.G.W. 1982, in Essays in Nuclear Astrophysics, ed. C. Barnes, D.
    Clayton, and D. Schramm (Cambridge Univ. Press: Cambridge), p. 23.
Clayton, D.D. 1968, Principles of Stellar Evolution and Nucleosynthesis (McGraw-
    Hill: New York).
Danziger, I.J., et al. 1987, Astr. Ap., **177**, L13.
Dufour, R.J. 1984, in IAU Symp. No. 108: Structure and Evolution of the Magellanic
    Clouds, ed. S. van den Bergh and K. de Boer (Dordrecht: Reidel), p. 353.
Fuhr, J.R., Martin, G.A., Wiese, W.L., and Younger, S.M. 1981, J. Phys. Chem.
    Ref. Data, **10**, 305.
Hamuy, M., Suntzeff, N.B., Gregory, B.A., and Elias, J.H. 1987, A. J., in press.
Lamb, S.A., Howard, W.M., Truran, J.W., and Iben, I. 1977, Ap. J., **217**, 213.
Lindgard, A., and Nielsen, S.E. 1977, Atomic Data and Nuclear Data Tables, **19**, 533.
Menzies, J.W., et al. 1987, M.N.R.A.S., **227**, 39P.
Mihalas, D. 1969, Stellar Atmospheres (W.H. Freeman: San Francisco), ch. 11.
Phillips, M.M. 1979, Ap. J. Suppl., **39**, 377.
Prantzos, N., Arnould, M., and Arcoragi, J.-P. 1987, Ap. J., **315**, 209.
Reader, J., Corliss, C.H., Wiese, W.L., and Martin, G.A. 1980, Wavelengths and
    Transition Probabilities for Atoms and Atomic Ions, NBS Publ. 68.
van Genderen, A.M., van Driel, W., and Greidanus, H. 1986, Astr. Ap., **155**, 72.
Wiese, W.L., Smith, M.W., and Glennon, B.M. 1966, Atomic Transition
    Probabilities, NSRDS-NBS 4.
Williams, R.E. 1987, Ap. J. (Letters), **320**, L117.
Woosley, S.E., Pinto, P.A., Martin, P.G., and Weaver, T.A. 1987, Ap. J., **318**, 664.

# ON THE ORIGIN OF SUPERNOVAE OF TYPE Ib


David Branch

Department of Physics and Astronomy, University of Oklahoma

Norman, Oklahoma 73019, U.S.A.


## I. INTRODUCTION

The issue I want to discuss is this: supernovae observed during their photospheric phases display three distinct kinds of spectra - but there are four distinct chemical compositions that should be considered for the outer layers of exploding stars. What is the correspondence between the observed supernova type and the chemical composition?

The three kinds of observed spectra are Type II, which by definition shows optical hydrogen lines, Type Ia, which has neither hydrogen nor helium lines, and Type Ib, which has helium but not hydrogen.

The compositions to be considered are hydrogen-rich, or, speaking loosely, "solar"; helium-rich, or, loosely, "Wolf-Rayet"; deflagration - initially a mixture of heavy elements from carbon to radioactive nickel, which decays through cobalt to iron; and detonation - initially a mixture of just helium and radioactive nickel.

By definition, a Type II spectrum is to be associated with a hydrogen-rich composition. Type Ia, lacking hydrogen and helium, is identified with the deflagration composition. Detailed hydrodynamical models of carbon deflagrations in accreting white dwarfs lead to good agreement between theoretical and observed "early-time" photospheric spectra, "late-time" nebular spectra, and light curves. In spite of the successes, the association of Type Ia with deflagrations should be tentative, in view of some unresolved issues such as whether the requisite number of white dwarfs can grow to the Chandrasekhar mass and whether the high predicted gamma-ray flux is really emitted. At present, however, no other model predicts the appropriate heavy-element composition in the outer layers.

The question becomes - which composition, helium-rich or detonation, corresponds to Type Ib?

## II. TYPE Ib: MASSIVE-STAR CORE-COLLAPSE OR WHITE-DWARF DETONATION?

Three fundamental constraints on the nature of Type Ib are the presence of strong optical lines of He I during the photospheric phase (Harkness et al. 1987); the presence of strong forbidden lines of oxygen ions during the nebular phase (Gaskell et al. 1986, Filippenko and Sargent

1986); and the association with regions of recent star formation (Porter and Filippenko 1987). These constraints point immediately to the possibility that Type Ib result from the collapsing cores of massive stars that have lost their hydrogen envelopes, i.e., Wolf-Rayet stars (Wheeler and Levreault 1985, Begelman and Sarazin 1986, Chevalier 1986, Filippenko and Sargent 1986, Gaskell et al. 1986, Schaeffer, Casse, and Cahen 1987). The existence of an oxygen layer beneath a helium layer is natural, and massive stars undoubtedly will die in regions of star formation. Uomoto (1986) has discussed the closely related possibility that Type Ib are core-collapses of helium-rich stars that have lost their hydrogen via Roche-lobe transfer to close companions. Single Wolf-Rayet stars evidently come from stars whose initial mass exceeded 40 $M_\odot$ (Abbott et al. 1986) while the loss of hydrogen in a binary can happen to less massive stars.

An alternative way to match the fundamental constraints, involving a thermonuclear explosion in an accreting white dwarf, was suggested by Branch and Nomoto (1986). The motivation was not only to account for Type Ib but also to solve the problem of the "double detonations". The outcome of idealized one-dimensional (spherically-symmetric) models of accreting carbon-oxygen white dwarfs depends strongly on the mass accretion rate (Nomoto 1987, Woosley and Weaver 1986). At rates of $10^{-8}$ - $10^{-6}$ $M_\odot$ $yr^{-1}$ central ignition of degenerate carbon leads to a carbon deflagration and a Type Ia supernova. For rates below $10^{-9}$ $M_\odot$ $yr^{-1}$ hydrogen shell ignition probably produces a classical nova. At intermediate rates of $10^{-9}$ - $10^{-8}$ $M_\odot$ $yr^{-1}$, ignition of degenerate helium leads to an outwards detonation wave and, for most parameter choices (Woosley, Taam, and Weaver 1986), an inwards detonation through the CO core; together the two detonations disrupt and eject the entire star as a helium-nickel mixture. With more than a solar mass of radioactive nickel this double-detonation explosion would be bright and easy to see, yet no intrinsically bright supernova with only helium, nickel, and its decay products has been observed. Nature must have a way of suppressing the double-detonations.

Branch and Nomoto speculated that when two-dimensional calculations are carried out, to treat the realistic case of ignition at a point at the base of the helium layer rather than simultaneous ignition all around the spherical base of the layer, the double-detonations might not occur. Instead, the inwards detonation may fail, and the CO core may be left behind as a white dwarf or it may be ejected unburned. The latter case would correspond to a helium-nickel mixture in the outermost layers with a carbon-oxygen mixture beneath - possibly corresponding to Type Ib. Iben et al. (1987) have discussed a similar model in which the accreted mass is helium, rather than hydrogen. Kokhlov and Ergma (1987) have discussed a failed detonation in a low-mass helium white dwarf, but this model wouldn't provide the oxygen layer beneath the helium.

## III. OBSERVATIONAL CONSTRAINTS

In this section the Wolf-Rayet and white-dwarf models are discussed in the light of some of the observational constraints. Not all constraints are considered here; for example, the Type Ia infrared light curve shows a distinct dip a few weeks after maximum that is not seen in Type Ib (Elias et al. 1985), but since the cause of the dip in Type Ia is not known its absence in Type Ib does not now provide a useful constraint.

### i) The Association with Star-Formation Regions

Type Ib supernovae appear in or near regions of star formation in spiral galaxies (Porter and Filippenko 1987). However, Huang (1987) has found that Type II supernovae, most of which probably have main sequence masses between 8 and 15 $M_\odot$, also occur in star-formation regions. Therefore this constraint is consistent with but does not require the hypothesis that Type Ib have more massive progenitors than Type II. Type Ib also could have somewhat less massive progenitors than Type II. In the white-dwarf model, the initial masses are not expected to be low. If the white dwarf accretes at $10^{-8}$ $M_\odot$ yr$^{-1}$ at low efficiency (a few percent) from the wind of a companion, and the white dwarf needs to accumulate a few tenths of a solar mass of helium before it explodes, then the companion must be able to lose several solar masses in its wind. It would need to be initially a star of at least intermediate mass ($\geq$ 5 $M_\odot$).

### ii) The Radio Emission

The Type Ib supernovae 1983n and 1984l were observed as radio sources (Sramek, Panagia, and Weiler 1984, Panagia, Sramek, and Weiler 1986). Chevalier (1984a) was able to account for the radio emission from SN 1983n using the circumstellar-interaction theory he had introduced to account for Type II supernovae (Chevalier 1984b). Matching the theory to observation gives an estimate of the density in the circumstellar shell, or equivalently an estimate of the ratio of the pre-supernova mass-loss rate to the wind velocity, $\dot{M}/V_w$. Chevalier's estimate for SN 1983n was $5 \times 10^{-7}$ $M_\odot$ yr$^{-1}$/km s$^{-1}$. In the white-dwarf model the wind of the red-giant donor might be 10 km s$^{-1}$, so the wind mass-loss rate would have to be $5 \times 10^{-6}$ $M_\odot$ yr$^{-1}$, which is plausible. In the Wolf-Rayet model the wind velocity might be 2000 km s$^{-1}$, so the mass-loss rate would have to be $10^{-3}$ $M_\odot$ yr$^{-1}$. Observed mass-loss rates for Wolf-Rayet stars are less than $10^{-4}$ $M_\odot$ yr$^{-1}$ (Abbott et al. 1986). Considering the uncertainties, this discrepancy does not exclude the Wolf-Rayet model, but contrary perhaps to the prevailing wisdom the white-dwarf model has less difficulty with the radio emission than the Wolf-Rayet model, not more.

### iii) The Shape of the Light Curve

The shape of the Type Ib light curve is similar to that of Type Ia. Since the width of the light curve is sensitive principally to the ejected mass, and the carbon-deflagration model gives a good account of the Type Ia light curve with an ejected mass of 1.4 $M_\odot$ (Graham 1987), it is not surprising that the predicted light curve for a detonation model that leaves a CO white dwarf behind is too fast (Woosley, Taam, and Weaver 1986). Whether the predicted light curve will be slowed down when the velocity-sensitive line opacity and the point explosion are correctly treated remains to be seen. Similarly, it is not surprising that the predicted light curve for an 8 $M_\odot$ Wolf-Rayet model (Schaeffer, Casse, and Cahen 1987), which ejects 6.6 $M_\odot$, is much too broad for Type Ib; at late times the gamma rays from $^{56}$Co decay are trapped and the optical light curve follows the $^{56}$Co decay (Figure 1). If Wolf-Rayet stars produce Type Ib, they must reduce to 4 $M_\odot$ or less before they explode!

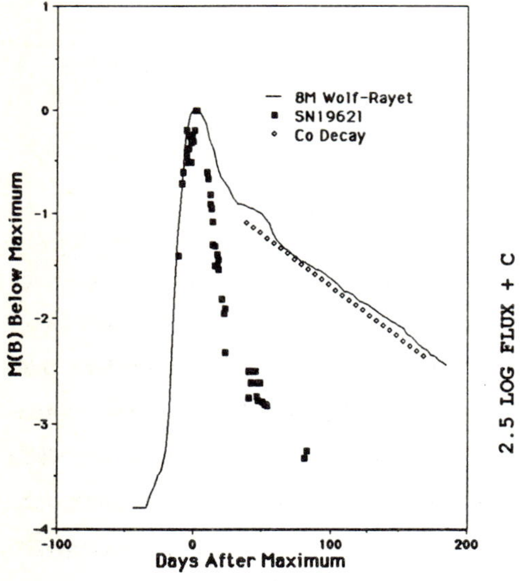


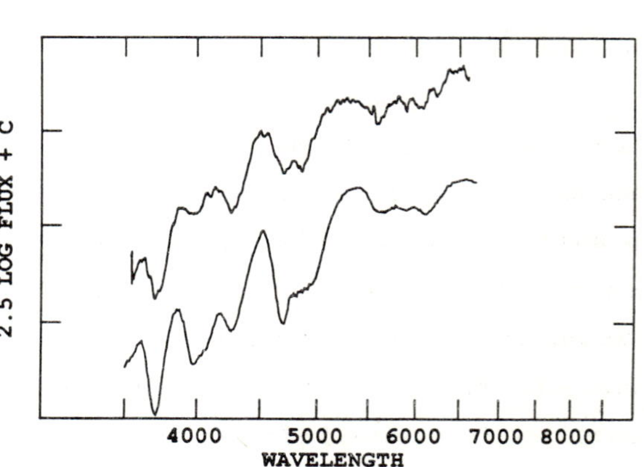


Fig. 1 - The theoretical light light curve for the explosion of an 8 $M_\odot$ Wolf-Rayet star (Schaeffer, Casse, and Cahen 1987) is compared to the observed light curve of the Type Ib SN 1962$\ell$ (Rust 1974), and to the rate of energy released by $^{56}$Co decay. Vertical normalization is arbitrary.

Fig. 2 - The observed pre-maxmimum spectrum of the Type Ib SN 1983v obtained at the AAT by Axon, Cannon, and McDowell (top) is compared to a synthetic spectrum consisting of lines of Fe II, He I, and Ca II and having a velocity at the photosphere of 18,000 km s$^{-1}$ (bottom).

iv) Spectra During the Photospheric Phase

Harkness et al. (1987) have argued that the optical spectra of Type Ib during the photospheric phases are consistent with a helium-rich composition but not with a detonation composition. I don't think the spectrum is sufficiently well understood to discriminate between the two possibilities. I have calculated the predicted LTE line optical depths at the photosphere for both the helium-rich and detonation compositions, as described in Branch (1987) for the hydrogen-rich composition. Neither composition predicts He I lines to be as strong as observed, so a strong non-LTE excitation of He I needs to be invoked. The reason for a large non-LTE effect in the Wolf-Rayet model is not obvious; a possibility that needs to be checked for the white-dwarf model is non-LTE excitation caused by the decay products of the radioactive cobalt that coexists with the helium.

Strong Fe II lines are predicted for both compositions. One might think that it would be easy to discriminate between the two compositions on the basis of the the lines that accompany the Fe II and He I lines. The problem is that the Fe II lines are ubiquitous - they are all over the optical spectrum and it is not easy to be sure what other lines are present.

Fe II lines also are predicted to dominate the near ultraviolet, for both compositions. A choice between compositions could be made based on the near infrared, where Mg II and Ca II lines are stronger than Fe II in the helium-rich case, but no infrared spectra have yet been observed for Type Ib.

An important constraint from the optical photospheric spectra that does not depend on the quantitative interpretion of line strengths is the presence of high-velocity matter just before maximum light. A spectrum of the Type Ib SN 1983 in NGC 1365, obtained at the AAT about 5 days before maximum light by Axon, Cannon, and McDowell, is compared in Figure 2 with a synthetic spectrum based on a velocity at the photosphere of 18,000 km s$^{-1}$. From the fact that the kinetic energy of the matter above the photosphere scales as $t^2 v^4$, it can be shown that this high-velocity phase must have been shortly (a week or less) after the explosion. Otherwise the kinetic energy would be much greater than $10^{51}$ ergs. This adds to the evidence that the ejected mass is low; not only is the light-curve peak narrow, the rise time is short. If the absorption near 3700 A is due to Ca II, as it is in the synthetic spectrum, then the detonation would need to falter (Khokhlov and Ergma 1987) in the outermost layers, to leave a small amount of calcium.

v) Spectra During the Nebular Phase

At this meeting Axelrod and Fransson both have shown synthetic nebular spectra based on the Wolf-Rayet model, and both find good qualitative agreement with observation. The main quantitative problem seems to be that since in this model the oxygen is concentrated into a shell, the line profiles come out to be box-shaped and broad compared to the narrow centrally-peaked observed lines. Whether this is a fundamental problem with the model or whether it is evidence that mixing of oxygen to deeper slower layers occurs during the explosion is not yet clear. Axelrod and Fransson both have expressed doubt that synthetic nebular spectra for the white-dwarf detonation model would match the observations, although the calculations have not yet been carried out. One virtue of the white-dwarf model is that the oxygen distribution is centrally peaked. One problem is that CO white dwarfs may contain too much carbon to be consistent with the relatively weak carbon lines in the the observed spectrum (C. Fransson, private communication). Another may be that the inner oxygen region will not absorb enough of the radioactive decay energy to keep the light curve bright and the spectrum dominated by oxygen lines (P. Pinto, private communication).

## III. SUMMARY

It is not evident which (if either) of the two models discussed here really produces Type Ib. The main problem with the Wolf-Rayet model appears to be that the light-curve width and the high-velocity pre-maximum spectrum demand that the ejected mass be small. Since the consequences for the white-dwarf model of a point ignition have not yet been explored, the white dwarf idea is really a speculation rather than a model. The main problem may be with the nebular spectrum. Although I have argued that the spectrum is not sufficiently well understood to provide a clear choice between the helium-rich and detonation compositions, I think that the situation will improve in the near future. The rapid development of quantitative spectroscopy of supernova photospheric spectra that has been stimulated by the high-quality observations of SN 1987a will lead before long to stringent tests of both models.

## ACKNOWLEDGEMENTS

I have enjoyed lively and informative discussions with many colleagues about the observations and the nature of Type Ib supernovae. I am especially grateful to Russell Cannon for sending me the spectrum of SN 1983v. This work has been supported by NSF grant AST 8620310.

## REFERENCES

Abbott, D., Beiging, J. H., Churchwell, E., and Torres, A. 1986, Ap. J., 303, 239.
Begelman, M. C., and Sarzin, C. L. 1986, Ap. J. (Lett.), 302, L59.
Branch, D., and Nomoto, K. 1986, Astr. Ap., 164, L13.
Chevalier, R. A. 1984a, Ap. J. (Lett.), 285, L63.
Chevalier, R. A. 1984b, Ann. N. Y. Acad. Sci., 422, 215.
Chevalier, R. A. 1986, Highlights Astron., 7, 599.
Elias, J. H., Mathews, K., Neugebauer, G., and Persson, S. E. 1986, Ap. J., 296, 379.

Filippenko, A. V., and Sargent, W. L. W. 1986, A. J., 91, 691.
Gaskell, C. M., Capellaro, E., Dinerstein, H., Garnett, D., Harkness, R. P., and Wheeler, J. C.
      1986, Ap. J. (Lett.), 306, L77.
Graham, J. R. 1987, Ap. J., 315, 588.
Harkness, R. P., Wheeler, J. C., Margon, B., Downes, R. A., Kirshner, R. P., Uomoto, A., Barker,
          E. S., Cochran, A. L., Dinerstein, H. L., Granett, D. R., and Levreault, R. M. 1987, Ap.
J., 317, 355.
Huang, Y.-L. 1987, P.A.S.P., 99, 461.
Iben, I., Jr., Nomoto, K., Tornambe, A., and Tutukov, A. V. 1987, Ap. J., 317, 717.
Khokhlov, A. M., and Ergma, E. V. 1986, Sov. Astr. Lett., 12, 152.
Nomoto, K. 1987, in Proc. 13th Texas Symp. Rel. Ap., ed. M. P. Ulmer (World Scientific Press,
      Singapore), in press.
Panagia, N., Sramek, R. A., and Weiler, K. W. 1986, Ap. J. (Lett.), 300, L55.
Porter, A. C., and Filippenko, A. V. 1987, A. J., 93,1372.
Rust, B. W. 1974, Oak Ridge Nat. Lab., Rept. 4953.
Schaeffer, R., Casse, M., and Cahen, S. 1987, Ap. J. (Lett.), 316, L31.
Sramek, R. A., Panagia, N., and Weiler, K. W. 1984, Ap. J. (Lett.), 285, L59.
Wheeler, J. C., and Levreault, R. 1985, Ap. J. (Lett.), 294, L17.
Woosley, S. E., Taam, R. E., and Weaver, T. A. 1986, Ap. J., 301, 601.
Woosley, S. E., and Weaver, T. A. 1986, Ann. Rev. Astr. Ap., 24, 205.

# MODEL CALCULATIONS AND
# SPECTROSCOPIC CONSTRAINS FOR SN1987A


P.Höflich

Max-Planck-Institut für Physik und Astrophysik, Institut
für Astrophysik, Karl Schwarzschild Str. 1, 8046 Garching, FRG


## Summary


We present synthetic spectra for atmospheres in order to interpret the observed spectra of
the supernova 1987A during the first few months after the initial event. Spherical symmetry
and density profiles are assumed which are given by the homologous expansion of the stellar
structure of a B3 supergiant. For hydrogen, up to eight levels and, for helium, 16 levels
are allowed to deviate from LTE. The radiation transport is calculated consistently with the
rate equations both for the continua and for the lines. Radiative equilibrium is assumed.
The observed spectra and colours in the optical wavelength range can be well reproduced
by pure hydrogen models during the first few weeks. The behaviour of the UV flux is due
to changes in the effective temperature and the photospheric radius. For later stages, the
influence of elements other than hydrogen and helium must be taken into account in order
to compare the calculated and observed spectra in the optical wavelength range. Reasonable
agreement between calculations and observations can be obtained with the assumption of
half solar abundances for all elements but for the s-process elements Sc,Ti and Ba which
are overabundant. To explain the small changes in the spectra after about 3 weeks up to 4
months, we need a total hydrogen mass of about 8 to 10 $M_\odot$.


## 1. Introduction

Supernova 1987A in the LMC is the first type II supernova which has been observed
simultaneously from the X-ray to the radio wavelength range beginning from the very early
stages. In principle, these measurements allow very detailed checks of hydrodynamic models
for supernova explosions and tests of stellar evolution calculations (Arnett, 1987; Woosley et
al., 1987; Hillebrandt et al. 1987; Maeder, 1987; Truran, 1987). However SN1987A shows
some pecularities with respect to observations of other type II supernovae: i) the very early
occurrence of H lines and strong lines of heavier elements and the fast development of the flux
spectrum at all wavelengths during the first few weeks, and ii) the relatively small changes
in the spectra after about 4 weeks during several months after the explosion.

Detailed analyses of the observed spectra are helpful in order to answer questions concerning the hydrodynamic models, the mass of the envelope, the chemical composition as a
function of radius and therefore the stellar evolution of the progenitor, etc. . A critical test
for atmospheric calculations is the observed time dependences of the features.

To address these questions and to allow an interpretation of the observed spectra of
SN1987A, we have modeled photospheres of type II supernovae by using a computer code for

the construction of spherical extended non-LTE models (Höflich et al., 1986) which has been modified in order to allow for the treatment of high velocity fields (Höflich, 1987a, 1987b).

We assume stationarity and radiative equilibrium for the energy balance because the radiative timescales are short in respect to the hydrodynamic timescales soon after the initial increase in luminosity. Spherical symmetry is assumed. According to detailed numerical models (Falk and Arnett, 1977; Müller, personal communication, 1987; Nomoto, 1987; Nomoto et al., 1987) and also analytical solutions for strong shock waves in spherical expanding envelopes (Sedov, 1959) density profiles are taken which are given by the self-similar expansion of an initial structure, i.e.

$$\rho\,(r) = \lambda^{-3}\,\rho(R) \text{ and } r = \lambda(t)\,R \,,$$

where r is the distance of a given mass element at time t and R its corresponding distance at time 0. The expanding matter of the envelope is mainly accelerated during the very first stages of the evolution. Consequently the expansion parameter $\lambda$ is a linear function of time. The envelope is assumed to consist of hydrogen for models corresponding to the early stages. For later stages, helium and heavier elements are also included. For hydrogen, five to eight levels are allowed to deviate from the local thermodynamical equilibrium (LTE). All bound-bound and bound-free transitions have been included in the rate equations. In addition, two-quantum decay of the second energy level is taken into account. Helium is represented by a 16 level model atom (He I singlet: 1s, 2s, 2p, 3s, 3p, 3d, 4, 5; He I triplet: 2s, 2p, 3s, 3p, 3d, 4, 5; He II: 1). The bound-bound and bound-free opacities of the same transitions are also included in the radiation transport equation which is solved by the comoving frame method developed by Mihalas et al. (1975) for the first momentum of the intensity. To calculate the emitted flux in the observers frame and to archive energy balance, we use the integral method of Schmid-Burgk (1975) which has been modified in order to allow the treatment of high velocity fields. Thomson scattering, free-free opacities and the opacities due to the higher members of the Balmer series (up to the main quantum number of 30) are taken into account. These line opacities and the corresponding source functions are calculated under the assumption that populations of the upper levels are determined by LTE. This is a reasonable approximation because these populations are mainly dominated by collisional processes. In order to analyse later stages, line blanketing due to heavier elements is included for line formation calculations. Pure scattering and LTE is assumed for the level populations for all these lines. The ionisation balance of elements other than H and He is calculated under the assumption that photoionisation occurs from the ground states only. Here the radiation field is given by a H/He non-LTE model which is blanketed by lines of heavier elements (about 2000) in the UV. The ionisation balance of most of the ions is not very sensitive to this treatment as test calculations have shown in which the LTE approximation was used. Note that the assumption of LTE for the relative populations inside an ionisation state should be regarded as a first order approximation. This allows the identification of observed features, a representation of the blanketing due to the large number of weak lines and to give first estimates of the element abundances, if a lot of lines of a certain ion is taken into consideration. However the element abundances determined by this treatment may show major errors if only a few strong features are used (see below).

## 2. Discussion of the models

Our models are described by the following free parameters: i) the initial density profile which is given by the stellar structure of a B3I star as calculated by A. Weiss (personal communication), ii) the expansion parameter $\lambda$, iii) the photospheric expansion velocity, iv) a statistical component of the velocity field, and v) the total luminosity. The total luminosity is determined by the observed flux in the visual wavelength range assuming the distance of SN1987A to be 46 kpc (Höflich, 1987a). Note that the velocity at the photosphere and the expansion parameter $\lambda$ are free parameters for only one model at a certain time. For all other stages these two parameters are determined by the assumption of constant kinetic energy (see above).

The comparison of the synthetic and the observed spectra at a specific time allows us to deduce the photospheric radius $R_{5000}$, the corresponding effective temperature $T_{eff}$, the particle density $N_o$ and the velocity field v at $R_{5000}$, the distance $R_{HII}$ at which most of the hydrogen becomes neutral, and the mass fraction of the envelope M in solar units which has passed the photosphere. Please note that for extended geometries the photospheric radius is a definition. Here $R_{5000}$ is the distance at which the optical depth for true absorption at 5000 Å equals 1 (from outside). This layer corresponds to the innermost layers from which photons can be observed because of the dominance of Thomson scattering $\tau_{sc}$ over bound-free opacities for all wavelengths but the far UV $(\tau_{sc}(R_{5000}) \geq 10)$. In table 1 we give besides these quantities the value of n which approximately corresponds to a density profile $\propto r^{-n}$ at $R_{5000}$, and the dates corresponding to SN1987A.

During the first week, the photospheric radius is strongly increasing due to the fact that the expansion of the photosphere is mainly coupled to the expanding matter already a few hours after the initial event. Simultaneously, the effective temperature decreases rapidly. This explains the rapid drop in the observed UV flux as a consequence of the changes in $T_{eff}$ and the increase of the IR-flux due to an increase in the photospheric radius. After about one to two weeks, the geometrical dilution of the material and the recombination of hydrogen outside a certain distance $R_{HII}$ result in a much slower change in $R_{5000}$ and, consequently, in smaller changes in time of the spectral energy distribution as long as the continuum is determined by electron scattering, i.e. the hydrogen envelope can be observed at the continuum forming region. The importance of an exact NLTE-treatment of H should be noted in respect to the electron density.

Firstly, pure hydrogen models are discussed in some detail. This allows an analysis of the optical spectra observed during the first few weeks. A comparison of the observed with the synthetic spectra show good agreement (figure 2). The early occurrence of broad absorption components can be understood as a result of the steeper density gradients and lower optical depths of electron scattering at the line forming region in comparison to normal type II supernovae, i.e. as a direct consequence of the fact that the progenitor star was a B3 supergiant. The increasing of the $H_\alpha$ emission is due to the growth of the line forming region and optical depth effects. The too low emission component of $H_\beta$ on Feb. 24 may be a hint that the assumed density structure is somewhat too steep at this stage. Additional discrepancies, mainly in the spectrum corresponding to Mar.01, are a consequence of the omission of line blanketing due to elements other than hydrogen.

Table 1: The distance $R_{5000}$ at which the optical depth is 1 for true absorption at 5000 Å, the effective temperature $T_{eff}$, the particle density $N_o$, the velocity field v in km/sec and the n corresponding to a density profile $N(r) \propto r^{-n}$ at the distance $R_{5000}$ are given for models with a density structure which is obtained by a homologous expansion of structure of a B3I star ($\dot{\lambda} \approx 10^{-3}$). All quantities are given in CGS. In addition the mass $\dot{M}$ in solar masses which has gone through the photosphere, the radius $R_{HII}$ of the HII-region and the corresponding date are given for SN1987A. Please note that we assumed a statistical velocity of 2000, 1000 and 500 km/sec for model I-III, IV-V and VI.

| No. | $R_{5000}$ | $T_{eff}$ | $N_o$ | v | n | $R_{HII}$ | Date | $\dot{M}$ |
|---|---|---|---|---|---|---|---|---|
| I | 1.28E14 | 12400 K | 3.8E12 | 20000. | 13. | - | Feb25 | 7.E-3 |
| II | 2.4E14 | 9150 K | 1.3E12 | 16500. | 10. | 3.9E14 | Feb26 | 1.2E-2 |
| III | 4.9E14 | 6500 K | 4.8E11 | 12000. | 7.2 | 7.8E14 | Mar02 | 2.E-1 |
| IV | 9.1E14 | 5100 K | 3.6E11 | 8000. | 5.6 | 1.15E15 | Mar09 | 8.E-1 |
| V[1] | 1.05E15 | 4800 K | 3.3E11 | 6000. | 5.2 | 1.25E15 | Mar16 | 1.3 |
| VI[2] | 1.15E15 | 5000 K | 4.0E11 | 4500. | 3.8 | 1.32E15 | Apr02 | 3.8 |

[1] H and He model

[2] H, He, C, N, O, Na, Mg, Si, S, Sc, Ca, Ti, V, Cr, Mn, Fe, CO, Ni, Sr and Ba model

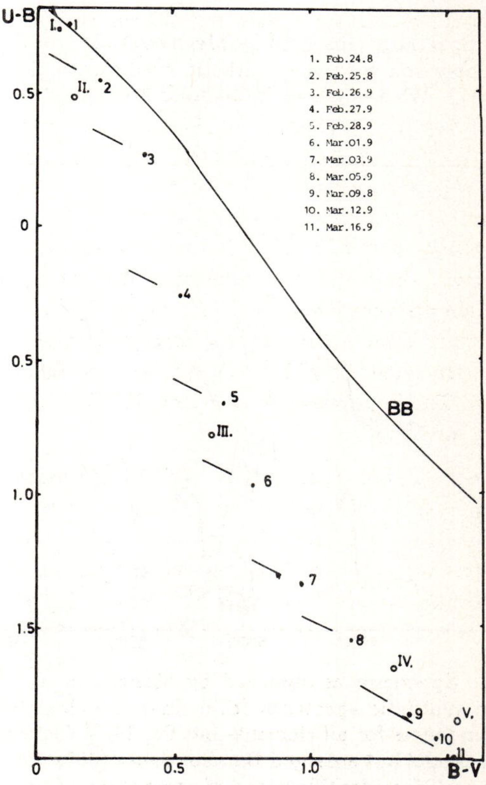


Figure 1: The colour index $E_{U-B}$ as a function of $E_{B-V}$ in the filter system of Johnson (1966). The indices as derived from a Planck function (BB; thin line), those as observed by Menzies et al. (1987) without (dots with numbers) and including interstellar reddening correction ($A_{B-V} = 0.08...0.18$ mag ,Wampler et al., 1987; bars), and those as calculated (open circles) by the models I-VI (see table 1) are given.

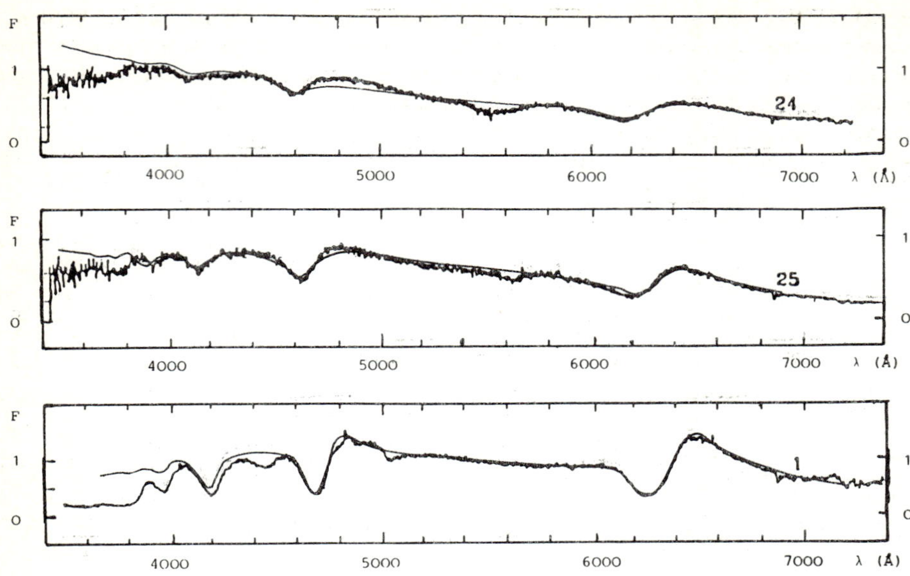

Figure 2: Spectra as observed by Menzies et al. (1987) at February 24.9, February 25.9 and March 1.8 in comparison with the synthetic spectra as calculated by the corresponding models I to III (see table 1). We assume an interstellar reddening $A_{B-V}$ of 0.08 mag.

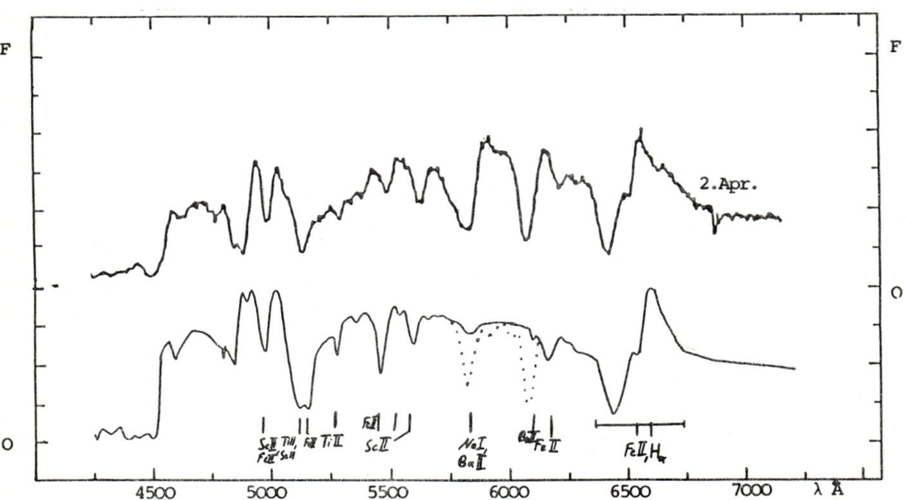

Figure 3: Spectrum as observed by Menzies et al. (1987) at Apr. 2. in comparison with the reddened synthetic spectrum (thin line) as calculated by model VI (see table 1) assuming half solar abundances for all elements but Sc, Ti, V,Cr,Sr and Ba (2.5*solar). The flux as calculated by the same model but enlarged Ba abundance (50*solar) is shown between 5600 and 6200 Å (dotted line). In addition identifications of some strong features are given.

The comparison between the calculated and the observed colour index $E_{U-B}$ as a function of $E_{B-V}$ shows good agreement (figure 1). The early departure from the slope expected from a black body fit, which is a good approximation for normal type II supernovae, is mainly due to H-line blanketing. Discrepancies between the observed and the calculated colour relation after about 2 weeks are due to the neglect of line blanketing by elements other than hydrogen (and helium) for these models. Therefore, a quantitative analysis of the observed spectra using pure hydrogen (and helium) atmospheres is restricted to about the first few weeks.

In order to analyse later stages and to get first estimates for the chemical abundances we have carried out line formation calculations by using some simple approximations with respect to the treatment of the heavier elements (see above). A comparison between the synthetic spectrum and the spectrum observed on Apr.02 in the optical wavelength range is shown in figure 3. In addition, some strong features are marked by ions which give the main contribution to the opacity. Note, however, that the calculated spectral slope is mainly determined by blanketing due to weak lines (about 1000), especially those of ScII, TiII and FeII. We use half solar abundances for all elements but the s-process elements Sc,Ti,V,Cr,Sr and Ba ($2.5_*$solar) to get the best fit. Features due to the s-process elements Sc, Ti and Ba can be identified. A reduction of the abundances with respect to the sun is needed in order to reduce line blanketing effects mainly due to iron at the longer wavelengths, and an overabundance of Sc and Ti is needed to fit some of the observed features and to explain the spectral slope below 5300 Å. In principle, the observed spectrum can be well reproduced with the exception of two strong features at about 5800 and 6080 Å which can be attributed to a blend of NaI and BaII and to a single BaII line, respectively. In order to get quantitative agreement, Na and Ba have to be choosen overabundant by a factor of about 10-15 and 50, respectively, in relation to the sun. However these estimates are based on a LTE-analysis of only few strong features and should be regarded as very uncertain. In fact test calculations show that the overabundances of Na and Ba may be reduced to a factor of 5-8 and 10-20 by NLTE-effects.

Note that the early occurrence of strong absorption features in the optical wavelength range is due to a low density in the outer region and therefore may be a hint for a low mass loss of the progenitor in the near past. The spectra observed later than Apr.02 up to about 4 months after the initial event show only small changes. This behaviour is to be expected from the models as long as the continuum is dominated by Thomson scattering, i.e. as long as the continuum region is formed in hydrogen rich layers. Therefore, the total mass of the hydrogen shell can be estimated to be 8-10 $M_\odot$. Note in addition that hydrogen features are visible for some more months because parts of the H lines are formed at much larger radii than $R_{5000}$.


## 3. Results

The observations can be understood by models with a density profile which is given by the holomogous expansion of a B3I star.

For the early stages pure hydrogen models can be used. The flux variations with time of the continua in the ultraviolet, the optical and the infared wavelength range are due to an early rapid decrease in the effective temperature and an increase in the photospheric radius. The smaller changes in the following are a result of the slow increase of the photospheric radius mainly due to a geometrical dilution effect and due to the recombination of hydrogen outside a certain radius. The early occurrence of the absorption features due to hydrogen and the rapid development of the spectrum in contrast to type II supernovae other than SN1987A can be understood as due to the steep density gradient and is therefore a direct consequence of the spectral type of the progenitor. The increase in the emission component of the hydrogen lines is due to the growth of the line forming region. For later stages, the line absorption due to heavier elements must be taken into account. The spectra can be understood by models with half solar abundances for all elements but the s-process elements (Sc, Ti and Ba) and Na which are overabundant. From the relatively small changes in the observed spectrum during several months after a few weeks, a mass of 8 to 10 $M_\odot$ can be infered for the H-rich envelope.

However we have also to stress the limits of the models. Because of the LTE-treatment of the heavier elements the given abundances should be regarded as a first estimate. Especially NLTE-treatment is needed in a more accurate determination of Na and Ba abundances. These models are restricted to stages at which the continua are dominated by electron scattering, i.e. to the first few months after the initial event.


### References


Arnett, W.D. *Astrophys.J.* **319** 136 (1987)

Falk,S.W.; Arnett, W.D. *Astrophys.J.Suppl.* **33** 515 (1977)

Hillebrandt,W.; Höflich,P.; Truran,J.W.; Weiss,A. *Nature* **327** 597 (1987)

Höflich,P. *Proceedings of the 4th workshop on Nuclear Astrophysics* Ringberg (1987a)

Höflich,P. *Proceedings of the ESO-workshop on SN1987A* Munich (1987b)

Höflich,P.; Wehrse,R.; Shaviv,G. *Astron. Astrophys.* **163** 105 (1986)

Johnson,H.L. *Ann.Rev.Astron.Astrophys.* **4** 197 (1966)

Maeder, A. *Proceedings of the ESO-workshop on SN1987a* (1987)

Menzies,J.W. et al. *Spectroscopic and photometric observations of SN1987A. The first 50 days* preprint (1987)

Mihalas,D.; Kunasz,R.B; Hummer,D.G. *Astrophys.J.* **202** 465 (1975)

Nomoto, K. personal communication (1987)

Nomoto, K.; Shigeyama,T.; Hashimoto,M. *Proceedings of the ESO-workshop on SN1987A* Munich (1987)

Schmid-Burgk,J. *Astron.Astrophys.* **40** 149 (1975)

Sedov,L.I. *Similarity and Dimensional Methods in Mechanics* , Academic Press, New York, p. 260 (1959)

Truran, J.W. *Proceedings of the ESO-workshop on SN1987A* Munich (1987)

Wampler, E.J.; Truran, J.W.; Lucy, L.B.; Höflich, P.; and Hillebrandt, W. *Astron. Astrophys.* **182** L51 (1987)

Woosley,S.E.; Pinto, P.A.; Martin,P.G.; Weaver,T.A. *Astrophys.J.* **318** 664 (1987)

TYPE II SUPERNOVA PHOTOSPHERES AND THE DISTANCE TO SUPERNOVA 1987A

Manorama Chilukuri and Robert V. Wagoner
Department of Physics, Stanford University
Stanford, California 94305-4060, USA

## I.  Introduction and Summary

Among the many historic opportunities provided by the recent supernova in the LMC is that to improve our understanding of the physical conditions in the neighborhood of supernova photospheres, even though 1987A was initially characterized by radial and time scales smaller (by a factor 5-10) than "standard" more luminous SNII. Two consequences of this understanding, which we shall focus on in this contribution, are (a) an estimate of the (frequency-dependent) location and thickness of the photosphere and (b) the only direct determination of the distance of the supernova (via the generalized Baade method). We find that the photosphere is sharp enough to allow the use of plane-parallel geometry in the calculation of the emergent continuum spectral flux, if we confine our attention to those epochs (temperature T ~ 5000-6000 K) at which hydrogen is recombining at the photosphere. We also find that the distance to this supernova is 43 ± 4 kpc. The reliability of this determination should improve when accurate spectrophotometric data for dates other than March 1 become available to us.

## II.  Model Photospheres

We have extended our previous calculations of radiative transfer within low-density photospheres (Hershkowitz, Linder, and Wagoner 1986; Hershkowitz and Wagoner 1987) to conditions appropriate to SN 1987a. Our only major assumption is that the photosphere is spherically symmetric and relatively sharp (steep gradient of optical depth). Although we will see that recombination does sharpen the photosphere (as expected), the possible presence of a second bright image associated with the supernova (Nisenson, Papaliolios, Karovska, and Noyes 1987; Meikle, Matcher, and Morgan 1987) might force us to eventually abandon spherical symmetry.

In addition to this geometrical assumption, our models employ the following approximations, most of which have been validated by previous calculations:

a) The velocity gradient is not strong enough to induce either an explicit time dependence of the physical conditions affecting an emerging photon or a significant violation of radiative detailed balance in the hydrogen lines. This last result, coupled with the fact that collisional excitations are negligible at these densities ($n_H \sim 10^{10} - 10^{12}$ cm$^{-3}$), means that the (non-LTE) level populations of hydrogen are governed solely by a balance between photoionization and direct recombination. This induces a cancellation which reduces the effective bound-free absorptive opacity of hydrogen.

b) The Eddington factor $f_\nu = K_\nu/J_\nu$ and the outer boundary condition on $H_\nu/J_\nu$ are taken to be that of a scattering-dominated (plane-parallel) atmosphere (Mihalas 1978). The absorptive opacity is usually less than the scattering opacity, especially in the more luminous (lower density) Type II supernovae.

c) The effective continuum scattering opacity $\chi_\nu(sc)$ is dominated by electron scattering and by the combined effect of very many velocity-gradient broadened spectral lines $[\chi_\nu(bb)]$. We have used the results of Karp, Lasher, Chan, and Salpeter (1977) [incorporating 260,000 lines] to estimate this contribution to the opacity. We have fitted their full grid of frequency-binned results (kindly provided by Alan Karp) to smooth functions of mass density $\rho$, matter temperature T, and inverse velocity gradient $t-t_o \cong r/v$ (where $t_o$ is the time when the mass element began freely coasting after the passage of the shock). Although this contribution of lines cannot be rigorously described by either a true opacity or emissivity, and Karp et al. (1977) assumed that the level populations were governed by LTE, we believe that the approximate equality of $J_\nu$ and $B_\nu$ at the photosphere and the above-mentioned dominance of radiative excitations allow one to use their results as a first approximation to a <u>scattering</u> opacity. Since their calculation employed a standard Pop. I abundance (heavy element mass fraction Z = 0.02, two to four times greater than that in the LMC) and their line list may be incomplete, as another first approximation we have rescaled their line opacity by a factor $Z_*/0.02$.

d) The continuum absorptive opacity $\kappa_\nu$ is dominated by hydrogen (H and $H^-$) bound-free and free-free processes. The constraint of statistical equilibrium allows us [using the result mentioned in (a) above] to eliminate the explicit appearance of the departure coefficients of the hydrogen level populations in the transfer equations, which results in increased numerical accuracy and stability (Hershkowitz, Linder, and Wagoner 1986).

Our set of model photospheres, as defined by the above conditions, can be essentially characterized by three free parameters (in addition to $t-t_o$), in much the same way as stellar photospheres. These are an effective temperature $T_e$, a scale height $\Delta R$, and the effective abundance $Z_*$, defined by the relations

$$\int_0^\infty F_\nu d\nu = \sigma_R T_e^4 \,, \qquad \int_r^\infty n_H dr = n_H(r)\Delta R \,, \qquad \chi_\nu(bb) \propto Z_* \,. \qquad (1)$$

Here $F_\nu$ is the spectral flux and $n_H$ is the total hydrogen number density. Although we have strictly assumed an exponential density profile $\rho(r)$, if the photosphere is sharp only the local value of $\Delta R(r)$, as defined by equation (1), is important.

In Figure 1 is shown one of the key relations which characterize the physical conditions near the photosphere for a model which we will see produces a reasonably good fit to the observed spectrum near the beginning of the recombination phase of SN 1987a (March 1). Plotted is the total optical depth

$$\tau_\nu = \int_r^\infty \chi_\nu dr = \int_r^\infty [\chi_\nu(sc) + \kappa_\nu]dr \tag{2}$$

at frequencies in the infrared, visible, and near ultraviolet. Note that the $n_H$ axis is a linear measure of radial distance. We choose to define the extent of the photosphere in the following way. Let the outer extent be the point where the flux $F_\nu$ reaches within 3% of its emergent value. Let the inner extent be the point where the mean intensity $J_\nu$ reaches within 3% of the Planck function $B_\nu(T)$.

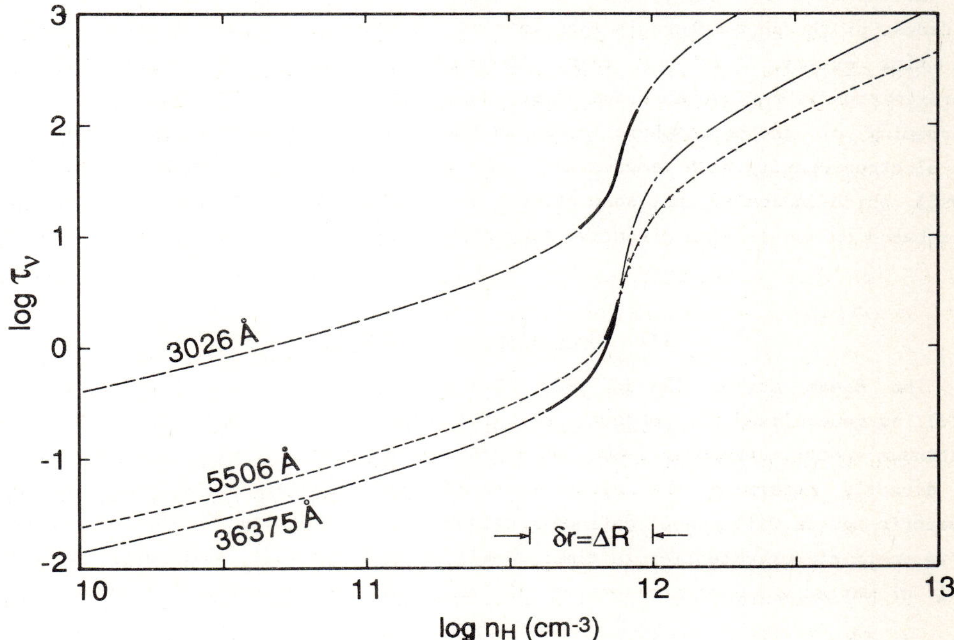


Figure 1. The total optical depth at three wavelengths is plotted versus the total hydrogen number density. The extent of the photosphere is indicated by the solid portion of the curves. A radial distance equal to the scale height is also shown. The model parameters are $T_e$ = 5750 K, $\Delta R = 4 \times 10^{13}$ cm, $Z_* = 0.010$.

From these results, we construct Table 1, which provides a quantitative measure of the location and thickness of the photospheric region. Of special interest are the last two rows, giving the radial extent of the photosphere as a fraction of the scale height $\Delta R$ and the radius of the photosphere on March 1 (as determined in the next section to be $R_p = 4.3 \times 10^{14}$ cm).

Table 1.  <u>Thickness of Photosphere</u>

| $\lambda(\overset{\circ}{A})$ | <u>3026</u> | <u>5506</u> | <u>36375</u> |
|---|---|---|---|
| $\tau_\nu(min)$ | 11.6 | 1.34 | 0.27 |
| $\tau_\nu(max)$ | 126 | 2.30 | 2.56 |
| $\log n_H(min)$ | 11.74 | 11.84 | 11.63 |
| $\log n_H(max)$ | 11.95 | 11.88 | 11.88 |
| $\Delta r/\Delta R$ | 0.48 | 0.09 | 0.58 |
| $\Delta r/\Delta R_p$ | 0.045 | 0.008 | 0.054 |

It should be pointed out that the true extent of the photosphere is given by the greatest outer radius (at the longest wavelength) and the smallest inner radius (at the shortest wavelength), giving $(\Delta r)_{max} = 0.069\ R_p$. We note, however, that the thickness of the photosphere is much less at the visible wavelengths, where most of the flux exists.  In any case, these results indicate that the plane-parallel approximation is a reasonable one under these conditions of recombination.  This sharpening of the photosphere is due in large part to the precipitous reduction of the electron density with temperature. The results of Höflich, Wehrse, and Shaviv (1986), which indicated that some effects of spherical extension could be important, were based on models with effective temperatures well above those of recombination.


### III.  <u>Determination of Distance</u>

The Baade method (Baade 1926; Branch and Patchett 1973; Kirshner and Kwan 1974), as generalized to objects of arbitrary redshift emitting an arbitrary continuum spectrum (Wagoner 1980, 1987) provides a powerful, relatively reliable way to directly determine the distance to any expanding (or collapsing) spherically symmetric source with a well defined photosphere.  As indicated in the last section, if we restrict our attention to times t well after the time $t_o$ when any particular shell of matter was last accelerated, the radius of the photosphere is given by

$$R_p = v_p(t-t_o) \qquad [\gg R(t_o)] \ , \qquad\qquad (3)$$

where $v_p(t)$ is the velocity of the matter at the photosphere at that particular time.  The (radial) angular size $\theta$ of the photosphere is given by

$$\theta^2 = (1 + z)^3\ e^{\tau_I}\ f_\nu/F_{\nu'} \ , \qquad\qquad (4)$$

where z is the source redshift, $\tau_I(\nu)$ the optical depth due to intervening matter, and $f_\nu$ the observed flux. The frequency $\nu'$ is that measured in the center-of-mass (c.m.) frame of the supernova, taken to be the same as that of the parent galaxy. The (proper motion) distance is then $D = R_p/\theta$.

The only complete (UV→IR) absolute spectrophotometry that we have thus far been able to obtain was acquired on March 1, 1987 (Kirshner et al., 1987; Bouchet et al. 1987; Danziger et al. 1987). Below we shall use that data to determine the angular size of the photosphere at that time. But first we shall determine the photospheric radius.

It is easily seen that there must be a discontinuity in slope of the P-Cygni line profiles at a frequency corresponding to the velocity of the matter at the photosphere, if the photosphere is sharp. However, as Branch (1980) has shown, this discontinuity is most easily observed if the line Sobolev optical depth $\tau_{ij} \lesssim 5$, in which case the break appears at the frequency of maximum absorption. Using the results of Blanco et al. (1987) and Phillips (1987) [assuming that the LMC velocity had been removed], we do indeed find that the weakest lines (FeII $\lambda$5018, FeII $\lambda$5169, B$\gamma$ $\lambda$21656) do give the same velocity, $v_p = (8.5 \pm 0.4) \times 10^3$ km s$^{-1}$ for March 1.1-1.4 (UT). As expected, the velocities obtained from the frequencies of maximum absorption of the stronger lines are greater than this.

Allowing for uncertainties in the shock breakout time (relative to the neutrino burst on Feb. 23.316) and the spread of observation times, we obtain $t-t_o = 5.8 \pm 0.3$ days. From equation (3), these values then give

$$R_p = (4.26 \pm 0.3) \times 10^{14} \text{ cm} \qquad \text{(March 1.1-1.4)} \quad . \qquad (5)$$

In order to obtain the angular radius from equation (4), we must first transform our emergent fluxes $F_\nu$, calculated in the frame moving with the matter, to the c.m. frame of the supernova. [The galactic redshift correction $(1 + z)^3$ is negligible because the velocity of the LMC is 280 km s$^{-1}$.] Approximating the angular distribution of the emergent intensity by that of a pure scattering atmosphere, we obtain for this Doppler correction

$$\Delta \log F_\nu \cong 0.434\beta[1.732 - 0.711(\text{dlog } F_\nu/\text{dlog } \nu)] \quad , \qquad (6)$$

with $\beta = v_p/c = 0.028$ for this epoch.

Next we must consider the correction for intervening absorption, $\tau_I$. To do this, we assume that the total amount is given by $A_\nu = E_{B-V} \, \psi(\nu)$ magnitudes, with $\psi(\nu)$ the same function for our galaxy and the LMC. The existing evidence (Clayton and Martin 1985) indicates that this is not a bad approximation at the wavelengths $\lambda > 2600$ Å where appreciable flux exists during the recombination era. If we consider any two frequencies $\nu_*$ and $\nu \gg \nu_*$, then the function

$$\widetilde{F}(v,v_*) - \widetilde{f}(v,v_*) = 0.4\ E_{B-V} \quad , \tag{7}$$

where $\widetilde{Q}(v,v_*) \equiv [\log Q(v) - \log Q(v_*)]/[\psi(v) - \psi(v_*)]$, should be independent of frequency (as well as distance) for any particular source. Therefore, plotting $\widetilde{F}$ – $\widetilde{f}$ versus $v$ with fixed $v_*$ for various models ($T_e$, $\Delta R$, $Z_*$) allows one to identify the best model atmosphere as that one which produces the smallest frequency dependence. The magnitude of the resulting (approximate) constant then provides the amount of absorption. We choose $\lambda_* = 5.01\ \mu m$, the longest wavelength at which data are available, since of course the intervening absorption is smallest there [$\psi(\lambda_*) = 0.07$].

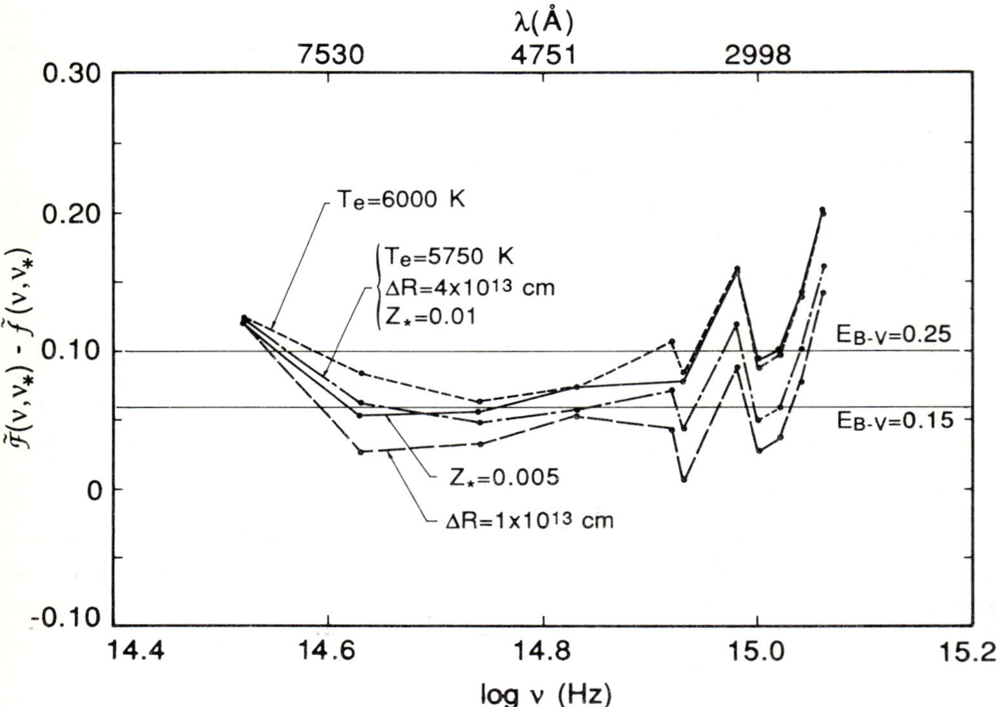


Figure 2. The absorption corrected combination of emitted and observed relative fluxes [eq. (7)] versus frequency. Lines corresponding to two choices of color excess are also shown. Results from our "fiducial model", as well as models in which only the parameter indicated has been varied, are shown.

In Figure 2 we plot the left-hand side of equation (7) at frequencies $v$ where the best values of the observed continuum flux are available and the denominator is large enough to be accurate. Results for our "fiducial model" (Fig. 1) and three others are shown. From these results and those for many other values of the parameters we can draw the following conclusions.

a)  The lack of an observed Pashen jump ($\lambda' = 8203$ Å) puts an upper limit on the photospheric density, corresponding to $\Delta R \gtrsim 2 \times 10^{13}$ cm.  It was also found that excess  UV  flux  was  produced  unless  $\Delta R \lesssim 6 \times 10^{13}$ cm.

b)  The UV flux was most sensitive to the parameter $Z_{*}$, since it controlled the dominant opacity at those wavelengths.  It has been found that values in the range $Z_{*} \cong 0.01 - 0.02$ were required to reduce the UV flux sufficiently.

c)  The shape of the spectrum in the visible region was most sensitive to the effective temperature.  Values in the range $5600 \lesssim T_e \lesssim 5900$ K gave reasonable fits.

d)  Those computed spectra which best matched the observed continuum at the wavelengths indicated (they all matched well in the infrared) were consistent with the expected amount of absorption indicated by the limiting values of $E_{B-V}$ shown on Fig. 2.

In addition to these estimates of the allowed ranges of our parameters, we have computed  the sensitivity of the flux at our normalization wavelength ($\lambda_o = 3.16$ µm) to changes in these parameters.  We obtain $\partial \log F_{\nu_o} / \partial \log X = -0.03$ (X = $\Delta R$), +0.05  (X = $Z_{*}$), and +1.7 (X = $T_e$).  Using all these results, we arrive at a value

$$\log F_{\nu_o} \cdot (\text{erg cm}^{-2}\ \text{s}^{-1}\ \text{Hz}^{-1}) = -4.61 \pm 0.03 \quad (\lambda_o = 3.16\ \mu\text{m}) \tag{8}$$

for  the Doppler-corrected emitted flux.  (The Doppler correction at this wavelength is $\Delta \log F_{\nu} \cong +0.01$.)  The spectrum of our fiducial model, absorbed by the amount indicated, is compared to the observed spectrum and a blackbody in Figure 3.

At  our  normalization  wavelength, the absorption corresponding to $E_{B-V} = 0.18$ (Sonneborn et al. 1987) gives us a flux correction $\Delta \log f_{\nu} = 0.013$.  We adopt a de-absorbed  observed  flux  $\log f_{\nu_o}$ (erg cm$^{-2}$ s$^{-1}$ Hz$^{-1}$) $= -21.60 \pm 0.03$ at $\lambda_o = 3.16$ µm, using the data plotted by Bouchet et al. (1987).  We note that  the  calibration of  their  spectrophotometry  is  claimed  to  agree  with their broad band infrared photometry to better than 10%.  Their K photometry in turn appears  to  agree  with that  of  the  South  African group (Menzies et al. 1987) to within 4% at this time.

This observed flux gives an angular radius

$$\log \theta = -8.495 \pm 0.02 \tag{9}$$

from equations (4) and (8).  Combining this result with equation (5) then  leads  to the distance

$$D = 43.3 \pm 4\ \text{kpc} \quad . \tag{10}$$

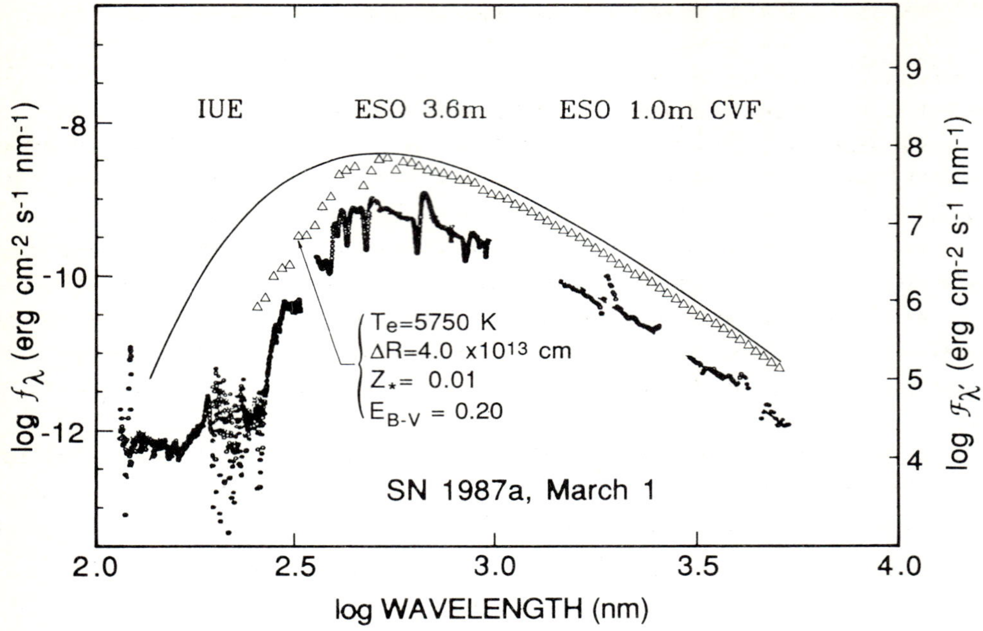


Figure 3.    A  composite  spectrum  of  SN  1987a  observed  on  March  1  (Danziger et al.
1987)  is  compared  with  one  of  our  best-fitting  (absorbed)  computed  spectra,
corresponding   to   the   parameters   indicated.    Both   the   theoretical   spectrum
(triangles)  and  that  of  an  (unabsorbed)  blackbody  at  the  same  effective   temperature
have  been  displaced  by  the  same  amount  for  clarity  (see  separate  scales).   No
theoretical  points  are  shown  at  wavelengths  shorter  than  that   at   which   the   other
stars  dominate  the  flux.

Using  essentially  the  same  method,  but  <u>assuming</u>  blackbody  emission  (in

disagreement with our calculated spectra),  Branch  (1987)  obtained  a distance of

55 ± 5 kpc.  This is another example of effects such as flux dilution and distortion

produced   by   a   scattering   dominated   photosphere   (Wagoner   1981).

The   distance   to   the   LMC   has   been determined by various classical methods.

Extensive JHK photometry of Cepheids has yielded a very recent value of 51.8  ±  1.2

kpc  (Welch  et  al.   1987),  in agreement with that obtained from optical photometry

(Caldwell and Coulson 1986).   However,  the  largest value of the distance,  58.1 ± 1.9

kpc (Visvanathan 1985),  was obtained from 1.05 μm photometry of Cepheids.   The  use

of  RR Lyrae variables has given 48.3 ± 2.2 kpc (Walker 1985),  while photometry of B

stars  has  yielded 45.7 ± 4.2 kpc (Shabbrook and Visvanathan 1987).   The lowest value

of the distance,  43.7 ± 4.0  kpc  (Schommer,  Olszewski,  and  Aaronson  1984),  was

obtained  from  main-sequence  fitting of two globular clusters; although Chiosi and

Pigatto  (1986)  have  obtained  a  larger  distance  by  including  convective  overshooting.

It should be noted that all these results refer to the <u>mean</u> distance  of  the  LMC,  and are subject to effects of the uncertain deficiency in heavy elements.

Our direct determination of the distance to the supernova, unlike these determinations of the distance to the LMC, involves no distance ladder calibrations or selection effects. In addition, the assumptions involved in this method are independently tested by a) the match between the frequency dependence of the computed and observed spectra, b) the requirement that the ratio of the two time dependent quantities ($R_p$ and $\theta$) that determine the distance remains constant, and c) the predicted break in the (weaker) line profiles. It is especially important that we acquire accurate data for other dates so that we can invoke test (b). We believe that the distance to any Type II supernova is most reliably determined during that period of time when the photosphere lies within the hydrogen recombination shell, because this gives a long time base with which to more accurately determine $R_p$ as well as a sharper photosphere to more accurately determine $\theta$.


## IV.   Acknowledgments

We wish to thank Stephen Hershkowitz for his help, and especially Alan Karp for kindly making available to us his extensive opacity tables. This research was supported in part by the National Science Foundation (grant PHY 86-03273) and by the National Aeronautics and Space Administration (grant NAGW-299). Many of the computations utilized the San Diego Supercomputing Center, supported in part by the National Science Foundation through a grant to Stanford University.


## References

Baade, W. 1926, Astr. Nach., 228, 359.
Blanco, V.M., Gregory, B., Hamuy, M., Heathcote, S.R., Phillips, M.M., Suntzeff, N.B., Terndrup, D.M., Walker, A.R., Williams, R.E., Pastoriza, M.G., Storchi-Bergmann, T., and Matthews, J. 1987, Ap. J., 320, 589.
Bouchet, P., Stanga, R., Le Bertre, T., Epchtein, N., Hamann, W.R., and Lorenzetti, D. 1987, Astron. Ap., 177, L9.
Branch, D. 1980, in Supernova Spectra, ed. R. Meyerott and G.H. Gillespie (New York: Am. Inst. Phys.).
Branch, D. 1987, Ap. J., 320, L23.
Branch, D., and Patchett, B. 1973, M.N.R.A.S., 161, 71.
Caldwell, J.A.R., and Coulson, I.M. 1986, M.N.R.A.S., 218, 223.
Chiosi, C., and Pigatto, L. 1986, Ap. J., 308, 1.
Clayton, G.C., and Martin, P.G. 1985, Ap. J., 288, 558.
Danziger, I.J., Fosbury, R.A.E., Alloin, D., Cristiani, S., Dachs, J., Gouiffes, C., Jarvis, B., and Sahu, K.C. 1987, Astron. Ap., 177, L13.
Hershkowitz, S., Linder, E., and Wagoner, R.V. 1986, Ap. J., 301, 220.
Hershkowitz, S., and Wagoner, R.V. 1987, Ap. J., 322, 967.
Höflich, P., Wehrse, R., and Shaviv, G. 1986, Astron. Ap., 163, 105.
Karp, A.H., Lasher, G., Chan, K.L., and Salpeter, E.E. 1977, Ap. J., 214, 161.
Kirshner, R.P., and Kwan, J. 1974, Ap. J., 193, 27.
Kirshner, R.P., Sonneborn, G., Crenshaw, D.M., and Nassiopoulos, G.E. 1987, Ap. J., 320, 602.
Meikle, W.P.S., Matcher, S.J., and Morgan, B.L. 1987, Nature, 329, 608.

Menzies, J.W., Catchpole, R.M., van Vuuren, G., Winkler, H., Laney, C.D., Whitelock,
    P.A., Cousins, A.W.J., Carter, B.S., Marang, F., Lloyd Evans, T.H.H., Roberts,
    G., Kilkenny, D., Spencer Jones, J., Sekiguchi, K., Fairall, A.P., and
    Wolstencroft, R.D. 1987, M.N.R.A.S., 227, 39P.
Mihalas, D. 1978, Stellar Atmospheres (San Francisco: W.H. Freeman).
Nisenson, P., Papaliolios, C., Karovska, M., and Noyes, R. 1987, Ap. J., 320, L15.
Phillips, M.M. 1987, Bull. Am. Astron. Soc., 19, 722.
Schommer, R.A., Olszewski, E.W., and Aaronson, M. 1984, Ap. J., 285, L53.
Shabbrook, R.R., and Visvanathan, N. 1987, M.N.R.A.S., 225, 947.
Sonneborn, G., Altner, B., and Kirshner, R.P. 1987, Ap. J., in press.
Visvanathan, N. 1985, Ap. J., 288, 182.
Wagoner, R.V. 1980, in Physical Cosmology, ed. R. Balian, J. Audouze, and D.N.
    Schramm (Amsterdam: North-Holland).
Wagoner, R.V. 1981, Ap. J., 250, L65.
Wagoner, R.V. 1987, in Proceedings of the Vatican Observatory Conference on Theory
    and Observational Limits in Cosmology, in press.
Walker, A.R. 1985, M.N.R.A.S., 212, 343.
Welch, D.H., McLaren, R.A., Madore, B.F., and McAlary, C.W. 1987, Ap. J. 321, 162.

# SPECTROSCOPIC DIAGNOSIS OF SN1987A AND LESSER LIGHTS


J. Craig Wheeler, Robert P. Harkness
Department of Astronomy
University of Texas at Austin

Zalman Barkat
Department of Physics
Hebrew University of Jerusalem


## ABSTRACT


SN 1987A gives a unique chance to study both the precursor star and the subsequent dynamical evolution of the explosion. Comparison of the light curves shows that either $H_0 \sim 100$ km/s/Mpc, or SN 1987A ejected significantly less $^{56}$Ni than ordinary Type II supernovae. Investigation of the stellar structure pertinent to SK -69 202 reveals multiple solutions. For given luminosity, effective temperature, core mass and core radius, there are two families of envelope mass, one with large envelope mass and one with small envelope mass. The small envelope mass solutions can be ruled out by considerations of kinematics and the light curve. Envelopes of moderate mass may avoid each of these problems, but must be helium rich to be structurally self-consistent.

The spectrum in both the optical and the ultraviolet at about two days is fairly well represented by a hydrogen envelope with a power law density profile $(\rho \propto r^{-11})$ of one-quarter solar metallicity in LTE. Theoretical spectra at this early epoch tend to favor luminosities on the high side of observational estimates in order to ionize Ca II and prevent excessively strong lines at H and K and the infrared triplet, with some ramifications for distance estimates.

The spectra of SN1987A present an interesting contrast to other SN II events. A McDonald Observatory spectrum of SN 1985H in NGC 3359 of uncertain epoch shows a very close resemblance to that of SN 1987A at about two month's age, including the strong line at 607 nm attributed by Williams to Barium. SN 1985H may have been of the same class of event as SN 1987A.

# 1. INTRODUCTION

SN 1987A has confirmed some of the most basic predictions of supernova theory, but has also presented unique aspects which project supernova research in exciting new directions guided by an impressive wealth of detailed observations. This situation provides a marvelous laboratory to test a wide variety of theories, assumptions, tech-

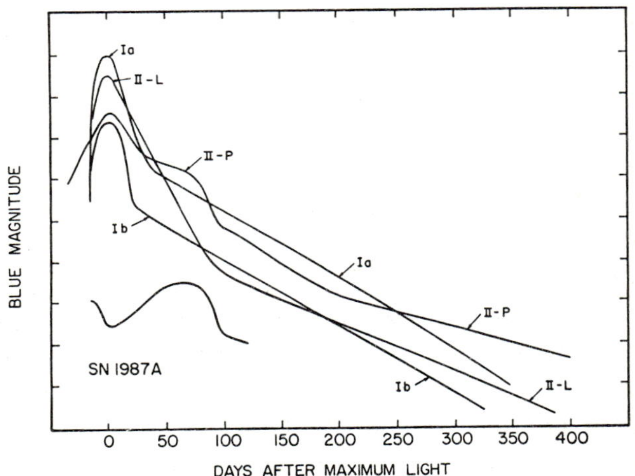



Figure 1. Schematic B magnitude light curves are presented for supernovae of Type Ia, Ib, II linear, and II plateau in contrast with that of SN 1987A. The absolute magnitude of SN 1987A is relatively well known. The displacement with respect to the other types assumes a distance scale corresponding to $H_0 = 50$ km/s/Mpc.

niques, and relations. In this paper we discuss some aspects of the structure of the progenitor star, including multiple solutions of the stellar structure equations and the controversy over the mass of the hydrogen envelope, the potential of supernova atmosphere calculations to give a new dimension to the analysis of Type II supernovae, and a comparison of SN 1987A with other Type II supernovae.

To put SN 1987A in perspective, the evolving B magnitude light curve is presented in Figure 1 with schematic light curves of the other classes of supernovae adapted from Doggett and Branch(1986), but normalized to show the differences in relative peak magnitudes. Type Ia are the brightest with some Type II linear events approaching similar luminosity, but also showing a dispersion downward to the magnitudes characteristic of Type II plateau and Type Ib events. The long term tails in this figure are not to be taken too literally, since they are statistical, and may contain bias from background galaxy light. In Figure 1, SN 1987A has been dereddened by 0.6 m. This plot suggests that there is little relation between SN 1987A and the other classes of supernovae. This conclusion is in striking contrast to that reached by Peter Wood (private communication) and others, that the declining part of the light curve of SN 1987A matches rather precisely onto the plateau and tail phase of Type II plateau supernovae.  There are some minor differences in assumptions and one major one that lead to these different conclusions.  One is that there is some uncertainty in the zero point of time, and this can be adjusted to make the fit look as good as possible.  In addition, Wood plotted V magnitude light curves.  SN 1987A shows a strong UV deficiency, discussed in Section 3, which may account for some of the disparity in magnitude shown in Figure 1.  The major difference is that while the "standard" light curves in Figure 1 are plotted only relative to one another, an implicit choice of Hubble constant of about 50 km/s/Mpc was made for them in plotting the relative amplitude of SN 1987A.  Wood, in contrast, implicitly assumed a Hubble constant of 100 km/s/Mpc. This exercise thus leads us to one of two important conclusions.  Either the congruity of light curves noted by Wood is accepted and the Hubble constant is about 100 km/s/Mpc, or "normal" Type II-P events are intrinsically brighter than SN 1987A on the tail, suggesting that they eject about 4 times as much $^{56}$Ni as SN 1987A.  Both of these possibilities deserve careful consideration.


## 2. PROGENITOR STRUCTURE

The advent of SN 1987A and the very probable identification of its progenitor, SK - 69 202, give an unprecedented opportunity to explore the structure of a supernova progenitor.  The spatial agreement of SK -69 202 and the supernova, and its disappearance in the UV (Walborn, et al. 1987, Kirshner et al. 1987) strongly argue that this blue supergiant was the progenitor star.  Furthermore, models of the light curve constrain the luminosity and radius of the progenitor to be very similar to those observed for SK -69 202 (Arnett 1987a, b, Woosley et al. 1987, Woosley, Pinto and Ensman 1987, Shigeyama, et al. 1987, Grassberg et al. 1987, Hillebrandt et al. 1987).  There is still considerable uncertainty about the mass of the hydrogen envelope, with some authors arguing for about 10 $M_\odot$ (e.g. Arnett 1987b, Woosley, Pinto and Ensman 1987), and others advocating a rather small envelope, of order 0.1 $M_\odot$ (Wood and Faulkner 1987, Maeder 1987).  Within the range of assigned luminosities for SK -69 202, depending on the distance to the LMC and extinction, plausible progenitors have original main sequence mass in the range roughly 15 to 20 $M_\odot$ with helium cores in the range 4 to 6 $M_\odot$.  We emphasize, however, that only the luminosity and radius (or effective temperature) are directly observationally constrained.  The mass of the star and that of the inner core are not known without recourse to solution of the equations of stellar structure and evolution.

In order to understand the physical structure of the progenitor of SN 1987A,  models have been constructed which match the radius, effective temperature, and luminosity of SK -69 202 by integrating  the equations of stellar structure inward from the surface. The composition is assumed to be  X = 0.70, Y = 0.295, and Z = 0.005 for most of the models, although the sensitivity to changes in X, Y , and Z has been explored.  Details will be presented in Barkat and Wheeler (1988).

In the current models constructed by inward integration, the composition is specified so that the place where the composition changes at the helium core, either suddenly or gradually, is a free parameter. The specification of the helium core mass implies that the temperature at larger Lagrangian mass may not exceed the threshold temperature for hydrogen burning ($\sim 4 \times 10^7$ K). This criterion will be used as a measure of self-consistency of the models. The bolometric luminosity of SK -69 202 depends on the adopted distance and reddening, with estimates ranging from L = 60,000 to 100,000 $L_\odot$ (Arnett 1987a, Woosley et al. 1987). Models with $M_{He}$ = 4 and 6 $M_\odot$ have been explored in some detail to span a reasonable range in allowed parameters. For the models constructed here we take L = 55,000 $L_\odot$ with a helium core mass of 4 $M_\odot$ and L = 110,000 $L_\odot$ with $M_{He}$ = 6 $M_\odot$.

Evolutionary calculations tend to show that once a core forms, its structure and subsequent evolution are rather independent of the outer envelope. This provides the rationale for studying evolving helium cores (Arnett 1977) without the complication of the envelope. The success of such undertakings gives rise to the intuitive notion that one may construct models of the progenitor of SN 1987A by adopting a core from calculations of evolution at constant mass, and then omitting arbitrary amounts of envelope mass to seek structures that reproduce observations. Woosley, Pinto, and Ensman (1987) have used this technique effectively to explore a large range in parameter space in order to constrain the structures that will reproduce the light curve of SN 1987A. This procedure tacitly assumes that both the core mass and the core radius may be held constant while the luminosity and envelope mass are varied.

Another way of parameterizing the problem of the progenitor structure of SN 1987A is to specify the luminosity, the total mass, and the core mass, and then to examine the behavior of the core radius, $R_{He}$, as a function of $T_{eff}$ (or R). Figure 2 shows the result for models with the standard envelope composition, M = 15 $M_\odot$ and $M_{He}$ = 4 $M_\odot$. For fixed L, M, $M_{He}$, *and* $R_{He}$, if there is a solution with large $T_{eff}$ then there are, in fact, *three* solutions to the stellar structure equations, with different values of $T_{eff}$ and R. A similar pattern exists for models with cores of 6 $M_\odot$. The result is not particularly sensitive to the specific parameters assumed.

The reason for the behavior shown in Figure 2 is illustrated in Figure 3 which gives the plot of the radius, r, as a function of the mass, $M_r$, for the series of models with L = 55,000 $L_\odot$, M = 15 $M_\odot$, and $M_{He}$ = 4 $M_\odot$ for a range in effective temperatures. Note that for large $T_{eff}$ the gradient in r is rather flat through most of the extent of the mass down to the chosen core mass. For lower $T_{eff}$ an inflection point is reached before the core is encountered. This causes the curves for lower $T_{eff}$ to cross those corresponding to higher values, so that two choices of $T_{eff}$ can give the same value of $R_{He}$. The reason for this is that the gradient in r with respect to $M_r$ gets larger as $T_{eff}$ decreases and $r^2 \rho$ decreases. As $T_{eff}$ decreases even further, the envelope develops convection which flattens out the gradient in r, yielding a larger core radius again, and providing the third solution in $T_{eff}$ for a given value of $R_{He}$.

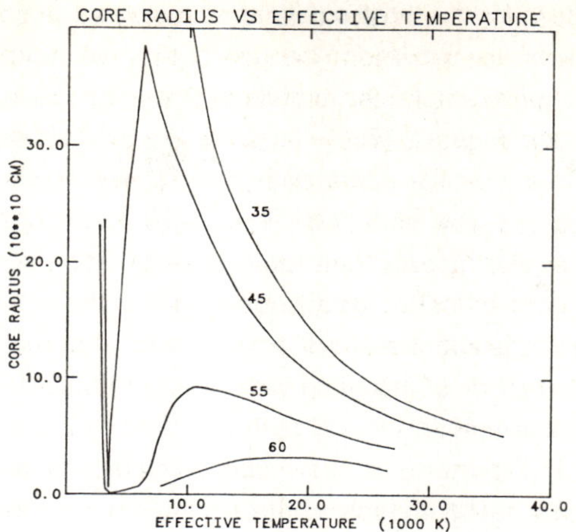


Figure 2.  The radius of the core in units of $10^{10}$ cm is given as a function of effective temperature for a  total mass of 15 M$_\odot$ and a helium core mass of 4 M$_\odot$.  The curves are labeled with the luminosity in units of 1000 L$_\odot$.  Note that for fixed total mass, core mass, core radius, and luminosity, there are three solutions with different effective temperature.  These three solutions have different density structures.

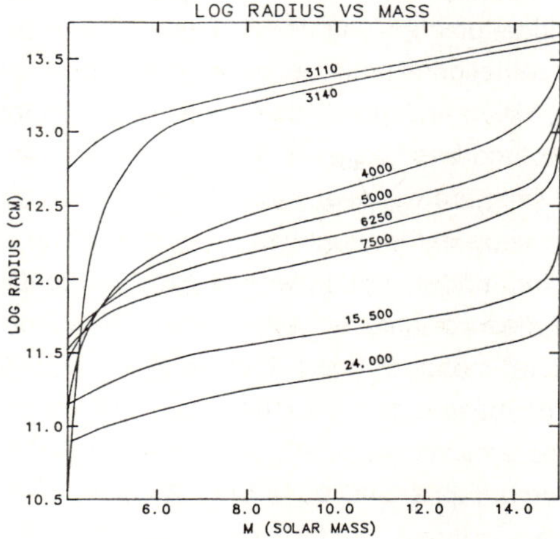


Figure 3.  The radius is given as a function of Lagrangian mass for a series of models with total mass 15 M$_\odot$, and helium core mass 4 M$_\odot$, for a range of values of the effective temperature.   Note that as the effective temperature declines the temperature gradient steepens giving two solutions with the same core radius, eg, the curves for 15,500 and 4000 K.  As the effective temperature continues to decline, the envelope becomes fully convective, and the temperature gradient becomes more shallow, providing the third solution. A model with effective temperature between 3110 and 3140 K would also have the same core radius as the 15,500 and 4000 K models.

The envelope structure for the three solutions at a single core radius are different.  In general, the core solutions are similar for the two coolest solutions, and rather different for the bluest solution.  For example, for models with L = 55,000 $L_\odot$ and $R_{He}$ ~ 8 x $10^{10}$ cm the two cooler models in Figure 2 ($T_{eff}$ = 3,046 K, R ~ 5 x $10^{13}$ cm, and $T_{eff}$ = 9,500 K, R ~ 5 x $10^{12}$ cm)  have very similar cores, with  the  coolest model being on the Hyashi track.  The hottest model ($T_{eff}$ = 15,500 K, R ~ 2 x $10^{12}$ cm) has a central core more nearly equal to the other two models than for a corresponding series with smaller core radius, but the outer core is hotter and denser, as is the envelope, and this may significantly alter the propagation of a shock between the core and the envelope.

To account for the number of red supergiants in the LMC, the typical massive star must spend an appreciable time in the red before turning to the blue (Maeder 1987).  If SK -69 202 followed this pattern, it must have evolved from the red to the blue.  Depending on the details of the change of the core radius, an evolutionary calculation could end up on the hotter or the cooler side of the peak of one of the constant luminosity curves in Figure 2.    Note also that two different evolutionary or structure codes could give very different solutions, depending on the exact input physics, and perhaps the numerical treatment.  If the solution falls near one of the peaks in Figure 2, then one code could give a blue solution, and another could give  a  slightly  smaller luminosity, or slightly  larger core radius, and find  only  a  solution on the Hyashi track.   Similarly, variation of input physics in a single code, i. e. composition, could cause swings from the red to the blue as the critical points illustrated in Figure 2 are encountered.  This may account in part for the presentation of some  evolutionary  tracks  for SK -69 202 that stay in the blue (Arnett 1987a, Hillebrandt et al 1987) while others swing to the red and back to the blue (Woosley, Pinto and Ensman 1987, Wood and Faulkner 1987).

Consideration of the structure equations also gives an interesting constraint on the mass of the hydrogen envelope of the progenitor.  Figure 4 gives the temperature profile as a function of mass for models with the envelope composition, L = 55,000 $L_\odot$ corresponding to a core mass of $M_{He}$ = 4 $M_\odot$, $T_{eff}$ = 15,500 K (R = 33.4 $R_\odot$), and a range of values of the total mass, M, from 7 to 18 $M_\odot$.  Note that  the temperature gradient is shallow for large masses but that it becomes ever steeper as the total mass is decreased.  For a given core mass, $M_{He}$, this has the result that the temperature at that point, $T(M_{He})$, has a maximum as a function of the total mass.  This is because for large M the temperature gradient is small , and for sufficiently small M, the temperature has insufficient range to rise, despite the steep gradient, before $M_{He}$ is encountered.

Many of the models demand that the temperature at some mass cut above the helium core be in excess of the ignition temperature of hydrogen, $T_H$, and hence, the structure is self-inconsistent.  For a helium core mass of 4 $M_\odot$, models with total mass in the range 4.05 to 14 $M_\odot$, are excluded.  This conclusion does not depend sensitively on the specific value adopted for $T_H$, since the temperature profiles tend to rise so steeply in the excluded range.  Similar conclusions follow for models with larger luminosities and

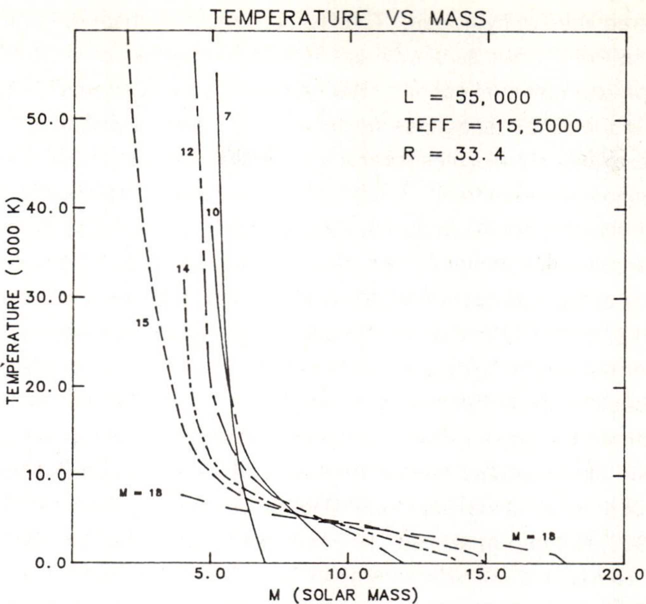


Figure 4. The temperature in units of 1000 K is given as a function of Lagrangian mass for models with L = 55,000 $L_\odot$, Teff = 15,500 K, R = 33.4 $R_\odot$ and envelope composition X = 0.7, Y = 0.295, Z = 0.005, for a range of total masses. This luminosity corresponds to original main sequence mass of about 15 $M_\odot$ and a helium core mass of 4 $M_\odot$. Note that for smaller total mass and hence smaller envelope mass the temperature rises rapidly to exceed the hydrogen ignition temperature well outside the core.

larger associated core masses. For L =110,000 $L_\odot$ and helium core mass of 6 $M_\odot$, the excluded range of mass is about 6.2 to 17 $M_\odot$.

We thus reach the conclusion that for the standard composition the mass of the envelope of SK -69 202 must fall into two extreme ranges. Either the envelope mass is relatively large, corresponding to the star having reached the endpoint of its evolution with virtually all of its main sequence mass intact, or it is rather small, having suffered considerable mass loss. Intermediate values, which might be invoked by adopting arbitrary amounts of the ill-known mass loss, are excluded. This bifurcation in the allowed envelope masses for given core mass accounts for the extreme range of models which have been proposed. In Arnett (1987a) the envelope mass is about 11 $M_\odot$ for a core of 4 $M_\odot$, and in Wood and Faulkner (1987) the envelope mass is only about 0.2 $M_\odot$ for a core of 5.3 $M_\odot$. The present conclusions may depend to some extent on the assumption of strict thermal equilibrium which may fail in realistic evolution calculations. The range of excluded mass is smaller if the envelope is helium rich, and indeed any envelope mass is allowed by this criteria if Y exceeds 0.5 (Barkat and Wheeler 1988). This point will be the subject of further investigation.

The proper structural solution for SK -69 202 can, in principle, be obtained from evolutionary calculations, or from observations of the light curve and spectra of the supernova itself. A basic constraint from the spectra is the observed Doppler shifts of the optical and infrared lines of hydrogen. The Balmer lines are observed to move as slowly as 5000 km/s and the IR lines as slowly as 2000 km/s (Menzies et al. 1987, Blanco et al. 1987, Phillips, private communication). We have computed dynamical models for stars of 15-20 $M_\odot$ with their main sequence mass virtually intact and kinetic energy $\gtrsim 10^{51}$ erg/s which give minimum velocities for the hydrogen envelopes $\sim 2000$ km/s, in accord with the observations. In contrast, models which reproduce the observed properties of SK -69 202 with small mass hydrogen envelopes produce velocities in the envelope far in excess of the observed minima. For induced kinetic energy of 6.6 x $10^{50}$ ergs, the minimum velocity of the hydrogen in a model with a core of 6 $M_\odot$ and an envelope of 0.1 $M_\odot$ is 6510 km/s, on the high side of the observations. For a kinetic energy of 0.95 x $10^{50}$ ergs the minimum hydrogen velocity is 2100 km/s, in keeping with the observations, but the early luminosity is very low. A model with a large hydrogen envelope to tamp the core expansion with a reverse shock gives a very slowly expanding core. This releases the radioactive heat slowly and can give a reasonable representation of the rise to maximum and decline to the radioactive tail (Woosley, Pinto, and Ensman 1987, Nomoto et al. 1987 ). For instance, we have calculated a model with a total mass of 15 $M_\odot$ and a kinetic energy of 1.4 x $10^{51}$ ergs which gives a core energy of only 0.59 x $10^{50}$ ergs. The light curves of both the low mass envelope models with 6 $M_\odot$ cores just mentioned rise too quickly to maximum due to the rapid expansion of the core, despite the low induced energy. These models can thus be ruled out on the basis of the high hydrogen velocity, the rapid expansion, or both. The same is presumably true for the model presented by Wood and Faulkner (1987) with a core mass of 5.2 $M_\odot$ and envelope mass of 0.2 $M_\odot$.


## 3. MODEL ATMOSPHERES

SN 1987A also gives an important opportunity to explore the process of radiative transfer in supernovae. There are features in the spectrum that are special to SN 1987A, and others that are general to Type II supernovae. A careful study of the spectra of SN 1987A should provide a deeper understanding of that particular event, and of the principles to be applied to other supernovae. Excellent spectral coverage has been provided by observatories throughout the world. Optical spectra have been presented by Tyson and Boeshaar (1987), Danziger et al. (1987), Menzies et al., (1987), Blanco et al. (1987), Ashoka et al. (1987), Catchpole, et al. (1987) and Hanuschik and Dachs (1987). Ultraviolet spectra have been presented by Wamsteker et al. (1987), Cassatella, et al. (1987) and by Kirshner et al. (1987). There is also a growing body of infrared data, but that will not be considered in detail here.

The optical spectra for the first 4 days while the supernova was rising to an initial local maximum were rather simple, showing broad Balmer lines. After two days (circa February 25), the spectrum showed He I $\lambda5876$ but no sign of Ca H and K or the IR triplet, due, presumably, to the high temperature that ionizes Ca II. After four days (circa

February 27) Ca H and K were readily apparent, but the He I line had faded.  After about 10 days (circa March 5) structure due to Fe II and Na D appeared.  A few days later a line appeared at 6070 Å which Williams (1987) attributes to Ba II.  Up to this time, the lines of the Balmer series became deeper, shifting to lower velocities, with H alpha showing a growing net emission.  There was no discernible Balmer jump in absorption or emission (Dopita, Menzies, private communication).  After about 10 days the Balmer lines began to fade rapidly.  Hβ and Hγ were barely discernible after 20 days.  The optical spectrum did not change qualitatively for the next four months as the supernova passed through maximum in mid-May.  Hβ did re-emerge with a P-Cygni profile around the end of April (65 days) (Catchpole et al. 1987).

The evolution of the spectrum of SN 1987A in the ultraviolet was extremely rapid and unique for an SN II.  For the first day or so the spectrum was dominated by broad features from 1200 to 3200 Å which were unlike any previous observed supernova (Wamsteker et al. 1987, Kirshner et al. 1987).  By the third or fourth day, the spectrum resembled that of previously observed SN Ia and SN Ib.  Wheeler et al. (1986) attribute the UV spectra of SN I to resonant line scattering of Fe II, and hence a similar effect is presumably at work in SN 1987A.  SN 1987A faded rapidly until it could no longer be seen at short wavelengths with the IUE satellite.

Various aspects of the formation of the optical and UV spectra of SN 1987A have been discussed by Fransson et al. (1987), Williams (1987), Branch (1987b), and Lucy (1987).  Details of the current discussion will be given by Harkness and Wheeler (1988).  Here we focus on the relatively simple spectrum when the supernova was two days old.  As a simplifying assumption, we adopt a power law density structure ($\rho \propto r^{-n}$) for the atmosphere.  Branch (1987a) and Dopita et al. (1987) assume a constant opacity and hence that the optical depth and density both follow the same power law structure to relate the power law index to the rate of recession of the velocity of the matter at the photosphere.  They derive n ~ 11 - 15 for the first four or five days and n ~ 4 - 5 during the subsequent few weeks.    Models are calculated with n = 5, 7 and 11.  Homologous expansion is assumed and an inner velocity and hence radius is assigned corresponding to the selected epoch, eg, 2 days.  The envelope composition is taken to be X = .7438, Y = .2520, and heavier elements 1/4 of the solar distribution.  The atmosphere models are calculated in the approximation of LTE.  The calculation is first done in the locally co-moving frame in which the opacities can be solved exactly without concern for "expansion opacity" effects, and then transformed to the observer frame (Harkness 1985, 1986).  All sources of continuum opacity are included and lines are treated by resonance scattering.  Of order 200 lines are included in the calculations.

The velocities of the absorption minima of the Balmer lines are basically set by the density at a given velocity.  The absorption minima are essentially independent of the luminosity of the model for reasonable values of the luminosity.  At higher luminosity and temperature more hydrogen is ionized at lower radii and hence velocities, but more of the remaining hydrogen is excited, and these two effects are found to offset one another.  Several aspects of the spectra do change significantly with the luminosity.   One is the

H$\alpha$ line profile. The blue edge of the feature, which samples the highest velocity matter, is distinctly rounded in the observations. The brightest models reproduce this aspect rather well, whereas the dimmer models give sharper, abrupt blue edges. Other features sensitive to the luminosity are the Ca II lines, at H and K and the infrared triplet. These are not observed at two days, but appear in many of the theoretical spectra which are sufficiently dim or extended that appreciable Ca II exists. Satisfactory models must have the Ca II ionized away at this epoch.

No models with n = 5 at two days are satisfactory. They are all plagued by very strong Ca II lines and give colors that deviate considerably from those observed. The same is true to a lesser extent for the models with n = 7. Of this series of models, the best agreement occurs for n = 11. Figure 5 gives three models with n = 11, a normalizing density of $1.5 \times 10^{-14}$ gm/cm$^3$ for the matter moving at 20,000 km/s, at different luminosities. At two days, Menzies et al. (1987) give a bolometric luminosity for SN 1987A of $4 \times 10^{41}$ erg/s for a distance modulus of 18.5 and. $A_v = 0.6$. At $4 \times 10^{41}$ erg/s the Ca II lines are still rather strong in the models, whereas at $5 \times 10^{41}$ erg/s, they have nearly disappeared. The higher luminosity is thus favored in this series, but a somewhat steeper density gradient would probably accommodate the lack of Ca II

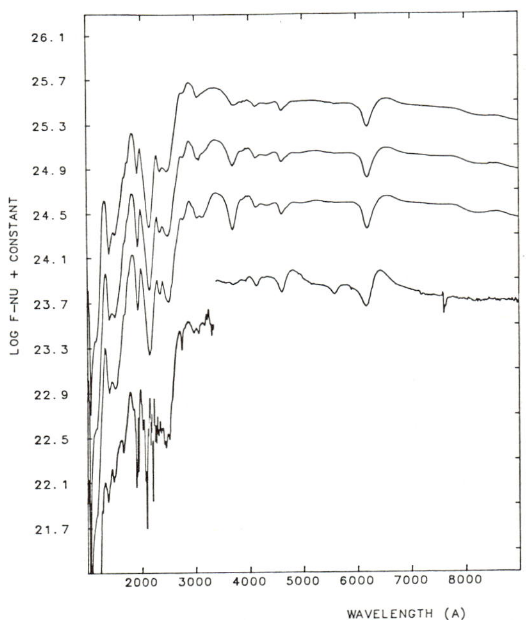


Figure 5. Three model atmospheres are given corresponding to a density profile of $r^{-11}$, with $\rho = 1.5 \times 10^{-14}$ gm/cm$^3$ at 20,000 km/s at two days. The three curves correspond to luminosities of 4.9, 3.9, and $3.2 \times 10^{41}$ erg/s, from top to bottom, respectively. The lower curves are optical spectra from Cerro Tololo provided by Mark Phillips and UV spectra provided by Bob Kirshner. Note the diminishing Ca II features with increasing luminosity, and the strong UV deficit which is unprecedented in a Type II supernova.

features with a somewhat lower luminosity. At this epoch, the lack of the Ca II features thus serves as an intrinsic luminosity indicator.

As anticipated, the ultraviolet spectra are formed by resonant scattering of many Fe II lines.   The resulting peaks are not emission lines, but rather the absence of  blended P-Cygni absorption troughs in wavelength ranges where the density of Fe II lines is less. This is particularly true of the peak at 1800 Å which occurs at about 2000 Å in the somewhat lower velocity environment of SN Ia and SN Ib (Wheeler, et al. 1986).  Mg II $\lambda$2797 also makes a significant contribution to the minimum at 2400 Å.   The success of these models in reproducing the UV spectrum suggests that "ordinary" Type II supernovae, which do not display such a UV deficit, have a more significant circumstellar nebula in the immediate environment of the supernova.  This matter can dynamically affect the density gradient in the supernova ejecta, and produce a separate source of UV radiation.


## 4. COMPARISON TO OTHER SUPERNOVAE

The McDonald Observatory archives of supernova spectra have been examined to compare and contrast SN 1987A with other Type II supernovae. SN 1979C and SN 1980K were two well-studied Type II events with "linear" light curves.   There is controversy over whether linear SN II represent a distinct class or a part of a continuum for which well defined "plateau" events are another extreme.   Near maximum light, SN 1979C and 1980K showed nearly pure continua rather than the strong Balmer lines which characterized SN 1987A for the first several weeks.   At about two months, SN 1980K, in particular, still displayed strong Balmer lines which had nearly disappeared from SN 1987A at a similar epoch.  SN 1979C and 1980K show no sign of the "barium" line at 6070 Å.  Whether these various differences mean that Type II linear events differ from SN 1987A in mass, composition, or excitation in the envelope is an important topic for future investigation.

SN 1986I in M 99 was the first supernova discovered by the Berkeley supernova search, and it has proved itself to be a classical plateau event.  The spectra of SN 1987A do not at any epoch resemble the McDonald spectra of 1986I  at about one month. There are some important similarities between the two events when they are about two months old, including some evidence in SN 1986I for the line at 6070 Å in a spectrum obtained by C. Foltz which is distinctly  absent in the earlier McDonald spectra. Most of the similarities are due to ubiquitous features of Fe II, and hence they alone do not provide a solid link between SN 1986I and SN 1987A.

The most intriguing spectrum in the McDonald archives in this regard is of SN 1985H in NGC 3359 obtained at an uncertain epoch.  This spectrum is shown in Figure 6 along with a  spectrum of SN 1987A on April 28 kindly provided by Mike Dopita.  There are strong similarities between these two spectra, most of which are again due to Fe II. The H$\beta$ feature is strong  in this spectrum of SN 1985H.  In SN 1987A it was just

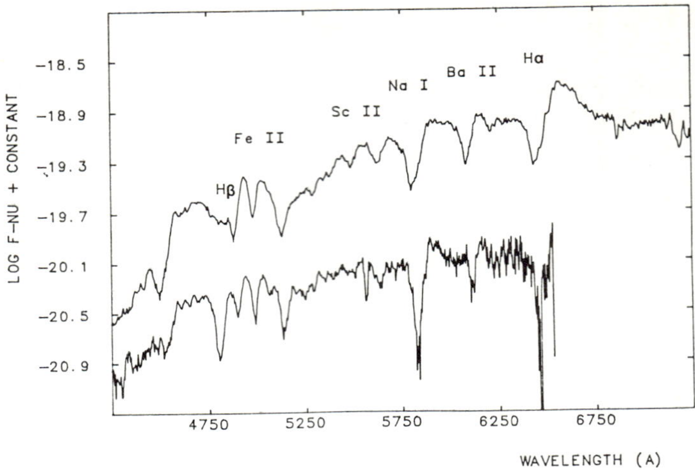



Figure 6. The spectrum of SN 1985H in NGC 3359 obtained by E. Barker and A. Cochran with the 2.7 meter telescope at McDonald Observatory on April 17, 1985 is shown in comparison with a spectrum of SN 1987A on April 28,1987 obtained at Siding Springs and provided by Mike Dopita. Note the similarity of the Hα absorption, Na D, the many Fe II features in the blue and especially the strong absorption at 6070 Å.

beginning to strengthen at the epoch shown in Figure 6 and reached a strength comparable to that displayed here for SN 1985H about a month later, at the end of May. In general, the features in SN 1985H are narrower than those of the April 28 spectrum of SN 1987A, suggesting an epoch for the spectrum of SN 1985H of greater than two months. The most interesting feature is the strong line at 6070 Å, suggesting that SN 1985H is closely related to SN 1987A.

At this writing, nothing is known about the photometric history of SN 1985H, so it is not known how long after the explosion this spectrum was obtained, nor what luminosity should be expected. The supernova was at V = 16.4 on April 12, 1985 (IAUC 4053). The distance modulus of NGC 3359 is about 30.21 according to de Vaucouleurs (private communication), so at the time the spectrum was obtained the absolute magnitude was about $M_V$ = -13.8 on this distance scale, neglecting reddening. The spectra of SN 1985H suggest that it is of order two months old (IAUC 4058), or older, as discussed above. Type II plateau events can remain on the plateau in V for about three months during which they decline ~ 0.2 magnitudes (Doggett and Branch 1985). On the plateau they are about $M_V$ ~ -16 to -17 on the short distance scale and 1.5 magnitudes brighter on the long scale. If SN 1985H is related to such events, it must have declined off the plateau by 2 to 3 magnitudes and hence be of order 150 days after maximum, independent of the distance scale. Over the temporal range of interest, 60-150 days, SN 1987A varied between $M_V$ ~ -16 and -14.5 for m-M = 18.5 and $A_V$ = 0.6. SN 1985H would have been significantly dimmer than this on the short distance scale, but within this range on the long scale. Uncertainties in the epoch and distance of SN 1985H prevent any firm conclusion, but one can not on this basis rule out the notion that SN 1985H may have been of the same class of event as SN 1987A.

## 5.  CONCLUSIONS

Three different structures are found for stellar models with identical  luminosity, total mass, core mass, and core radius as the effective temperature is varied.  Two of these solutions can correspond to effective temperatures in the range pertinent to SK -69 202. This implies that models for the progenitor of SN 1987A will prove very sensitive to physical assumptions and numerical treatment in structural and evolutionary calculations.

Self-consistent models which match the observational constraints of SK -69 202 with envelopes of normal hydrogen and helium abundance suggest that the progenitor of SN 1987A must have undergone essentially no mass loss, or must have lost nearly all the hydrogen-rich envelope leaving ~ 0.1 $M_\odot$.  The models with smaller mass envelopes are shocked to excessive velocities in the core or envelope even with very low kinetic energy, and hence may be ruled out.  The envelope composition may not be normal, but may be helium enriched.  This  avoids the problems of structural self-inconsistency, and may allow a range of envelope masses that are consistent with basic observational constraints.

The LTE models with $\rho \propto r^{-11}$ give a reasonable representation of the optical and the "unique" UV spectra of SN 1987A at about 2 days.   A lower limit to the luminosity is given by the Balmer line profiles and the absence of Ca II features.  It is difficult to see how the current models could be reconciled with a distance modulus of 18.3 and/or an extinction significantly smaller than $A_V = 0.6$, but this important issue deserves further study.  Comparison of the light curve of SN 1987A to previous SN II plateau events suggests that either $H_0 \sim 100$ km/s/Mpc or that SN 1987A has ejected about 4 times less $^{56}$Ni than ordinary SN II plateau events.

At the close of this conference, we do not seem to have a completely self-consistent model in which the basic features of the light curve and spectra are satisfied by calculations  of the theoretical spectra and light curves at all epochs.  There is much left to be learned about the structure of the progenitor and the manner in which it exploded.

This research is supported in part by NSF Grant 8413301, by the R. A. Welch Foundation, and by an allotment of computer time at the Center for High Performance Computing at the University Texas.  We are grateful to Mike Dopita, Mark Phillips, and Bob Kirshner for providing us with optical and UV data in numerical form, and to Stan Woosley for some of his dynamical models.  We have benefitted from conversations with numerous colleagues who have shared their excitement and insights into this remarkable event.

## REFERENCES

Arnett, W. D. 1977, *Ap. J. Suppl.*, **35**, 145.
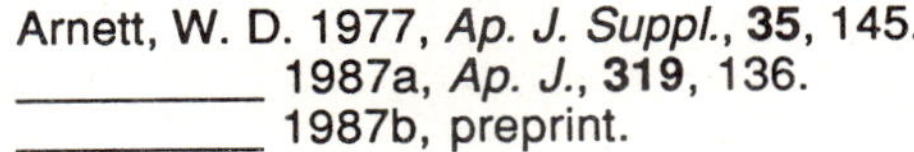 1987a, *Ap. J.*, **319**, 136.
__________ 1987b, preprint.

Ashoka, B. N., Anupama, G. C., Prabhu, T. P., Giriidhar, S., Ghash, K. K., Jain, S. K., Pati, A. K., and Kameswara Rao, N. 1987, *J. Astron. Astr.*, **8**, 195.

Barkat, Z. and Wheeler, J. C. 1988, in preparation.

Blanco, V. M., et al. 1987, *Ap. J.*, **320**, 589.

Branch, D. 1987a, *Ap. J. (Letters)*, **320**, L23.

___________ 1987b, *Ap. J. (Letters)*, **320**, L121.

Cassatella, A., Fransson, C., van Santvoort, J., Gry, C., Talavera, A., Wamsteker, W. and Panagia, N. 1987, *Astr. & Ap.*, **177**, L29.

Catchpole, R. M. et al. 1987, *M.N.R.A.S.*, in press.

Chevalier, R. A. and Fransson, C. 1987, *Nature*, **328**, 44.

Danziger, I. J., Fosbury, R.A.E., Alloin, D., Cristiani, S., Dachs, J., Gouiffes, C., Jarvis, B., and Sahu, K. C. 1987, *Astr. & Ap.*, **177**, L13.

Doggett, J. B., and Branch, D. 1985, *Ap. J.*, **90**, 2303.

Dopita, M. A., Achilleos, N., Dawe, J. A., Flynn, C. and Meatheringham, S. J. 1987, preprint.

Fransson, C., Grewing, M., Cassatella, A., Panagia, N. and Wamsteker, W. 1987, *Astr. & Ap.*, **177**, L33.

Grassbergh, E. K., Imshennik, V. S., Nadyozhin, D. K., and Utrobin, V. P. 1987, preprint.

Hanuschik, R. W. and Dachs, J. *Astr. and Ap.*, **182**, L29.

Harkness, R. P. 1985, in *Supernovae as Distance Indicators*, ed. N. Bartel (Berlin: Springer-Verlag), p. 183.

Harkness, R. P. 1986, in *Radiation Hydrodynamics in Stars and Compact Objects*, ed. D. Mihalas and K.-H.A. Winkler (Berlin: Springer-Verlag), p. 166.

Harkness, R. P. and Wheeler, J. C. 1988, in preparation.

Hillebrandt, W., Höflich, P., Truran, J. W., and Weiss, A. 1987, *Nature*, **327**, 597.

Kirshner, R., Sonneborn, G., Crenshaw, D. M., and Nassiopoulos, G. E. 1987, *Ap. J.*, **320**, 602.

Lucy, L. B. 1987, *Astr. & Ap.*, **182**, L31.

Maeder, A. 1987, *Proceedings of the ESO Conference on SN 1987A*, in press.

Menzies, J. W., et al. 1987, *M.N.R.A.S.*, **227**, 39p

Nomoto, K., Shigeyama, T., and Hashimoto, M. 1987, in *Proceedings of the ESO Workshop on SN 1987A*, in press.

Shigeyama, T., Nomoto, K., Hashimoto, M., and Sugimoto, D. 1987, *Nature*, **328**, 320.

Tyson, J. A. and Boeshaar, P. C. 1987, preprint.

Walborn, N. R., Lasker, B. M., Laidler, V. G., and Chu, Y.-H. 1987, *Ap. J. (Letters)*, **321**, L41.

Wamsteker, et al. 1987, *Astr. & Ap.*, **177**, L21.

Wheeler, J. C., Harkness, R. P., Barkat, Z. and Swartz 1986, *P.A.S.P.*, **98**, 1018.

Williams, R. E. 1987, *Ap. J. (Letters)*, **320**, L117.

Wood, P. R. and Faulkner, D. J. 1987, preprint.

Woosley, S. E., Pinto, P. A., Martin, P. J., and Weaver, T. A. 1987, *Ap. J.*, **318**, 664.

Woosley, S. E., Pinto, P. A., and Ensman, L. 1987, *Ap. J.*, in press.

# LIGHT CURVE MODELS FOR SN 1987A AND DIAGNOSIS OF SUPERNOVA INTERIOR


Ken'ichi Nomoto[1], Toshikazu Shigeyama[1], and Masa-aki Hashimoto[2]

[1] Department of Earth Science and Astronomy, University of Tokyo, Tokyo
[2] Max-Planck-Institut fur Astrophysik, Garching


## Abstract


Presupernova evolution of the progenitor of SN 1987A, hydrodynamics of explosion (shock propagation, explosive nucleosynthesis), optical light curve due to shock heating and $^{56}$Co decay, and X-ray and $\gamma$-ray light curves are calculated and compared with the observations of SN 1987A. Constraints on the mass of the hydrogen-rich envelope $M_{env}$ (i.e., mass loss history) and the helium abundance in the envelope are obtained from the progenitor's blue-red-blue evolution as well as from the light curve. The explosion energy $E$ and the mass and distribution of $^{56}$Ni are inferred from the light curves. Models and observations are in reasonable agreement for $E/M_{env} = 1.5 \pm 0.5 \times 10^{50}$ erg/$M_\odot$, $M_{env} = 5$ - $10$ $M_\odot$, and $M_{Ni} \sim 0.07$ $M_\odot$. Mixing of $^{56}$Ni into the envelope is indicated.

Light curves of exploding bare helium stars are also calculated to see whether the observed Type Ib supernova light curves can be accounted for.


## 1. INTRODUCTION

SN 1987A in the LMC is directly related to the three issues discussed at our Colloquium: chemical peculiarity, mass loss, and explosion. This supernova is providing us with valuable information on the three major uncertainties involved in the current theory of massive star evolution and explosion, namely, 1) the mechanism that tranforms collapse into explosion, 2) mass loss, and 3) convection (material mixing, in general). Because of the lack of clear understanding of these processes, the explosion energy $E$, mass and distribution of $^{56}$Ni, and the mass of the hydrogen-rich envelope $M_{env}$ in SN 1987A are highly uncertain.

From the comparison between observations and the theoretical models of stellar evolution, explosion, and the optical, X-ray, $\gamma$-ray light curves, we will examine:
1) why was the progenitor of SN 1987A blue,
2) how much $M_{env}$ was retained in the progenitor after mass loss,

3) how large is $E$,

4) what kind of heavy elements are synthesized and how they are distributed in the ejecta.

In addition to SN 1987A, explosion of bare helium stars is calculated to examine whether and under what conditions Type Ib-like light curves are realized.

These consideration will lead us deeper understanding of the mixing, mass loss, and explosion during the evolution of massive stars.

## 2. WHY WAS THE PROGENITOR BLUE ?

The most likely progenitor of SN 1987A, Sk-69 202, was surprisingly a B3 blue supergiant. Its luminosity is about $1.3 \times 10^5$ $L_\odot$ which corresponds to the presupernova luminosity of a helium core of $M_{core} \sim 6$ $M_\odot$ (Woosley 1988; Nomoto, Shigeyama, and Hashimoto 1987; Nomoto and Hashimoto 1988). Its main-sequence mass is $M_{ms} = 17 - 22$ $M_\odot$. From the luminosity and the spectral type, the radius of Sk-69 202 is estimated to be about $3 \times 10^{12}$ cm.

The occurrence of a Type II supernova from such a blue supergiant progenitor is not known before. It is very likely that the progenitor of SN 1987A evolved first to become a red supergiant and then came back to the blue as indicated by the following observations. The UV observations have shown the existence of a circumstellar shell which is nitrogen overabundant and whose expansion velocity is less than 30 km s$^{-1}$ (Panagia *et al.* 1987). The existence of a dust shell is also inferred from the IR observations (Chalavaev *et al.* 1987). The existence of the bright red supergiants in the LMC corresponding to stars up to $\sim 50$ $M_\odot$ (Humphreys and Davidson 1978) is consistent with the above scenario.

Saio, Kato, and Nomoto (1988) recently examined under what conditions a massive star undergoes a blue-red-blue evolution. The evolution of a star of initial mass 20 $M_\odot$ star in the HR diagram is shown in Figure 1 from the zero-age main-sequence through carbon ignition at the center. The metallicity in the envelope was assumed to be $Z = 0.005$ and the Schwarzschild criterion was adopted. The star shows the three types of evolutionary path (A, B, C) depending on the mass loss, metallicity, and the change in the helium abundance $Y$ in the envelope.

Such a behavior can be understood from the fact that the blue- supergiant envelope solution with only a certain combination of $(M_{env}, Y)$ fits to the evolved core model of $\sim 6$ $M_\odot$ for a given luminosity (Fig. 2a). For example, for $L = 1 \times 10^5$ $L_\odot$ and $Y = 0.25$, only two solutions with $M_{env} \sim 14$ $M_\odot$ (almost no mass loss) and $0.05$ $M_\odot$ are allowed to exist (see also Barkat and Wheeler 1987, Wheeler *et al.* 1988).

Among the three cases of A, B, and C in Fig. 1, the progenitor of SN 1987A must have evolved along the path C as follows.

**Evolution from the Blue to the Red:** The progenitor was $\sim 20$ $M_\odot$ on its main-sequence. During early helium burning, the star was a blue supergiant. Whether the star remains blue or moves to the red depends on mass loss. If $M_{env}$ would be sufficiently large (i.e., almost no mass loss) for $Y \sim 0.25$ (Fig. 2a), the star would have remained blue without undergoing extensive redward evolution. (This corresponds to the models by Brunish and Truran 1982, Hillebrandt *et al.* 1987, and Arnett 1987.) However, the progenitor of SN 1987A should have undergone mass loss and significantly reduced its $M_{env}$. Then the star moved redward to become a red supergiant. This is because the blue supergiant envelope solution with the reduced $M_{env}$ does not exist for its $L$ and $Y$. Thus the mass loss is the driving mechanism of the redward evolution. Before reaching

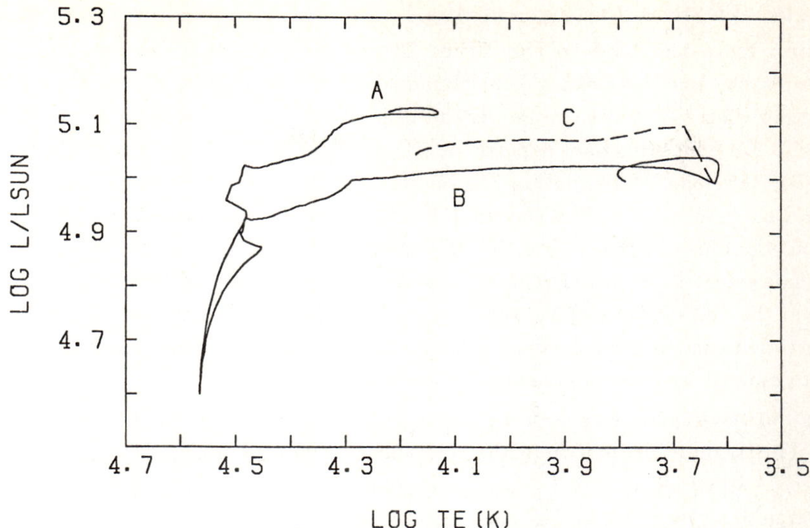


**Figure 1:** Evolutionary tracks in the HR diagram of the initially 20 $M_\odot$ star. Cases A, B, and C correspond to models with no mass loss (A), with no artificial enhancement of helium in the hydrogen-rich envelope (C), and with enhancement of helium up to $Y = 0.4$ (B).

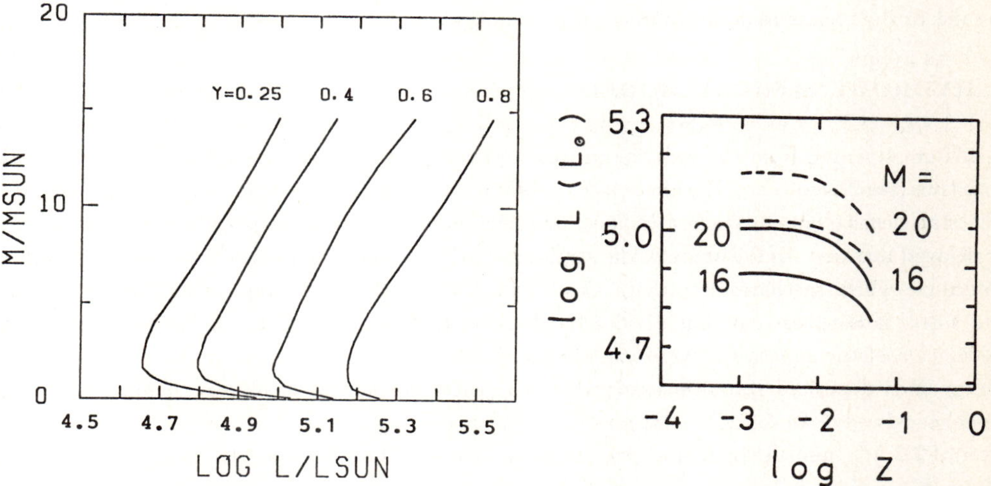


**Figure 2a:** Envelope solutions with the outer boundary of $\log T_{\text{eff}} = 4.2$ and the inner boundary conditions that fit to the core at $M_r = 5.45\ M_\odot$ and $r = 0.6\ R_\odot$. The ordinate is the mass of the hydrogen-rich envelope ($M_{\text{env}} = M - 5.45\ M_\odot$) given as a function of luminosity $L$ and the mass fraction of helium $Y$.

**Figure 2b:** Envelope solutions for different metallicity. For $\log T_{\text{eff}} = 4.2$, the lines for the constant total mass ($M = M_{\text{env}} + 5.45\ M_\odot = 20$ and $16\ M_\odot$) are given as a function of the luminosity $L$ and the mass fraction of heavy elements $Z$. The solid and dashed lines are for helium abundance $Y = 0.255 - Z$ and $0.4 - Z$, respectively.

the Hayashi line, the star lost as large as 5 $M_\odot$.

**Evolution from the Red to the Blue:** During the red supergiant phase, envelope mass was still decreasing and the mean helium abundance $Y$ in the envelope increased. A blueward evolution began when C+O core contracted and helium shell burning became active. Whether the star returns to the blue supergiant depends on $M_{env}$ and $Y$. If $M_{env}$ and $Y$ would not match the blue supergiant envelope solution in Fig. 2a, the star would not move blueward in the HR diagram as case B in Fig. 1; for case B, $Y$ is too small for $M_{env} \sim 9$ $M_\odot$. For case C, on the contrary, the star moves to the blue because $Y$ is artificially enhanced to 0.4. In the progenitor of SN 1987A, such an enhancement of $Y$ have occurred probably due to mixing of helium and nitrogen-rich material from the deepest layer of the envelope to the surface by the meridional circulation. (The surface convection zone may not be deep enough.) Such a material mixing drove the progenitor of SN 1987A to move to the blue just before the explosion.

Another important point for case C evolution is that $M_{env}$ is as large as 10 $M_\odot$ at the supernova explosion. This is quite distinct from the solutions obtained by Maeder (1987) and Wood and Faulkner (1987) where $M_{env} \ll 0.1$ $M_\odot$. Their solutions and case C solution correspond to the small and large $M_{env}$ branches in Fig. 2a, respectively.

If the above interpretation is correct, the star with only a narrow range of $(M_{env}, Y)$ can become blue and end up its life as a blue supergiant. It is interesting to note that the envelope solution depends on its metallicity $Z$ as seen in Fig. 2b. For a given $L$, the solution with larger $Z$ has a larger $M_{env}$ or larger $Y$ to compensate the larger opacity with larger $Z$. In other words, more enhancement of $Y$ is required for a given $M_{env}$ if $Z$ is larger. Therefore, the evolution from the red to the blue is more likely to occur for smaller $Z$.

## 3. HYDRODYNAMICAL MODELS

After returning from the red, the progenitor evolved to form an onion-skin like composition structure (see Nomoto and Hashimoto 1987, 1988). The hydrogen-depleted core was about 6 $M_\odot$. The star then collapsed, made a bounce, and generated a strong shock wave either in a prompt or delayed manner. Unfortunately, the statistics of the observed neutrinos is not good enough to determine which mechanism (prompt shock or delayed neutrino heating) worked for SN 1987A and where is the mass cut that divides the neutron star and the ejecta. In the hydrodynamical calculation of the supernova explosion, therefore, the mass cut is assumed to be 1.4 $M_\odot$ and an energy $E$ is deposited instantaneously there to generate a strong shock wave. Thus the initial model consisted of an 1.4 $M_\odot$ point mass neutron star, heavy element layer of 2.4 $M_\odot$, helium-rich layer of 2.2 $M_\odot$, and the hydrogen-rich envelope of $M_{env}$. The total mass of the ejecta is $M = 4.6$ $M_\odot + M_{env}$.

The subsequent propagation of the shock wave, the expansion of the star, and the optical light curve are calculated for models 11E1Y4, 11E1Y6, 11E1.5, 11E2, 7E1Y4, where 11 and 7 denote the ejecta mass of $M = 11.3$ $M_\odot$ ($M_{env} = 6.7$ $M_\odot$) and 7.0 $M_\odot$ ($M_{env} = 2.4$ $M_\odot$), respectively; E1, E1.5, and E2 denote explosion energy of $E = 1$, 1.5, and 2 $\times 10^{51}$ erg, respectively; Y4 and Y6 stand for a surface composition of $Y = 0.40$ and 0.6, respectively. For all models the initial radius of the progenitor is assumed to be $R_0 = 3 \times 10^{12}$ cm (see Nomoto *et al.* 1987, Shigeyama *et al.* 1988 for details).

### 3.1. Explosive Nucleosynthesis

As the shock wave propagates through Si and O-rich layers, explosive nucleosynthesis takes place behind the shock. The peak temperature at the deep layer with $M_r < 1.67\ M_\odot$ exceeds 5.5 $\times 10^9$ K so that the materials, originally composed of Si, S and Ca, are burned into almost nuclear statistical equilibrium (NSE) elements. Because of low density and high temperature, such a layer undergoes alpha- rich freezeout being composed of mostly $^{56}$Ni and $^4$He (Thielemann *et al.* 1986). In the inner O-rich layer, oxygen is burned into mostly Si, S, Ca, and some $^{56}$Ni. The resulting composition structure after the shock passage is shown in Figure 3. The masses of $^{56}$Ni produced by silicon burning at $M_r = 1.60$ - $1.67\ M_\odot$ and by oxygen burning at $M_r > 1.67\ M_\odot$ are 0.05 $M_\odot$ and 0.025 $M_\odot$, respectively (Hashimoto *et al.* 1988).

As shown in § 4, the total amount of $^{56}$Ni that power the optical light curve of SN 1987A is $\sim 0.07\ M_\odot$. This is consistent with the above composition structure if the mass cut is 1.60 $M_\odot$. The resulting neutron star has a gravitational mass of $\sim 1.45\ M_\odot$, which is consistent with the observations of neutrinos (e.g., Totsuka 1988; Wilson 1988).

### 3.2. Shock Propagation and Material Mixing

The shock wave propagates through the hydrogen-rich envelope. When the expansion of the inner core is decelerated by the low-density envelope, a reverse shock forms and produces a density inversion (Nomoto *et al.* 1987). Because of Rayleigh-Taylor instability, the core material will be mixed during early stages (see also Woosley *et al.* 1988). Since the time scale and the extent of mixing are uncertain, we calculated several models 11E1Y6 assuming mixing up to $M_r/M_\odot =$ 0.07 (no mixing), 3.4, 4.6 (outer edge of helium layer), and 6.0. For the mixed layer, we assumed uniform composition. Rayleigh-Taylor instability may also form clumpy medium and could cause a leak of $\gamma$-ray and X-rays.

The shock wave arrives at the original surface of the star at $t_{\mathrm{prop}}$, which is approximated for different values of the initial radius $R_0$ and the ejected mass $M$, and for different explosion energies $E$ as (Shigeyama *et al.* 1987, 1988)

$$t_{\mathrm{prop}} \sim 2\mathrm{hr}(R_0/3 \times 10^{12}\mathrm{cm}) \left( \frac{M/10M_\odot}{E/1 \times 10^{51}\mathrm{erg}} \right)^{1/2} \tag{1}$$

The condition $t_{\mathrm{prop}} < 3$ hr (Kamiokande time is used) rules out all the models with too large $R_0$ and too low $E/M$ (Shigeyama *et al.* 1987). A progenitor radius larger than 4.5 $\times 10^{12}$ cm is ruled out for Model 11E1 simply from this condition. If $R_0 = 3.5 \times 10^{12}$ cm, $E/M$ should be larger than 0.6 $\times 10^{50}$ erg/$M_\odot$.

### 3.3. Velocity Profile

After the shock wave reaches the surface, the star starts to expand and soon the expansion becomes homologous. In Fig. 4, the velocity distribution for the homologous expansion is shown for 11E1 and 11E2. The velocity gradient with respect to the enclosed mass, $M_r$, is very steep near the surface, while it is almost flat in the helium layer and the heavy element core. This is because the core material is decelerated and forms a dense shell due to the reverse shock. The

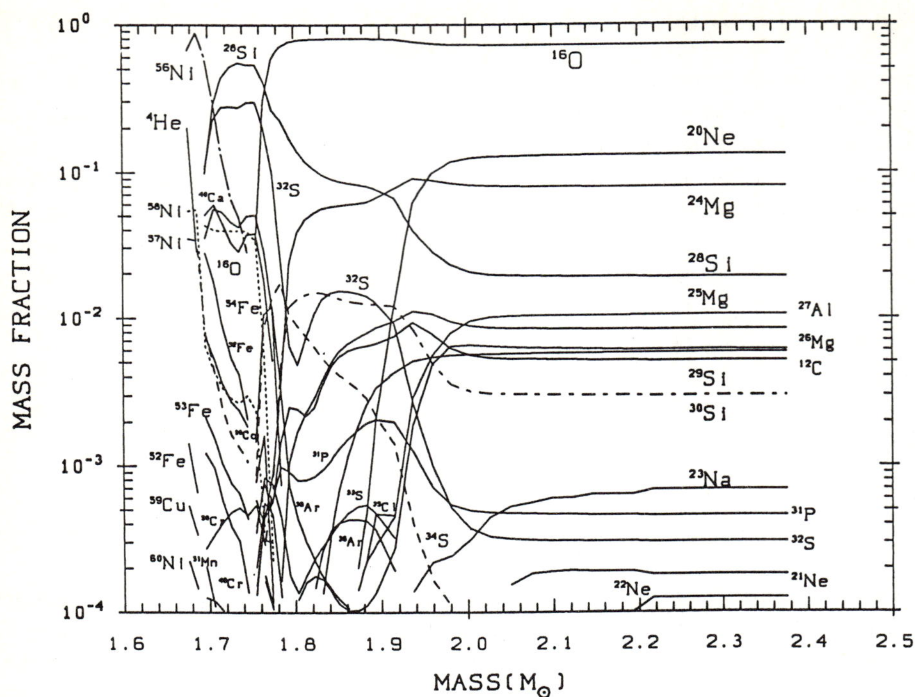

**Figure 3:** Abundances after explosive oxygen burning at $M_r \geq 1.67\ M_\odot$ for the 6 $M_\odot$ core.

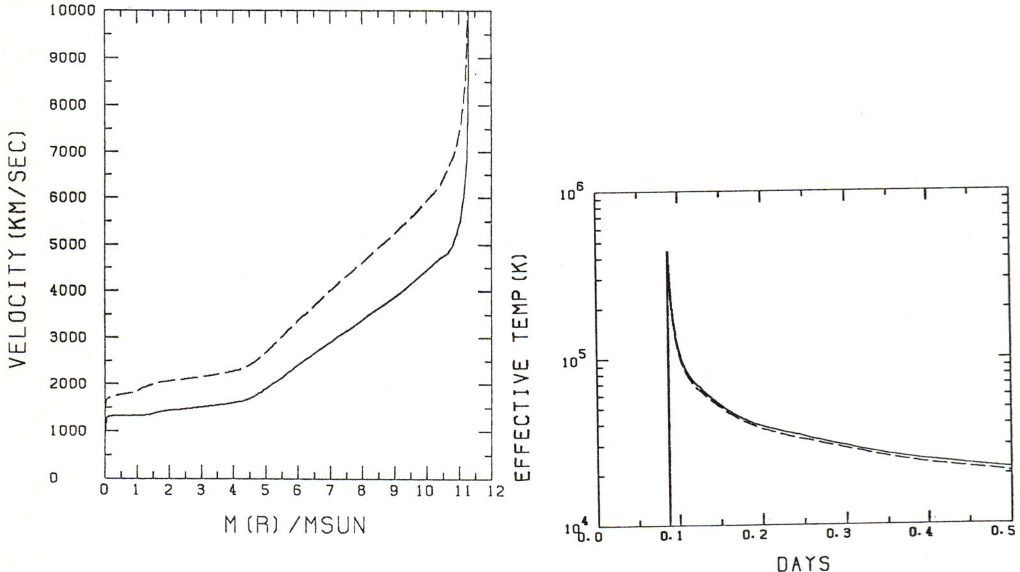

**Figure 4:** Velocity distribution for Models 11E1 (solid: $M = 11.3\ M_\odot$, $M_{\rm env} = 6.7\ M_\odot$, $E = 1 \times 10^{51}$ erg) and 11E2 (dashed: $E = 2 \times 10^{51}$ erg) at $t = 116$ d.

**Figure 5:** Change in the effective temperature near the shock break out at the surface for 11E1Y6 (solid) and 11E1Y4 (dashed).

expansion velocities of the helium and heavy element layers are so small that the kinetic energy of these layers is only 10 percent of the total kinetic energy for 11E1.

Expansion velocities have been measured to give important constraints on the above hydrodynamical models (Phillips 1988). As the star expands, the photosphere moves inward in $M_r$. Because of the steep velocity gradient near the surface, the velocity of the material at the photosphere decreases. Good agreement between the observed and theoretical material velocities at the photosphere (Nomoto and Shigeyama 1988; Shigeyama *et al.* 1988) suggests $E/M_{\mathrm{env}}$ should not be so different from $\sim 1.5 \times 10^{50}$ erg$/M_\odot$. The lowest velocity of the hydrogen lines would provide an information about the mixing between the hydrogen-rich envelope and the core as suggested from the optical, X-ray, and $\gamma$-ray light curves.

## 4. LIGHT CURVE DUE TO SHOCK HEATING

The light curve of SN 1987A shows quite unique features, which give important constraints on the hydrodynamical model of explosion. In the theoretical model, the shock wave initially establishes the radiation field with energy of roughly a half of explosion energy $E$; e.g., $4.4 \times 10^{50}$ erg for 11E1. The early light curve up to $t \sim 25$ d can be accounted for by diffusive release of this energy.

When the shock wave arrives at the photosphere, the bolometric luminosity reaches $2 \times 10^{44}$ erg s$^{-1}$ and the effective temperature becomes as high as $5 \times 10^5$ K (Fig. 5) for Model 11E1. Hence, most of the radiation is emitted in the UV band. Total energy of radiation during the first two days amounts to $10^{47}$ erg, which is enough to ionize the circumstellar matter.

Afterwards, the ejected gas and the radiation field expands rapidly so that the interior temperature decreases as $r^{-1}$, i.e., almost adiabatically (Fig. 5). Most of the radiation field energy is lost by $PdV$ work. As a result of decreasing photospheric temperature, the intensity peak shifts rapidly to the optical wavelength. Two unique features of SN 1987A at this optical flare-up are: 1) It took only 3 hours for the visual magnitude to reach 6.4 magnitude after the neutrino burst (McNaught 1987). 2) At the subsequent plateau, the optical light was much dimmer than typical Type II supernovae. The steep rise in luminosity in 3 hr and a less luminous plateau require relatively large $E/M$ and small $R_0$ (Shigeyama *et al.* 1987).

Figure 6 shows the changes in the calculated V magnitude for 11E1Y4 (dashed) and 11E1Y6 (solid). To account for the observed points, $E = 1 \times 10^{51}$ erg is required for $M = 11.3\ M_\odot$. For the envelope with smaller $Y$, the luminosity is lower and hence a larger $E$ is required because of a larger scattering opacity. Moreover, the model with $Y = 0.6$ is in better agreement with the overall shape of the observed light curve than $Y = 0.4$ especially at $t > 3$ d. Such an enhancement of helium abundance may be consistent with the restriction from the envelope solution as seen in Fig. 2a.

In the theoretical visual luminosity, there remain uncertainties. The supernova atmosphere is scattering dominated so that the color temperature may be significantly higher than the effective temperature (Shigeyama *et al.* 1987; Hoflich 1988), i.e., diluted black body radiation is emitted. The bolometric correction is sensitive to the color temperature because it is as high as $4 - 6 \times 10^4$ K. More careful calculations would be required to obtain accurate constraints on $R_0$ and $E/M$.

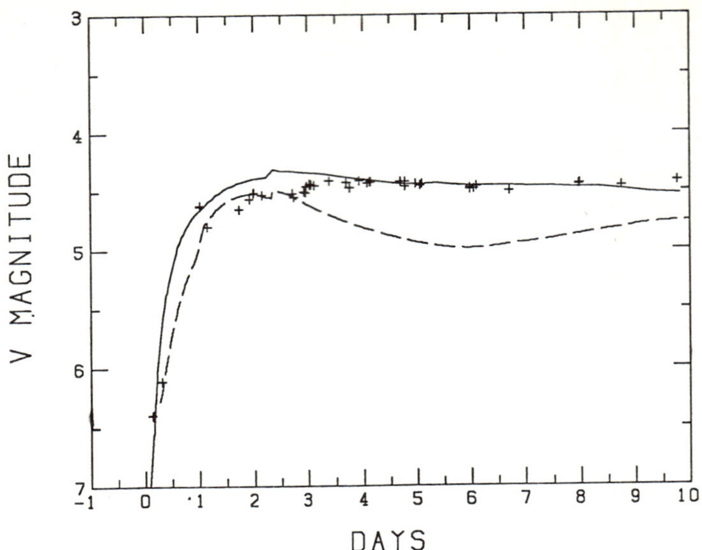

Figure 6: The calculated visual magnitude for Models 11E1Y6 (solid) and 11E1Y4 (dashed). Observed data are taken from ESO (Cristiani et al. 1987), CTIO (Blanco et al. 1987), and SAAO (Menzies et al. 1987).

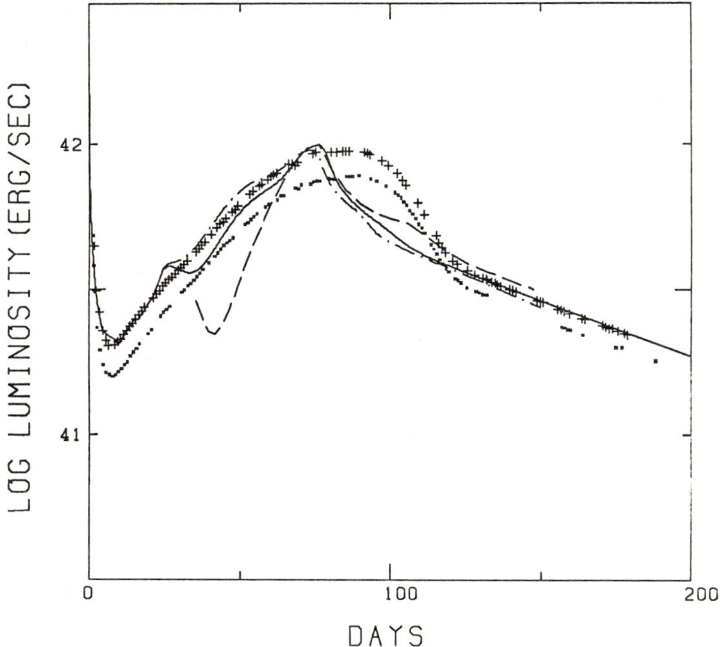

Figure 7: Bolometric light curve for several cases of mixing (11E1Y6). Production of 0.07 $M_\odot$ $^{56}$Ni is assumed. For the dashed curve $^{56}$Ni is assumed to be confined in the innermost layer while the solid curve assumes that $^{56}$Ni is mixed uniformly up to $M_r = 4.6$ $M_\odot$ (solid) and 6.0 $M_\odot$ (dash-dotted). Observed points are taken from SAAO (+; Catchpole et al. 1987) and CTIO (x; Hamuy et al. 1988).

# 5. LIGHT CURVE POWERED BY RADIOACTIVE DECAYS

After these early stages, the observations show an increase in the bolometric luminosity. The energy source that continually heats up the expanding star is certainly the decaying $^{56}$Co as evident from the light-curve tail after 120 d.

The line $\gamma$-rays from $^{56}$Co (847 keV, 1238 keV, 2599 keV, 1771 keV, ...) undergo multiple Compton scatterings and are degraded into X-rays. The resulting X-rays and fast electrons ionize atoms, which recombine emitting UV and optical radiation. In this way, the line $\gamma$-rays are thermalized to heat up the expanding materials until X-rays and $\gamma$-rays emerge from the supernova due to the decrease in the column depth of the ejecta.

The theoretical light curve with radioactive decays of $^{56}$Ni and $^{56}$Co was calculated assuming the production of 0.07 $M_\odot$ $^{56}$Ni. Figures 7 and 8 compare the calculated bolometric light curve with observations (Catchpole *et al.* 1987; Hamuy *et al.* 1988) for several models. The light curve shape is sensitive to the hydrodynamics and thus is a useful tool to infer 1) the distribution of the heat source $^{56}$Ni, 2) the mass of the hydrogen-rich envelope $M_{\rm env}$, and 3) the explosion energy $E$.

As the star expands, the photosphere becomes deeper as the recombination front proceeds through the hydrogen-rich envelope deeper in mass. At the same time a heat wave is propagating out from the interior. At a certain stage, energy flux due to radioactive decays exceeds that from shock heating. The dates when the radioactivity starts to dominate and when the luminosity reaches its peak depend on the above three factors as follows.

1) **Mixing**: Mixing of $^{56}$Ni into outer layers is likely to occur due to Rayleigh-Taylor instability (§ 3.2). Figure 7 shows how the light curve depends on the distribution of $^{56}$Ni by comparing the cases with mixing of $^{56}$Ni up to $M_r/M_\odot = 0.07$ (no mixing; dashed line), 4.6 (outer edge of helium layer; solid), and 6.0 (dashed-dotted) for 11E1Y6. If we assume that $^{56}$Ni is confined in the innermost layer of the ejecta, the increase in luminosity due to radioactive heating is delayed to $t = 42$ d, a dip appears in the curve, and the light curve shape in the rising part is too steep as compared with the observations. On the other hand, if $^{56}$Ni is mixed into outer layers, heat is transported to the envelope earlier. As a result the optical light increases earlier, i.e., $t = 33$ d and 26 d for $M_{\rm mix} = 4.6$ and 6.0 $M_\odot$, respectively, and the light curve shape is less steep, being in better agreement with the observations.

This suggests that the heat source had actually been mixed into outer layers and its effect began to dominate the light curve from $t \sim 26$ d. From the observational side, Phillips (1988) noted that the changes and kinks of the color started from $t = 25$ d, which may indicate the appearance of heat flux due to radioactive decays.

2) **Envelope mass**: For Model 7E1 with a hydrogen-rich envelope of 2.6 $M_\odot$, a peak luminosity of $\sim 10^{42}$ erg s$^{-1}$ is reached too early at about $t = 50$ d (Fig. 8), because the expansion velocities of the helium and heavy element layers are larger than in 11E1 and the photosphere approaches the helium layer earlier. Therefore $M_{\rm env}$ should be larger than $\sim 3$ $M_\odot$. Probably we need $M_{\rm env} = 5$ - 10 $M_\odot$ for a good agreement with the observations.

3) **Explosion Energy**: For Model 11E1 5Y4 ($E = 1.5 \times 10^{51}$ erg), the luminosity peak of $1.3 \times 10^{42}$ erg s$^{-1}$ is reached at $t = 63$ d which is a little too bright and too early compared with the observation. The cases with $E = 1.0$ - 1.3 $\times 10^{51}$ erg for $M_{\rm env} = 6.7$ $M_\odot$ show a better fit to the observations. This constraint is consistent with the condition obtained from the photospheric velocity.

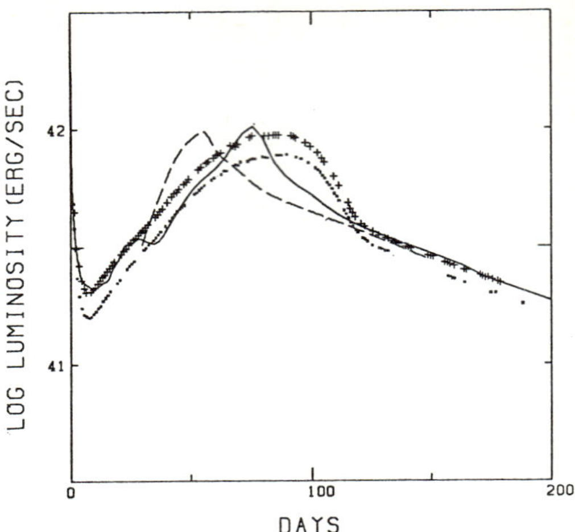



**Figure 8:** Dependence of the light curve on $M_{env}$. The solid line is Model 11E1Y6 ($M = 11.3\ M_\odot$, $M_{env} = 6.7\ M_\odot$; mixing of $^{56}$Ni at $M_r < 3.4\ M_\odot$) and the dashed is 7E1Y6 ($M = 7\ M_\odot$, $M_{env} = 2.4\ M_\odot$, $E = 1 \times 10^{51}$ erg).

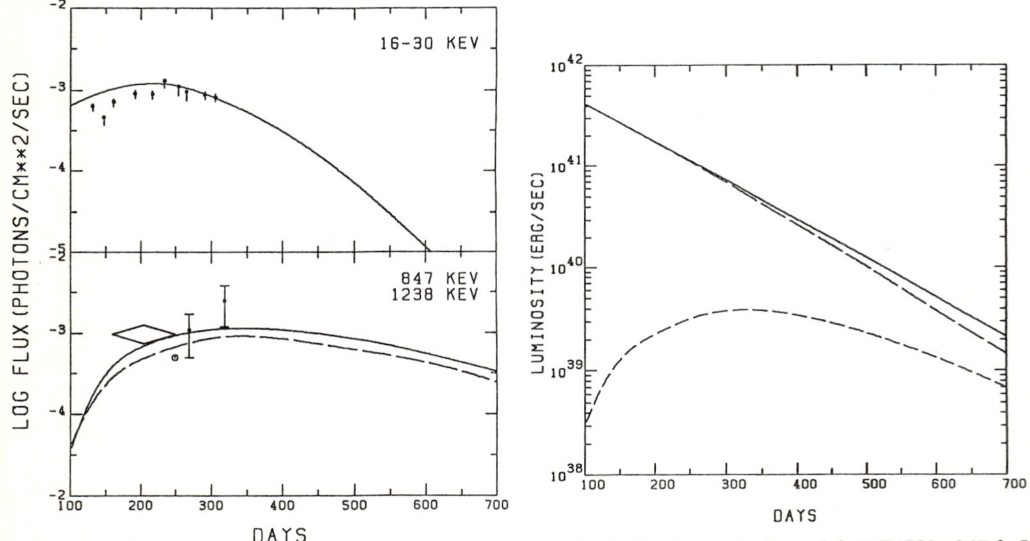



**Figure 9:** Calculated X- and $\gamma$-ray light curves for the hydrodynamical model 11E1Y6. 847 keV and 1238 keV line fluxes are shown by the solid and the dashed line, respectively. Time is measured from the beginning of explosion. $^{56}$Co and other heavy elements are mixed uniformly up to the outer edge of the original helium layer ($M_r = 4.6\ M_\odot$) and inhomogeneously in the hydrogen-rich envelope up to $M_r = 10\ M_\odot$ (see text). The crosses are the X-ray light curve observed by Ginga (Makino 1988). Gamma-ray observations by SMM (Matz et al. 1988) and balloon borne observations (Sandie et al. 1988; Cook et al. 1988; Rester et al. 1988) are indicated.

**Figure 10:** Changes in the X-ray and $\gamma$-ray luminosity above 5 keV is shown by the dashed line. The associated changes in the optical bolometric luminosity (dashed) are compared with the energy generation rate due to $^{56}$Co decay (solid line).

For Model 11E1, the bolometric luminosity reaches a peak value of $L_{pk} = 1 \times 10^{42}$ at $t = 77$ d. After the peak, the luminosity decreases more rapidly than SN 1987A which formed a broad peak. Finally the light curve enters the radioactive tail (Fig. 7).

It should be noted that the observed peak luminosity is higher than the energy generation rate due to Co-decay which gives $L = 5.4 \times 10^{41}$ erg s$^{-1}$ at $t = 70$ d for $M_{\mathrm{Co}} = 0.07\ M_\odot$. This implies that previously deposited energy from Co-decay is also radiated away during the broad peak of the light curve (Woosley $et\ al.$ 1988). How this additional energy is radiated away is rather sensitive to the dynamical behavior and to the opacity. The difficulty to reproduce the broad peak may require the improvement of the flux-limited diffusion approximation used in the calculation.

Despite this difficulty, both the peak luminosity and the tail can be consistently accounted for by the radioactive decay of $^{56}$Co whose mass is $\sim 0.07\ M_\odot$.

## 6. HARD X-RAYS AND GAMMA-RAYS

As the column depth of the supernova ejecta decreases as $t^{-2}$, $\gamma$-rays and hard X-rays suffer from less Compton scattering and photoelectric absorption and eventually emerge from the surface. The timing of their emergences and light curves are sensitive to $E$, $M_{\mathrm{env}}$, and the distribution of elements. In other words, X-ray and $\gamma$-ray observations provide another good diagnosis of the supernova interior.

X-ray and $\gamma$-ray light curves for model 11E1Y6 were calculated by Itoh $et\ al.$ (1987) and Ebisuzaki and Shibazaki (1988). They have shown that $^{56}$Co needs to be mixed into hydrogen-rich envelope (i.e., the column depth to the cobalt layer is $\sim 20$ g cm$^{-2}$ at $t = 200$ d) to account for the emergence of the X-rays as early as observed by Ginga (Dotani $et\ al.$ 1987) and Kvant (Sunyaev $et\ al.$ 1987). (See also Pinto and Woosley 1988 and Sutherland $et\ al.$ 1988). Recently, it is reported that SMM has been observing $\gamma$-rays since August 1987 (Matz $et\ al.$ 1988). Such an early $\gamma$-ray emergence requires the column depth to the $^{56}$Co layer as small as $\sim 2.5$ g cm$^{-2}$ at $t = 200$ d (Itoh $et\ al.$ 1988; Kumagai $et\ al.$ 1988). Figure 9 shows the calculated light curves of 16 - 30 keV X-ray and the 847 keV and 1238 keV line $\gamma$-rays. They are consistent with the observations. This model assumes a step-like distribution of $^{56}$Co where the mass fraction of $^{56}$Co in the layers at $M_r$ $\leq 4.6\ M_\odot$, 4.6 - 6 $M_\odot$, 6 - 8 $M_\odot$, and 8 - 10 $M_\odot$ are $X_{\mathrm{Co}} = 0.0128$, 0.0035, 0.0021, and 0.0011, respectively. The X-ray spectrum and its time evolution are in good agreement with observations (see Itoh $et\ al.$ 1988 in this volume).

This model predicts:

1) The X-ray flux at 20 - 30 keV will decrease below the detection limit of Ginga around $t \sim 400$ d while harder X-rays remain observable until t $\sim 2$ years.

2) The line $\gamma$-rays will be observable until $t \sim 2$ yr (Fig. 9).

3) With increasing high energy flux, the deviation of the optical bolometric luminosity from the exponential decline will become significant as seen in Figure 10. The luminosity above 5 keV reaches 5 and 8 % of the energy generation rate at $t = 300$ d and 350 d, respectively.

Further satellite and balloon-borne observations will test the predicted spectral evolution and light curves and provide new information on the distribution of the mixed heavy elements in the supernova ejecta.

# 7. LIGHT CURVE MODELS FOR TYPE Ib SUPERNOVAE

Now that some details of SN 1987A has been analyzed, the relation between SN 1987A and Type Ib supernovae (SN Ib) is worth studying. Massive helium stars have been the most popular progenitor model for Type Ib supernovae (e.g., Wheeler *et al.* 1987), while a helium detonation in accreting white dwarfs could be a possible alternative model (e.g., Branch and Nomoto 1986). If massive helium stars are the SN Ib progenitor, there would be a similarity in the core structure, explosion energy, and nucleosynthesis between SN 1987A and SN Ib. Then the same parameter could reproduce the light curve of SN Ib because of the difference in the envelope structure.

By applying the same physics to the same 6 $M_\odot$ helium core model as used in the modeling of SN 1987A, light curves powered by the decay of 0.07 $M_\odot$ $^{56}$Co are calculated and compared with observations. The $\gamma$-ray opacity of 0.03 cm$^2$ g$^{-1}$ (Sutherland and Wheeler 1984) and the optical opacity of 0.1 cm$^2$ g$^{-1}$ are assumed. For the 6 $M_\odot$ helium star with $E = 1 \times 10^{51}$ erg (where the masses of the ejecta $M_{\rm ej}$ and the neutron star residue are assumed to be 4.4 $M_\odot$ and 1.6 $M_\odot$, respectively), the calculated bolometric light curve after the peak declines at nearly the $^{56}$Co decay rate because almost all $\gamma$-ray energies are deposited in the star (see also Schaeffer *et al.* 1987; Ensman and Woosley 1988). This is slower than even the slowest SN Ib, SN 1984L (Harkness *et al.* 1987).

The decline rate of the light curve depends largely on how fast $\gamma$-rays and X-rays escape from the star. The column depth to the cobalt layer at time $t$ is roughly proportional to (mass of the overlying layer)$^2/(Et^2)$ (e.g., Chan and Lingenfelter 1987). Accordingly the optical light curve declines faster, if the ejecta mass is smaller, $^{56}$Co is mixed closer to the surface, and $E$ is larger.

Possibilities that $M_{\rm ej}$ is smaller than 4.4 $M_\odot$ include: 1) the helium star itself is smaller than 6 $M_\odot$, e.g., 4 - 5 $M_\odot$, and 2) the helium star undergoes mass loss and loses most of its helium envelope before the explosion. To explore the most extreme case of mass loss, we removed the helium envelope to expose the C+O layer. Then $M_{\rm ej} = 2.2\ M_\odot$. (This also corresponds to the less massive helium star of 3.6 - 3.8 $M_\odot$.) Its light curve is shown by the dashed curve in Fig. 11 and is consistent with SN 1984L. However, it is still slower than the bolometric light curve of SN 1983N (Panagia 1987) and the photographic light curve of SN 1964L (Bertola *et al.* 1965).

The other possibility is the mixing of $^{56}$Co into outer layers. The solid line in Fig. 11 shows a light curve for rather extreme case of mixing where $^{56}$Co is uniformly distributed in the whole ejecta of 2.2 $M_\odot$. The light curve near the peak is consistent with SN 1983N but its decline after $\sim 35$ d is still slower than the observations (see also Ensman and Woosley 1988). Even for $E = 3 \times 10^{51}$ erg with uniform mixing of $^{56}$Ni, the decline shown by the dashed curve in Fig. 12 is slower than SN 1983N. Since the column density of the ejecta scales as $M_{\rm ej}{}^2/Et^2$, the dashed curve in Fig. 12 suggests $M_{\rm ej}$ is as small as the white dwarf mass if $E \sim 1 \times 10^{51}$ erg.

Finally, we tried a numerical experiment where $^{56}$Ni exists only in the outermost 0.2 $M_\odot$ layer beneath the surface. The escape of X-rays and $\gamma$-rays are significant and the resulting optical light curve (the dash-dotted curve in Fig. 11) is as narrow as SN 1964L. Such a composition inversion is not realistic for helium stars but a natural outcome from the off-center detonation in accreting white dwarfs.

The above light curve modeling is meant to explore rather extreme cases and quite preliminary, but suggestive.

1) The slowest light curve of SN Ib such as SN 1984L may be accounted for by the helium star

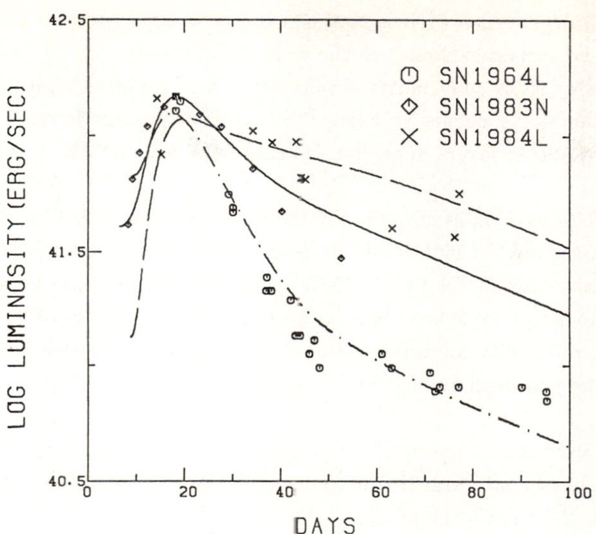

**Figure 11:** Calculated bolometric light curves for the ejecta of 2.2 $M_\odot$ and $E = 1 \times 10^{51}$ erg. The mass of $^{56}$Ni and the optical opacity are assumed to be 0.07 $M_\odot$ and 0.1 cm$^2$ g$^{-1}$, respectively. Shown are the cases where $^{56}$Ni is confined in the central region (dashed), uniformly distributed throughout the ejecta (solid), and confined in the outer layers of 0.2 $M_\odot$ below the surface (dash-dotted). The observed light curve of 1983N is bolometric, while those of 1984L and 1964L are visual and photographic, respectively.

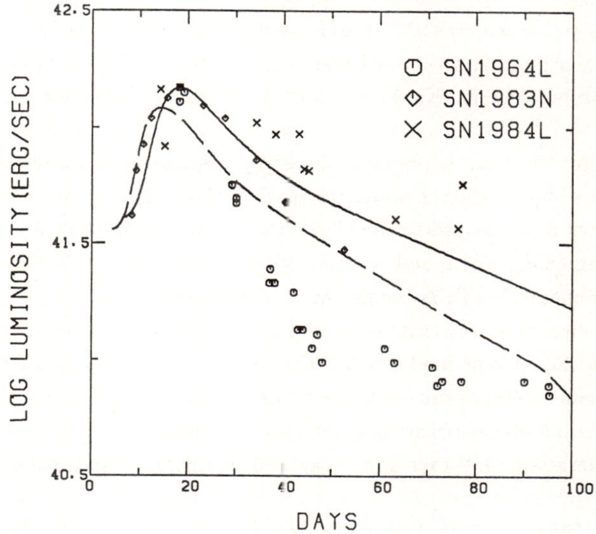

**Figure 12:** Same as FIg. 11 but for different explosion energy $E = 1 \times 10^{51}$ erg (solid) and 3 $\times 10^{51}$ erg (dashed). $^{56}$Ni is unifcrmly distributed in the ejecta of 2.2 $M_\odot$.

models of $M_{ej} \sim 2$ - $3$ $M_\odot$ even if $^{56}$Ni is confined in the central region. The origin of such a helium star is likely to be a mass exchange in the interacting binary.

2) The light curve of SN 1983N may indicate that $M_{ej}$ is as small as white dwarf mass and a large scale mixing in the ejecta occurs to bring $^{56}$Ni into fairly outer layers. Such a mixing of radioactive materials into outer layers might be an origin of a large non-LTE excitation of helium lines (Branch 1988).

3) If the width of the bolometric light curve of SN 1964L is even narrower than SN 1983N as is the case for the blue magnitude, SN 1964L could be essentially different from SN 1983N and 1984L and belongs to a separate class of SN I (i.e., SN Ic). The progenitor could be an accreting white dwarf having some composition inversion. It is interesting that the spectra of SN 1964L and 1962L are similar to SN 1983I and 1983V for which helium lines are weak (Wheeler *et al.* 1987). Also the expansion velocity at the photosphere of 1983V observed near maximum light is as high as 18,000 km s$^{-1}$ (Branch 1988).

To clarify whether a separate class of SN Ic exists, observations and modeling of nebula phase spectra of SN Ic would be crucial. Also the mechanism of extensive mass loss and mixing required from the helium star model need to be studied.

## 8. CONCLUDING REMARKS

Comparison between the presupernova evolutionary model, hydrodynamical models and the observations imposes several interesting constraints on $M_{env}$ and $E$, i.e., on the presupernova mass loss history, the explosion mechanism, and material mixing before and after the explosion. These are summarized as follows:

1) The optical light curve at very early phase and near maximum gives a lower bound and an upper bound of $E/M_{env}$, respectively. Reasonable agreement between the model and observations is obtained for $E/M_{env} = 1.5 \pm 0.5 \times 10^{50}$ erg/$M_\odot$ (see also Woosley 1988).

2) The near maximum light curve sets a lower bound of $M_{env}$ which is about 3 $M_\odot$. $M_{env} \sim 5$ - 10 $M_\odot$ would be more reasonable. Kirshner (1988) and Dopita (1988) have also estimated that $M_{env}$ is fairly large.

3) Another constraint on $M_{env}$ is obtained from the presupernova evolutionary model where the star evolves from the blue to the red when the mass of the hydrogen-rich envelope significantly decreases and comes back from the red to the blue if the helium abundance is sufficiently enhanced by mass loss and mixing (Saio, Kato, and Nomoto 1988). The observed N/C ratio, which is $\sim 40$ times the solar ratio (Panagia 1988), is consistent with $M_{env} \sim 7$ - 11 $M_\odot$.

Enhancement of helium and nitrogen may require mixing due to meridional circulation because the convection in the current model is too shallow. The observed enhancement of s-process elements, Sr, Ba (Williams 1988) might be related to this mixing.

4) Mixing of $^{56}$Co into hydrogen-rich envelope may be required to account for the optical light curve and the early emergence of X-rays and $\gamma$-rays. Mechanism of mixing needs to be explored.

We would like to thank Drs. R. Catchpole, J. Danziger, M. Dopita, R. Kirshner, and R. Williams for sending us data taken at SAAO, ESO, MSSSO, IUE, and CTIO, respectively, prior to publication. A part of this work is based on the collaboration with Drs. H. Saio, M. Kato, M. Itoh, and Ms. S. Kumagai. We are grateful to them for their contribution. This work has been supported in part by the Grant-in-Aid for Scientific Research (62540183) of the Ministry of Education, Science, and Culture in Japan and by the Space Data Analysis Center, Institute of Space and Astronautical Sciences.

## References

Arnett, W.D.: 1987, *Astrophys. J.*, **319**, 136.

Barkat, Z., and Wheeler, J.C. 1988, *Astrophys. J.*, in press.

Bertola, F., Mammano, A., Perinotto, M.: 1965, Contr. Asiago Obs., No. 174.

Blanco, V.M., *et al.*: 1987, *Astrophys. J.*, **320**, 589

Branch, D.: 1988, in *IAU Colloquium 108, Atmospheric Diagnostics of Stellar Evolution*, ed.
K. Nomoto, Springer-Verlag, p. 281.

Branch, D., and Nomoto, K.: 1986, *Astron. Astrophys.*, **164**, L13.

Brunish, W.M. and Truran, J.W. 1982, *Astrophys. J. Suppl.*, **49**, 447.

Catchpole, R. *et al.*: 1987, *Monthly Notice Roy. Astron. Soc.*, **229**, 15p.

Chalabaev, A., Perrier, C., and Mariotti, J.M. 1987, *IAU Circular 4481.*

Chan, K.W., Lingenfelter, R.E.:1987, *Astrophys. J. Letters*, **318**, L51.

Cook, W.R., Palmer, D., Prince, T., Schindler, C., Starr, C., Stone, E.: 1988, *IAU Circular 4527.*

Cristiani, S., *et al.*: 1987, *Astron. Astrophys.*, **177**, L5

Dopita, M.: 1988, *Nature*, **331**, 506.

Dotani, T. *et al.*: 1987, *Nature*, **330**, 230.

Ebisuzaki, T., Shibazaki, N.: 1988 , *Astrophys. J. Letters*, **327**

Ensman, L., Woosley, S.E.: 1988, preprint.

Hamuy, M., Suntzeff, N.B., Gonzalez, R., Martin, G.: 1987, *Astron. J.*, **95**, 63.

Harkness, R.P. *et al.*: 1987, *Astrophys. J.*, **317**, 355.

Hashimoto, M., Nomoto, K., Shigeyama, T.: 1987, in preparation.

Hillebrandt, W., Hoflich, P., Truran, J.W., Weiss, A.: 1987, *Nature*, **327**, 597.

Hoflich, P.: 1988, in *IAU Colloquium 108, Atmospheric Diagnostics of Stellar Evolution*, ed.
K. Nomoto, Springer-Verlag, p. 288.

Humphreys, R.M., Davidson, K.: 1978, *Astrophys. J.*, **232**, 409.

Itoh, M., Kumagai, S., Shigeyama, T., Nomoto, K., Nishimura, J.: 1987, *Nature*, **330**, 233.

Itoh, M., Kumagai, S., Shigeyama, T., Nomoto, K., and Nishimura,J.: 1988, in *IAU Colloquium
108, Atmospheric Diagnostics of Stellar Evolution*, ed. K. Nomoto, Springer-Verlag, p. 446.

Kirshner, R.: 1988, in *IAU Colloquium 108, Atmospheric Diagnostics of Stellar Evolution*, ed.
K. Nomoto, Springer-Verlag, p. 252.

Kumagai, S., Itoh, M., Shigeyama, T., Nomoto, K., Nishimura, J.: 1988, in *Supernova 1987A in
the LMC*, ed. M. Kafatos, Cambridge University Press, in press.

Maeder, A.: 1987, in *SN 1987A*, ed. I.J. Danziger, ESO, Garching, p. 251.

Makino, F.: 1988, *IAU Circular 4532.*

Matz, S.M., Share, G.H., Leising, M.D., Chupp, E.L., Vestrand, W.T., Purcell, W.R., Strickman,
M.S., Reppin, C.: 1988, *Nature*, **331**, 416.

Menzies, J.M., *et al.*: 1987, *Monthly Notice Roy. Astron. Soc.*, **227**, 39

McNaught, R.H.: 1987, IAU Circ. No. 4389.

Nomoto, K., Hashimoto, M.: 1986, *Prog. Part. Nucl. Phys.* **17**, 267.

Nomoto, K., Hashimoto, M.: 1987, in *Chemical Evolution of Galaxies with Active Star Formation*, ed. K. Takakubo., Universal Acad. Press. p. 93.

Nomoto, K., Hashimoto, M.: 1988, in *Theory of Supernovae*, ed. G.E. Brown, *Physics Report*, in press.

Nomoto, K., Shigeyama, T.: 1988, in *Supernova 1987A in the LMC*, ed. M. Kafatos, Cambridge University Press, in press.

Nomoto, K., Shigeyama, T., Hashimoto, M.: 1987, in *SN 1987A*, ed. I.J. Danziger, ESO: Garching, p. 325.

Panagia, N.: 1987, in *High Energy Phenomena Around Collapsed Stars*, ed. F. Pacini, D. Reidel, p. 33.

Panagia, N. *et al.*: 1987, *IAU Circular* 4514.

Phillips, M.M.: 1988, in *Supernova 1987A in the LMC*, ed. M. Kafatos, Cambridge University Press, in press.

Pinto, P., Woosley, S.E.: 1988, *Astrophys. J.*, **329**, in press.

Saio, H., Kato, M., Nomoto, K.: 1988, *Astrophys. J.*, **331**, in press.

Sandie, W.G., Nakano, G.H., Chase, L.F., Jr., Fishman, G.J., Meegan, C.A., Wilson, R.B., Paciesas, W., Lashe, G.P.: 1988, preprint.

Schaeffer, R., Casse, M., Cahen, S.: 1987, *Astrophys. J.*, **316**, L31.

Shigeyama, T., Nomoto, K., Hashimoto, M.: 1988, *Astron. Astrophys.*, in press.

Shigeyama, T., Nomoto, K., Hashimoto, M., Sugimoto, D.: 1987, *Nature*, **328**, 320.

Sunyaev, R. *et al.*: 1987, *Nature*, **330**, 227.

Sutherland, P., Xu, Y., McCray, R., Ross, R.P.: 1988, in *IAU Colloquium 108, Atmospheric Diagnostics of Stellar Evolution*, ed. K. Nomoto, Springer-Verlag, p. 394.

Sutherland, P., Wheeler, J.C.: 1984, *Astrophys. J.*, **280**, 282.

Thielemann, F.-K., Nomoto, K., and Yokoi, K.: 1986, *Astron. Astrophys.*, **158**, 17.

Totsuka, Y.: 1988, in *IAU Colloquium 108, Atmospheric Diagnostics of Stellar Evolution*, ed. K. Nomoto, Springer-Verlag, p. 335.

Wheeler, J.C., Harkness, R.P., Barker, E.S., Cochran, A.L., Wills, D.: 1987, *Astrophys. J. Letters*, **313**, L69.

Wheeler, J.C., Harkness, R.P., Barkat, Z.: 1988, in *Supernova 1987A in the LMC*, ed. M. Kafatos, Cambridge University Press, in press.

Williams, R.: 1988, in *IAU Colloquium 108, Atmospheric Diagnostics of Stellar Evolution*, ed. K. Nomoto, Springer-Verlag, p. 274.

Wilson, R.B., Fishman, G., Meegan, C., Paciesas, W., Sandie, W., Chase, L., Nakano, G.: 1988, *Nature*, submitted.

Wilson, J.R.: 1988, in *IAU Colloquium 108, Atmospheric Diagnostics of Stellar Evolution*, ed. K. Nomoto, Springer-Verlag, p. 348.

Wood, P.R., Faulkner, D.J.: 1988, in *IAU Colloquium 108, Atmospheric Diagnostics of Stellar Evolution*, ed. K. Nomoto, Springer-Verlag, p. 410.

Woosley, S.E.: 1988, *Astrophys. J.*, in press.

Woosley, S.E., Pinto, P., Ensman, L.: 1988, *Astrophys. J.*, **324**, 466.

Zoltowski, F.: 1987, *IAU Circular* 4389.

Observation of a Neutrino Burst from  the Supernova SN1987a

The  KAMIOKANDE-II Collaboration

presented by Yoji Totsuka
ICEPP, Faculty of Science
University of Tokyo

A neutrino burst was observed in the KAMIOKANDE-II detector on 23 February, 7:35:35 UT ($\pm 1$ minute) during a time interval of 13 seconds. The signal consisted of 11 electron events of energy 7.5 to 36 MeV, of which the first 2 point back to the Large Magellanic Cloud (LMC) with angles $18° \pm 18°$ and $15° \pm 27°$.

A new analysis of the special low threshold trigger counts strongly indicates that KAMIOKANDE did not observe any signals at 2:52:36 which is the time claimed by the Mont Blanc group.

1. Introduction

Following the optical sighting on 24 February 1987 of the supernova [1], now called SN1987a, a search was made of the data taken in the detector KAMIOKANDE-II during the period from 1609h, 21 February 1987 to 0731h, 24 February 1987.  The results of that search has been published elsewhere. [2]

The KAMIOKANDE detector has been continuously upgraded primarily for detection of low energy neutrino interactions, especially for solar neutrinos. The low energy background which comes mainly from radioactive elements $^{214}$Bi in water has been reduced by a factor of almost 10000 since 1985. The remaining background due to unstable elements fragmented from $^{16}$O by cosmic ray muons has been systematically studied and a program to remove those background was ready. Then came the information of SN1987a. It was thus quite easy to find a neutrino burst, since necessary programs were all available, and the background rate was quite low. A burst of low energy electron events was clearly seen on the print-out already in the morning of February 28.

## 2. Detector

The KAMIOKANDE-II detector, directed primarily at nucleon decay and solar $^8$B neutrino detection, has been operating since the beginning of 1986. It is described in detail elsewhere, [3] but its salient features are shown schematically in Fig.1. The inner detector fiducial volume containing 2140 tons of water [4] is viewed by an array of 20" diameter photomultiplier tubes (PMT), on a 1m x 1m lattice on the surface. The photocathode coverage amounts to 20% of the total inner surface. The attenuation length of the water for Cerenkov light is in excess of 45m. The inner detector is completely surrounded by a water Cerenkov

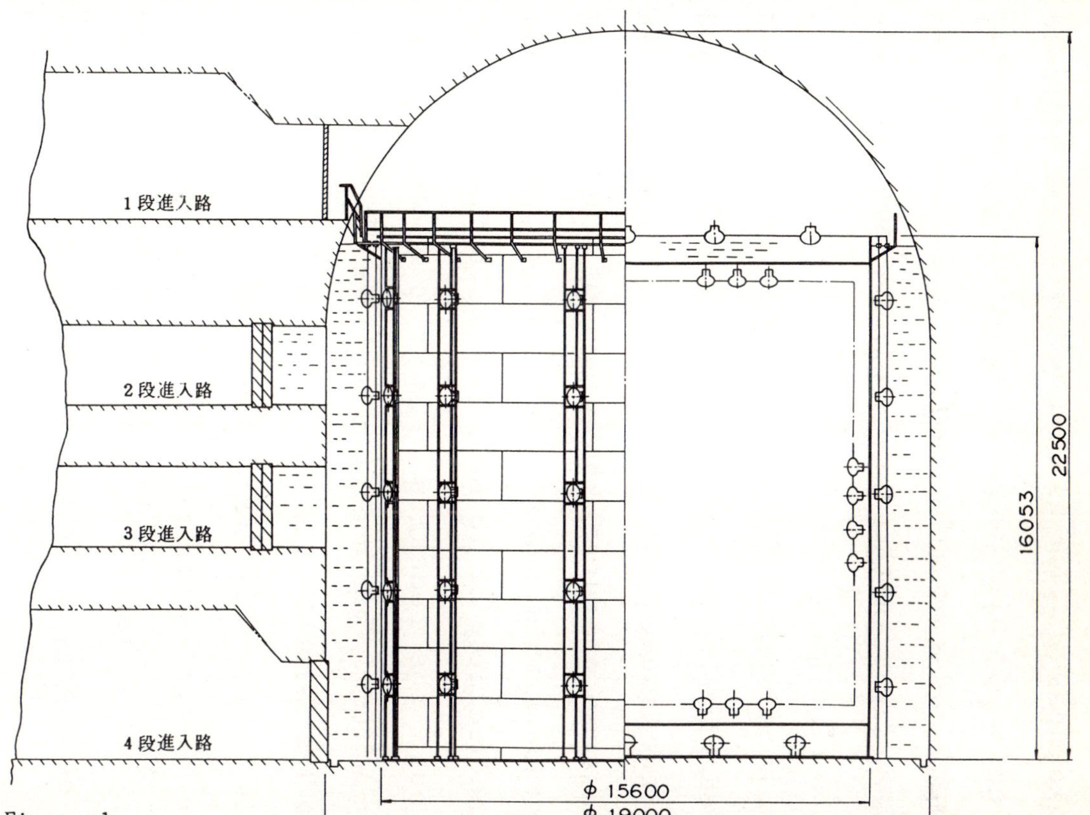


Figure 1.

Schematic view of the KAMIOKANDE-II detector. The inner detector contains 3000 tons of water of which 2140 tons are fiducial volume. It is viewed by 948 20" diameter PMTs mounted on a 1 meter grid on the inner surface. The outer(veto) counter surrounds the inner detector and is viewed by 123 PMTs. Dimensions in the figure are in millimeters.

counter of thickness $\geq 1.4$ m to ensure the containment of low energy events. It also is an absorber of $\gamma$ rays from surrounding rock and a monitor of slow muons which do not produce Cerenkov light in the inner detector but may decay there.

Neutrinos of different flavors are detected through the scattering reaction $\nu e \rightarrow \nu e$. The kinematics of this reaction and the subsequent multiple scattering of the recoiling electron preserve knowledge of the incident neutrino direction within approximately $28°$ rms at electron energies in the vicinity of 10 MeV. In addition, $\bar{\nu}_e$ are detected through the reaction $\bar{\nu}_e p \rightarrow e^+ n$ on the free protons in the water. This reaction produces $e^+$ essentially isotropically. The Cerenkov light of a 10 MeV electron gives on average 26.3 hit PMT's (NHIT) at 1/3 photoelectron threshold. The energy calibration is obtained by observing $\mu \rightarrow e$ decays and by using the Compton scattered electrons from $\gamma$ rays of energy up to 9 MeV from (n+Ni) using a Cf neutron source.

The detector is triggered by 20 PMT discriminators firing within 100 ns. Charge and time information for each channel above threshold is recorded. The trigger accepts 8.5 MeV electrons with 50% efficiency, and 14 MeV electrons with 90% efficiency over the volume of the detector [4]. The raw trigger rate is 0.60Hz of which 0.37Hz is cosmic ray muons.   The remaining 0.23 Hz is largely due to radioactive contamination in the water.

The electronics consists of ADC and TDC for each PMT. Every TDC and ADC has fast memories of depth 4.  Therefore the electronics dead time is essentially determined by the gate width of the circuits, namely 400 ns.  The data are stored on magnetic tapes and sent to the University of Tokyo, where they are analysed with the computer M380 stationed at the International Center for Elementary Particle Physics, ICEPP, Facutly of Science.

Reconstruction of the vertex of low energy events is performed with an algorithm based on the time and position of hit PMTs.  After the vertex is established, a separate fit is used to obtain the angle of the electron.  The distribution of the events presented here is consistent with an uniform volume distribution.

3. Search of a Neutrino Burst

The search for a neutrino burst from SN1987a was carried out  on the data of Run #1892, which continuously covered  the period from 16:09, 21 February  to 07:31, 24 February in Japanese Standard Time, which is UT plus 9 hours. Events satisfying the following four criteria were selected: 1) the total number of photoelectrons per event in the inner detector had to be less than 170, corresponding to a 50 MeV electron; 2) the total number of photoelectrons in the outer detector had to be less than 30, ensuring event containment; and 3) the time interval from the preceding event had to be longer than 20 µ sec, to exclude electrons from muon decay. 4) Events must not be correlated spacially and timely  with preceding muons to exclude a burst of events in a short time interval that were the production of an energetic nuclear cascade by an incident cosmic ray muon.

The short time correlation of these low energy contained events was investigated and the event sequence as shown in Fig.2 was observed at 16:35:35JST(7:35:35UT) of 23 February. The event sequence during 0 to 15 sec is shown expanded in Fig.3. The properties of the  events in the burst ( numbered 1 to 12 in Fig.3) are summarized in Table I. Event number 6 has NHIT<20 and has been excluded from the signal analysis. A scatter plot of event energy vs cosine of the angle between the measured electron direction and the direction of the Large Magellanic Cloud(LMC), known to contain SN1987a, is shown in Fig.4.  It is seen that the earliest two events point back to LMC with angles $(18 \pm 18)^{\circ}$ and $(15 \pm 27)^{\circ}$. The angular distribution of the remainder of the events is consistent with isotropy.  A search was also made  on a larger data sample of 42.9 days, 9 January 1987 – 25 February 1987, and  no other burst candidates were found. Figure 5 shows that from the extended period  the  number of events per 10 second time interval is well described by a

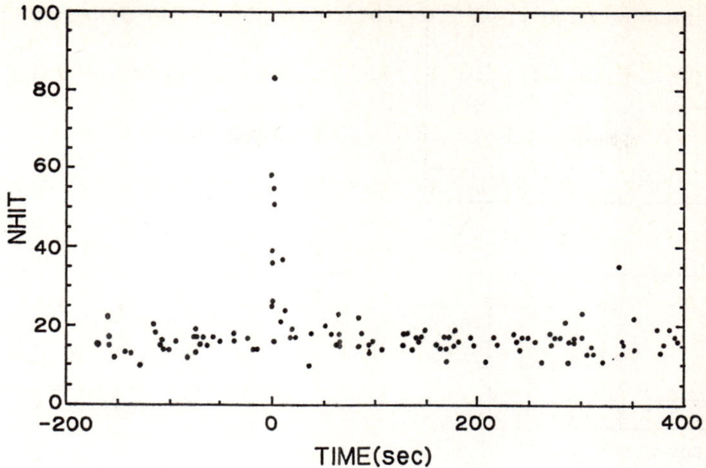

Figure 2.

The time sequence of events in a 600 second interval centered on UT 07:35:35, 23 February 1987.  Dots represent low energy electron events in units of the number of hit PMT,NHIT.

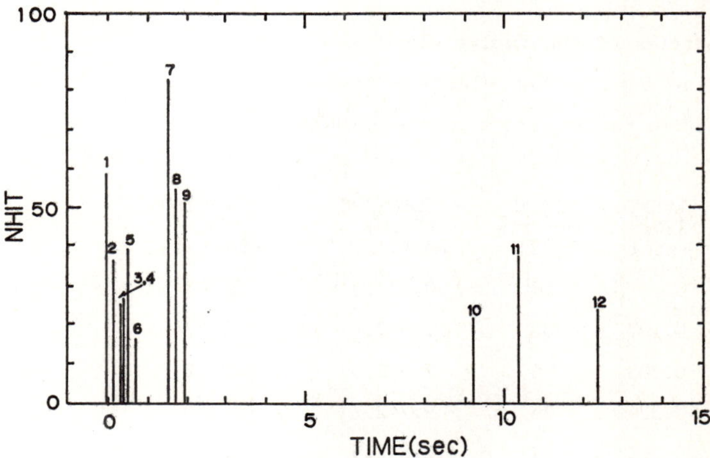

Figure 3.

The time sequence of events during 0 to 15 sec.  Solid lines represent low energy electron events in units of NHIT.

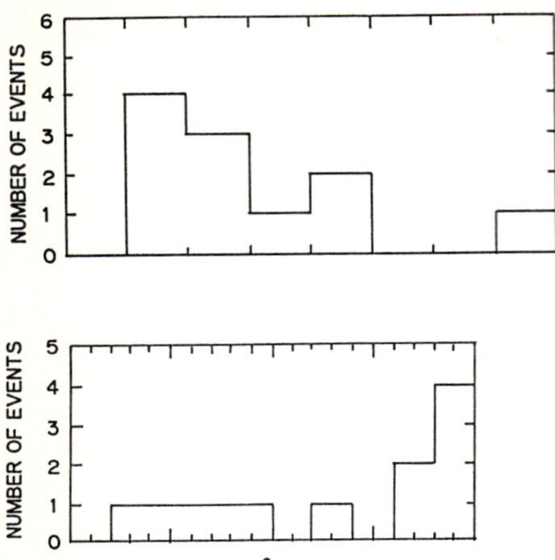
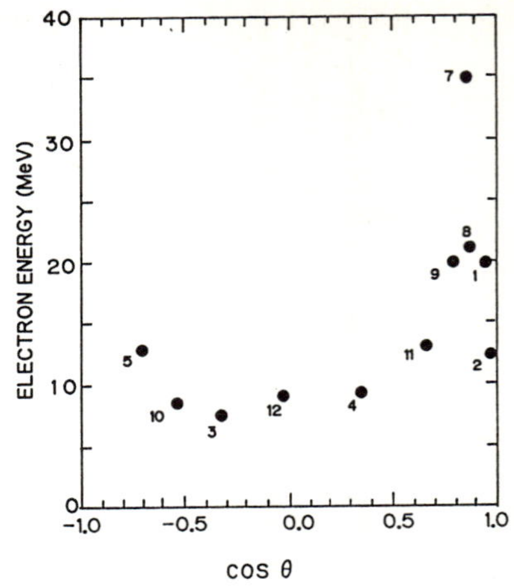


Figure 4.

Scatter plot of the detected electron energy in MeV and the cosine of the angle between the measured electron direction and the direction of the Large Magellanic Cloud.  The number to the left of each entry is the time sequential event number from Table I.  The two projections of the scatter plot are also displayed.

TABLE I.

Measured properties of the twelve electron events detected in the Neutrino burst.  The electron angle in the last column is relative to the directrion of SN1987a.

| EVENT NUMBER | EVENT TIME (sec) | NUMBER OF PMTs (NHIT) | ELECTRON ENERGY (MeV) | ELECTRON ANGLE (Degrees) |
|---|---|---|---|---|
| 1 | 0 | 58 | $20.0 \pm 2.9$ | $18 \pm 18$ |
| 2 | 0.107 | 36 | $13.5 \pm 3.2$ | $15 \pm 27$ |
| 3 | 0.303 | 25 | $7.5 \pm 2.0$ | $108 \pm 32$ |
| 4 | 0.324 | 26 | $9.2 \pm 2.7$ | $70 \pm 30$ |
| 5 | 0.507 | 39 | $12.8 \pm 2.9$ | $135 \pm 23$ |
| 6 | 0.686 | 16 | $6.3 \pm 1.7$ | $68 \pm 77$ |
| 7 | 1.541 | 83 | $35.4 \pm 8.0$ | $32 \pm 16$ |
| 8 | 1.728 | 54 | $21.0 \pm 4.2$ | $30 \pm 18$ |
| 9 | 1.915 | 51 | $19.8 \pm 3.2$ | $38 \pm 22$ |
| 10 | 9.219 | 21 | $8.6 \pm 2.7$ | $122 \pm 30$ |
| 11 | 10.433 | 37 | $13.0 \pm 2.6$ | $49 \pm 26$ |
| 12 | 12.439 | 24 | $8.9 \pm 1.9$ | $91 \pm 39$ |

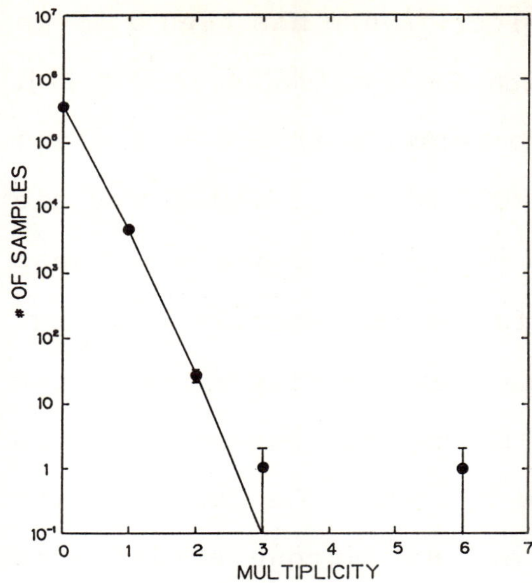

Figure 5.

Distribution of number of event with NHIT $\geq$ 30 per 10 sec
time interval for a data sample of 42.9 days. 9 January 1987
- 25 February 1987.   A burst at multiplicity 6 is the
observed neutrino burst.   The line corresponds to a Poisson
distribution with mean 0.012.

Poisson distribution of mean $\bar{n}$ =0.0121 for events with NHIT
$\geq$30.  Hence the rate of occurrence of 6 events in a 10
second time interval due to a statistical fluctuation is
less than one  in $7 \times 10^7$ years in our experiment.

4.  2:52:36 UT

M. Aglietta et al. (the Mont Blanc group) have reported
the observation of five events in  7 seconds beginning from
2:52:36 UT. [6]  This time is about 3:40 earlier than our
time.  Therefore if this observation is indeed true, the

supernova must have banged twice. There has been a large
debate about its consistency with the KAMIOKANDE result [7].
It is true that we observed two events near the Mont Blanc
time, namely one event with NHIT=30 at 2:52:40 UT and the
other with NHIT=20 at  2:52:47 UT.  Some people take these
two events very seriously and further arbitrarily shift our
energy normalization, concluding that the rate of these two
events is consistent with Mont Blanc.  We, however, wish to
stress that occurrence of these two events is just expected
from background.  Thus they are not strong evidence to
support Mont Blanc.

We present here the other argument against the
consistency between Mont Blanc and KAMIOKANDE.  The
KAMIOKANDE trigger system possesses a special low level
discriminator output, the rate of which is counted with an
online scaler.  The low level discriminator has a threshold
of approximately 6.4 MeV at  50% efficiency.  The scaler
count is read out at every event occurrence.  The result is
shown in Fig.6.  The count rates between 2:52 UT and 2:54 UT
are compared with the average rate of 9.96 Hz.  They are all
consistent with background.  Since the threshold level is
similar with Mont Blanc, one can easily calculate the event
rate inferred from the Mont Blanc result:

$$( 5 \pm \sqrt{5}) \times 2140/90/1.3/10 = 9.2 \pm 4.1 \text{ Hz},$$

where 5 is the number of events observed by the Mont Blanc

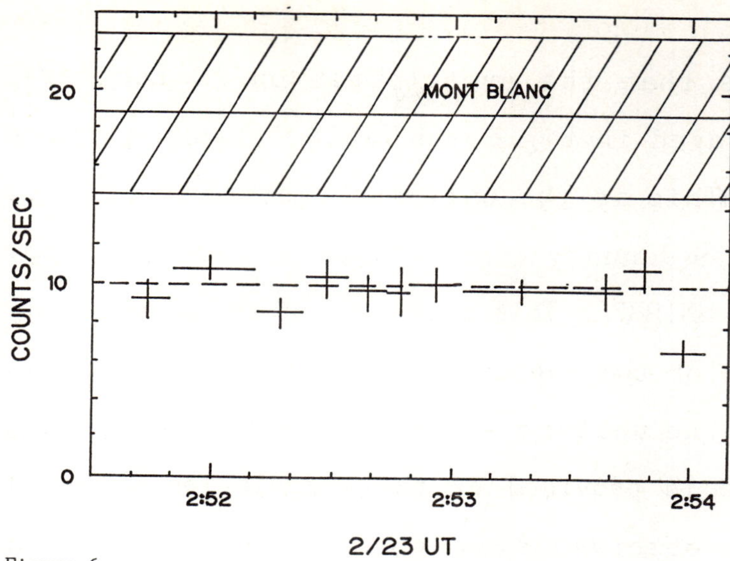


Figure 6.

Scaler count rates with 6.4 MeV threshold as a function of
time.  Hatched region is the count rate expected from the
Mont Blanc burst ($5 \pm 5^{1/2}$ events, burst period = 10 sec).
Our data are well below the burst rate, thus failed to
confirm the Mont Blanc result.  The dashed line is the
expected background rate.


group;  2140 and 90 are the fiducial volumes for KAMIOKANDE
and Mont Blanc, respectively; 1.3 is the relative number of
protons in unit mass of the two detectors ($H_2O$ and $H_2C$); a
burst period of 10 sec is assumed.  If the Mont Blanc
observation is indeed true, our scaler must have counted
this extra rate above background.  As shown in Fig.6, this
is clearly not consistent with our result.  Unless the
threshold energy of Mont Blanc is much lower than 6.4 MeV,
the KAMIOKANDE data does not confirm the Mont Blanc result.

345

## 5. Conclusion

We conclude that the event burst on 23 February, 7:35:35 UT, displayed in Fig.2 and Table 1, is a genuine neutrino burst. This is the only such burst found by us during the period 9 January to 25 February. We therefore associate it with SN1987a. This association is supported by the time structure of the events in the burst, their energy distribution and the uniform volume distribution. Strong additional support is provided by the correlation in angle of the first two observed events with the direction to SN1987a. The event burst occurred roughly 18 hr prior to the first optical sighting. [1] Correcting for energy dependent detection efficiency and assuming 9 of the 12 events are due to $\bar{\nu}_e p \rightarrow e^+ n$ we obtain an integral flux of $1.0 \times 10^{10}$ cm$^{-2}$ for the burst for $\bar{\nu}_e$ energy above 8.8 MeV. This in turn leads to the $\bar{\nu}_e$ output of SN1987a of $8 \times 10^{52}$ erg. for an assumed average energy of 15 MeV. Our data between 2:52 UT and 2:54 UT did not confirm the Mont Blanc result, unless the Mont Blanc threshold energy is much lower than 6.4 MeV.

## References


[1] IAU Circular No.4316.

[2] K. Hirata et al., Phys. Rev. Lett. 58 (1987) 1490 .

[3] E.W. Beier, Proceedings of 7th WOGU/ICOBAN'86, April, 1986. Toyama, Japan.

[4] This is for the entire volume inside  the PMT array. The detection efficiency for a fiducial volume of 780 ton,    2m inside  the PMT array is 90% at 10 MeV and 50%    at 7.6 MeV.

[5] A. Suzuki, Proceedings of the 12th Int. Conf. on Neutrino Physics and Astrophysics, page 306, June, 1986, Sendai,  Japan.

[6] M. Aglietta et al., Europhys. Lett. 3(1987) 1315.

[7] A. de Rujula, Phys. Lett. 193(1987) 514.

SN1987a: CALCULATIONS VERSUS OBSERVATIONS

James R. Wilson
Lawrence Livermore National Laboratory
Livermore, California

In this report the results of old calculations (Mayle 1985; Woosley, Wilson, Mayle 1986; Mayle, Wilson, Schramm 1987) of collapse driven explosions and new calculations of the Kelvin-Helmholtz proto-neutron star cooling will be compared with the neutrino observations of supernova 1987a.  The calculations are performed by a modern version of the computer model of Bowers and Wilson 1982.  (See Mayle 1985 for more recent improvements).

First we give the results of the old calculations.  In the collapse, bounce and cooling of the central iron core of a massive star, about 0.1% of the binding energy of the eventual neutron star is emitted in a short deleptonization burst as the bounce shock passes through the photosphere; 5% is emitted in the total deleptonization process; and 95% is released as thermal emission in all neutrino species.  In a survey of a wide range of stellar masses, stars in the range 20 to 30 $M_\odot$ are found to have the most energetic antineutrino spectra ($<\epsilon_{\bar{\nu}_e}> \sim 15$ MeV).  In calculations where black holes were formed (Woosley, Wilson, Mayle 1986 and Wilson 1971) very little neutrino emission was found associated with black hole formation.  The neutrinos associated with BH formation also have low energies.  The time history of the neutrino pulse is sensitive to the explosion mechanism.  If the mechanism is a prompt exiting through the star of the bounce shock wave, the pulse has a high peak as the shock wave passes near the photosphere.  It falls rapidly for the next first few tenths of a second and then declines slowly over several seconds to effectively zero.  If no prompt explosion occurs then the shock becomes an accretion shock and matter continues to fall onto the proto-neutron star keeping up the luminosity.  After the late time mechanism ejects the envelope the luminosity drops several fold to the luminosity associated with the bare proto-neutron star.  During the accretion phase the average antineutrino energy is about 10 MeV while during the cooling phase the energy rises to about 15 MeV.

We have made three new calculations following the cooling of proto-neutron stars until the luminosity fell below an observable level.  In the first model a soft equation of state (EOS) was used (gravitational critical mass 1.50 $M_\odot$). The proto-neutron star was selected by taking a post bounce calculation of the core of a 25 $M_\odot$ star and removing all the mass but for the inner 1.64 $M_\odot$.  The second model was made with a stiffer EOS using the same core as the first model.  The third model was made by following the collapse of a 15 $M_\odot$ model of Woosley that had a 1.27 $M_\odot$ iron core. This model gave a prompt explosion with an energy of $3\times10^{50}$ ergs. In Figure 1 we see the energy emitted in antineutrinos versus time for these three models. We note that only the high mass soft EOS model can sustain an appreciable luminosity for the

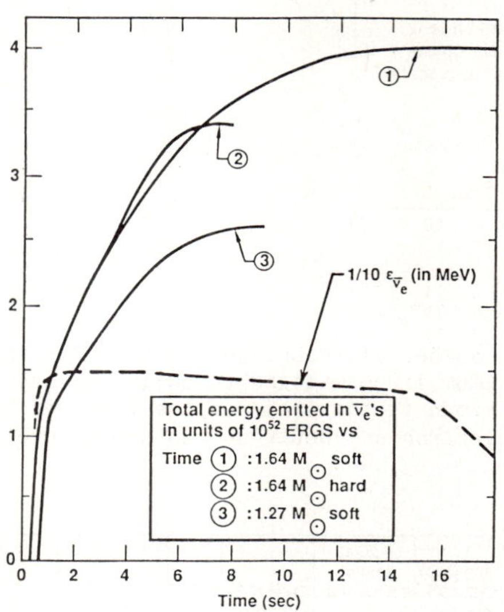


Figure 1. Cooling curves for the three models studied. Also shown is the average antineutrino energy for the 1.64 $M_\odot$ core evolved with the soft equation of state.

12 seconds needed to explain the observations (Hirato et al. 1987 and Bionta et al. 1987). The antineutrino energy peaks a few tenths of a second after bounce and only falls slightly until the neutron star has almost stopped emitting neutrinos. At the photosphere the antineutrino energy is always rising except near the very end. The red shift at the photosphere is 1.30 for the 1.64 $M_\odot$ soft EOS model at the end of the calculation. The mean energy of the antineutrinos is considerably greater than 3.15 times the matter temperature at the photosphere at late times because the antineutrinos exchange energy with matter principly through electron scattering since the density of protons near the photo is quite low. In Figures 2a, b the time integrated number spectrum for the 1.64 $M_\odot$ soft EOS model is given along with the spectra multiplied by the detector responses, Q, and the capture cross section for antineutrinos, $\sigma$. We see that our expected energy for Kamiokande is somewhat high and for IMB it is somewhat low. Statistical studies by the Kolmogorov-Smirnov method show that our spectra can account for both the Kamiokande and IMB results with good confidence.

Table 1 gives a summary of results of the three model calculation and Table 2 gives the expected observational results from our models. We estimate the neutron

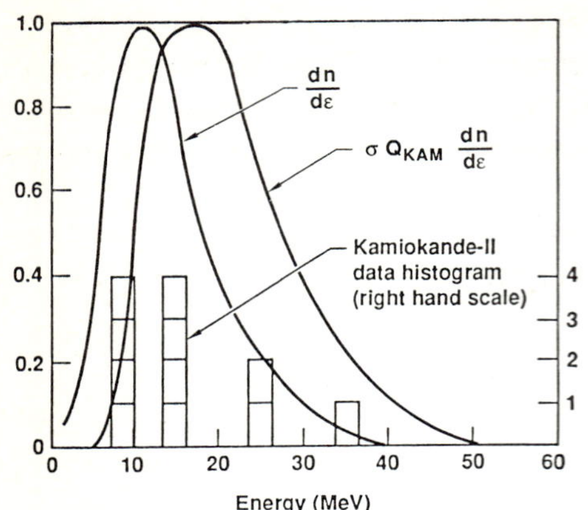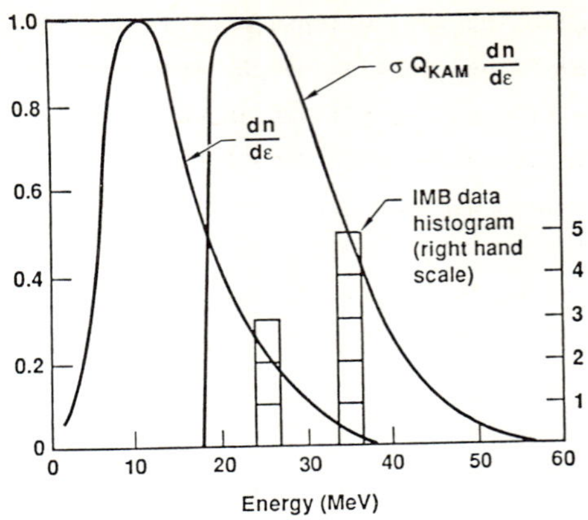


Figure 2a, b.  Antineutrino number distribution function for the 1.64 $M_\odot$ core evolved with the soft EOS.  The maximum is normalized to unity, as only the shape of the curve is being considered.  Also shown is the shape of the expected positron number spectrum that would be produced by electron antineutrino capture on protons taking into account detector characterestics.

| $M/M_\odot$ | EOS | $\rho_c$ $(10^{15}$ g/cc) | r (km) | Time (sec) $L \geq 5 \times 10^{51}$ erg/sec | B.E. at end of calculation $(10^{53}$ erg) | Total B.E. $(10^{53}$ erg) | $\varepsilon_{\nu_e}$ (MeV) | $\varepsilon_{\bar{\nu}_e}$ (MeV) | $\varepsilon_\mu$ (MeV) | % B.E. in $\nu_e$'s | % B.E. in $\bar{\nu}_e$'s | % B.E. in $\nu_{\mu,\tau}$'s |
|---|---|---|---|---|---|---|---|---|---|---|---|---|
| 1.27 | Soft | 1.3 | 10.7 | 7.7 | 1.65 | 2.3 | 10.5 | 13.9 | 23.5 | 15 | 17 | 68 |
| 1.64 | Hard | 0.74 | 20.0 | 6.5 | 2.17 | 2.7 | 7.2 | 12.0 | 18.4 | 15 | 16 | 69 |
| 1.64 | Soft | 1.5 | 9.8 | 13.2 | 3.08 | 4.1 | 10.5 | 14.0 | 23.3 | 12 | 14 | 74 |

TABLE 1

| $M/M_\odot$ | EOS | # $\bar{\nu}_e$ Absorb. Kam. | # $\bar{\nu}_e$ Absorb. IMB | # $e^-$ Scatt. Kam. | # $e^-$ Scatt. IMB | $\varepsilon_{e^+}$ Kam. | $\varepsilon_{e^+}$ IMB | # $\nu_e$ Absorb. $C_2Cl_4$ |
|---|---|---|---|---|---|---|---|---|
| 1.27 | Soft | 9.5 | 5.7 | 1.3 | 0.45 | 20.3 | 28.7 | .10 |
| 1.64 | Hard | 10.4 | 3.9 | 1.5 | 0.49 | 17.4 | 26.0 | .09 |
| 1.64 | Soft | 14.8 | 8.9 | 2.1 | 0.84 | 20.2 | 27.9 | .16 |

TABLE 2

star binding energy by comparing the observed counts to the number predicted by our calculation.  We do this both for Kamiodande and IMB and average the two results.  We add 20% to this number because when the luminosity for the 1.64 $M_\odot$ soft EOS model had fallen to the point it would be very difficult to observe, it still contains 20% of the total binding energy appropriate to the EOS used.  We assumed a distance to the

LMC of 52 Kps.  We arrive at an estimate of the binding energy of $(3.0 \pm 1.0) \times 10^{53}$ ergs.  The uncertainty arises from Poisson statistics and distance uncertainties.  We may also estimate the binding energy by the core mass and EOS required to account for a 12 second signal.  Our 1.64 $M_\odot$ soft EOS has a binding energy of $4.0 \times 10^{53}$ ergs and was the only model to produce an antineutrino signal that lasted as long as 12 seconds.  The largest uncertainty in the calculation is the shape of the EOS function above nuclear density.  Because a soft EOS is indicated, a black hole remnant is still a possibility.

If we plot the Kamiokande and IMB data together as in Figure 3, we see that for the first two seconds a fairly high luminosity is followed for 10 seconds by a much lower luminosity.  We infer from this fact that there may have been a 2 second period of accretion, followed by the explosion and subsequent Kelvin-Helmholtz cooling of the proto-neutron star.

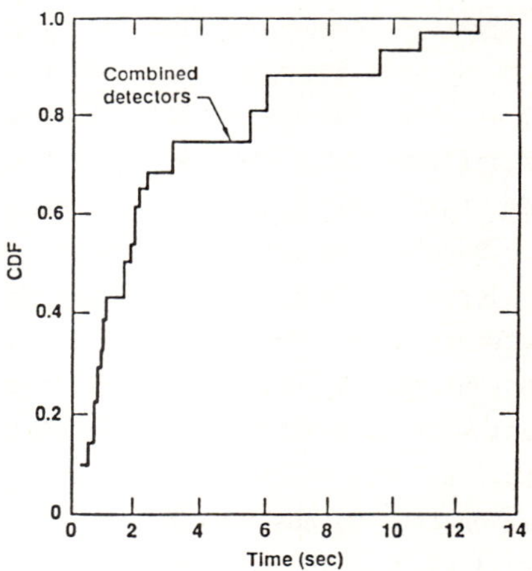


Figure 3. The stair step curve is the detected sample cumulative distribution function (CDF) for the combined detectors. We have weighted the two detectors equally so that the height of an IMB detection is 12/8 the height of a Kamiokande detection (note we have included the count at .686 seconds rejected by the Kamiokande group as being too close to their threshold).  If millions of counts had been seen, the CDF would be smooth and directly proportional to the number luminosity emitted by the supernova.

An estimate of the iron core mass before collapse is found as follows.  We need 1.6 to 1.7 $M_\odot$ to explain the 12 second signal and a few tenths of a solar mass for the accretion phase luminosity.  From a series of stellar evolution calculations of stars producing different iron core masses by Weaver and Woosley (private communication), we find that for models with iron core masses less than 1.5 $M_\odot$ the density exterior to the core falls so rapidly with radius that appreciable accretion could occur not in a few seconds.  For high mass iron cores the density doesn't fall off

rapidly with radius.  We would get too large a final mass since the initial core mass
would be augmented by a sizeable accretion mass.  We thus arrive at 1.50 $M_\theta$ as our
estimate for the initial iron core mass.

Convection will not drastically alter the results of our calculations.  Smarr et
al. (1981) found that with complete core overturn the luminosity increased at most by
30%.  Mayle (1985) in a series of calculations with mixing length theory found about
a 20% enhancement of the luminosity.  Our models are unstable by the LeDoux and salt
finger criteria but not the Schwartzschild criteria.  Future calculations will
include convection.

Our models produce luminosity and spectra that change smoothly with time, hence,
we can put no limits on the neutrino mass below the present laboratory limets.

Since our calculations agree fairly well with the standard Weinberg-Salam neutrino theory we can put some limits on competing neutrino theories.  We repeated the
cooling calculation of the 1.64 $M_\theta$ soft EOS model with additional lepton families.
We find the time after bounce at which the antineutrino luminosity had fallen to $10^{50}$
ergs/sec to be 11 seconds for 3 flavors, 7.4 seconds for 5 flavors, and 5.6 seconds
for 7 flavors.  Thus it appears that SN1987a will give an appreciable restriction on
the number of neutrino types.  Double beta decay experiments have indicated the possibility that neutrinos are majoran particles.  We have put the Gelmini-Roncadelli
neutrino model into our computer program.  The result of calculations is that the
majoran-neutrino coupling constant must be small or the antineutrino energies will
be so low as to be inconsistant with observations.  A calculation was done also with
the inclusion of axion cooling.  A limit on the axion coupling constant of $1 \times 10^{12}$
GeV was found in order that axion cooling not remove the antineutrino signal.  All
these above particle models will be done in the near future with much more care.  We
only want to point out at this time the possibilities to be derived from SN1987a.

In conclusion, we mention the following statistically weak observational oddi-
ties; the 7 sec time gap in Kamiokande data, the lack of late time events in IMB,
the occurrence of two possible electron scattering events very early in time, and
the discrepancy of the Kamiokande and IMB average neutrino energies.  The work on
particle models is being pursued with G. Fuller, R. Mayle, K. Olive, D. Schramm.

References


Bionta et al. 1987, "Physical Review Letters (Phys. Rev. Lett.)", <u>58</u>, 1494.

Bowers, R. L., and Wilson, J. R. 1982, "The Astrophysical Journal Supplement Series
   (Ap.J. Supp. Ser.)", <u>50</u>, 115.

Hirata et al. 1987, "Physical Review Letters (Phys. Rev. Lett.)", 58, 1490.

Mayle, R. W. 1985, Lawrence Livermore National Laboratory, preprint # UCRL-53713.

Mayle, R. W., Wilson, J. R., and Schramm, D. N. 1987, "The Astrophysical Journal (Ap.J.)", 318, 288.

Wilson, J. R. 1971, "The Astrophysical Journal (Ap.J.)", 163, 209.

Woosley, S. E., Wilson, J. R., and Mayle, R. W. 1986, "The Astrophysical Journal (Ap.J.)", 302, 10.


This work was performed under the auspices of U.S.D.O.E. through LLNL under contract number W-7405-ENG-48.

SUPERNOVA MECHANISMS: BEFORE AND AFTER SN1987a

Sidney H. Kahana
Physics Department
Brookhaven National Laboratory
Upton NY 11973
U S A

The impact of SN1987a on theoretical studies of the specific mechanism generating Type II supernovae is examined. The explosion energy extracted from analysis of the light curve for SN1987a is on the edge of distinguishing between a prompt explosion from a hydrodynamic shock and a delayed, neutrino-induced, explosion. The detection of neutrinos from 1987a is also reanalysed.

## Introduction

The observation of Supernova (Shelton) 1987a [1,2] on February 23 of this year in the Large Magellanic Cloud, has provided astrophysicists and neutrino-physicists with a magnificent opportunity to test out their equipment and theories. Some surprise was clearly created, especially for the presupernova evolvers who had not seriously expected blue giants to explode; on the other hand, the neutrino observations might be said to be just what one might expect for a gravitational collapse, Type II supernova (SNII). It is the collapse and explosion phase for such a supernova that concerns me here. I wish to report on new calculations [3] by my colleagues and myself, and to remind you of our earlier description [4] of the prompt mechanism for Type IIs. I will also present results [5] from the analysis of the neutrinos seen by Kamiokande (KII) [6] and by IMB [7].

There is a long history [8-11] of the difficulties inherent in producing a prompt explosion in an SNII progenitor, first thought to be red supergiants and now clearly also blue giants. It appeared early on that the analysis of the SN1987a light curve would help to distinguish between possible mechanisms. This may still prove to be correct. For the moment, however, both Nomoto and coworkers [12], as well as Woosley and coworkers [13], and others [14], have analysed the light curve and have arrived at explosion energies which may conceivably be generated by either a prompt or a delayed shock mechanism. The energies suggested, at this Symposium, from total luminosities, are 1.0 to 1.5 $\times 10^{51}$ ergs [12] and 0.5 to 1.5 $\times 10^{51}$ ergs [13], whose lower limits are perhaps too low to rule out delayed explosions. There remain problems, however. Nomoto and co-workers [12] finds an energy of as much as $2.5 \times 10^{51}$ ergs may be demanded by consistency with the colour temperature, while their description of the later time light curve analysis (from

---

This work has been supported by the U. S. Department of Energy under contract no. DE-AC02-76CH00016.

75 days to 125 days) is probably improved by increasing the ejected mass $(M_E)$ and thus also the explosion energy $(E_{expl})$. There is always a degree of ambiguity associated with $M_E$ and $E_{expl}$, for example the initial velocities and initial propagation time for the shock to reach the photosphere are given by [12]

$$t_{prop} \propto R_0 \left( \frac{M_{E'}}{E_{expl}} \right)^{1/2} . \tag{1}$$

Increasing the mass at the same time as mixing the radioactive sources throughout the star will both keep the peak in the light curve near the observed 90 day point after explosion, and broaden the curve. Both of the features would improve the fit in Ref. 12 to the observed light curve.

There are two presently extant mechanisms for producing explosions in the collapsing core of SNII progenitors. In a prompt explosion the hydrodynamic shock produced just after bounce is sufficiently energetic to expel the stellar mantle and envelope. In a so–called delayed explosion a stalled, accretion shock is resuscitated by neutrino emission from a hot interior. Wilson, Mayle and Bethe [15] have discussed the neutrino-induced mechanism and in general produce low explosion energies $\lesssim 1.0 \times 10^{21}$ ergs for progenitor main sequence masses initially below $25 M_\odot$. Baron, Cooperstein and Kahana were the first to generate successful prompt explosions for Woosley, Weaver [16,17] models and can obtain explosion energies up to 2.5 or $3.0 \times 10^{51}$ ergs. These energies include corrections from nuclear burning in the mantle and from the gravitational binding of the mantle and envelope.

Perhaps the most novel aspect of SN1987a is the detection [6,7] of neutrinos from the production and cooling of a compact remnant. One hopes this is only the beginning of a new field of astronomy. The analysis I present here [5], parallel to the analysis of many other authors [23-28], finds remnant binding energy $\sim 2.0 \pm 0.5 \times 10^{53}$ ergs and remnant mass 1.2 to 1.7 $M_\odot$ consistent with what one expects for neutron star generation. An upper limit of 10–15 eV may also be inferred for the electron neutrino mass.

## Elements in the Success of a Prompt Explosion

### A. Core characteristics

Theoretical modelling of the brief period from fuel exhaustion to collapse, bounce and shock formation in SNIIs is one of the more difficult problems in astrophysics. Early hydrodynamic simulations [8] at the beginning of this decade found that neutrino and dissociation losses stalled the shock inside the "iron" core, and that eventually accretion of mantle material would drive the core towards a black hole. An important factor in this demise of the hydrodynamic shock was the excessive core mass, $M_{core} \sim 1.50$ $M_\odot$ to 1.55 $M_\odot$, predicted in presupernova modelling of massive stars. The Chandrasahkar limit $M_{Ch} \sim 5.7 Y_e^2$ pretty well determines the core mass at collapse, but details of the presupernova calculations have led to an evolution of $M_{core}$ from the early Woosley, Weaver, Zimmerman [16] value 1.51 $M_\odot$ to later Woosley, Weaver [17] level of $M_{core} \sim 1.36$ $M_\odot$ for a star with the main sequence masses $M_{ms} \sim 12, 15$ $M_\odot$, and more recently to $M_{core} \approx 1.18$ $M_\odot$ for a Nomoto–Hashimoto [18] model with main sequence mass 13 $M_\odot$. Should initial core masses at collapse stay as low as those in these recent calculations, prompt explosions will be hard to avoid and may obtain for progenitors in mass up to perhaps 20 $M_\odot$, i.e., for helium cores in the neighborhood of 6 $M_\odot$.

The core characteristics do not change appreciably in these diverse models with central densities near $10^{10}$ g/cm$^3$, a central electron fraction $Y_e^c \approx 0.42$, and temperature $T_c \approx 0.5$ MeV. The central entropy per baryon rises and falls with the core mass, with high values inimical to healthy shocks.

B. Background to Calculations: Input Physics To understand the delicacy of the collapse simulation, one should recall the virial theorem for the non-relativistic core matter supported by a relativistic electron gas, for which

$$E_{gravitational} + E_{internal} \approx 0 \quad \text{for} \quad P \sim \rho^{4/3} \, . \tag{2}$$

The adiabatic collapse preserves this relation but drives the total gravitational energy to above $10^{53}$ ergs. Thus, the shock energy for a viable prompt explosion, $\sim 2 \times 10^{51}$ ergs, is small in comparison to the internal or gravitational energy. Simulation must be accurate both in computation and in the input physics.

A second important feature to keep in mind is the small radius of the homologously collapsing core at maximum density, $R_{core} \approx 15$ fm, in comparison to the Schwarzschild radius $R_s \approx 2$ km for the 0.7 or so solar masses inside this core. The sensitivity in physical effects determining the final shock energy extends also to the treatment of gravitation. General relativity seems necessary, both helps and hurts the prompt mechanism, but in the end is crucial to producing explosions.

The elements of physics input that are marked out for careful treatment, then, include:

1. $\beta$–capture, neutrino transport: effect on $Y_e$.

2. Hadronic equation of state at high density, $\rho = \rho_0$ to $4\rho_0$, $\rho_0 \approx 2.4 \times 10^{24}$ g/cm$^3$ the saturation density for the asymmetric nuclear matter found in the collapsing core. A moderate softening of nuclear matter under these conditions allows the core to collapse further into the gravitational well and thus increases the shock energy at formation.

3. General relativity magnifies this effect, but at the same time reduces the mass of the homologous core at bounce and forces the shock to unfavourably traverse more stellar material on its way out of the core. The equation of state used in our astrophysical simulations has the simple form [19]

$$P_N\left(\rho\right) = \frac{K_0\left[Z/A\right] \; \rho_0\left[Z/A\right]}{9\Gamma} \left[\left(\frac{\rho}{\rho_0}\right)^{\Gamma} - 1\right] \tag{3}$$

for the cold, hadronic pressure. The incompressibility $K_0$ (and density $\rho_0$) are, of course, functions of the charge to mass ratio $Z/A$, the latter reduced by $\beta$–capture from 0.42 to nearer 0.32 in the collapse environment. A good measure of the softening so helpful to the prompt mechanism then follows from the behaviour of the incompressibility in asymmetric matter $(N = 2Z)$ [19,4]

$$K_0\left(Z/A\right) = K_0^{symm} \left[1 - 2\left(\frac{Z}{A} - \frac{1}{2}\right)^2\right] \, . \tag{4}$$

The combination of a softer equation of state with relativistic gravitation is the key to the prompt mechanism.

4. Laboratory constraints on the hadronic equation of state do exist and must be respected.

a. The incompressibility at normal nuclear density $\rho_0$ is reasonably well extracted from the energy of the breathing mode in heavy nuclei to be [20]

$$K_0^{symm} = 210 \pm 30 \text{ MeV, at } Z/A = 1/2 \, . \tag{5}$$

b. In contrast, early analysis of relativistic heavy ion experiments at LBL [21] pointed to a considerably stiffer nuclear matter at high density. This is still an open question, but further investigations [22] suggest the neglect in the analysis of heavy ion collisions of the velocity (and hence density) dependence of nuclear forces as the culprit in the discrepancy between heavy ion collisions and breathing mode equations of state.

## Results of Prompt Mechanism Simulations

The parameters of explosions resulting from our hydrodynamic simulations have been extensively reported. Confirmation of these results are seen in the recent work of Bruenn [23], some of which is as yet unpublished. A key component of any such calculations is the $\nu$–transport scheme which in our calculations has only recently been carried out in complete detail to the level found in Bruenn [23] or Wilson and Mayle [15].

It suffices to quote some results (Table I) for the 12, 15 $M_\odot$ initial models of Woosley–Weaver and for the 13 $M_\odot$ model of Nomoto–Hashimoto. The much lower core masses found by the latter group for even 20 $M_\odot$ ($M_{He} \approx 6.0\ M_\odot$) progenitors presage viable explosions in these cases as well. A vital factor in our introduction of full transport is once again the strong role played by symmetry effects. A high symmetry energy suppresses free proton number (and hence $\beta$–capture) in the collapse environment and thus sustains a high $Y_e$ throughout the collapsing core. This in turn leads to shock formation at a favourably large radius.

From Table I one can conclude that within reasonable ranges for $K_0\ (Z/A)$ and the adiabatic index $\Gamma$, prompt explosion energies can describe SN1987a or perhaps even more energetic supernovae. A close examination of the work of Shigeyama, Nomoto and Hashimoto [12] leads me to suspect they may have underestimated the energy in 1987a.

## Neutrino Detection

The 'sightings' of simultaneous, bunches of neutrinos in the KII [6] and IMB [7] detectors some three or four hours before optical observations of SN1987a is surely as good a demonstration of the existence of gravitational collapse supernovae as we can desire. The very short time between neutrinos and optical visibility is a surprise, speaking to the small size and unusual nature of the progenitor. We have performed [5] one of the many parallel analyses of these neutrinos [23-28].

At the heart of our analysis are the equations

$$\frac{dE}{dt} = L = (0.011) \left(\frac{T}{MeV}\right)^4 \left(\frac{R}{10\ \text{km}}\right)^2 \times 10^{51}\ \text{ergs/sec.} \tag{6}$$

$$T = T_0\ e^{-t/\tau} \tag{7}$$

describing the compact remnant emission as black body from a sharp neutrinosphere and with Newtonian cooling of the remnant. The analysis must take account of the neutrino cross sections and detector efficiencies [5].

The observed mean energy more or less determines the initial cooling temperature $T_0$; the total number of neutrinos in the detectors then determine the binding energy $B$ and compact remnant mass $M_{ns}$. The chronology of detected events, divorced from their energy distribution, yields a cooling time $\tau$. We find an acceptable range for $T_0$ between 4.5 and 5.5 MeV and then extract $B = 2.0 \pm 0.5$ MeV and $M_{ns} = 1.2$ to 1.6 $M_\odot$. Further, there is a complete consistency in the numbers extracted separately from KII and IMB data within the stated range of initial temperature.

Finally, I turn to the question of limits on neutrino masses. One can obtain a good limit on $m_{\bar{\nu}_e}$ by examination of Table II, constructed by mapping the Kamiokande data back to its source at the supernova, after imposition of a finite mass. The KII data itself corresponds to $m_{\bar{\nu}_e} = 0$; I

Table I  Explosion Energies from the Prompt Mechanism, for Various Collapse Models taken from Refs. 3 and 4.  Equation of state parameters $K_0^{symm}, \Gamma$ are the equivalent incompressibility at saturation for symmetric matter $Z/A = 1/2$, and the adiabatic index, as given in equations (3) and (4).  All calculations take full account of general relativity except for model #38*, which is Newtonian.  The maximum central density reached in the calculation $\rho_{max}^c$ is in units of the saturation density appropriate to the asymmetric matter $Z = 1/3$ relevant to bounce, which is $\rho_0(1/3) = 2.4 \times 10^{14}$ g/cm$^3$.  The precollapse models for #48-45 are the main sequence $M = 12$, 15 M$_\odot$ models of Ref. 17, while #61-63 are from the 13 M$_\odot$ model of Ref. 18.  The explosion energy $E_{expl}$ was obtained from the estimated shock energy by correcting for oxygen burning in the mantle and gravitational binding of mantle and envelope. Models are further distinguishable by the symmetry energy $W_s$ which we believe is experimentally closer to the higher values in the table and by a trapping density which is set at $0.4 \times 10^{12}$ g/cm$^3$ in the first six models in the table, and at the more realistic $1 \times 10^{12}$ g/cm$^3$ for #62, 63.

| Model number | Mass | $K_0^{symm}$ | $\Gamma$ | $W_s$ | $\dfrac{\rho_{max}^c}{\rho_0(0.33)}$ | $E_{expl}$ |
|:---:|:---:|:---:|:---:|:---:|:---:|:---:|
| | M$_\odot$ | MeV | | meV | | $10^{51}$ ergs |
| 38* | 12 | 180 | 2 | 29.3 | 2.3 | 0.1 |
| 40 | 12 | 180 | 2 | 29.3 | 12.0 | 3.2 |
| 41 | 12 | 180 | 3 | 29.3 | 3.1 | 0.8 |
| 43 | 15 | 180 | 2.5 | 29.3 | 4.1 | 1.7 |
| 45 | 15 | 90 | 3 | 29.3 | 4.0 | 0.8 |
| 61 | 13 | 180 | 2.5 | 29.3 | 4.1 | 2.4 |
| 62 | 13 | 180 | 2.5 | 36.0 | 4.1 | 2.6 |
| 63 | 13 | 180 | 2.5 | 34.0 | 4.1 | 1.9 |

have also included the mapping for $m_{\bar{\nu}_e} = 10$ and $m_{\bar{\nu}_e} = 15$ eV. It is my opinion that the first eight KII events in 1.92 sec. constitute the 'real' cooling pulse. The width in time of this pulse starts at 1.92 sec. and is stretched to 5.4 sec. for a 10 eV mass, to 10.8 sec. for a 15 eV mass. Moreover, the spectrum itself becomes very hard for either of these masses, with neutrinos of low energies being emitted well before those of high energies as the neutron star cools. To select a mass of 30 eV and claim this is a more conservative [28] limit for the neutrino mass seems unreasonable; one is then dealing with almost unconnected events at the supernova. The energy versus time sequencing and pulse width become unacceptable near $m_{\bar{\nu}_e} = 15$ eV, especially since the source pulse width is then $\sim 10$ sec. somewhat larger than the cooling time $\tau/4 \lesssim 4$ sec. we find from the data itself.

The above limit is comparable to or slightly better than that obtainable from the presently best laboratory experiments [29]. How does one evaluate this estimate statistically? The weakest link in the argument is I think the dependence of the mapped pulse width on the time of arrival of the lowest energy event #3. The probability that both #3 and #6 (rejected as background) are background events determines the level of confidence in our conclusions. This probability is roughly 5%. Otherwise, one would rely on event #4 (9.5 MeV electron energy) and extract a limit closer to 20 eV for the mass upper bound.

The late time KII events, after 10 seconds, are potentially quite interesting. It would seem the probability of all three of these events #10, 11, 12 being background is slight [6]. They could represent some reheating of the compact remnant due to, say, material falling back into the core or, more speculatively, a phase change in the core. However, the falling back of matter not ejected in the explosion should not take as long as 10 seconds.

<u>Table II  Limit on the neutrino mass</u>. The KII event sequence is mapped back in time to the source as a function of neutrino mass. The pulse width for the first eight events becomes increasingly long as the neutrino mass $m_{\bar{\nu}_e}$ increases, and the spectrum becomes increasingly hardened. Knowledge of the supernova and neutron star cooling strongly suggest a mass of 15 eV as an upper limit.

| Event Number | $t_{detection}$ (secs) | Energy (MeV) | $(t - t_1)$ source $m_{\bar{\nu}_e} = 10$ eV | $(t - t_1)$ source (secs) $m_{\bar{\nu}_e} = 15$ eV |
|---|---|---|---|---|
| 1 | 0.0 | 20 | 0 | 0 |
| 2 | 0.107 | 13.5 | -0.63 | -1.55 |
| 3 | 0.303 | 7.5 | -3.50 | -8.27 |
| 4 | 0.324 | 9.2 | -2.00 | -4.90 |
| 5 | 0.507 | 12.8 | -0.38 | -1.49 |
| 6 (rejected) | 0.686 | 6.5 | | |
| 7 | 1.541 | 35.4 | 1.97 | 2.51 |
| 8 | 1.728 | 21.0 | 1.80 | 1.89 |
| 9 | 1.915 | 19.8 | 1.92 | 1.89 |
| 10 | 9.2 | 8.6 | 6.45 | 7.39 |
| 11 | 10.4 | 13.0 | 9.55 | 8.49 |
| 12 | 12.4 | 8.9 | 9.87 | 6.71 |
| Pulse width at source (first 8 events) | 1.92 | | 5.42 | 10.8 |

There are a number of other interesting limits to be drawn on neutrino properties by somewhat more sophisticated use of the supernova dynamics. Putting another neutrino–antineutrino pair [30], i.e., another two species, into any calculation of the neutron star cooling would probably accelerate this process unacceptably. Further, one can place an upper limit [5] of 45 eV on the mass of any species mixing with the electron neutrino, else no supernova mechanism would succeed, delayed or prompt.

# 1. References


1.  I. Shelton (Las Campanas Observatory, Chile): International Astronomical Circular No. 4316, 24 February 1987, communicated by W. Kunkel and B. Madore.

2.  R. H. McNaught: IAU Circular No. 4389 (1987);
    R. Catchpole, et al: M.N.R.A.S. (submitted 1987);
    R. Gilmozzi, et al: Nature <u>328</u>, 318 (1987);
    R. Kirshner, G. Sonneborn, D. M. Crenshaw, and G. E. Nassiopoulos: Astrophys. J. Lett. (in press).

3.  E. Baron, H. A. Bethe, G. E. Brown, J. Cooperstein, and S. Kahana: Phys. Rev. Lett. <u>59</u>, 736 (1987).

4.  E. Baron, J. Cooperstein, and S. Kahana, Phys. Rev. Lett. <u>55</u>, 126 (1985).

5.  S. Kahana, in Proceedings of "La Structure Elementaire de la Matiere," (Les Houches, France, March 1987);
    S. Kahana, J. Cooperstein, and E. Baron: Phys. Lett. (in press).

6.  R. Hirata et al: Phys. Rev. Lett. <u>58</u>, 1490 (1987).

7.  R. M. Bionta et al: Phys. Rev. Lett. <u>58</u>, 1494 (1987).

8.  T. Mazurek, J. Cooperstein, and S. Kahana: in "Dumand '80", edited by V. J. Stenger (Hawaii Dumand Center, Honolulu, 1981) and in "Supernovae: A Summary of Current Research," edited by M. Rees and R. J. Stoneham (Reidel, Dordrecht, 1982).

9.  J. R. Wilson: Ann. N. Y. Acad. Sci. $\underline{336}$, 358 (1980).

10. W. D. Arnett: in "Supernovae: A Summary of Current Research," edited by M. Rees and R. J. Stoneham (Reidel, Dordrecht, 1982).

11. W. Hillebrandt: in "Supernovae: A Summary of Current Research." edited by M. Rees and R. J. Stoneham (Reidel, Dordrecht, 1982).

12. T. Shigeyama, K. Nomoto, M. Hashimoto, and D. Sugimoto: Nature (submitted April 1987); and in Proceedings of ESO Workshop on SN1987a, (Garching, July 1987); and proceedings of this Symposium.

13. S. E. Woosley, P. A. Pinto, and L. Ensman: Astrophys. J. (submitted April 1987); and proceedings of this Symposium.

14. W. Hillebrandt, P. Höflich, J. W. Truran, and A. Weiss: Nature (submitted April 1987).

15. J. R. Wilson, in Numerical Astrophysics, eds. J. Centrella, J. Leblanc and R. Bowers (Jones and Bartlett: Boston 1985);
    H. A. Bethe and J. R. Wilson: Astrophys. J. $\underline{295}$, 14 (1985);
    R. Mayle: University of California at Berkeley, PhD thesis (1984) unpublished.

16. T. A. Weaver, B. Zimmerman, and S. E. Woosley: Astrophys. J. $\underline{225}$, 1021 (1978).

17. S. E. Woosley and T. A. Weaver: Bull. Am. Astr. Soc. $\underline{16}$, 971 (1984); and private communication.

18. K. Nomoto and M. Hashimoto: Prog. in Part. & Nucl. Phys. $\underline{17}$, 2670 (1986); and private communication.

19. S. H. Kahana: Prog. in Part. & Nucl. Phys. $\underline{17}$, 231 (1986);
    S. H. Kahana: in Conf. Proc. Weak and Electromagnetic Interactions in Nuclei (Heidelberg, FRG, July 1986);
    E. Baron, J. Cooperstein, and S. H. Kahana: Nucl. Phys. $\underline{A440}$, 744 (1985).

20. J. P. Blaizot: Physics Reports $\underline{64}$, 171 (1980).

21. H. A. Gustafsson et al: Phys. Rev. Lett. $\underline{52}$, 1590 (1984);
    J. W. Harris et al: Phys. Lett. $\underline{153}$, 377 (1982).

22. T. Ainsworth, E. Baron, G.E. Brown, J. Cooperstein, and M. Prakash: Nucl. Phys. (in press).

23. S. Bruenn: Phys. Rev. Lett. $\underline{59}$, 938 (1987).

24. J. N. Bahcall and S. L. Glashow: Nature $\underline{326}$, 376 (1987);
    J. N. Bahcall, T. Piran, W. H. Press, and D. N. Spergel: Phys. Rev. Lett. (submitted April 1987).

25. J. Lattimer: Proceedings of the XI Particle and Nuclei Conf. (Kyoto, Japan, April 1987);
    A. Burrows and J. Lattimer: "Neutrinos from SN1987a" University of Arizona preprint (April 1987).

26. W. D. Arnett and J. L. Rosner: Phys. Rev. Lett. $\underline{58}$, 1906 (1987).

27. K. Sato and H. Suzuki: Phys. Rev. Lett. $\underline{58}$, 2722 (1987).

28. E. W. Kolb, A. J. Stebbins, and M. S. Turner: Phys. Rev. D (in press).

29. R. G. W. Robertson et al: Reported in Telemark IV (March 1987): $m_\nu < 27$ eV (95% cl);
    M. Frischi et al: Phys. Lett. $\underline{173B}$, 485 (1986): $m_\nu < 18$ eV (95% cl).

30. D. N. Schramm: Proceedings of XXII Recontre de Moriond (March 1987).

# SN 1987a: THEORETICAL CONSIDERATIONS

S. E. Woosley
Board of Studies in Astronomy and Astrophysics
Lick Observatory, University of California at Santa Cruz
Santa Cruz CA 95064

## I. THE STAR THAT EXPLODED

There is now general agreement that the presupernova star for 1987a was indeed SK-69-202, a star that had visual magnitude 12.36 (Walborn *et al.* 1987), bolometric correction $1.15^m$ (Humphreys and McElroy 1984), and extinction $0.5^m$ (Woosley *et al.* 1987). The distance modulus to the LMC is presently controversial, but most astronomers prefer a value near 18.5 (*e.g.*, Walker 1985; Chiosi and Pigatto 1986) with an uncertainty of perhaps $0.2^m$. Putting it together, one arrives at a bolometric magnitude for SK-69-202 of $-7.8$ with a probable range $-7.5$ to $-8.2$, or a luminosity of $4 \times 10^{38}$ erg s$^{-1}$ with a range of 3 to $6 \times 10^{38}$ erg s$^{-1}$. At the time of the supernova the hydrogen burning shell contributes negligible energy generation so the critical quantity determining the luminosity is the helium core mass. Inspection of a variety of current stellar models indicates that SK-69-202 had a helium core mass of $6 \pm 1$ M$_\odot$ from which we may infer that, on the main sequence, it had a mass of $19 \pm 3$ M$_\odot$. A B3-I supergiant has a surface temperature of $\sim$16,000 K (Humphreys and McElroy 1984). For the above range of luminosities and assuming a temperature in the range 15,000 to 18,000 K one obtains a radius for SK-202-69 of $3 \pm 1 \times 10^{12}$ cm.

Unfortunately observations of the presupernova star do not constrain the mass of the hydrogen envelope. An unknown amount of mass loss could have occurred leaving anywhere from 14 M$_\odot$ to as little as a few tenths M$_\odot$. Since the envelope mass greatly affects the dynamics of the explosion, the light curve, and the spectroscopic history of the supernova, its determination is of high priority. Based upon observations of the supernova one conclusion of this paper will be that the envelope mass was in the range 5 to 10 M$_\odot$.

Another major and as yet unresolved issue centers upon precisely why SK-202-69 was a blue supergiant, and not a red one. This issue has been recently reviewed by Woosley (1987) and will be briefly summarized here. The essential problem is that there exist multiple solutions to the structure equations for the stellar atmosphere (see also Wheeler, this volume). Two stars having the same helium core mass and only slightly different luminosities, for example, can have radically different envelope structures, either a convective red supergiant or one that is radiative and blue (Woosley, Pinto, and Ensman 1987). There are several physical parameters that may break this symmetry and cause the star to chose one solution and not the other. Among them are metallicity, (extreme) mass loss, and the theory of convection used in calculating the stellar model.

A number of groups have invoked the reduced metallicity of the LMC as the probable cause for the small radius and provided evolutionary calculations to justify their contention.

My own contribution to this is in Fig. 1. A major difference with respect to Arnett (1987) and Truran and Weiss (1987), for example, is that their stars never became red supergiants, while the low metallicity models here burned helium as red supergiants and moved back to the blue (as recently as 20,000 years ago) just in time to explode. The reason underlying this different behavior is use of the LeDoux criterion for convective instability in the present models rather than the less restrictive Schwarzschild criterion. Observations of the 30 Doradus region show the existence of many red supergiants. So at least a portion of massive stars in the LMC must evolve to or through that state. Also recent observations of low velocity ($V \lesssim 200$ km s$^{-1}$) nitrogen-rich gas (Kirshner, this volume), presumably a circumstellar shell, require SK-202-69 itself to have experienced considerable mass loss. This would be easier to understand if the star spent a portion of its life as an extended red supergiant.

An alternative means of obtaining a blue supergiant progenitor, while passing along the way through a red supergiant stage, is extreme mass loss. Well before 1987a it had been recognized that a massive star which lost most of its hydrogen envelope would evolve back from the red to the blue in the HR-diagram, exploding as a blue supergiant or, in the extreme limit of complete hydrogen evaporation, a Wolf-Rayet star (Chiosi and Maeder 1986). Models for SK-202-69 of this sort have been computed by Maeder (1987, see also this volume) and Wood and Faulkner (1987). The problem with these solutions is that so much mass must be lost before the star becomes blue again (less than 1 $M_{\odot}$ remains on the presupernova star). Prior to this year it had been believed that stars would need to be massive than 20 $M_{\odot}$, perhaps 40 $M_{\odot}$ (Humphreys 1984) in order to lose most of their envelope. More importantly, as we shall see, the explosion of a 6 $M_{\odot}$ helium core tamped by less than a few solar masses of envelope will give hydrogenic velocities, an optical light curve, and an x-ray light curve incompatible with observations of 1987a.

## II. MODELS

Because the explosion energy and hydrogen envelope mass are not known *ab initio*, the strategy here has been to calculate a variety of models based upon the explosion of a 6 $M_{\odot}$ helium core (extracted from a previous 20 $M_{\odot}$ presupernova model; Woosley and Weaver 1986) capped by envelopes of various masses. The envelopes were constructed separately in thermal and hydrostatic equilibrium with a radius and luminosity appropriate to SDK-202-69. Explosion was simulated in the hybrid configuration by removing the collapsed iron core and replacing it with a piston of specified trajectory. Rapid motion of the piston initiated a shock wave that ejected all exterior matter. After a time, when the expansion had become homologous, the total kinetic energy, hereafter referred to as the "explosion energy," could be sampled. A collection of such models is given in Table 1 which gives, besides the explosion energy and envelope mass, the time when the shock wave broke through the surface of the star, $t_{break}$; the column depth to the center of the star (actually the outer edge of the $^{56}$Ni mass) when the supernova was $10^{6}$ s old, $\phi_o$; how long the energy contained in the hydrogen envelope following shock passage could power the supernova at the observed luminosity, $\tau_H$; and the velocity of the slowest moving hydrogen in the ejecta, $v_{slow}$. This last quantity is a powerful constraint upon the models. Elias and Gregory (1987) have determined that the slowest hydrogen ejected was moving no faster than and probably close to 2100 km s$^{-1}$. This immediately suggests that the favored model will resemble some subset of 3VL, 5L, 10H, and 14VH.

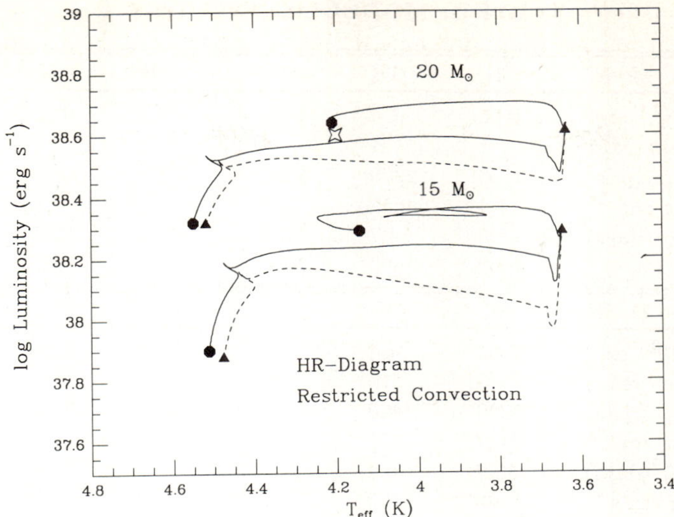


Fig. 1 - Hertzsprung-Russell diagram for stars of 15 and 20 $M_\odot$ and composition appropriate to the LMC ($Z_\odot/4$) (solid lines) and to the sun (dashed lines) evolved through hydrogen, helium, and carbon burning. The location of the presupernova stars are indicated. The four-pointed star indicates the best estimated properties of SK-202-69.

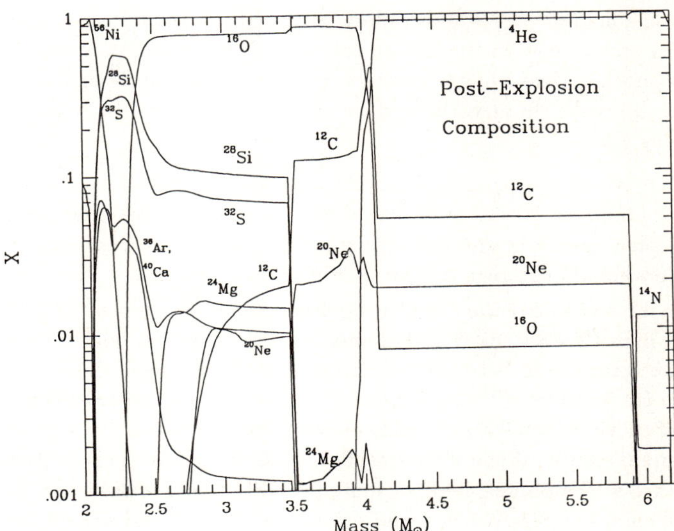


Fig. 2 - Composition of the 6 $M_\odot$ helium core used in all studies. Interior to about 3 $M_\odot$ the composition is a result of explosive nucleosynthesis. Farther out the fossil remnants of previous burning stages are ejected.

## TABLE 1: MODEL DEFINITIONS

| Model | 1L | 1H | 3L | 3H | 5L | 5H |
|---|---|---|---|---|---|---|
| Expl. KE ($10^{51}$ erg) | 0.65 | 1.4 | 0.65 | 1.4 | 0.65 | 1.3 |
| Envel. Mass ($M_\odot$) | 1 | 1 | 3 | 3 | 5 | 5 |
| $t_{break}$ (s) | 3100 | 2200 | 5200 | 3700 | 5500 | 3900 |
| $\phi_o/10^4$ (g cm$^{-2}$) | 1.4 | 0.52 | 2.3 | 1.0 | 4.6 | 1.9 |
| $\tau_H$ (days) | 12.5 | 14.7 | 20.5 | 23.4 | 23.9 | 26.7 |
| $v_{slow}(H)$ km s$^{-1}$ | 4000 | 5600 | 2400 | 3500 | 1800 | 2800 |

| Model | 10L | 10H | 14H | 14VH | 3VL | 1VVL |
|---|---|---|---|---|---|---|
| Expl. KE ($10^{51}$ erg) | 0.65 | 1.4 | 1.3 | 2.6 | 0.41 | 0.34 |
| Envel. Mass ($M_\odot$) | 10 | 10 | 14 | 14 | 3 | 1 |
| $t_{break}$ (s) | 9000 | 6400 | 7600 | 5500 | 6300 | 4000 |
| $\phi_o/10^4$ (g cm$^{-2}$) | 15.4 | 7.1 | 17.1 | 7.8 | 5.1 | 8.0 |
| $\tau_H$ (days) | 35.2 | 40.2 | — | — | — | — |
| $v_{slow}(H)$ km s$^{-1}$ | 1100 | 1700 | 1200 | 1700 | 1800 | 2900 |

The composition within the helium core following shock wave passage is given in Fig. 2.

## III. THE EARLY LIGHT CURVE

After approximately one minute the shock initiated by iron core collapse arrives at the outer edge of the helium core whose radius is typically $5 \times 10^{10}$ cm. The hydrodynamic interaction with the envelope slows the helium core down, the deceleration propagating into the core as a "reverse shock." Meanwhile the outgoing shock continues though the hydrogen. The time when the shock breaks through the surface of the envelope can be estimated (Shigeyama et al. 1987; Woosley 1987),

$$t_b \sim 2500 \left( \frac{M_{env}/M_\odot}{E_{51}} \right)^{1/2} \text{ s,} \tag{1}$$

with $E_{51}$ the explosion energy in units of $10^{51}$ erg. This result is in very good agreement with the calculated values for shock break out given in Table 1.

As the shock breaks out the electromagnetic display commences (Fig. 3). Initially the temperature is so high ($2 - 3 \times 10^5$ K) that most of the radiation will be in the ultraviolet. Figure 4 shows the comparison between the first two days of optical data and the calculated visual magnitudes for 4 models (Table 1) having a variety of explosion energies and envelope masses. This set of curves was calculated using electron scattering opacity only and a bolometric correction based upon simple, single temperature blackbody model. It is well known, however, that the temperature at the photosphere, or more properly "surface of last scattering", is not a good match to the color temperature in situations where electron scattering opacity dominates. Instead the radiation and electron temperature fall out of equilibrium and the star radiates a *dilute* blackbody spectrum having a color temperature, $T_c$, approximately equal to the local temperature where the generalized optical depth, $(\kappa_{tot}\kappa_{abs})^{1/2}\phi$, is unity. Here $\kappa_{tot}$ and $\kappa_{abs}$ are respectively the total opacity (approximately the electron scattering opacity) and that portion of the opacity in which interaction does not preserve photon energy and $\phi$ is column depth. A

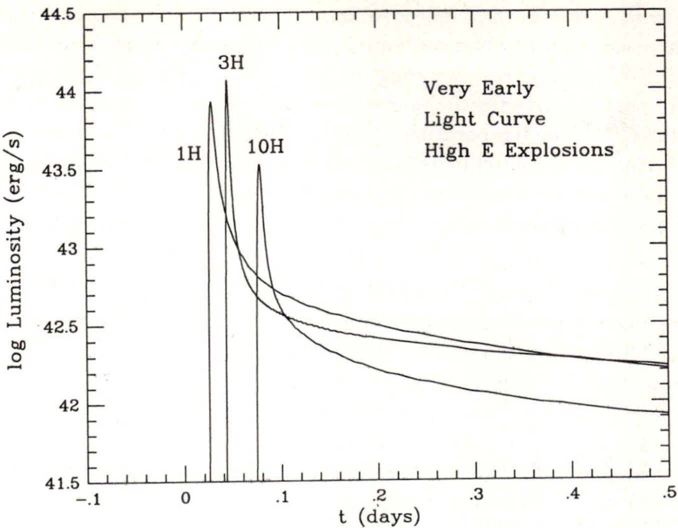



**Fig. 3 -** Bolometric luminosity during the first half day of several explosions (Table 1). The decline during the first hour is especially rapid. A total of about $10^{47}$ erg is emitted as hard UV-radiation during the first day of the supernova.

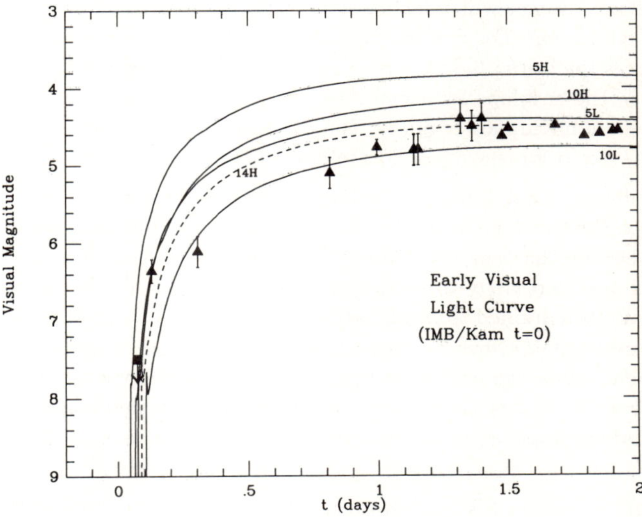



**Fig. 4 -** Visual light curve during the first two days. Time zero is defined by the Kamiokande - IMB neutrino signal. Light curves from four models (Table 1) are shown as solid lines. A distance modulus of 18.5 and visual extinction of $0.5^m$ have been adopted. Shown for comparison are observational data points.

365

photon originating from deeper in the supernova than this will undergo enough scattering events to encounter several energy non-conserving events and be (approximately) thermalized. On the other hand the radiative diffusion equation may be solved for a region of constant luminosity to yield $(T_c/T_e)^4 = \tau_e$ with $T_e$ the effective emission temperature at the scattering photosphere (what the code edits) and $T_c$ the temperature at electron scattering optical depth $\tau_e$. Combining equations one finds that the color temperature will be approximately $(\kappa_{tot}/\kappa_{abs})^{1/8}$ times the effective emission temperature evaluated at the scattering photosphere. The problem then is one of determining the non-conservative opacity as a function of temperature, density, composition, and velocity shear.

The relatively weak dependence on the ratio $\kappa_{tot}/\kappa_{abs}$ suggests that the modification to our calculated results will not be great except at very early times. The effective temperature calculated for Model 10H, for example, is, without modification, within 15% of the values inferred from the spectrum (Suntzeff, private communication) on days 1.14 (13,600 K), 1.51 (12,700 K), and 1.85 (11,690 K). Figure 5 illustrates the effect for $\kappa_{tot}/\kappa_{abs} = 1$, 0.3, and 0.1. The latter corresponds to a color temperature one third greater than the effective emission temperature. Karp *et al.* (1977) have considered the effect of Doppler broadened lines on the bound-bound opacity. For typical photospheric densities ($10^{12}$ g cm$^{-3}$) and temperatures (5000 K to 50,000 K) the line opacity is approximately 20% to 200% that of electron scattering (see their Table 3). This should keep the color temperature within about 20% of the effective emission temperature.

## IV. THE LATER LIGHT CURVE

Figure 6 shows the bolometric light curves for the first 200 days for 4 of the models defined in Table 1. Following shock break out and its associated high temperatures, the supernova enters a "plateau" stage, which lasts about one month, with energy released (though not provided) by recombination. For greater explosion energies, larger presupernova radii, and smaller hydrogen masses the initial light curve is brighter. This reflects both the greater internal energy deposited in the envelope by a more energetic explosion and the greater expansion velocity given an envelope having lower mass. The duration of this plateau ($\tau_H$ in Table 1) is also determined by the explosion energy and envelope mass.

Following hydrogen recombination the luminosity rises at a rate that is very sensitive to the explosion energy, the envelope mass, and to the opacity in the helium core. The ultimate source of the energy here is the decay of $^{56}$Co to $^{56}$Fe, a reaction that has powered the light curve since late March and especially through the peak and tail. Because the amount of $^{56}$Ni synthesized is artificially constrained to be the same (0.07 M$_\odot$; §V) in all our models, the peak luminosity does not vary greatly with envelope mass in Fig. 6. In a more realistic case the $^{56}$Ni mass would depend upon the explosion energy and would be greater for more energetic explosions. We see from Fig. 6 that the favored model, for an explosion energy near $1.4 \times 10^{51}$ erg, has about 10 M$_\odot$ of envelope, consistent with Fig. 4 and restrictions on the slowest moving hydrogen (§II).

All of the light curves in Fig. 6, however, suffer from an obvious deficiency - they go through a period of decline near the end of hydrogen recombination that is not reflected in the observations. Indeed, quite the opposite is observed - after the first week the light curve *increases*, steadily and smoothly, all the way to its peak. This smoothness of the bolometric light curve has been one of the most perplexing aspects of 1987a. In Fig. 7 Model 10H has been recalculated using i) an opacity, in addition to electron scattering, of 0.001 cm$^2$ g$^{-1}$ for

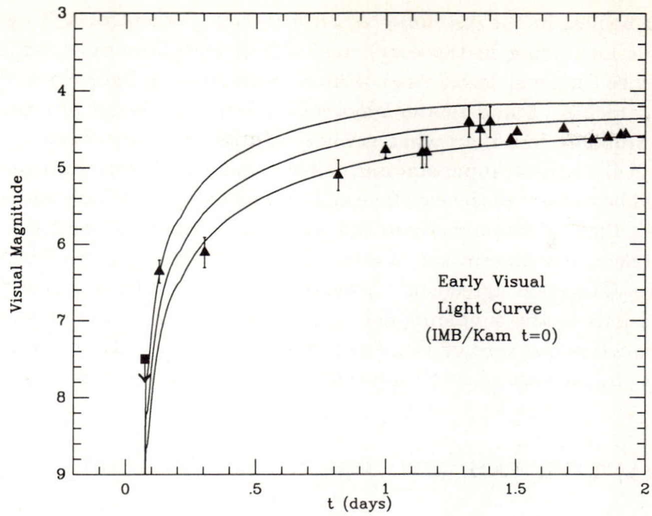



**Fig. 5 -** Correction to the early visual light curve of Model 10H for the fact that the color temperature does not equal the effective emission temperature for an atmosphere whose opacity is dominantly due to electron scattering. The three curves from top to bottom have the non-conservative opacity equal to 1, 0.3, and 0.1 of the electron scattering opacity.

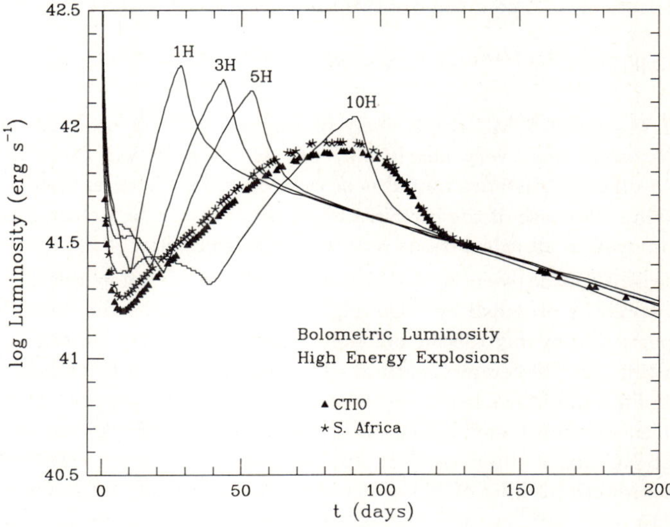



**Fig. 6 -** Bolometric luminosities during the first 200 days for 4 of the models defined in Table 1 compared to data from Catchpole *et al.* (1987) and Hamuy *et al.* (1987). All models employed the same 6 $M_\odot$ helium core capped by hydrogen envelopes of various masses. The opacity, chiefly due to electron scattering while the gas remains ionized, was given a lower floor of 0.02 cm$^2$ g$^{-1}$ for elements heavier than helium.

both hydrogen and helium in the envelope; ii) an artificial gradient of hydrogen and helium in the envelope (but no mixing in the core); and iii) an additive opacity of 0.025 cm$^2$ g$^{-1}$ within the helium core (internal to 6.2 M$_\odot$). Effect i) reduces the light curve during the first $\sim$10 days compared to Fig. 6 and relates physically to the opacity of broadened lines. The value employed is probably less than what actually characterizes the outer layers with large velocity shear. Effect ii) is most important and reflects, partly, the real gradient of helium and hydrogen that would be present in the envelope and mixing that would have occurred during the explosion. Effect iii), the additive opacity in the core, affects the width and timing of the light curve peak and is taken to represent the Doppler-broadened line opacity of helium and heavy elements in the core. The good agreement shows that the observations can be fit very well by reasonable modifications to the simple model. Also shown in Fig. 7 are two light curves that used the same parameters but zero or twice as much radioactivity. Both are clearly excluded by the observations. In the case of zero radioactivity the light curve dies at approximately $\tau_H$ for Model 10H.

## V. THE RADIOACTIVE TAIL AND THE NICKEL BUBBLE

Unlike Type Ia supernovae, the slow expansion of the core of Type II's renders them optically thick to $\gamma$-radiation for a period of about 2 years. During the first year at least this $\gamma$-radiation is degraded chiefly into optical and ultraviolet radiation which, after the peak of the light curve, diffuses out in a time short compared to the elapsed time. During this interval the UV-optical light curve should track exactly the $^{56}$Co decay rate. Figure 8 shows the bolometric luminosity of Catchpole et al. (1987) and Hamuy et al. (1987) during the first 188 days based upon the same assumptions regarding distance modulus and visual extinction as the authors (18.5 and 0.6$^m$ respectively). Also given are two lines generated by the equation

$$\dot{S}_{nuc} = 3.90 \times 10^{10} e^{-t/\tau_{Ni}} + 7.21 \times 10^9 \left( e^{-t/\tau_{Co}} - e^{-t/\tau_{Ni}} \right) \ \text{erg g}^{-1}\text{s}^{-1} \qquad (2)$$

multiplied by 0.07 M$_\odot$ and 0.2 M$_\odot$ respectively of radioactive $^{56}$Ni produced initially in the explosion. *It is apparent that very nearly 0.07 M$_\odot$ of mass 56 has been produced in the explosion,* though a different distance modulus or correction for visual extinction would give a slightly different value. Because of the certainty with which it can be determined this value of $^{56}$Co mass was employed in all calculations reported in this paper.

As an interesting aside, we note that the near match of the bolometric light curve on the tail to that which would be provided by $^{56}$Co (eq. 2) places limits on the possible contribution from a pulsar. In particular on day 188 the bolometric luminosity of the supernova was $1.8 \times 10^{41}$ erg s$^{-1}$ (Nick Suntzeff, private communication). The contribution of a second source having a very different mean life would have led to discernable deviations from eq. (2) if the background source contributed more than a small fraction of this, say 10%. If a pulsar exists, similar to the one in the Crab Nebula, its luminosity at this stage would be $L \sim 4 \times 10^{43} B_{12}^2 P_o^{-4}$ erg s$^{-1}$ with $B_{12}$ the field strength in units of $10^{12}$ gauss and $P_o$ the period in ms (Ostriker and Gunn 1969). Limiting $L$ to $\sim 2 \times 10^{40}$ erg s$^{-1}$ thus implies (for $B_{12} = 4.3$ as in the Crab; Manchester and Taylor 1977), that the period of the pulsar is presently greater than 14 ms (*i.e.*, 1/2 the Crab).

The energy released by the decay of $^{56}$Ni ($2.96 \times 10^{16}$ erg g$^{-1}$) and by $^{56}$Co ($6.41 \times 10^{16}$ erg g$^{-1}$) is comparable to the kinetic energy density in the iron and overlying layers. Following the passage of the reverse shock, all of the elements heavier than helium are typically moving at 1000 to 2000 km s$^{-1}$ corresponding to a kinetic energy density of $\sim 2 \times 10^{16}$ erg $^{-1}$. Even

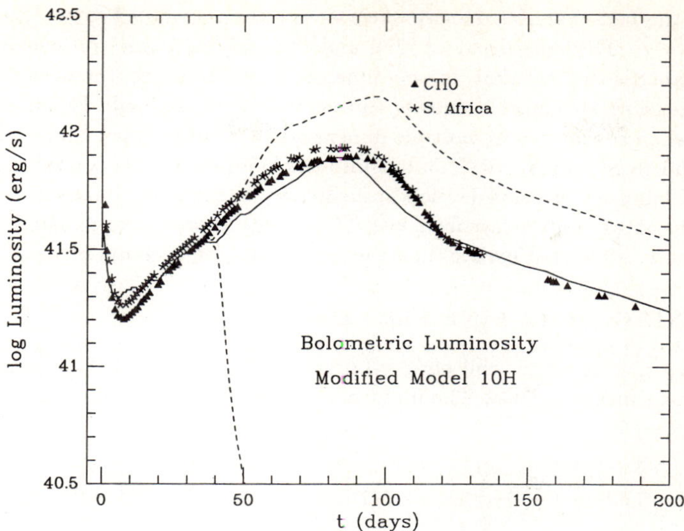

**Fig. 7 -** Comparison of the observed bolometric light curve to several modified versions of Model 10H. Each employed an artificial gradient of helium in the envelope to aid in replicating the rising nature of the light curve during the first 40 days. Each also adopted a floor to the hydrogen and helium opacities in the envelope of 0.001 cm$^2$ g$^{-1}$ and to the opacity within the helium core of 0.025 cm$^2$ g$^{-1}$ (including helium itself). The two dashed lines show the effect of varying the mass of $^{56}$Co produced in the explosion. The upper curve used twice as much $^{56}$Ni (0.14 M$_\odot$); the lower curve, no radioactivity.

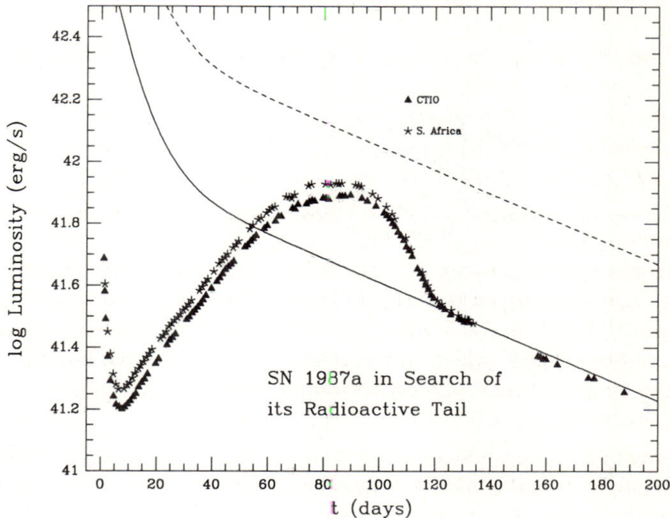

**Fig. 8 -** The observed bolometric light curve (Catchpole *et al.* 1987 and Hamuy *et al.* 1987) compared to that which would result from 100% optical conversion and escape of energy from the radioactive decay of $^{56}$Ni and $^{56}$Co. The upper curve is for 0.20 M$_\odot$ of mass 56 and the lower curve is for 0.07 M$_\odot$.

more relevant is the fact that the velocity *shear* across the region of heavy elements is only a few *hundred* km s$^{-1}$. Thus the decay of $^{56}$Ni and $^{56}$Co has dynamic consequences. Of great potential significance is the fact that, in one dimensional calculations, the decay of $^{56}$Co at the center of the supernova produces a density inversion that is Rayleigh Taylor unstable. It is thus likely that when considered in multiple dimensions the bubble may "pop," that is develop "fingers" that lead to the mixing of $^{56}$Co out into the helium core. This mixing might also be accompanied by clumping. If voids develop in the heavy elements overlying the core, this might allow us to see the x-rays and $\gamma$-rays from the $^{56}$Co decay, as well as any manifestations of the central neutron star, somewhat earlier than the simple one-dimensional models would predict.

## VI. X-RAY AND GAMMA-LINE FLUXES

The only species which will emit $\gamma$-line radiation at a level that might possibly be detected in the near future is $^{56}$Co. The flux from a mass, $M_{56}$ of $^{56}$Co in solar masses located in the LMC (50 kpc) is

$$F = 0.602 \left( \frac{M_{56}}{0.1 M_\odot} \right) \exp \left( -t/113.6\,\mathrm{d} - \kappa_\gamma \, \phi_o \, (t_o/t)^2 \right) \mathrm{cm}^{-2}\, \mathrm{s}^{-1} \qquad (3)$$

where $t$ is the elapsed time since the explosion, $t_o$ some fiducial time at which the column depth to the edge of the $^{56}$Co layer, $\phi_o$, is to be determined, and $\kappa_\gamma$ is the opacity to 1 MeV $\gamma$-rays. Here $F$ is the flux of some line, such as 847 keV, through which all decays proceed and homologous expansion has been assumed. An appropriate value of $\kappa_\gamma$ is 0.06 cm$^2$ g$^{-1}$ and a reasonable time to evaluate the column depth is $t_o = 10^6$ s (Table 1).

This flux will have a maximum at time

$$t_{max} = (2\tau_{Co}\, \kappa_\gamma \, \phi_o \, t_o^2)^{1/3} = 263\,(\phi_o/10^4)^{1/3} \text{ days.} \qquad (4)$$

The maximum flux for models in Table 1 is easily obtained by evaluating eq. (3) at time $t = t_{max}$,

$$F_{max} = 0.602 \left( \frac{M_{56}}{0.10 M_\odot} \right) \exp \left[ 3 \left( \frac{\kappa_\gamma \, t_o^2 \, \phi_o}{4\tau_{Co}^2} \right)^{1/3} \right]$$

$$= 0.602 \left( \frac{M_{56}}{0.10\, M_\odot} \right) \exp \left( -0.161\, \phi_o^{1/3} \right) \mathrm{cm}^{-2}\, \mathrm{s}^{-1}, \qquad (5)$$

a result which is extremely sensitive to the column depth at $t_o$, *i.e.*, to the expansion rate. Models 3VL, 5L, and 10H, which have $\phi_o/10^4$ g cm$^{-2}$ = 5.1, 4.6, and 7.1 respectively would have peak fluxes of 1.1, 1.4, and 0.5 $\times 10^{-3}$ cm$^{-2}$ s$^{-1}$ at days 450, 440, and 510 respectively. Model 1H, on the other hand, which unfortunately (for $\gamma$-line astronomers) is disallowed by comparison to the light curve and photospheric velocity history, would have presented a flux of about 0.026 cm$^{-2}$ s$^{-1}$ on day 210.

The $\gamma$-ray optical depth at maximum emission is 1.4 $(\phi_o/10^4$ g cm$^{-2})^{1/3}$, which is in the range 2-3 for any reasonable model. Thus the continuum will be quite strong. Indeed well before the $\gamma$-lines become visible at a detectable level, the supernova should be quite bright in hard x-rays. At the meeting Tanaka (see this volume) announced the discovery of the supernova in the 10 to 30 keV band by GINGA beginning in mid-August. Prior to the meeting Pinto and I had calculated the expected x-ray spectrum and light curve for Model 10H (as had several other groups). Without mixing or clumping, the x-rays would not have been detectable until

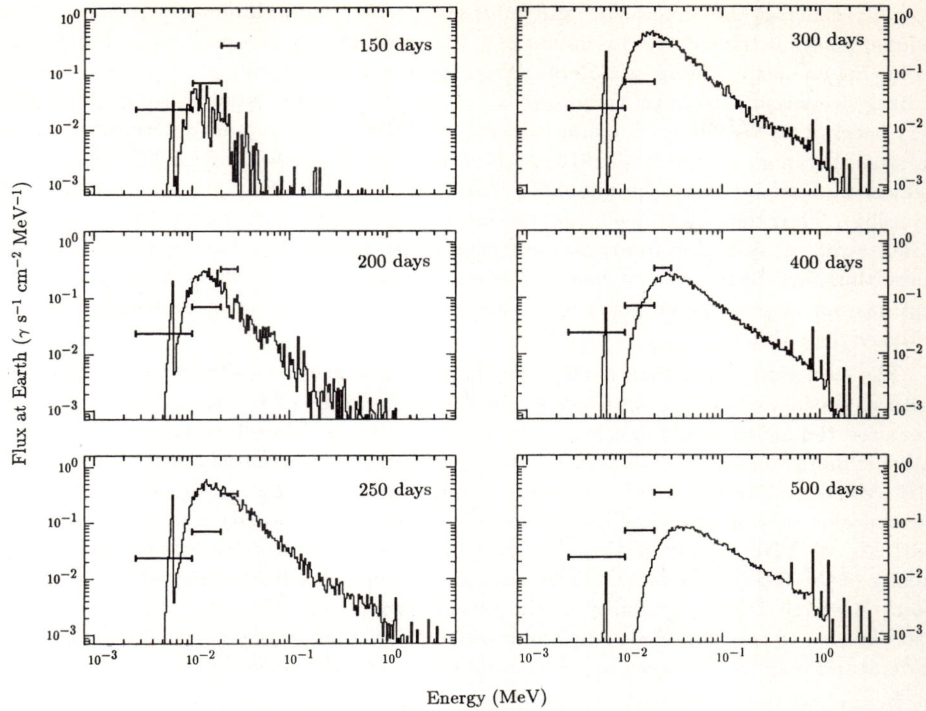



Figure 9. X-ray spectra at several times for Model 5LM evaluated at the dates shown from Monte Carlo calculations by Pinto and Woosley (1987). Horizontal error bars show the GINGA sensitivity at various energies.

about day 250 (Nov. 1), *i.e.*, a discrepancy of about 75 days. Immediately upon returning home we calculated the expected light curve of a more promising model, 5L, in which some outward mixing of the $^{56}$Co was presumed to have occurred (§V). In particular Model 5LM (Woosley 1987) was examined in which the $^{56}$Co abundance was artificially given a gradient such that it declined smoothly by a factor of 10 between the center of the supernova and the edge of the helium and was zero in the hydrogen envelope. The x-ray spectra and onset (Fig. 9) are in reasonable accord (perhaps bright by a factor of two and a little too soft a spectrum) with the data available in September. For further detail see Pinto and Woosley (1987).

## VII. SUMMARY

The star that exploded, SK-202-69, was, as theory required, a massive star. When it lived on the main sequence, it had a mass of 19 $\pm$ 3 $M_\odot$. At the time it exploded it had a helium core mass of 6 $\pm$ 1 $M_\odot$, a radius 3 $\pm$ 1 $\times 10^{12}$ cm, a luminosity 3 to 6 $\times 10^{38}$ erg s$^{-1}$, and an effective temperature 15,000 to 18,000 K. Further consideration of the stellar models (Woosley 1987; Nomoto, this volume) suggests that the iron core mass at the time of collapse was 1.45 $\pm$ 0.15 $M_\odot$. Adding $\sim$0.15 $M_\odot$ for matter between the iron core and the entropy jump which usually demarks the "mass cut" and subtracting 10% for the binding energy implies a gravitational mass for the collapsed remnant of $1.40 \pm 0.15$ $M_\odot$. In the particular case of a 20 $M_\odot$ star having an entropy jump at 1.55 $M_\odot$ (Woosley and Weaver 1987) where the "mass cut" is most likely to develop (see also Nomoto *et al.* 1987), the remnant would have gravitational mass very nearly 1.40 $M_\odot$. This compares favorably with the accurate mass determined for two neutron stars in binary pulsar PSR 1913+16 ($1.451\pm0.007$ $M_\odot$ and $1.378\pm0.007$ $M_\odot$; Taylor 1986) which are believed to be the remnants of stars in the 16 to 18 $M_\odot$ range (Burrows and Woosley 1986). Thus the object is almost certainly a neutron star, not a black hole. However, even if it is a pulsar, radiation from this neutron star has not contributed significantly to the light curve thus far. Therefore if it has a magnetic moment like that of the Crab pulsar and accretion has not choked the emission mechanism, the neutron star must be rotating with a period longer than 15 ms.

The explosion mechanism itself might have been due to the shock wave created by core bounce, especially for iron core masses in the lower range of the error bars, but more likely required the aid of neutrino energy transport, *i.e.*, was a delayed explosion. This latter alternative is more consistent with properties of the neutrino burst measured by Kamiokande and IMB (Mayle and Wilson 1987). A severe constraint on both the presupernova structure and the explosion mechanism (that has yet to be imposed and explored) is that the explosion eject 0.07 $M_\odot$ of $^{56}$Ni. Decay of this $^{56}$Ni and its daughter $^{56}$Co releases sufficient energy that mixing of the heavy elements (carbon through calcium) may have occurred following the explosion. Rayleigh-Taylor instability in the reverse shock (*e.g.*, Chevalier and Klein 1978) may also have led to mixing. Models having mixed compositions agree marginally better with observations and may be necessary to understand the x-ray light curve.

In any case the total kinetic energy of the explosion did not exceed $2\times10^{51}$ erg, and even this large a value is only allowed if the star had not lost much of its hydrogen envelope prior to exploding. Otherwise the early light curve would have been too bright (Fig. 4), the radioactive portion of the light curve would have peaked too early (Fig. 6), and the slowest hydrogen ejected would have had a velocity in excess of 2100 km s$^{-1}$ (Table 1; Woosley 1987) which is disallowed by observations (Elias and Gregory 1987). For explosion energies less than about $3 \times 10^{50}$ erg, on the other hand, large portions of the star would have failed to achieve escape velocity, especially in the case of large envelope masses which effectively tamp the expansion of the heavy element core. We know that this did not occur because of the radioactivity that is now powering the light curve. Even for energies as low as $4 \times 10^{50}$ erg the envelope mass cannot be less than about 3 $M_\odot$ or the light curve would have risen too rapidly and peaked too early . Putting it together we obtain Fig. 10 which summarizes the allowed range of explosion energy and envelope mass allowed by constraints coming from the light curve. The favored model emerges as one having a hydrogen envelope in the range 5 to 10 $M_\odot$ and thus an explosion energy in the range 0.8 to $1.5 \times 10^{51}$ erg.

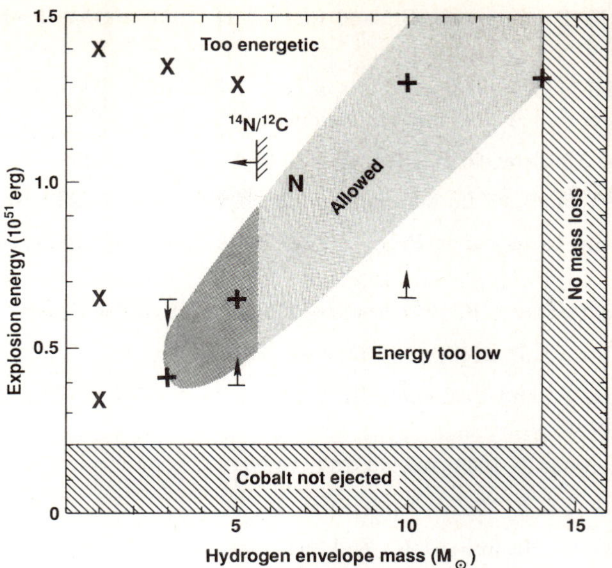


Figure 10. Allowed values of explosion energy and hydrogen envelope mass are broadly delineated for 1987a. Based upon the explosion of a 6 $M_\odot$ core (main sequnce mass 20 $M_\odot$), the atmosphere can be no greater than 14 $M_\odot$. Symbols "X" denote a model that can be excluded on the basis of one or more observational constraints; "+" indicates a moderately successful model; arrows indicate lower and upper bounds provided by three of the models; and "N" a successful model recently published by Nomoto et al (1987). Explosion energies below $3 \times 10^{50}$ erg lead to reimplosion of the core and loss of all $^{56}$Co.


This work has been supported by the National Science Foundation (AST-84-18185). I would like to express my appreciation to Ken Nomoto for his hospitality and for his role in organizing a productive and unually exciting meeting.


## BIBLIOGRAPHY


Arnett, W. D. 1987, *Ap. J.*, **319**, 136.

Burrows, A., and Woosley, S. E. 1986, *Ap. J.*, **308**, 680.

Catchpole, R. M., Menzies, J. W., Monk, A. S., Wargau, W. F. and 16 others. 1987, preprint, South Africa Observatory..

Chevalier, R., and Klein, R. I. 1978, *Ap. J.*, **219**, 994.

Chiosi, A., and Maeder, A. 1986, *Ann. Rev. Astron. and Ap.*, **24**, 329.

Chiosi, C., and Pigatto, L. 1986, *Ap. J.*, **308**, 1.

Elias, and Gregory 1987, in preparation for *Ap. J.*

Hamuy, M., Suntzeff, N. B., Gonzalez, R., and Martin, G. 1987, NOAO Preprint No. 102, submitted to *Ap. J.*

Humphreys, R. M. 1984, in *Observational Tests of Stellar Evolution Theory*, IAU Symposium 105, Ed. A. Maeder and A. Renzini, (D. Reidel: Dordrecht), p. 279.

Humphreys, R. M., and McElroy, D. B. 1984, *Ap. J.*, **284**, 565.

Karp, A. H., Lasher, G., Chan, K. L., and Salpeter, E. E. 1977, *Ap. J.*, **214**, 161.

Maeder, A. 1987, preprint to appear in *Proc. of ESO Workshop on SN1987a*, ed. J. Danziger, in press.

Manchester, R. N., and Taylor, J. H. 1977, *Pulsars*, (Freeman: San Francisco).

Mayle, R. W., and Wilson, J. R. 1987, preprint, submitted to *Ap. J.*

Nomoto, K., Shigeyama, T., and Hashimoto, K. 1987, in *Proc. ESO Workshop. on SN1987a*, ed. J. Danziger, in press.

Ostriker, J. P., and Gunn, J. E. 1969, *Ap. J.*, **157**, 1395.

Pinto, P. A., and Woosley, S. E. 1987, submitted to *Ap. J.*

Shigeyama, T., Nomoto, K., Hashimoto, K., and Sugimoto, D. 1987, *Nature*, **328**, 320.

Taylor, J. 1986 in *The Origin and Evolution of Neutron Stars*, ed. D. J. Helfand and J.-H. Huang, (D. Reidel: Dordrecht), p. 383.

Truran, J. W., and Weiss, A. 1987, Max Planck Inst. for Ap. preprint No. 303. To appear in *Proc. of 4th Workshop on Nuclear Astrophysics*.

Walborn, N. R., Lasker, B. M., Laidler, V. G., and Chu, Y.-H. 1987, *Ap. J. Lettr.*, in press.

Walker, A. R. 1985, *MNRAS*, **212**, 343.

Wood, P. R., and Faulkner, D. J. 1987, *Proc. Astron. Soc. Australia*, in press.

Woosley, S. E. 1987, submitted to *Ap. J.*

Woosley, S. E., Pinto, P. A., Martin, P., and Weaver, T. A. 1987, *Ap. J.*, **318**, 664.

Woosley, S. E., Pinto, P. A., and Ensman, L. 1988, *Ap. J.*, **324**, 000.

Woosley, S. E., and Weaver, T. A. 1987, *Physics Reports*, Proc. Bethe Birthday Workshop, submitted.

Woosley, S. E., and Weaver, T. A. 1986, in *Radiation Hydrodynamics in Stars and Compact Objects*, ed. D. Mihalas and K.-H. A. Winkler, (Springer Verlag: Berlin), p. 91.

# Calculated Late Time Spectra of Supernovae

*Timothy S. Axelrod*

Lawrence Livermore National Laboratory
Livermore, California
U.S.A.
October, 1987

## 1. Introduction

We consider here the nebular phase spectra of supernovae whose late time luminosity is provided by the radioactive decay of $^{56}$Ni and $^{56}$Co synthesized in the explosion. A broad variety of supernovae are known or suspected to fall in this category. This includes all SNIa and SNIb, and at least some SNII, in particular SN1987a. At sufficiently late times the expanding supernova becomes basically nebular in character due to its decreasing optical depth. The spectra produced during this stage contain information on the density and abundance structure of the entire supernova, as opposed to spectra near maximum light which are affected only by the outermost layers. A numerical model for nebular spectrum formation is therefore potentially very valuable for answering currently outstanding questions about the post-explosion supernova structure. As an example, we can hope to determine the degree of mixing which occurs between the layers of the ''onion-skin'' abundance structure predicted by current one dimensional explosion calculations.

In the sections which follow, such a numerical model is briefly described and then applied to SN1972e, a typical SNIa, SN1985f, an SNIb, and finally to SN1987a. In the case of SN1987a predicted spectra are presented for the wavelength range from 1 to 100 microns at a time 300 days after explosion.

## 2. Numerical Model

### 2.1. Optical Depths

As a rough approximation, an expanding supernova can be treated as nebular when the continuum optical depth to the center falls below unity. This occurs when

$$t > 228 \frac{\sqrt{\kappa M}}{u_9}$$

where $t$ is the time after explosion in days, $\kappa$ is the average opacity in $cm^2/gm$, $M$ is the mass in solar masses, and $u_9$ is the expansion velocity in units of $10^9 cm/sec$. As discussed by a number of authors, for example [1], the average opacity is quite uncertain due to the effects of dense thickets of overlapping lines that occur in rapidly expanding material with substantial amounts of heavy elements. A lower limit is obtained by considering only electron scattering, so that

$$\kappa \approx 0.4 \frac{\bar{Z}}{\bar{A}}$$

where $\bar{Z}$ is the average degree of ionization and $\bar{A}$ is the average atomic weight. Light curve studies, for example [2], show this value for kappa is probably not grossly in error.

As will be described in the sections that follow, all the supernova models we consider consist of one or more regions containing heavy elements, possibly surrounded by outer regions of helium and hydrogen. Consider, for example, the model for SN1987a discussed in Section 6. At times near 300 days, the inner ''metal'' region is roughly characterized by $\bar{Z} \approx .3$, $\bar{A} \approx 16$, $M \approx 2$, $u_9 \approx 0.1$. The outer ''helium'' region has $\bar{Z} \approx 0.1$, $\bar{A} \approx 4$, $M \approx 12$, $u_9 \approx 2$. If these $\bar{Z}$ values were constant in time, then equation 1 implies that the outer region is optically thin for $t > 40$, while the inner metal region becomes thin only for $t > 300d$. So for this model, the nebular approximation is (at least possibly) reasonable for times of 300 days and later. As the above reasoning makes clear, at these times what optical depth there is comes dominantly from the inner metal region, with the outer helium region forming a very thin atmosphere above it. The situation for the other model classes is similar, but they become optically thin at much earlier times, due to their lower mass and higher expansion velocity.

## 2.2. Energy Deposition

The power source for the optical spectrum is the radioactive decay of $^{56}$Ni and $^{56}$Co, whose role in supernovae has been extensively discussed for nearly 20 years[3]. At the times under discussion here, the Ni has all decayed and the only significant energy production is due to Co. The numerical model operates by first calculating the energy deposition from the Co decay products throughout the nebula. The gamma deposition is determined by solving a 4 group steady state transfer equation in spherical geometry. The complex Co gamma emission spectrum is approximated by emission at a single characteristic energy, usually taken to be 1.5 mev, and these photons then are scattered into lower energy groups. At a scattering event the incident photon deposits a fixed fraction of its energy (taken from the data of Plechaty et al [4]) and then scatters isotropically at a correspondingly lower energy. Comparison of this approximation with a full Monte Carlo treatment of the gamma transport has shown that the accuracy achieved is more than adequate for the supernovae considered here.

In addition to the gammas, $^{56}$Co emits a continuous positron spectrum accounting for 4% of the emitted energy on average. The positrons are expected to deposit nearly locally due both to their high energy deposition crossection and the large effect of even interstellar level magnetic fields on their effective path length. To allow a crude investigation of the effects of nonlocal deposition, the positrons are treated identically to the gammas, but with only a single group and with an energy deposition fraction which is always unity. The effective crossection can then be varied to simulate nonlocal deposition, with a large value resulting in local deposition.

## 2.3. Temperature and Ionization State

Energy deposition from $^{56}$Co decay products causes both ionization and heating throughout the nebula. The ionization state and temperature at each point is determined by iteratively solving for a steady state in which ionization by fast particles and reabsorbed recombination radiation is balanced by radiative and dielectronic recombination; and in which heating is balanced by radiative cooling, principally from collisionally excited forbidden lines. This latter calculation is made more rapid by the use of cooling tables precalculated for each ion as a function of temperature and electron density. For the most part spatial coupling between different regions of the nebula is weak. The one important exception is that recombination radiation emitted in one region may transport to another and strongly affect the ionization balance there by causing increased ionization. This is treated by solving a full transfer equation for the recombination radiation and coupling it to the ionization balance equation.

## 2.4. Spectrum Calculation

Once the temperature and ionization state has been determined as a function of radius throughout the nebula, calculation of the emergent optical spectrum is straightforward. For each zone in the nebula, level populations for each ion are calculated by solving steady state rate equations. Then, for each radiative transition, the contribution to the emergent flux is calculated by using the line profile appropriate for an optically thin shell moving at the given velocity.

The numerical model is set up to handle atomic data for an arbitrary number of elements, and for each of these an arbitrary number of ionization stages can be incorporated. The atomic model for each ion in general consists of two separate models used for different purposes. The "x-ray" model is utilized for computing radiative and dielectronic recombination rates and photon generation from the cascade which follows such events. It is also used for the total photoionization crossections required to calculate the transport of the cascade radiation and its effects on the ionization balance. The "optical" model, on the other hand, is used for calculation of cooling rates and for generating the optical spectrum. In general, the "optical" model contains detailed information on low lying levels and the collisional and radiative transitions that connect them, while the "x-ray" model contains a less detailed level structure that covers a much greater energy range.

The x-ray models used have been generated by a Hartree-Slater code [5, 6] of J. Scofield's. Radiative recombination rates are calculated from the photoionization crossections. Information for the optical models has been drawn from a wide variety of sources. The ions included in the present calculation are: FeI, II, III, IV, V, VI; CoI, II, III, IV, V; OI, II, III, IV; and HeI, II. Many important ions are not included, particularly in the intermediate Z range near Si. To minimize (but not eliminate) the effects of these omissions on the temperature and ionization structure, the missing ions are mapped to "glopium", a mythical element that cools and ionizes like O, but produces no actual optical spectrum.

## 3. Supernova Models

The supernova models utilized below have all been generated by S. Woosley using the Kepler [7] supernova code. The evolution is followed with Kepler until the expansion becomes essentially homologous. At this point the Kepler information is transferred to the spectrum code. It is important to emphasize that once the supernova model has been selected the predicted spectra are generated without adjustable parameters. The adjustable parameters of the model are on the whole those with direct astrophysical relevance, for example the initial stellar mass and amount of mass loss. It is,.of course, feasible and useful to use the comparison of the observed and predicted spectra to refine the supernova model. This, however, has not yet been done except in a rough way. At present, the nebular spectra have been used only to choose broad classes of feasible models and rule out others. For example, the He detonation models for SNIa have been ruled out on the basis of the spectrum[8]. The results presented, therefore, must be viewed accordingly as generic to a whole class of models whose details are still unresolved.

## 4. Application to SNIa

The model chosen for an SNIa is a carbon deflagration in a white dwarf of 1.4 $M_\odot$[9]. The "flame speed", which must currently be chosen somewhat arbitrarily, has been given an intermediate value. The explosion produces $1.0x10^{51}$ ergs and 0.51 $M_\odot$ of $^{56}$Ni. The optical spectrum from the numerical model is compared to the observation of SN1972e at a time 264 days after explosion[10]. The spectrum is dominated by [FeII] and [FeIII]. The feature near 6000 Angstroms has been previously identified with [CoIII] and shown to decay in a manner consistent with the radioactive decay of $^{56}$Co[11].

## 5. Application to SNIb

SNIb have only recently been recognized as a distinct class of supernovae [12]. At present there are only two SNIb for which observed spectra are available during the nebular phase. These are SN1983n [13] and SN1985f [14, 15]. The progenitors of these events are still uncertain. Here we consider as a possibility a Wolf-Rayet star of 4 $M_\odot$ that has evolved from a 15 $M_\odot$ main sequence star. The star was evolved with Kepler to the point of Fe core collapse. The dynamics of the core collapse and the subsequent explosion have not been followed for this model, but rather simulated with a mass cut and a piston that transfers $2.5x10^{51}$ ergs to the remaining star. 0.06 $M_\odot$ of $^{56}$Ni are synthesized and ejected in the resulting explosion. The optical spectrum from this model at a time 200 days after explosion are compared with the spectrum of SN1985f as observed in April, 1985 [15]. The time of explosion of SN1985f is not well known, so this choice of times is in itself speculative. The two spectra have more than a superficial resemblance, however. The dominant features in the calculated spectrum are due to OI and OII, and are matched well in intensity in the observation, although the widths are too great and the lineshape too flat-topped. The observed spectrum has a continuum which is absent in the calculation. This continuum probably results from the large number of allowed lines present in the Fe region, untreated in this model, which in the presence of large velocity gradients can generate a "quasi-continuum" across the entire optical wavelength region [11].

If SN1985f was indeed 200 days old at the time of the observed spectrum, the model spectrum can be used to infer a distance to M83. This distance is 7 Mpc, considered within the present uncertainties in the distance to this galaxy [13]. It is therefore worth noting that the model generates both a spectrum and a luminosity in reasonable agreement with observations in spite of the fact that it has only 0.6 $M_\odot$ of O compared with the 5 $M_\odot$ that has been previously claimed to be a requirement [16].

## 6. Application to SN1987a

Models for SN1987a have recently been extensively discussed by Woosley [17]. We have used model 10H from that paper for the spectrum calculation. As discussed there in detail, this model is the core of a 20 $M_\odot$ star which has lost 4 $M_\odot$ of mass in a wind. The core collapse and subsequent explosion produces ejecta with $1.4x10^{51}$ ergs and 0.08 $M_\odot$ of $^{56}$Ni. As discussed in an earlier section of this paper, this model is not thin in the continuum until roughly 300 days after explosion. In fact, the situation is worse than this, since the optical region remains obscured by allowed lines (principally from Fe) until considerably later times. In the infrared, however, this model is predicted to be quite transparent at this time, and we therefore present the predicted spectrum at 300 days. The major caveat is that for this model the neglect of allowed lines may result in considerable errors in the calculated temperature and ionization state. Nontheless, it is clear that the infrared spectrum should contain a wealth of information on the abundance, temperature, and ionization state of the ejecta. The late time spectrum of this model has also been calculated and discussed by Fransson and Chevalier[18]

## 7. Acknowledgements

I am grateful to Claess Fransson for making many valuable points on the late time spectra of supernovae. This work performed under the auspices of the U. S. Department of Energy by the Lawrence Livermore National Laboratory under contract No. W-7405-ENG-48.

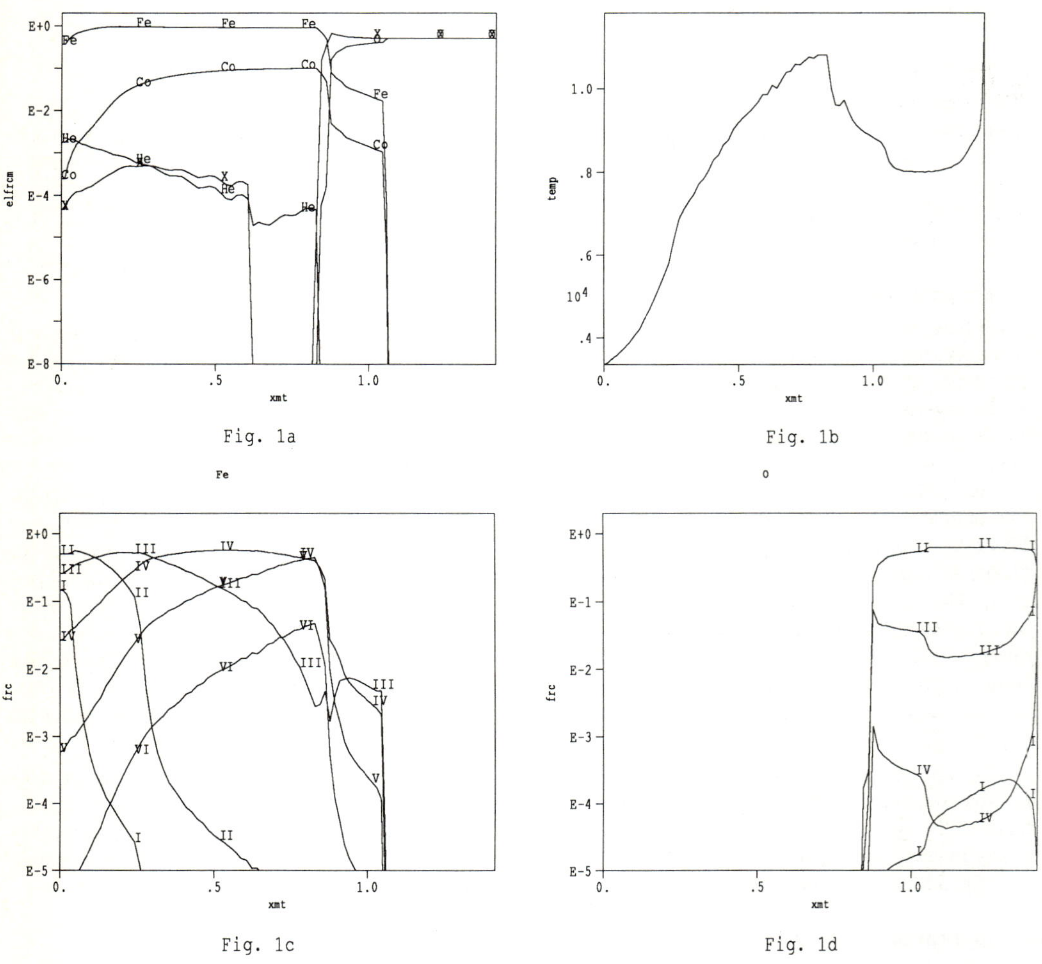

Fig. 1a

Fig. 1b

Fig. 1c

Fig. 1d

1.    Element mass fractions (Fig. 1a), temperature in degrees K (Fig. 1b), ionization state of Fe (Fig. 1c) and O (Fig. 1d) are shown for the SNIa model discussed in Section 4. The x axis is the mass coordinate in $M_\odot$

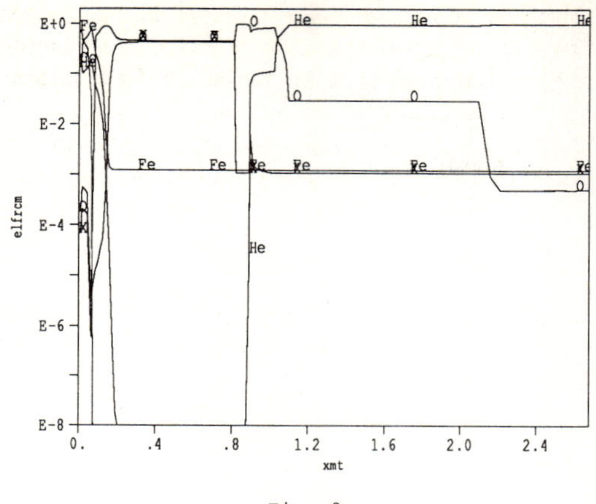

Fig. 2a

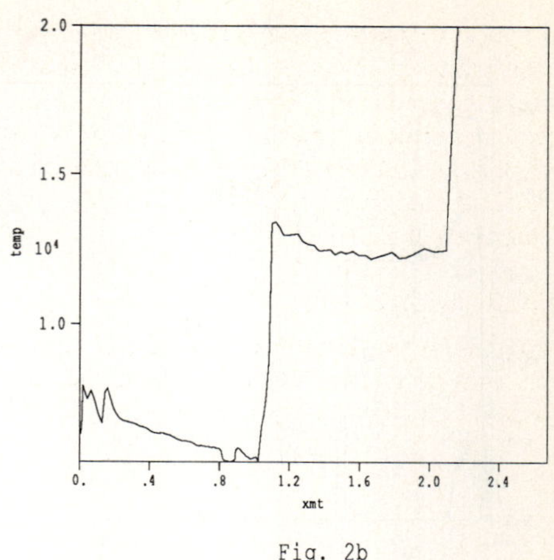

Fig. 2b

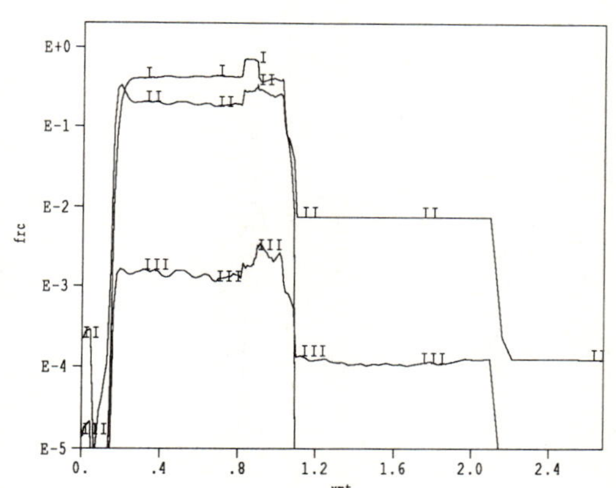

Fig. 2c

2.  Same as Fig. 1, but with Fe ionization state omitted, and for the SNIb model of Section 5.

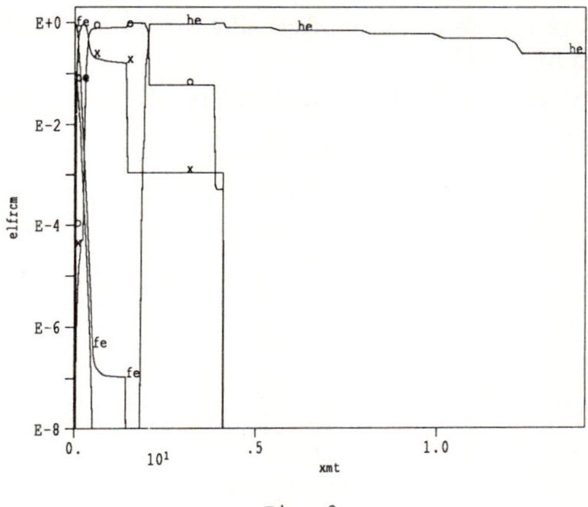

Fig. 3a

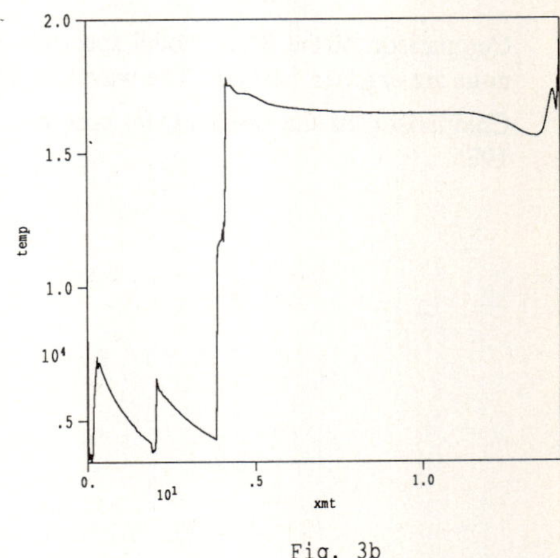

Fig. 3b

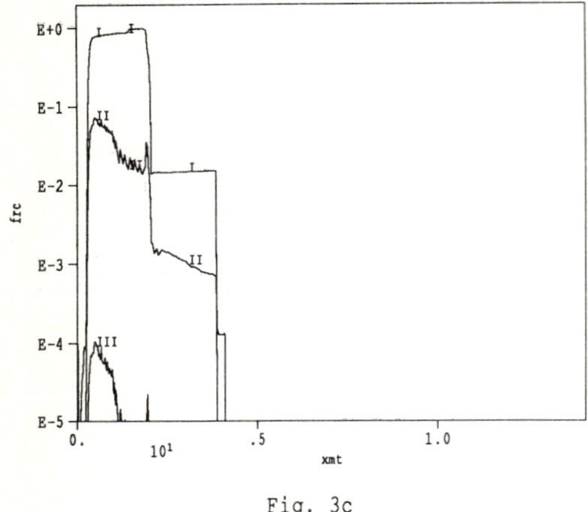


Fig. 3c

3.      Same as Fig. 2, but for the SN1987a model of Section 6.

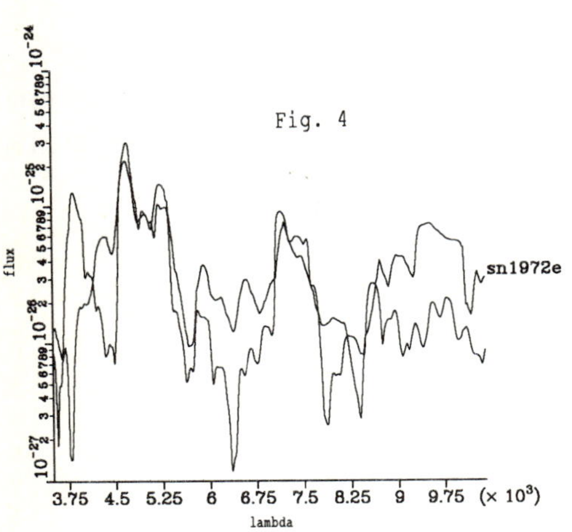


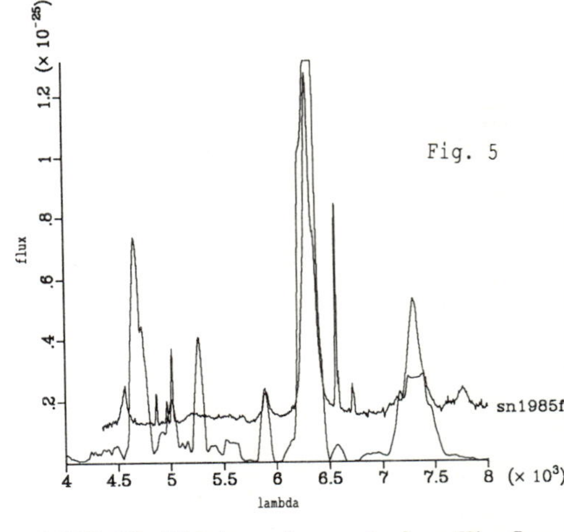


4.      Comparison of the SNIa model spectrum with that of SN1972e 264 days after explosion. The flux units are $erg/sec/cm^2/Hz$. The wavelength is in Angstroms.

5.      Comparison of the SNIb model spectrum 200 days after explosion with that of SN1985f in April 1985.

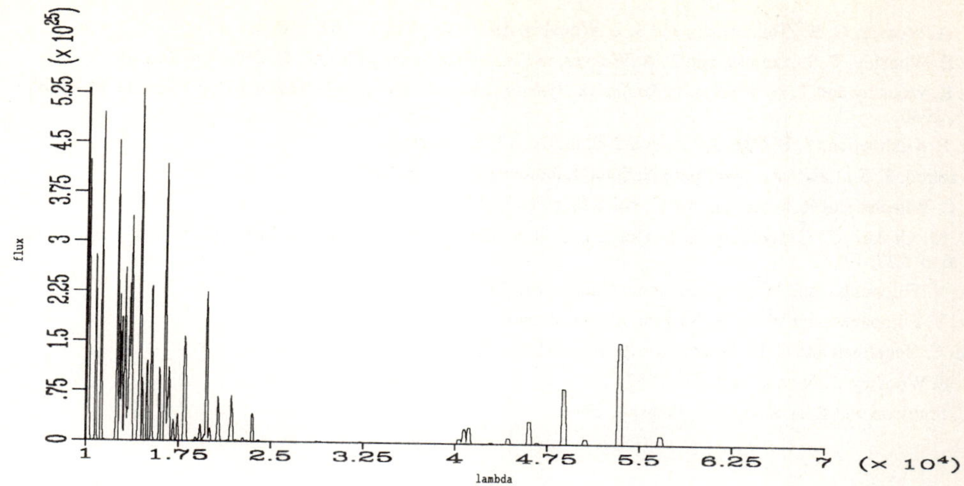

Fig. 6a

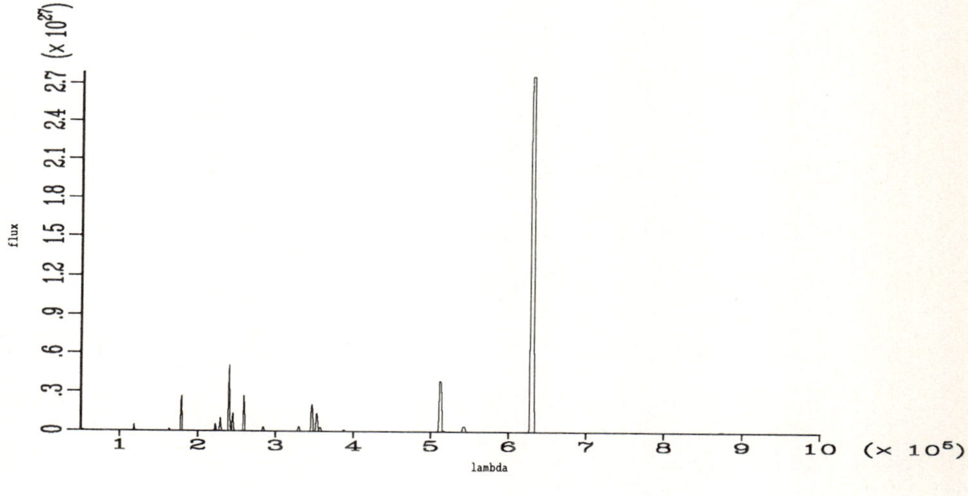

Fig. 6b

6.   Predicted IR spectrum of the SN1987a model at a time 300 days after explosion. The flux units are *erg/sec/Hz*.


**References**


1.   A. H. Karp, K. L. Chan, G. J. Lasher, and E. E. Salpeter, *Ap. J.*, vol. 214, p. 161, 1977.
2.   W. D. Arnett, *Ap. J.*, vol. 253, p. 785, 1982.
3.   S. A. Colgate and C. McKee, *Ap. J.*, vol. 157, p. 623, 1969.
4.   E. F. Plechaty, D. E. Cullen, and R. J. Howerton, Lawrence Livermore National Laboratory UCRL-50400, 1981.
5.   J. H. Scofield, *Phys. Rev.*, vol. 179, p. 9, 1969.
6.   J. H. Scofield, Lawrence Livermore National Laboratory UCRL-51231, 1972.

7.  T. A. Weaver, G. B. Zimmerman, and S. E. Woosley, *Ap. J.*, vol. 225, p. 1021, 1978.

8.  S. E. Woosley, T. S. Axelrod, and T. A. Weaver, in *Stellar Nucleosynthesis*, ed. C. Chiosi, p. 263, 1984.

9.  S. E. Woosley and T. A. Weaver, in *Radiation Hydrodynamics in Stars and Compact Objects*, ed. D. Mihalas, p. 91, 1986.

10.  R. P. Kirshner and J. B. Oke, *Ap. J.*, vol. 200, p. 574, 1975.

11.  Axelrod, T. S., Lawrence Livermore National Laboratory UCRL-52994, 1980.

12.  J. C. Wheeler and R. Levreault, *Ap. J.*, vol. 219, p. L17, 1985.

13.  C. M. Gaskell, E. Cappellaro, H. L. Dinerstein, D. R. Garnett, R. P. Harkness, and J. C. Wheeler, *Ap. J.*, vol. 306, p. L77, 1986.

14.  A. V. Filippenko and W. L. W. Sargent, *Nature*, vol. 316, p. 407, 1985.

15.  A. V. Filippenko and W. L. W. Sargent, *Astron. J.*, vol. 91, p. 691, 1986.

16.  M. C. Begelman and C. L. Sarazin, *Ap. J.*, vol. 302, p. L59, 1986.

17.  S. E. Woosley, *Submitted to Ap. J.*, 1987.

18.  C. Fransson and R. A. Chevalier, Preprint, 1987.

# OBSERVING THE NUCLEOSYNTHESIS FROM CORE COLLAPSE SUPERNOVAE

Claes Fransson
Stockholm Observatory
S-133 00 Saltsjöbaden, Sweden

## 1. INTRODUCTION

Since the appearance of the classical papers on stellar nucleosynthesis in the 1950's most of the observational tests have been through indirect sources of information. Even though this has been rather successful (cf. various contributions in this volume), it represents only an average over all sources, yielding little information about specific stars. The most direct evidence comes from observations of young galactic supernova remnants (age less than $\sim 10^3$ years). Unfortunately, the analysis of X-ray data are hampered by a lack of understanding of the detailed physics, eg. non-equilibrium and plasma effects, as well as by observational problems (Itoh and Nomoto, 1987). Optical observations (cf. Raymond, 1984) have given some valuable insight of eg. the oxygen-rich remnants. The analysis of these are, however, suffering from the fact that only a small fraction of the mass is seen in the optical.

Although being the most direct approach, comparatively little attention has been paid to direct observations of the nucleosynthesis in supernovae during the first years after the explosion. The main exception has been the modelling of the late Fe/Co spectra of Type I supernovae (Meyerott, 1980; Axelrod ,1980; this volume), and the early phases of Type I's by Branch et al. (1985). Both for Type I's and II's the early stages (before $\sim 200$ days) are complicated by NLTE effects, and the non-transparent nature of the envelope. Therefore, the phase most suitable for analysis is the epoch after which the envelope is transparent, and the density  low enough to make NLTE effects tractable. The most important condition is that the continua of the excited levels should be optically thin, so that a nebular approximation for the radiative transfer can be used. In this paper I will sketch the basic physics involved in this type of models, and then discuss some applications to Type Ib and Type II supernovae. For a more detailed account see Fransson and Chevalier (1987, 1988; in the following FC87 and FC88) and Fransson (1987).

## 2. RADIOACTIVE EXCITATION

Both from observations (Barbon et al. 1984) and from theory (eg. Woosley, this volume) there is strong evidence for the formation of $\sim$ 0.05-0.3 $M_\odot$ of $^{56}$Ni in the explosion of a massive star. This decays first to $^{56}$Co on a time scale of 8.6 days and then to $^{56}$Fe on 114 days, emitting 94% of the energy as 0.5-3 MeV $\gamma$-rays and 4% in positrons. The trapping and thermalization of these are determined mainly by the density and $\gamma$-ray optical depth of the oxygen core. By the oxygen core I mean all mass inside the oxygen shell, and similarly with the helium core. The important parameters are thus the mass, $M_c$, and expansion velocity,

$V_c$, of this. The mass of the oxygen core is relatively well determined, for a 15 $M_\odot$ star ~ 1.1 $M_\odot$, increasing to ~ 3.8 $M_\odot$ for 25 $M_\odot$ (Woosley and Weaver, 1987). The expansion velocity is less certain and depends on the deceleration of the core expansion by the hydrogen envelope. The total mass loss of the progenitor is therefore crucial. For stars more massive than ~20 $M_\odot$ mass loss may change the star from a red supergiant with an extended envelope, to a naked WR star (eg. Maeder, this volume). Calculations show that for a 15 $M_\odot$ star without mass loss the core velocity is ~ 900 km/s (Weaver and Woosley, 1980). For the same core mass, but without a H-envelope, the velocity increases to ~ 2500 km/s (Woosley and Weaver, 1987). Loss of part of the He-mantle may increase this further. The average density in the oxygen core is ~$5.7 \times 10^8$ ($M_c$/1 $M_\odot$) ($V_c$/$10^3$ km/s)$^{-3}$ $t_{yr}^{-3}$ cm$^{-3}$, and the energy averaged $\gamma$-ray optical depth $\tau_\gamma$~2 ($M_c$/1 $M_\odot$) ($V_c$/$10^3$ km/s)$^{-2}$ $t_{yr}^{-2}$. Hydrodynamical calculations show, however, that the density structure is highly non-uniform due to radioactive heating, which creates a bubble in the center, with most of the mass in a thin shell (Woosley and Weaver, 1987; Nomoto et al., 1987). The shell is likely to be subject to the Rayleigh-Taylor instability, smoothing the density distribution and causing a mixing of the elements.

The thermalization of the $\gamma$-rays proceeds in several steps: First the $\gamma$-rays loose their energy by Compton scattering off the bound and free electrons, producing fast non-thermal electrons with energies of 0.1-1 MeV. These loose their energy by ionizations, excitations and heating of the *thermal*, free electrons. In most cases direct excitations are unimportant. The ionizations are balanced by recombinations, and the heating by collisional excitations of low energy levels, mainly in the optical and near-IR. Each absorbed $\gamma$-ray photon thus produces ~ $10^6$ photons in the optical. The thermalization is treated in detail in FC 88, and for an Fe/Co plasma by Axelrod (1980). In this paper I will mainly discuss the application of these models to Type Ib and Type II supernovae, and will therefore only summarize the main ingredients.

The $\gamma$-ray and electron thermalization is modelled by Monte Carlo technique, giving the heating and ionization rates for a given composition and ionization state. This is done iteratively. The recombination and cooling include all important atomic processes. Most of the cooling is done by forbidden and semi-forbidden lines of neutral and singly ionized elements. For the line transfer we use the Sobolev approximation, and for the diffuse continuum a lambda iteration. For O I, and Ca II we take 9 and 4 levels into account. The elements included are  He, C, N, O, Ne, Na, Mg, Si, S, Ar, Ca, and Fe. In contrast to Axelrod (1980; these proceedings), who include only Fe, Co and O, we do not calculate the line emission of Fe and Co in detail. This is reasonable for massive stars, where most of the emission is dominated by the lighter elements, but of course insufficient for Type Ia's. Neither do we include hydrogen, since we are primarily interested in the emission from the region where nuclear processing has taken place.

## 3. GENERAL CONSIDERATIONS

The most interesting question is the correspondence between the abundance structure of the supernova ejecta and the resulting optical emission observed (FC87). To illustrate this we show in Fig. 1 the emission per unit mass in the various lines as a function of the mass from the center for the Woosley and Weaver (1987) 8 $M_\odot$ He-core (main sequence mass 25 $M_\odot$ ), 300 days after the explosion. The expansion velocity of the

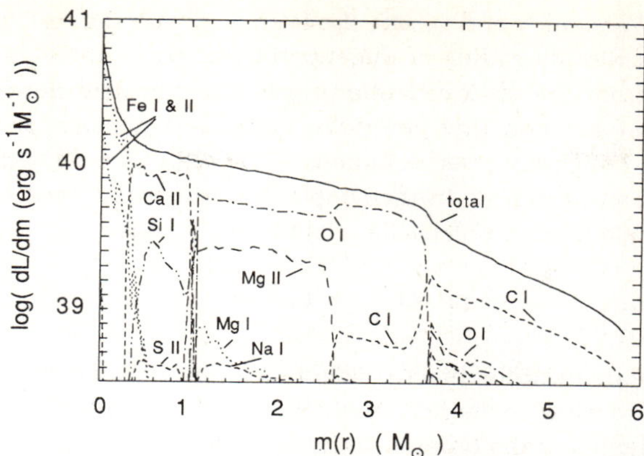


**Fig 1**. Luminosity per unit mass, dL/dm, of the most important lines 300 days after the explosion, for the Woosley and Weaver (1987) 8B model (ZAMS 25 M⊙ ). The spectrum from this model is shown in Fig. 3.

core was 5200 km/s, and the total energy $3 \times 10^{51}$ ergs. This model is discussed further in Sect. 4. Comparing the figure with the abundance structure (Woosley and Weaver, 1987), we see that there is a rough overall correspondence between the nuclear burning shells and the emission from the various zones. However, a closer examination shows that it is extremely important to take the detailed atomic physics into account. This is evident for the He-mantle, where He does is *not* contribute much to the emission, which is instead dominated by the trace elements, mainly C I and O I. The same thing can be seen in the inner part of the oxygen shell, where Mg I, Mg II and Na I lines are comparable to  O I, even though the total abundances are down by factors of ~25 and ~200, respectively. *Therefore, including the low abundance elements is necessary for a realistic comparison with observations.* Neglect of this may severely overestimate the emission from the dominant elements.It is important to realize that the presence of strong [O I] lines in the spectrum does *not* mean that the density has to be less than the critical density, $\sim 10^6$ cm$^{-3}$. While the thermalization cause the forbidden lines to radiate less efficiently compared to eg. the semi-forbidden, there are very few semi-forbidden or allowed transitions for neutral and singly ionized elements. This means that there are few channels other than the forbidden to do the cooling.  As shown by the figure most of the elements lighter than iron can be studied from observations of optical and near-IR lines. The most important exceptions are Ne and Ar, which have no lines in this region of the spectrum. There are, however, in the far-IR fine-structure transitions of these at 12.81 μ  ([Ne II]) and at 6.985 μ ([Ar II]), which should become prominent especially at late epochs (~450 days and later).

In general, the total emission from a particular burning zone is proportional to the γ-ray optical depth through the zone, $\Delta\tau_\gamma = \kappa_\gamma \int \rho(r)\, dr$. The density structure is thus important for the relative line strengths. Fortunately, both this and the core velocity may be obtained directly from observations of the line profiles at late epochs. The velocity field has then relaxed to a V∝r law, and for an optically thin line the emission per volume, j(r), is related to the intensity, I(ε) , of the line at the dimensionless frequency ε by

$$j(r=\varepsilon R) = \frac{1}{2\pi\varepsilon R^2}\frac{dI(\varepsilon)}{d\varepsilon} \tag{1}$$

Here $\varepsilon = (1-\nu/\nu_0)c/V_0$, and $\nu$ and $\nu_0$ are the frequency and the rest frequency of the line, respectively, R the maximum radius of the supernova and $V_0$ the velocity at this point (ie. $R/V_0 = t$). Therefore from the observed line profile one can determine the emissivity as a function of the radius $\varepsilon R$, and thus the $\gamma$-ray input and density. If the energy input is dominated by $\gamma$-rays, the heating per volume is $\sim 0.03\, \rho\, L_\gamma/(4\pi r^2)$, where $\rho$ is the density and r the distance from the center. In the simple, but important, case where all the cooling is dominated by a single line we obtain the relation (since $r=\varepsilon R$)

$$\rho(r=\varepsilon R) \quad \propto\ \varepsilon\ \frac{dI(\varepsilon)}{d\varepsilon} \tag{2}$$

Thus the shape of the line profile directly reflects the density distribution of the ejecta. It is, however, important to check whether the lines are really optically thin, since even the forbidden can have depths of the order of unity or larger. For [O I] $\lambda\,6300$ we have $\tau=2$ $(n(O\ I)/10^9\ cm^{-3})\,t_{yr} \sim 1.1\,M_c\,(V_c/10^3\ km/s)^{-3}\,t_{yr}^{-2}$.


## 4. TYPE Ib SUPERNOVAE

The main characteristics of the Type I b supernovae are: 1) Occurrence close to H II regions, 2) radio emission, presumably due to circumstellar interaction, 3) underluminous compared to Type Ia's, indicating a smaller mass of $^{56}$Ni, 4) after $\sim 200$ days the spectra are dominated by strong [O I] lines (see Panagia (1987) for a review). In Fig. 2 the beautiful spectrum of SN 1985f, obtained by Fillipenko and Sargent (1986) in March/April 1985, is shown. Unfortunately, for this supernova the exact time since the explosion is not known, but from a comparison with similar observations of SN 1983n Gaskell et al. (1986) argue that the epoch is $\sim 200$-300 days. Note the complete absence of any broad H$\alpha$ line, immediately showing that the gas has undergone advanced nuclear processing. Although not yet settled

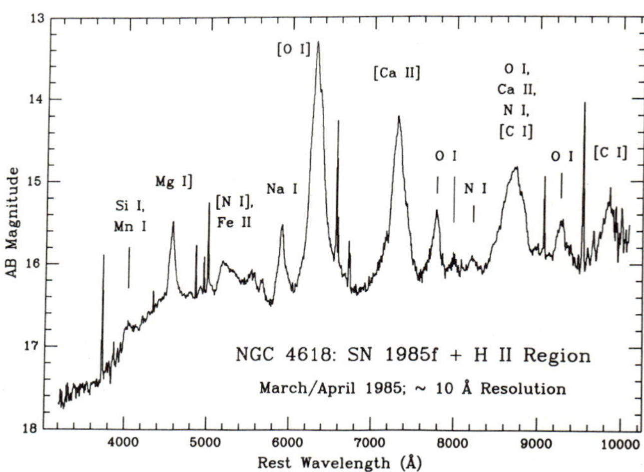


Fig. 2. Spectrum of SN 1985f in March/April 1985 from Fillipenko and Sargent (1986). Note the broad lines, the absence of a broad H$\alpha$ line and the peaked profiles of the lines.

(for a different view see Branch, this volume), most of these properties indicate a massive progenitor (Fillipenko and Sargent, 1985; Chevalier, 1986; Wheeler and Levreault, 1986; Fransson, 1986). From a nebular analysis of the O I and O II lines, Begelman and Sarazin (1986) proposed that SN 1985f was the result of the explosion of a ~ 50 $M_\odot$ Wolf-Rayet star, which had undergone a pair-instability collapse. Since the frequency of Ib's is comparable with the Type Ia's (Panagia, 1987), this origin is unlikely to apply for all Ib's. The main conclusion of their nebular analysis was that a minimum of ~ 5 $M_\odot$ of oxygen was needed to explain the observations, making a very massive progenitor necessary. There are, however, several loop-holes in their argument. The most obvious are the uncertainties in the distance and reddening, which affect the [O I] $\lambda$ 6300-64 flux, and thus also the oxygen mass. Taking these into account, as little as ~ 1 $M_\odot$ of oxygen is possible. Another complication is that the O I and O II zones in general do not coincide, meaning that the constraints from the O I recombination lines are further relaxed. It is thus difficult to make any definite conclusions from this type of analysis, although it is probably difficult to escape the conclusion that at least ~ 1 $M_\odot$ of oxygen is needed.

In addition to the oxygen lines, the late spectra of the I b's also display emission lines of several other ions, like C I, Na I, Mg I, and Ca II (Fig. 2), from which additional constraints can be obtained. For realistic conclusions one must, however, calculate a self-consistent model, with a density and abundance structure of the ejecta given by hydrodynamic calculations (or better, from observations of line profiles). As input models we have used the Woosley and Weaver (1987) models of exploding WR stars, without hydrogen envelope. Light curves for this type of models have been calculated by Ensman and Woosley (1986). Compared to the observed Type Ib light curves they, however, give too wide a peak. This may constrain the mass and energy severely for a WR origin. However, since the density distribution of these models do not reproduce the observed line profiles (see below), it is too early to make any definite conclusions. In this connection it should also be remembered that current models have difficulty in explaining the observed light curve for SN 1987a.

For a qualitative comparison we have calculated the spectrum of the exploded Woosley and Weaver (1987) 8 $M_\odot$ He core (ZAMS mass 25 $M_\odot$), model 8 B , with a velocity of ~ 5200 km/s at the O/He interface. In Fig. 3 the synthetic spectrum of this model is shown at 300 days. It should be stressed that there are very few free parameters in these models, except for the core mass, the $^{56}$Ni mass and the core velocity, both set by the observations. Since especially elements with low ionization potentials, like Na I, Mg I and Si I, are sensitive to details, such as the photoionization by diffuse emission and the density structure, the spectra presented here should be regarded as preliminary results. Given this, there is a remarkably good *qualitative* agreement between the observations in Fig. 2 and Fig. 3, both in terms of the lines present and their relative strengths. As noted by Fillipenko and Sargent, the identifications in Fig. 2 are uncertain due to the widths of the lines, and the feature at 4036 Å is eg. consistent with [S I] $\lambda$ 4069-76 within the errors. We note that most of the emission in the line at ~7296 Å is due to the Ca II] $\lambda\lambda$ 7291-7324 lines, with only a small contribution from [O II] $\lambda\lambda$ 7320-30. Also the wide feature at 8700 Å is well modeled by a blend of O I 8446, Ca II $\lambda\lambda$ 8498-8662, and most important [C I] $\lambda$ 8727. The main discrepancy is in the Na I $\lambda$ 5890 strength, which is severely underestimated. This is, however, probably the most uncertain line, both in terms of the ionization balance and also in the total Na abundance in the models. The forbidden [Si I] lines at 1.099 $\mu$ and 1.645 $\mu$ are not in the range observed for this supernova, but the 1.6068-1.6455 $\mu$ lines were probably present

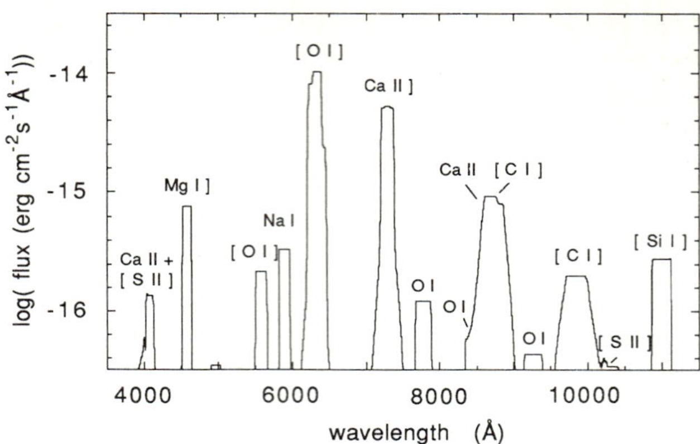


**Fig. 3.** Synthetic spectrum for the  25 M$_\odot$ (=8 M$_\odot$ He-core) model 300 days after the explosion. The expansion velocity of the O-core was ~5200 km/s. Note that several lines, eg  [O I] $\lambda\lambda$ 6300-64 and Ca II $\lambda\lambda$ 8498-8662 + [C I] $\lambda$ 8727 are blends. The large width of the [CI] $\lambda$ 9823-49 line is mainly due to the contribution of the high velocity He-mantle.

in the IR spectra of the Type Ib SN 1983n observed ~ 1 year after the explosion by Graham et al. (1986). These authors attributed the line to the [Fe II] 1.600-1.644-1.664-1.677 $\mu$ multiplet, but an equally likely interpretation, consistent with the expected strength, is due to the [Si I] line (Oliva, (1987); Fransson (1987)).

We have also tested WR models with smaller core masses, 4 - 6 M$_\odot$ , but find considerably worse agreement, both in absolute and relative line strengths. Especially the Ca II / [O I] ratio is sensitive to the core mass, and may in fact require a somewhat higher mass than 25 M$_0$ (see Sect. 5). Also an extreme white dwarf model, provided by Stan Woosley (priv. comm.) has been tested. In this the deflagration wave died out quickly, producing only ~0.2  M$_\odot$ of $^{56}$Ni, leaving ~0.5 M$_\odot$ of unburned O and ~0.5 M$_\odot$ of C, and ~0.2 M$_\odot$ of other elements. The main problems with this is the small absolute strengths of the [O I] $\lambda\lambda$ 6300-64 lines,  the far too large [C I] $\lambda\lambda$ 9823-49 / [O I] $\lambda\lambda$ 6300-64 ratio, and the large strengths of the [S I] lines. Most of the emission emerges as Fe emission. We thus find it unlikely that the Type Ib's can be explained by even extreme types of exploding white dwarf models.

The main problem with the WR models is that the line profiles in Fig. 3 are too flat, compared to the peaked profiles in Fig. 2. This is due to the shell structure of the core, due to the $^{56}$Ni heating, with nearly all the oxygen at one velocity. To estimate the required distribution of the ejecta, we have applied Eq. (2) to the blue wing of the [O I] $\lambda$ 6300 line in the spectrum of SN 1985f (Fillipenko and Sargent, 1985), where the emission indeed was dominated by the [O I] $\lambda\lambda$ 6300-64 line. In Fig. 4 we show the density in dimensionless units as a function of the velocity, and thus radius (V$\propto$r ), indicating a much wider distribution in velocity than  the models give. A possible explanation is that hydrodynamic instabilities during the first days may lead to mixing between different burning shells, and a smoothing of the distribution. Also, if clumps of the dense core material penetrate into the He-mantle and H-envelope, the line profiles of the metal lines will extend to higher velocities. Observations of Cas A (Chevalier and Kirshner, 1979) show evidence for this type of mixing.

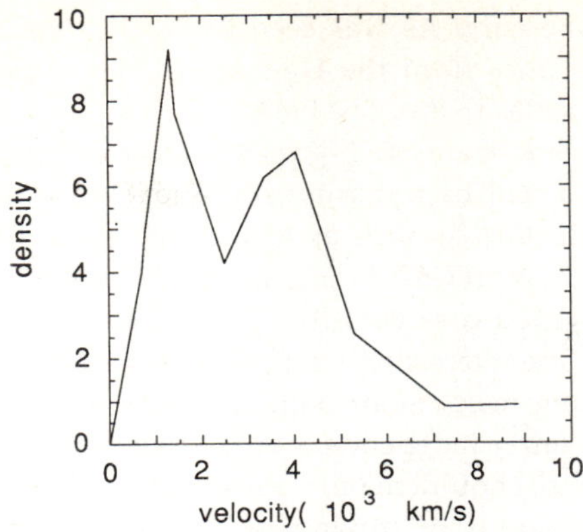

Fig. 4. Density as a function of expansion velocity ( $V(r) \propto r$ ) for the Type Ib SN 1985f, derived from the [O I] $\lambda\lambda$ 6300 line profile.

## 5. TYPE II SUPERNOVAE AND SN 1987a

Type II supernovae are thought to arise as the result of the explosion of stars more massive than ~10 $M_\odot$. As we have already discussed the main difference between Type II's and the explosion of a WR star is the presence of a massive hydrogen envelope for the former. For the late spectrum of Type II's this results in a very strong Hα line, dominating the spectrum (see eg. the spectrum of SN 1979c by Branch et al., 1981). Uomoto and Kirshner (1986) have shown that for SN 1980k the luminosity of this line decayed with a rate close to the $^{56}$Co decay time, as well as the total luminosity evolution observed by Barbon et al. (1982). This implies a γ-ray optical depth larger than unity. In that case one, however, expects the excited region to grow with the falling density, and the maximum velocity of the Hα line should increase. Since this is not observed, direct γ-ray excitation of the line is unlikely. The exact excitation is uncertain, but probably involves a combination of collisional excitation and ionizations by the Balmer continuum, followed by recombination. If the Balmer continuum flux follows the γ-ray input, as is likely, this may be reflected in the strength of Hα. The large Hα/Hβ ratio indicates a large optical depth in Hβ, transforming these photons into Hα and Pα.

For the understanding of the nucleosynthesis the Hα emission is of moderate importance. The expansion of the core is, however, sensitive to the presence of a hydrogen envelope. Because of the deceleration, the O-core velocity is only ~1000 km/s and that of the He-core ~2500 km/s (Woosley 1987). The lines are thus expected to be narrower and the γ-ray optical depth higher compared to Type Ib's, $\tau_\gamma \propto V_c^{-2}$. Therefore, the γ-ray trapping is expected to be efficient for a longer time, and the light curves of both SN 1979c and SN 1980k showed no deviation from a pure exponential, more than a year after the explosion. Except for the Hα emission surprisingly little is known about the spectra of these at late phases. Recent observations by Fillipenko (1987) of SN 1986i, ~9 months after the explosion, however, show a number of strong emission lines of O I, Ca II and Na I. The widths of these lines indicate an origin in the core. As was seen in Fig. 1, there may, however, also be a strong contribution to these lines from the He-mantle. This supernova is also interesting, since its early spectrum resembles that of SN 1987a.

Obviously the late spectrum of SN 1987a is of great interest for the understanding of the nucleosynthesis, and I will here discuss some results, which attempt to show what kind of observations are relevant, as well as to indicate the kind of information one can obtain from this type of analysis (FC87; Fransson 1987). In FC87 we studied two different models, one 15 $M_\odot$ model with a core velocity of ~1200 km/s, and one 25 $M_\odot$ model with core velocity of 2600 km/s. One interesting result was the increase in the [O I] $\lambda$ 6300-64 / Ca II] $\lambda$ 7300 ratio with increasing mass. Since both O I and Ca II are the dominant ionization stages, this ratio is probably fairly independent of the details of the ionization, in contrast to the Mg I and Na I lines, and should mainly reflect the nucleosynthesis. The ratio does, however, depend on the amount of mixing of Ca and O, which is uncertain in the explosion models. A mixing should also show up in the widths of the lines. Observations of these lines are therefore a useful probe of the nucleosynthesis.

As an example of this type of models, a calculation of a 20 $M_\odot$ (ZAMS) explosion model, specifically designed for SN 1987a (Woosley, 1987) has been done. Due to the $^{56}$Ni heating during the first days, most of the core mass is piled up in the high density peak outside the central $^{56}$Ni bubble. This density distribution will give rise to a flat profile, similar to those in Fig. 3, but with a half width of ~1000 km/s. High resolution observations can thus provide a badly needed test of the hydrodynamic calculations, as well as on the distribution of the individual elements. In Fig. 5 we show the spectrum in the optical and near-IR wavelength range one year after the explosion, and in Table 1 we give the line luminosities. This model is at the border line between the Ca II] dominated and [O I] dominated models, with [O I] $\lambda\lambda$ 6300-64 / Ca II] $\lambda$ 7300 ~ 0.9

**Table 1.**

Luminosities after one year relative to [O I] $\lambda\lambda$ 6300-64 of the strongest lines for the 20 $M_\odot$ model BF7 from Woosley (1987). L(6300-64) is the absolute luminosity of the [O I] $\lambda\lambda$ 6300-64 line, $V_c$ is the O-core velocity, and M($^{56}$Ni) the $^{56}$Ni mass.

| M($^{56}$Ni) ($M_\odot$) | | 0.075 | $V_c$ ( km/s) | | 1200 |
|---|---|---|---|---|---|
| L (6300-64) (erg/s) | | $2.1\times10^{39}$ | | | |
| [C I] | 8727 | 0.64 | Mg II | 2800 | 0.74 |
| [C I] | 9823-49 | 0.12 | [Si I] | 10991 | 0.49 |
| O I | 1356 | 0.04 | [Si I] | 16455 | 0.24 |
| [O I] | 5577 | 0.25 | [S II] | 4069-76 | 0.07 |
| [O I] | 6300-64 | 1.00 | [S II] | 10286-10371 | 0.05 |
| O I | 7774 | 0.21 | [S I] | 25.25 $\mu$ | 0.10 |
| [O I] | 63.15 $\mu$ | 0.04 | [Ar II] | 6.985 $\mu$ | 0.014 |
| [Ne II] | 12.81 $\mu$ | 0.006 | Ca II | 3934-68 | 0.17 |
| Na I | 5890-96 | 0.76 | Ca II] | 7291-7332 | 0.91 |
| Mg I] | 4571 | 0.43 | Ca II | 8498-8662 | 0.49 |

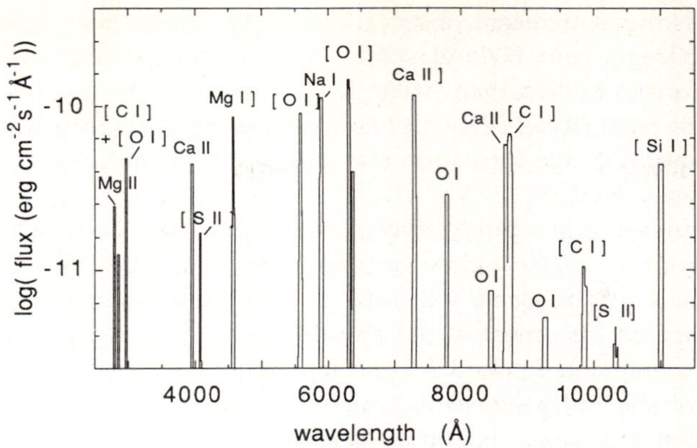



**Fig. 5.** Synthetic spectrum one year after the explosion of model BF7 from Woosley (1987), The ZAMS mass was 20 $M_\odot$  and the oxygen core velocity ~1200 km/s.

Although the He-mantle was included in the calculations, there is very little trace of He in the spectrum, which is due to the low cooling efficiency of this element. Most of the absorbed energy is emitted as [C I] $\lambda\lambda$ 8727, 9824-50 lines. Also the [O I] and Ca II] lines may contribute to the cooling of the He-mantle, depending on the relative abundances of C, O and Ca in the envelope. Lines from this region are expected to show a larger velocity width than lines coming solely from the core. Unless mixing is substantial, the iron lines are expected to come from the inner core. The emission from this region has not been calculated in detail, but the total Fe I-II emission is for both models less than ~20% of the total. Mixing of the iron throughout the core may, however, increase this substantially.

When comparing these results with observations it should be kept in mind that the line strengths in some cases are sensitive to, for example, the density distribution of the ejecta. This is especially true for the Na I, Mg I], and [Si I] lines, for which the ionization is dominated by photoionization by diffuse UV emission, mainly from O I recombination lines. This is sensitive to both the density and to resonance line blocking in the UV. At epochs less than ~450 days  the density in the core can be so high that forbidden lines like the [O I] and [Si I] lines are optically thick, resulting in P-Cygni type absorptions also for these lines.

The time dependence of the emission line luminosities is set by the decaying $\gamma$-ray input, proportional to $\tau_\gamma L_\gamma \propto t^{-2} e^{-t/114\,d}$ . As discussed in detail in FC87 and Fransson (1987), this may lead to a thermal instability at ~700 days, due to a transition from cooling by optical and near-IR transitions to cooling by far-IR fine structure lines. Most of the emission then emerges as far-IR lines, like [O I] 63.15 µ, [Ne II] 12.81 µ, [Si I] 1.645 µ, [Si I] 68.49 µ, [S I] 25.25 µ, and [Ar II] 6.985 µ. These may be strong also at earlier epochs, and are thus important to monitor. They also have the advantage of being easy to model and should thus yield

The thermal instability results in a sudden drop of the temperature from ~3500 K to less than 1000 K. In the case of a clumpy density distribution the instability may set in earlier, and may also trigger the formation of dust in this metal rich environment. In this connection it is interesting that recent observations of SN 1987a have shown evidence of CO in the ejecta (McGregor and Hyland, 1987). Preliminary calculations show that this requires the temperature to be less than ~4000 K, and the electron fraction to be less than ~0.1. Since the lines are fairly broad, they are, however, likely to be formed in the hydrogen envelope. At later stages molecule formation may also occur in the core regions, and be the first sign of dust condensation.

Finally, the presence of a pulsar may be important after about two years, when the radioactivity has decayed. A pulsar will be surrounded by a synchrotron nebula (a plerion), emitting ionizing radiation. The ejecta will thus be gradually ionized, as they expand. In contrast to the $\gamma$-ray heated case most of the absorbed energy is expected to be emitted as lines of highly ionized elements in the UV. The ionization of the envelope will also cause free-free absorption of the radio emission from the plerion, making a detection of this difficult during the first few years. For more details I refer to FC87 and Fransson (1987). Chevalier (1987) has proposed that the optical emission from SN 1986j (Rupen et. al., 1987) may be the result of pulsar excitation. Although the age of the supernova is not known, Chevalier argues from the evolution of the radio emission that it is ~ 4 years. The fact that the optical spectrum shows a number of strong He I lines with high excitation potentials may be consistent with photoionization by a central hard continuum source. Also the low velocities, FWHM~1000 km/s, argues for emission from the core. The modelling of this is, however, very sensitive to mixing, filamentation etc., known to be important for the Crab.


## 6. CONCLUSIONS

The discussion in this paper shows that we in the future can expect the understanding of the stellar nucleosynthesis from observations of supernovae to be considerably improved. Most of the physics in connection with the thermalization of the $\gamma$-rays is well understood, as well as most of the atomic data going into the calculations. There are, however, in this area some uncertain processes, most importantly the charge transfer reactions between the various ions, like O II + Na I $\rightleftharpoons$ O I + Na II. Also the ionization of the trace elements, Na I, Mg I and Si I, may be sensitive to the treatment of the UV radiation field. However, these problems are likely to be solved in the near future. Therefore, from a *given* explosion model of the density and abundance structure one can *predict* what the late spectrum should be, and compare this with the observations. Especially the line profiles are important, since they provide a test of the probably most uncertain part of the explosion calculations.

If the late spectra of SN 1987a do not correspond to the model predictions this is hardly surprising. However, using these observations I think that with our current knowledge of the spectral formation we can learn a lot about the structure of the supernova and of nucleosynthesis. It should also be remembered that SN 1987a is only one single event, and that it is necessary to test the models for a wide range of progenitors. It is thus important to get good spectral information about other more distant supernovae. In this respect we can expect much from the future, since the observational requirements are already within our capabilities.

Acknowledgements: Most of this work has been done in collaboration with Roger Chevalier. I am also grateful to Claes-Ingvar Björnsson for reading the manuscript and for discussions, to Stan Woosley for providing the models of SN 1987a, and to Alexei Fillipenko for the spectra of SN 1985f.


# REFERENCES


Axelrod, T.S.: 1980, Ph.D. thesis, Univ. of California, Santa Cruz.

Barbon, R., Ciatti, F., and Rosino, L.: 1982, Astron. Astrophys. **116**, 35.

Barbon, R., Cappelaro, E., and Turatto, M.: 1984, Astr. Ap. **135**, 27.

Begelman, M.C., and Sarazin, C.L.: 1986, Astrophys. J. (Letters) **302**, L59.

Branch, D., Falk, S.W., McCall, M.L., Rybski, P., Uomoto, A.K., Wheeler, J.C., and Wills, B.J.: 1981, Astrophys. J. **244**, 780.

Branch, D., Doggett, J.B., Nomoto, K., and Thielemann, F.-K.: 1985, Astrophys. J. **294**, 619.

Chevalier, R.A.: 1986, in *Highlights of Astronomy*, Ed. J.P. Swings, (D. Reidel: Dordrecht) p. 599.

Chevalier, R.A.: 1987, Nature, **329**, 611.

Chevalier, R.A., and Kirshner, R.P.: 1979, Ap. J. **233**, 154.

Chiosi, C., and Maeder, A.: 1986, Ann. Rev. Astron. Astrophys. **24**, 329.

Ensman, L.M., and Woosley, S.E. : 1986, B.A.A.S. **18**, 963.

Fillipenko, A.V.: 1987, in *Proc. of the 4th George Mason Conference on SN 1987A*, ed. M. Kafatos, in press.

Fillipenko, A.V., and Sargent, W.L.W.:1985, Nature, **316**, 407.

Fillipenko, A.V., and Sargent, W.L.W.:1986, Astron. J. **91**, 692.

Fransson, C.: 1986, in *Highlights of Astronomy*, Ed. J.P. Swings, (D. Reidel: Dordrecht) p. 611.

Fransson, C.: 1987, in *ESO Workshop on SN 1987A*, Ed. J.I. Danziger, in press.

Fransson, C., and Chevalier, R.A.: 1987, Ap. J. (Letters), Nov. 1 1987. (FC87)

Gaskell, C.M., Cappellaro, E., Dinerstein, H.L., Garnett, D.R., Harkness, R.P. and Wheeler, J.C.: 1986, Astrophys. J. (Letters) **306**, L77.

Graham, J.R., Meikle, P.S., Allen, D.A., Longmore, A.J., and Williams, P.M.: 1986, M.N.R.A.S. **218**, 93.

Itoh, H. , and Nomoto, K.: 1987, in *X-Ray Astrophysics*, Eds. A.C. Fabian and C.R. Canizares, in press.

McGregor, P.J., and Hyland, A.R.: 1987, I.A.U. Circ. No. 4486.

Meyerott, R.E.: 1980, Astrophys. J. **239**, 257.

Nomoto, K., Shigeyama, T., and Hashimoto, M.: 1987, in *ESO Workshop on SN 1987A*, Ed. J.I. Danziger, in press.

Oliva, E.: 1987, Astrophys. J. (Letters) **321**, L45.

Panagia, N.: 1987, in *High Energy Phenomena Around Collapsed Stars*, Ed. F. Pacini, Reidel (Dordrecht), p.33.

Raymond, J. C.: 1984, Ann. Rev. Astron. Astrophys. **22**, 75.

Rupen, M.P., van Gorkom, J.H., Knapp, G.R., Gunn, J.E., and Schneider, D.P.: 1987, Astron. J. **94**, 61.

Uomoto, A., and Kirshner, R.P.: 1986, Astrophys. J. **308**, 685.

Weaver, T.A., and Woosley, S.E.: 1980, in *Supernova Spectra*, ed. R. Meyerott and G.H. Gillespie, American Institute of Physics, p. 15.

Wheeler, J.C., and Levreault, R.: 1985, Ap. J. (Letters), **294**, L17.

Woosley, S.E.: 1987, in prep. for Ap. J.

Woosley, S.E., and Weaver, T.A.: 1987, preprint.

# X-rays and $\gamma$-rays from Supernova 1987a

P. Sutherland[1], Y. Xu[2], R. McCray[2], and R. Ross[3]

## SUMMARY

The observation of X-rays and $\gamma$-rays from SN 1987a can provide important constraints on parameters for models of this unique event. We present the results of detailed Monte Carlo calculations of the fluxes to be expected in several X-ray bands and for the strong line at 847 keV associated with the decay of $^{56}$Co. Our calculations use Model 10H of Woosley, Pinto, and Ensman(1988), with $0.075 M_{\odot}$ of radioactive material. If it is assumed that there is no mixing of this material with the layers above, then the X-ray fluxes do not become detectable as early as the observations made by the *Ginga* team in August, 1987. If these observations correspond to X-rays arising from $\gamma$-rays Compton scattered down in energy in the supernova ejecta, rather than the interaction of the ejecta with circumstellar matter, then they can only be explained by mixing outward of radioactive material or an envelope with some combination of less mass or greater kinetic energy per unit mass.

## INTRODUCTION

It has by now become abundantly clear from the bolometric light curve of SN 1987a that the ejecta contain $0.075 M_{\odot}$ of $^{56}$Co. The ejecta are almost entirely opaque to the $\gamma$-rays released in the decay of $^{56}$Co to $^{56}$Fe, and after several months it is the rate of deposition of their energy that directly powers the light curve because the diffusion time for optical photons is then shorter than the dynamical time. This permits the determination of the radioactive mass to the accuracy of the distance of the LMC. The $\gamma$-rays are not immediately absorbed, however, but are instead repeatedly Compton scattered and degraded in energy and subsequently they may be photoelectrically absorbed by heavy elements or they may escape as X-rays. As the ejecta continue to expand, less scattering and energy degradation occur and a higher fraction of the photons escapes. Many groups have calculated the X-ray and $\gamma$-ray fluxes expected in this scenario (see McCray, Shull and Sutherland, 1987; Chan and Lingenfelter 1987; Gehrels, MacCallum, and Leventhal 1987; Xu *et al.* 1988; Ebisuzaki and Shibazaki 1987; Pinto and Woosley, 1988; Itoh *et al.* 1987). We present and discuss results for a representative calculation done in mid-summer 1987 at the University of Colorado.

Before proceeding to the details, it is useful to estimate the critical factors that determine the fluxes. If the $\gamma$-rays were unimpeded by the ejecta then their flux at Earth would be:

$$f_{\gamma} = 1.6 \times (M/0.075 M_{\odot}) exp(-t/t_c) \text{ counts s}^{-1} \text{ cm}^{-2}; \qquad (1)$$

where $t_c = 113.6$ days is the mean life of $^{56}$Co. In each decay of a $^{56}$Co nucleus, on average 2.88 $\gamma$-rays are emitted with a mean energy of 1.24 MeV.

To estimate the critical epoch for the emergence of the X-ray flux we need to balance the effects of energy degradation through multiple-scattering and photoelectric absorption. A photon of energy $E_0$ has its energy reduced to

$$E_1 = E_0/[1 + (E_0/mc^2)(1 - cos\theta)] \qquad (2)$$

by Compton scattering off a cold electron. For purposes of estimation, we may set $cos\theta \sim 0$ and

---

[1]Physics Dept., McMaster University, Hamilton, Ontario, Canada and Visiting Member, JILA 1986-87

[2]JILA, University of Colorado and National Bureau of Standards, Boulder, Colorado, USA

[3]Dept. of Physics, College of the Holy Cross, Worcester, Mass., USA

then the energy loss equation may be iterated for $n$ scatterings to yield:

$$E_n \sim mc^2/n \sim mc^2/\tau_s^2 \tag{3}$$

independent of the initial $\gamma$-ray energy. For ejecta in the homologous expansion phase, $\tau_s = \tau_{s,0}(t_0/t)^2$ where $\tau_{s,0}$ is a fiducial value for the scattering depth at time $t_0$. The photoelectric optical depth, as a function of X-ray energy, is:

$$\tau_a = 1.5\zeta(E/10 \text{ keV})^{-3} \times \tau_s \tag{4}$$

where $\zeta$ is the metallicity relative to solar. At early times the characteristic energy of the "down-Comptonized" photons is very low, and these photons are absorbed with little chance of escape. The critical epoch for emergence of the X-ray flux occurs when the effective absorption optical depth $\tau_{a,eff} \sim [\tau_a(\tau_s + \tau_a)]^{1/2}$ falls below unity; $\tau_a$ is evaluated at the characteristic energy of the "down-Comptonized" photons. One readily finds that at this epoch $\tau_{s,crit} \sim 4\zeta^{-1/8}$ and the escaping X-rays have characteristic energy $\sim 30\zeta^{1/4}$ keV. The critical epoch is $t_{crit} \sim [\tau_{s,0}t_0^2/\tau_{s,crit}]^{1/2}$. Using naive scaling for the explosion itself, one finds $t_{crit} \propto M/K^{1/2}$ with M and K the total mass and kinetic energy of the explosion.

The emergent flux in a $\gamma$-ray line can be estimated with a simple escape probability: take equation (1) above and multiply by the angle-average of $exp(-\tau_l)$ where $\tau_l$ is the scattering optical depth at the line energy. This simplicity is a consequence of the fact that a single scattering will almost certainly remove the $\gamma$-ray from the line because of the large energy loss to recoil. The $\gamma$-ray lines will emerge from the continuum after the peak in the X-rays since the latter depends upon multiple-scattering and the former is suppressed by it.

For Model 10H of Woosley, Pinto and Ensman (1988) the radial optical depth to the radioactive shell is $\sim 10$ at $\sim 1$ year and this implies that it would be unlikely for X-rays to become detectable until spring, 1988.

It is clear from the above simple analysis that quantitative predictions demand a Monte Carlo calculation, for the following reasons: (i) for the first few scatterings when the photon energy is still relatively large, the Klein-Nishina cross-section is appropriate and this is considerably smaller than the Thomson value, (ii) the scatterings at high energy are strongly forward-peaked, (iii) the ejecta are very inhomogeneous and compositionally stratified so that absorption cannot be treated with a single, global parameter, and (iv) crucial to confrontation with observations is the precise determination of the distribution of X-ray flux with energy. The results presented below are similar to those given in Xu *et al.* (1988) but have been augmented with results for the 847 keV $\gamma$-ray flux and with results for a model in which fractions of the central core have been mixed in an attempt to match the recent observations from *Ginga* (Tanaka, 1987; Dotani *et al.* 1987). The code used was a modification of one developed by Ambwani and Sutherland (1988) and implemented the scheme of Pozdnyakov, Sobol, and Sunyaev (1983) which is crucial when there is a high probability of photon interaction and a low probability of escape. A derivative of this code is used by Pinto and Woosley (1988).


## RESULTS


The results of our calculation for the Woosley, Pinto, and Ensman (1988) Model 10H are shown in Figure 1. The continuum flux below 10 keV is very low because to lower a photon's energy that much requires very many scatterings, and below 10 keV absorption strongly dominates scattering. There is, however, a not insignificant line flux at 6.4 keV due to fluorescence of Fe by photons well above the Fe K edge. The fluxes in the 10-20 and 20-30 keV bands are predicted to be marginally detectable by the *Ginga* satellite, about 1 year after the explosion. Tanaka (1987) and Dotani *et al.* (1987) have reported that *Ginga* detected SN 1987a in August, 1987 at a level $\sim 2 \times 10^{-11}$ ergs cm$^{-2}$ s$^{-1}$ or $\sim 8 \times 10^{-4}$ counts s$^{-1}$ cm$^{-2}$. Although the flux level is approximately that expected, it was observed substantially earlier than we expected. If the observations indeed correspond to the multiple-scattering of the $\gamma$-rays from the radioactive shell then some aspect of the model must be modified.

The models for SN 1987a advocated by Woosley and his group (see Woosley, 1988, for a review of these models: the label 10H is first used there – this model was previously known as 2BF7) by Arnett (1987), by Nomoto, Shigeyama, and Hashimoto (1987), and by Hillebrandt, Höflich, Truran, and Weiss (1987) have in common, as their progenitors, $6M_\odot$ helium cores with primordial main sequence masses $\sim 15 - 20M_\odot$, as befits the identification of the star Sanduleak 69-202 as the site of the supernova. Beyond this, there are differences about the mass and composition of the envelope at the instant of explosion. Since the model explosions are intended to reproduce the observed light curve and the velocity distribution within the ejecta, with such uncertainties about the envelope must go uncertainties about the energy of the explosion. To get an earlier turn-on of the X-ray flux requires a less massive envelope and/or more energy per unit mass. This qualitative statement has been quantitatively confirmed by Pinto and Woosley (1988) where they evaluate the constraints placed on a variety of models by both the optical and the X-ray observations.

An alternative explanation for the early X-ray emergence is the possibility of mixing of the radioactive material with the overlying layers. As pointed out by Woosley, Ensman, and Pinto (1988) the thin shell of $^{56}$Ni will form a "bubble" that is Rayleigh-Taylor unstable with respect to the material above it. This is because the $^{56}$Ni shell is very slowly moving once the reverse shock from the envelope has moved through the core, and the energy per unit mass to be released through radioactive decay is comparable to the kinetic energy in the shell. No one has yet modelled this instability, and it is not known whether relatively thorough mixing will occur or whether a "salt-finger" instability will develop. We have made a very crude attempt at modelling the effects of mixing, and the results are shown if Figure 2. For the 3 calculations shown, we have taken a certain central mass in Model 10H and thoroughly mixed it, using momentum conservation to set the density, and then repeated our Monte Carlo calculations. We find that when $\sim 5M_\odot$ are mixed we are able to match the August 1987 *Ginga* observations. If there is any merit in these calculations, then it follows that the flux should increase by a factor of 2-3 by October-November 1987 after which it should show a gradual decline. Preliminary results from Dotani *et al.* do *not* seem to show a flux increase in late September 1987.

CONCLUSIONS

A flux of X-rays with a sharp cut-off near 10 keV and detectable by the *Ginga* satellite is a natural consequence of the "down-Comptonization" of $\gamma$-rays released by the radiaoactive material in the ejecta of SN 1987a. There is considerable confidence that the mass of this radioactive material has been accurately ($\pm 5\%$) determined by the bolometric light curve. The X-ray light curve can be used to constrain properties of the envelope of the exploding star and the degree of mixing of the radiaocative material within the core. The detection of the strong $\gamma$-ray lines will provide additional information as these sample the properties of the envelope in a different way. In principle, $\gamma$-ray line profiles could tell us a great deal about the structure and velocity distribution of the radioactive core (Chan and Lingenfelter, 1987; Pinto and Woosley 1988).

If the *Ginga* observations do not entirely fit the above view (and there is some evidence for a "soft" component in the spectrum: see Dotani, *et al.* 1987) then serious consideration has to be given to the possibility that we are seeing the interaction of the supernova with circumstellar matter (Nomoto *et al.* , 1987).

396

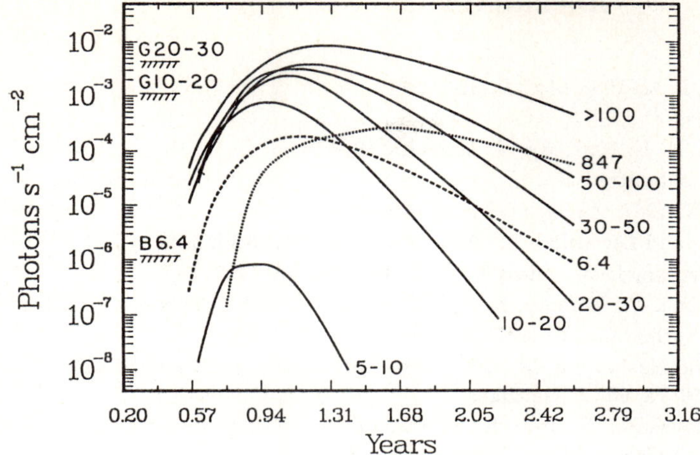


Figure 1: X-ray and $\gamma$-ray fluxes for Model 10H (no mixing). The numbers (with the solid curves) indicate the energy bands in keV; the dotted curve is for the $\gamma$-ray line at 847 keV and the dashed curve is for the Fe $K_\alpha$ line at 6.4 keV. The $5\sigma$ thresholds for detection by the *Ginga* satellite and a hypothetical detector of the BBXRT-class are also indicated.

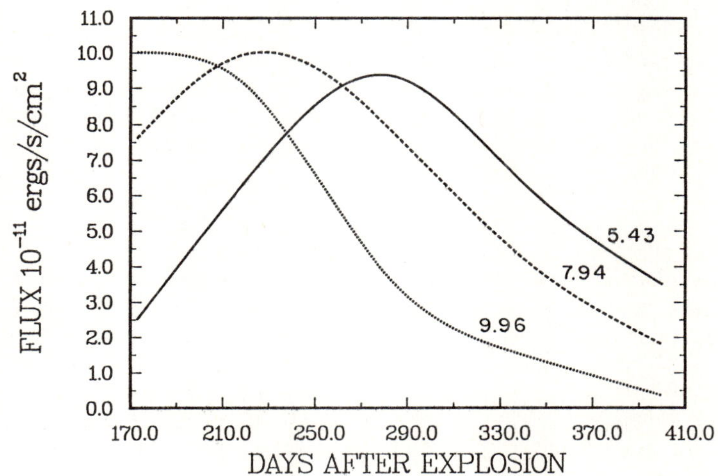


Figure 2: The effects of mixing the radioactive material with the overlying layers. The 3 curves correspond to different total masses (in $M_\odot$) into which the $0.075 M_\odot$ of radioactive material has been homogeneously mixed. The original model was 10H of Woosley, Pinto, and Ensman (1988) and momentum was conserved in the mixing. The flux is for the 10-20 keV band.

# REFERENCES

Ambwani, K., and Sutherland, P. G. 1988, *Ap. J.*, **325**, 000.

Arnett, D. 1987, *Ap. J.*, in press.

Chan, K. W., and Lingenfelter, R. E. 1987, *Ap. J. Letters*, **318**, L51.

Dotani, T. *et al.* 1987, *Nature*, submitted.

Ebisuzaki, T., and Shibazaki, N. 1987, *Ap. J. (Letters)*, submitted.

Gehrels, N., MacCallum, C. J., and Leventhal, M. 1987, *Ap. J. Letters*, **320**, L19.

Hillebrandt, W., Höflich, P., Truran, J. W., and Weiss, A. 1987, *Nature*, **327**, 597.

Itoh, M., Kumagai, S., Shigeyama, T., Nomoto, K., and Nishimura, J. 1987, *Nature*, submitted.

McCray, R., Shull, J. M., and Sutherland, P. 1987, *Ap. J. Letters*, **317**, L73.

Nomoto, K., Shigeyama, T., and Hashimoto, M. 1987, to be published in the Proceedings of the ESO Workshop on SN 1987a, ed. J. Danziger.

Nomoto, K., Shigeyama, T., Hayakawa, S., Itoh, H., and Masai, K. 1987, to be published in the Proceedings of the Fourth George Mason Workshop on SN 1987a.

Pinto, P. A., and Woosley, S. E. 1988, *Ap. J.*, submitted.

Pozdnyakov, L. A., Sobol, I. M., and Sunyaev, R. A. 1983, *Soviet Sci. Rev., Ap. and Space Phys.*, ed. R. A. Sunyaev, **2**, 189 (New York: Harwood).

Tanaka, Y. 1987, talk given at *IAU Colloquium 108*, Tokyo.

Woosley, S. E., Pinto, P., and Ensman, L. 1988, *Ap. J.*, **324**, 000.

Woosley, S. E. 1988, *Ap. J.*, submitted.

Xu, Y., Sutherland, P. G., McCray, R. A., and Ross, R. R. 1988, *Ap. J.*, in press.

# X-RAY OBSERVATION OF SN1987A FROM GINGA


Y. Tanaka
Institute of Space and Astronautical Science
3-1-1 Yoshinodai, Sagamihara
Kanagawa Prefecture 229, Japan



ABSTRACT.    An unusual hard X-ray source was discovered in  an  error
box  of  0.2° x  0.3° including  SN1987A  from  the  X-ray  astronomy
satellite  Ginga.  The energy spectrum is quite unusual for any  known
classes  of  X-ray  source, and apparently consists  of  two  separate
components,  a soft component and a very hard component. This  source
is  considered  to be identified with SN1987A.   The  X-ray  emergence
occurred in July, 1987, or possibly even earlier.  The soft  component
is  significantly time-variable and also showed a flare-like  increase
in  January,  1988,  while the intensity of  the  hard  component  has
remained relatively unchanged for more than 200 days.


## 1.    OBSERVATION OF SN1987A

We  began  search  for X-rays from the supernova 1987A  in  the  Large
Magellanic Cloud (LMC) from the X-ray astronomy satellite Ginga  since
February  25, 1987, right after the optical discovery  (Shelton  1987;
Nelson and Jones 1987).   Ginga  carries a set of proportional  counters
of  a total 4000 cm² effective area and a full field of view  of  2° x
4° .  (For more detail of Ginga, see Makino 1987.)  SN1987A  is  about
0.6°  away  from  LMC X-1.  This source is one of the  brightest  X-ray
sources in the LMC with an intensity of approximately 20 mCrab in  the
1 - 10 keV range, and is also time variable.  In order to observe  the
SN  separated  from LMC X-1, two different modes of  observation  have
been employed;  (1) slow scans along a path through the SN and LMC  X-
1, and  (2) pointing observations at a position about 1° offset  from
LMC  X-1  on  the side of the SN as shown in Fig. 1,  which  gives  an
exposure to the SN with approximately half the maximum sensitivity and
little contribution from LMC X-1.

## 2.    RESULTS

### 2.1. Discovery of a Hard Source Near SN1987A

In Fig. 2, we show the result obtained from the scanning  observations
on September 2 and 3 in the form of histograms of the count rate  with
time in different energy bands.   In the energy range below 10 keV, the

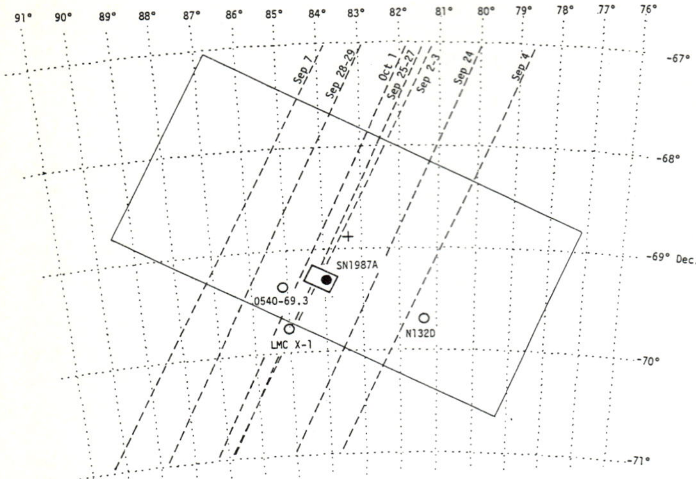


**Figure 1:**
The sky map of the LMC region. The scan paths (dashed lines), the target position (+) for the offset pointing mode and the full field of view (2° x 4°) are indicated. The thick solid rectangle shows the 90% confidence error box of the hard source.

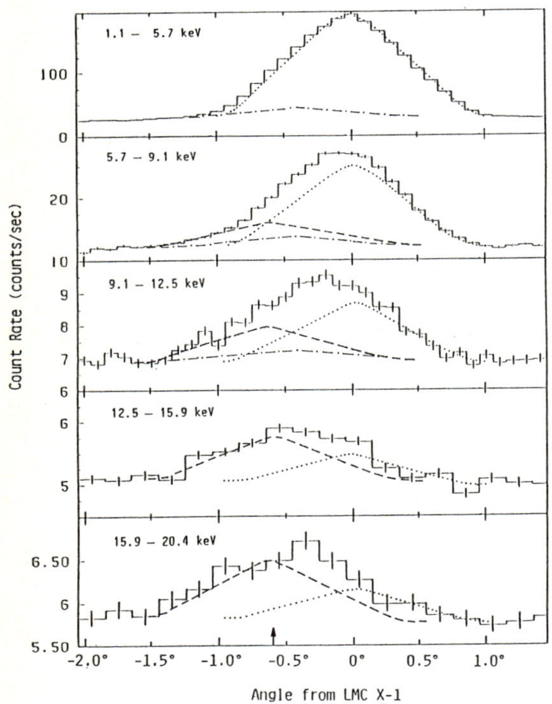


**Figure 2:**
Count rate histograms obtained from the scans on September 2 - 3 are shown in five different energy bands. The dotted lines (LMC X-1), dash-dotted lines (SNR0540-69.3) and dashed lines (the hard source) indicate the fits to the collimator response function. Position angle of SN1987A is marked by an arrow.

counts from LMC X-1 dominate. On the other hand, as the energy
increases, the peak position shifts towards the line position near the
SN.   The contributions of LMC X-1 and SNR0540-69.3 are minor in  the
range  above 10 keV and quickly diminish with increasing energy.   The
histograms   clearly  indicate the presence of a new source  having  a
very  hard spectrum separated from LMC X-1.  We are certain that  this
source  did  not exist in the March - April period.   By fitting  the
collimator response function to the observed count-rate histograms, we
determined  the  contribution of the variable source LMC X-1  and  the
intensity  as  well  as the line position of  the  hard  X-ray  source
separately. The contribution of the nearby supernova remnant 0540-69.3
is  also  taken  into account using the result  of  a  separate  Ginga
observation  of this remnant.  The result of the fit is shown  by  the
dashed  lines  in Fig. 2.  The line position of the hard  source  thus
determined  is in agreement with that of the  SN  within  $\pm 0.1°$ (90%
confidence limit).   The discovery of this hard source was announced on
an IAU Circular (Makino et al. 1987).

In  order  to determine the error box of the hard source, we  conducted
separate  scans on September 4 - 7 and September 24 - October 1  along
several  different  paths  which  were parallel  to  each  other  with
separations   of 0.3° to 05° .  These scan paths are shown in Fig.  1.
From the comparison of the peak count rate observed on each scan path,
we  determined the 90% confidence error box of the hard source of  the
size 0.2° x 0.3° as shown in Fig. 1.  This error box includes the  SN.
Furthermore,  as shown later, the spectrum of the new source is  quite
unusual  for  any of the known classes of X-ray  source.   Also,  this
source  was  not  present until April, 1987,  and  emerged  some  time
between  May  and July, 1987.  From these facts, we consider  it  most
plausible to identify the new source with SN1987A.  Once the  position
was  known,  pointng  observations give the source  flux  and  energy
spectrum  with much better statistics than scanning observations.   We
therefore  employed the pointing mode only, for later observations.

2.2   X-Ray Light Curve


The  light  curves  of  the  source  so  far  obtained  from  pointing
observations are shown for two different energy bands, 6 - 16 keV  and
16  -  28 keV, in Fig. 3.  We shall hereafter call  these  two  energy
bands the soft X-ray band and the hard X-ray band, respectively.   The
count  rates  are given after subtraction of  the  contributions  from
nearby  sources,  and  the  aspect correction.   In  addition  to  the
contributions from LMC X-1 and SNR0540-69.3, minor contributions  from
N132D,  N157B and two other Einstein sources (source No's. 26  and  83
from  the Einstein survey by Long, Helfand and Grabelsky (1981))  were
also  corrected for.   The  spectrum of each  of  these  sources  was
determined from separate Ginga observations. The other sources in  the
field  of view are much fainter and estimated to be  negligible.   The
contributions  from the nearby sources never exceeded 40%, except  for
the data on July 4, of the total counts in the soft X-ray band and 10%
in  the  hard  X-ray  band.   Therefore,  it is  beyond  doubt  that
significant  flux  is coming from SN1987A not only in the  hard  X-ray
band but also in the soft X-ray band.

In addition, the background subtraction is a source of systematic error. A background measurement is performed immediately before or after each SN observation. The systematic errors in the process of background subtraction was estimated to be about 1 % and is incorporated in the error estimation of the SN flux.

One notices in Fig. 3 that time variations in the two energy bands are qualitatively different. The first positive detection in the soft X-ray band was on July 20, 1987. The intensity in the soft X-ray band (6 - 16 keV) is found to vary significantly by a factor of two to three in 1987. The intensity on December 26 was the lowest so far

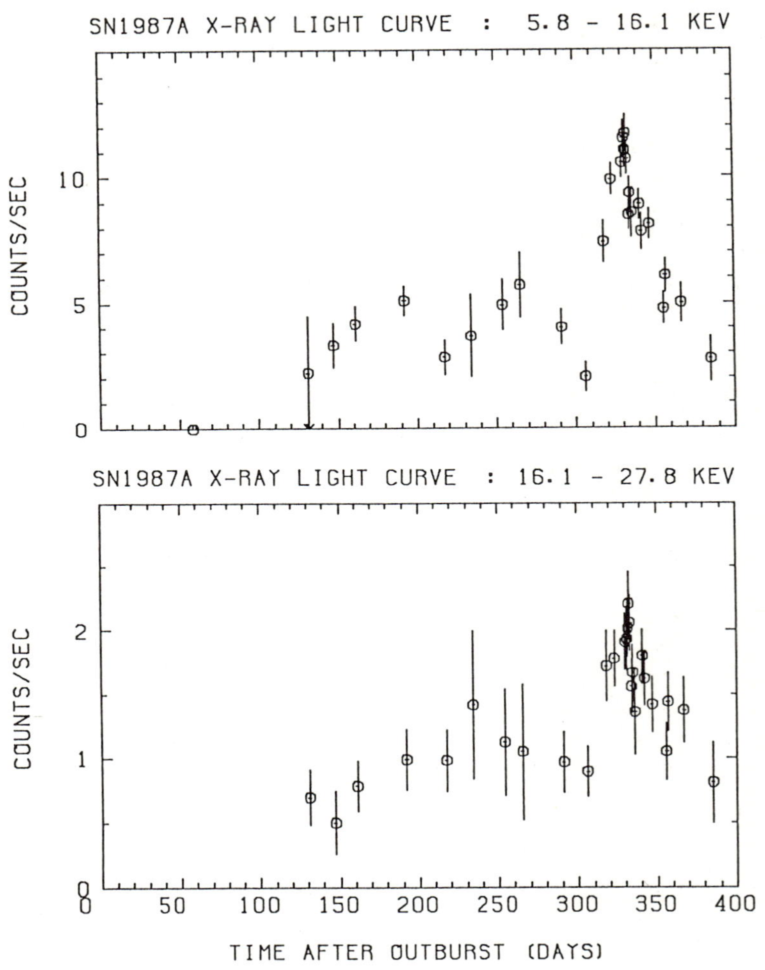


Figure 3: X-ray light curves in two energy bands.

observed.  After twelve days, on January 7, 1988, the sourcee showed a
dramatic intensity increase by nearly a factor of four.  The intensity
increased  further and stayed at the maximum level until  January  22,
which  was  about a factor of six of that on December 26.  Then,  on
January  23, it dropped by 30 %, which is  statistically  significant.
Since  then, it decreased slightly through February 5.  About 10  days
later, on February 14, the intensity was found to have returned to the
general  level  of  1987.  We shall hereafter  call  this  event  the
"Januray flare".

On the other hand, the X-ray intensity in the hard X-ray band (16 - 28
keV) exhibits much less change than those in the soft X-ray band.  The
first positive detection was on July 4, 1987.  It should be mentioned,
however,  that  the epoch when the hard X-ray intensity  exceeded  the
Ginga detection threshold remains somewhat uncertain.  This is because
the observing condition was unfavorable in May and June, 1987.  Since
the  detection,  the  intensity in the hard X-ray  band  has  remained
almost constant within the statistical uncertainties through December,
1987.  In  the  January flare, the  hard  X-ray  intensity  increased
simultaneously by a factor of two, which was however much smaller than
the factor of increase in the soft X-ray band.


2.3.  Energy Spectrum


The  energy  spectrum of SN1987A has a unique  shape,  suggesting  the
presence  of  two separate components (Dotani et al. 1987).  Fig.  4
shows the average spectra for three different intensity levels.  This
averaging is justified because the spectra observed on different  days
but  at the same intensity level are consistent to be the same  within
statistical  uncertainties.  The  spectrum A  shows  the  average  of
September  28 and December 26 for the lowest intensity group, and  the
spectrum B that of August 3, September 3 and November 4.  The spectrum
C  is  the  average from January 19 through 22 when the  intensity  was
highest during the January flare.  These spectra are corrected for the
energy-dependent detection efficiency.

The  shape of the spectra A and B would hardly be explicable in  terms
of  a  single  component.  In addition, while these  two  spectra  are
significantly  different  from  each  other below  15  keV,  they  are
essentially the same above 15 keV.  These facts suggest that there are
two  separate  components, a soft component and a hard  component,  of
which  the hard component remains essentially constant independent  of
the  soft  component.  This  explains  the  little  intensity  change
observed  in the hard X-ray band.  Besides, a flux minimum in the 10  -
15 keV range in the spectrum A suggests that the hard component is cut
off below 20 keV.

We  attempt to determine the spectral shape of the hard  component  by
assuming the form $E^{-1} \times \exp(-\sigma N_h)$.  However, the assumed  power-law
form  is  not critical for the present energy range limited  below  40
keV.  The  power-law  $E^{-1}$  was  simply  chosen  for  a  qualitative
representation  of the result of the Kvant observations in the  higher

energy range (Sunyaev et al. 1987) and also the numerical results of the Compton degradation model (e.g., Itoh et al. 1987, and references therein). The soft component is assumed to be expressed by the bremsstrahlung spectrum.

In order to determine the absorption column $N_h$, the above model was fitted to the average spectra. As the result, the spectra A and B gave $N_h$-values both approximately equal to $10^{25}$ H atoms/cm$^2$ for the cosmic abundances of element. The spectrum C is insensitive to determine the $N_h$-value uniquely. However, we assume the same form of the hard component also during the January flare, because the hard component is likely to be independent of the soft component. We therefore fixed the $N_h$ at $10^{25}$ H atoms/cm$^2$, and performed fitting to the spectrum C. The intensity of the hard component was dealt with as a free parameter.

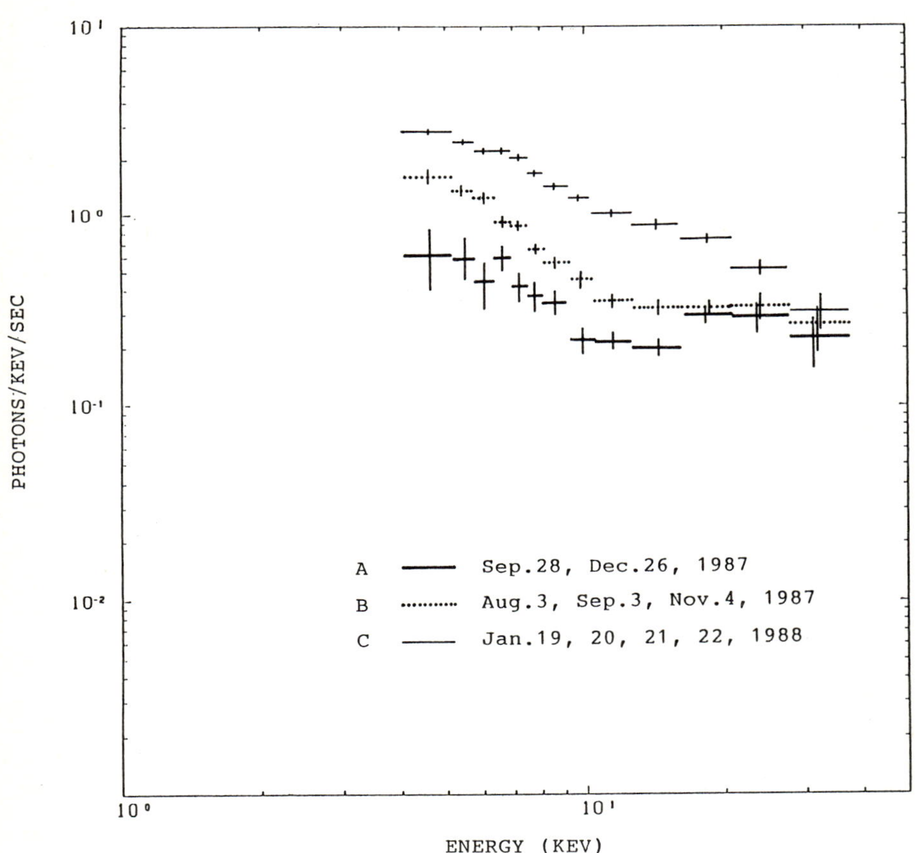


<u>Figure 4:</u>  Average spectra for three different intensity levels

Satisfactory fit was obtained for all cases. As the result, the intensity of the hard component turned out to be the same for all three spectra within the errors. This implies that the hard component remained essentially constant for more than 200 days. The temperature kT of the soft component is found to be about 10 keV for the spectra A and B but is higher than 50 keV for the spectrum C. The count rate increase in the hard X-ray band (16 - 28 keV) during the January flare (see Fig. 3) is thus interpreted as due to an enhanced contribution of the hardened soft component. The luminosity of the soft component reached $10^{38}$ ergs/sec at the flare peak, for the assumed distance of 50 kpc.

The absorption column of the soft component, $N_s$, is generally found to be about $10^{23}$ H atoms/cm² or less. The lower bound of $N_s$ is difficult to estimate because of the increasing systematic errors in the corrections for the nearby sources towards lower energies. For the same reason, whether or not $N_s$ changes with time remains uncertain.

An emission line of iron is significantly visible in the spectrum of the soft component when statistics is sufficiently good. This may be a supportive evidence for the thermal origin of the soft component. In particular, the iron line is clearly seen during the January flare. The energy of the iron line observed during the flare is determined to be $6.8 \pm 0.2$ keV with an equivalent width of about 200 eV. Therefore, the line is emitted most likely from helium-like or hydrogen-like iron ions.

What are the origings of X-rays from SN1987A? The spectral shape of the hard component is in general agreement with that expected from the model of the down-Comptonized $^{56}$Co gamma-ray lines. However, if it is the case, much earlier emergence of X-rays than expected and the essential constancy of the hard component for more than 200 days remain to be explained (see Itoh et al. 1987). Besides, the presence of an intense soft component was quite unexpected. A possible scenario is that the expanding ejecta interacts with fairly dense circumstellar matter and the shock-heated plasma emits thermal radiation (Masai et al. 1987). Very high luminosity, temperature close to $10^9$ K and a rapid change in intensity observed during the January flare might require rather exceptional conditions for this model. Search for regular pulsations is obviously very important. Thus far, no pulsation has been detected.


## REFERENCES

Dotani, T. et al. 1987, Nature 330, 230.
Itoh, M. et al. 1987, Nature 330, 233.
Long, K.S., Helfand, D.J. and Grabelsky, D.A. 1981, Ap.J. 248, 925.
Makino, F. 1987, Astrophys. Lett. & Communications. 25, 223.
Makino. F. et al. 1987, IAU Circular No.4447.
Masai, K. et al. 1987, Nature 330, 235.
Shelton, I. (reported by W. Kunkel and B. Madore); Jones, A., Nelson (reported by F.M. Bateson) 1987, IAU Circular No.4316.
Sunyaev, R. et al. 1987, Nature 330, 227.

# THE COMPOSITE IMAGE OF SANDULEAK −69° 202, CANDIDATE PRECURSOR TO SN 1987A IN THE LMC


*Nolan R. Walborn* and *Barry M. Lasker*
Space Telescope Science Institute
*Victoria G. Laidler*
Astronomy Programs, Computer Sciences Corporation
and
*You-Hua Chu*
Astronomy Department, University of Illinois


The image of Sk −69° 202 was scanned and analyzed on eight (of 32 available) blue through near-infrared photographic plates obtained at the prime focus of the Cerro Tololo Inter-American Observatory 4-meter telescope during 1974–1983. Both intensity syntheses of the image and density differences were derived by means of reference stars from the same plates, including the similar nearby object Sk −69° 203. Several of the density differences are shown in Figure 1. The analysis shows that the $12^m$ blue supergiant in Sk −69° 202 (Star 1) has two companions with $V$ magnitudes, position angles, and separations $15\overset{m}{.}3$, 315°, 3″ (Star 2) and $15\overset{m}{.}7$, 115°, 1″.5 (Star 3), respectively. Both companions appear to be early-type stars; there is no evidence for a bright red star in the system. The two companions are responsible for the spectra observed by the International Ultraviolet Explorer following the decline of the SN in the far UV, so that Star 1 has disappeared and was probably the progenitor. The most likely interpretation is that it was a post-red supergiant evolving blueward in the HR diagram.

The full paper appears in *The Astrophysical Journal Letters*, **321**, L41 (1987). The ST ScI is operated by AURA, Inc. under contract with NASA.

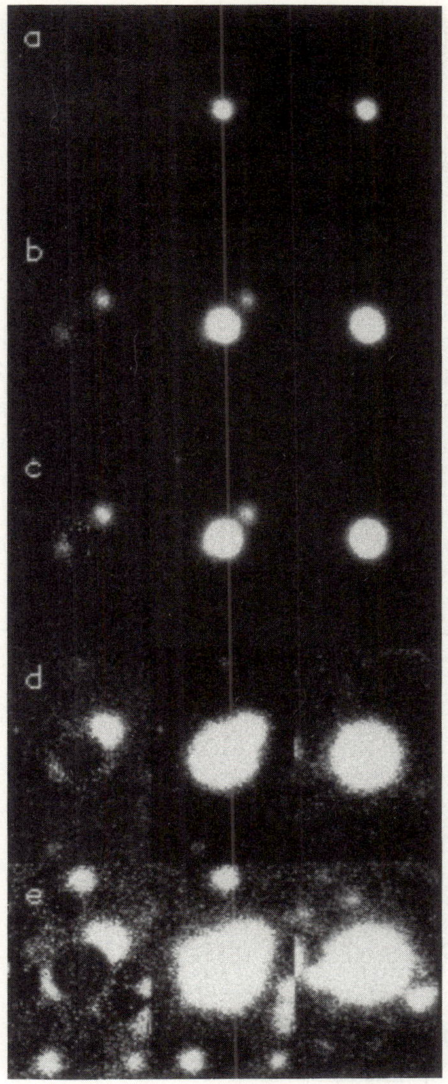

Figure 1 – Density differences from 5 plates, Sk −69°202 minus Sk −69°203. The left panel of each row shows the image subtraction from a given plate, while the center panel is the original image of Sk −69°202 and the right panel is that of Sk −69°203, from the same plate. North is up and east to the left in each case. Star 2 is at $3''$ in PA 315° and Star 3 is at $1\rlap{.}''5$ in PA 115°. (a) Plate No. 345 (4765 Å, 2 min—the apparent "nebulosity" is spurious), (b) No. 5973 (4765 Å, 10 min), (c) No. 5976 (5000 Å, 10 min), (d) No. 719 (6725 Å, 30 min), (e) No. 4858 (IV–N, 90 min).

# THE EVOLUTION OF THE PRECURSOR OF SN 1987 A

by

André M A E D E R

GENEVA OBSERVATORY


## SUMMARY

Ideally, the evolutionary models for the precursor of SN 1987 A should account for both the SN properties <u>and</u> the observational constraints for massive stars with relevant mass and composition.

Mass loss is an essential property of massive star evolution [1]. Recent parametrisations of mass loss rates $\dot{M}$ for galactic stars cover the whole HR diagram [2]. There are indications [3,4] that for given L and Teff values, $\dot{M}$ is lower at lower metallicity and therefore $\dot{M}$ is lower in the LMC than in the Galaxy, thus we take $\dot{M}_{LMC} = f \cdot \dot{M}_{Galaxy}$ with f < 1. Various models of an intitial 20 $M_\odot$ star with f=0.2, 0.4, 0.6 and 1.0 are constructed (cf. Fig. 1) with a metallicity


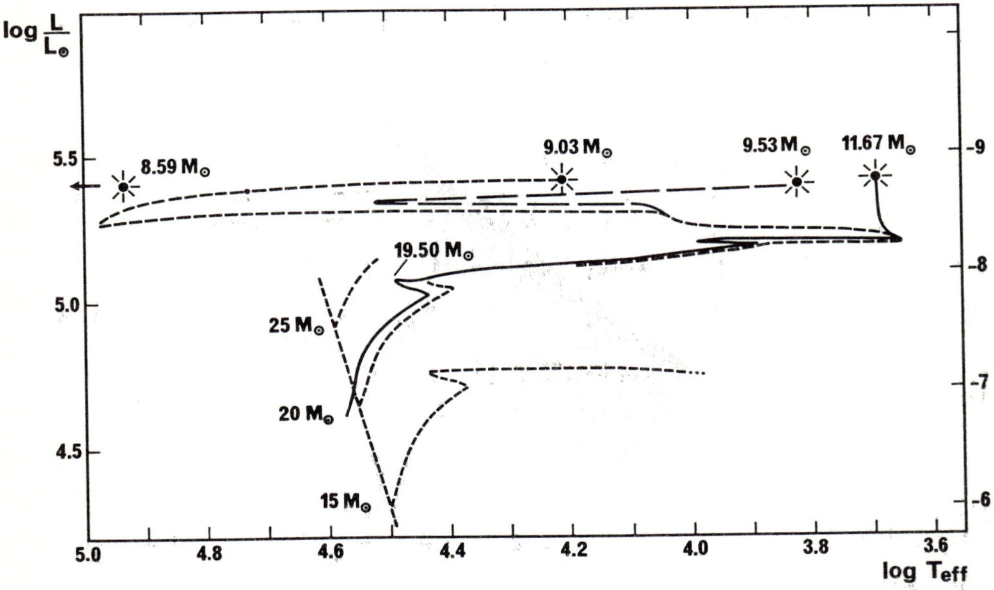



*Fig.1 : Evolutionary tracks in the HR diagram for a model with an intial mass of 20 $M_\odot$ and composition X=0.744 and Z=0.006. Various cases of mass loss in post-MS evolution are considered and the remaining final masses are indicated.*

Z=0.006 and a moderate overshooting $d_{over}=0.3$ $H_p$. From these models, we suggest an <u>initial mass</u> on the zero age sequence of 17 to 18 $M_\odot$. The pre-SN location in the HR diagram very much depends on the remaining stellar mass, or more precisely on the mass of the remaining H-rich envelope. A final location at log Teff $\simeq$ 4.2 is obtained for a final mass of about 9.0 $M_\odot$ (cf. Fig.1). Scaled to an initial value of 17 $M_\odot$, this corresponds to a <u>final mass of about 8 $M_\odot$</u> and a remaining H-rich envelope of a few tenths of a solar mass at most. The stellar surface exhibits CNO equilibrium values with C/N $\simeq$ 0.01 and O/N $\simeq$ 0.1 in mass fraction, and an hydrogen content X (surf) = 0.39. The blue progenitor is obtained for f=0.4, i.e. for $\dot{M}$-values in the LMC equal to 40% of the galactic values.

The models must also account for the frequency of red supergiants (RSG). The observations [5] suggest the following lifetime ratios $t_{RSG}/t_{tot}$ : 0.12 in the SMC, 0.08 in the LMC, 0.08 and 0.02 for the outer and inner galactic regions. As to the models, they indicate [1] the following relation : $\dot{M}$ $\nearrow$ $\Rightarrow$ $t_{RSG}/t_{tot}$ $\searrow$ . Since $\dot{M}$ is likely to increase along the sequence of the 4 galactic sites considered, the above observed lifetime ratios imply that the various considered galactic sites are located in the part of this last relation, where $t_{RSG}/t_{tot}$ is decreasing with increasing $\dot{M}$. Such a behaviour implies the existence of a substantial mass loss, even in the SMC and LMC. Interestingly enough, the observed lifetime ratio of 8% for the LMC is also reached for an f value of 0.40. Thus, the same model is able to simultaneously account for the blue SN precursor and for the observed RSG frequency as well as for the existing evidences of CNO processing in the spectrum of SN 1987 A [6,7].


## REFERENCES

1. Chiosi, C., Maeder, A. , 1986, *Ann. Rev. Astron. Astrophys.* <u>24</u>, 329.
2. de Jager, C., Nieuwenhuijzen, H., van der Hucht, K.A. , 1987, *Astron. Astrophys.,* in press.
3. Kudritzki, R.P., Pauldrach, A., Puls, J. , 1987, *Astron. Astrophys.* <u>173</u>, 293.
4. Azzopardi, M., Lequeux, J., Maeder, A. , 1987, *Astron. Astrophys.,* in press.
5. Meylan, G., Maeder, A. , 1982, *Astron. Astrophys.* <u>108</u>, 148.
6. Cassatella, A. , 1987, *ESO Workshop on SN 1987 A, Ed. J. Danziger,*in press.
7. Kirshner, R.P., 1987, this meeting.

**EVOLUTION OF THE PRECURSOR OF SN1987A**


P.R. Wood and D.J. Faulkner
Mount Stromlo and Siding Spring Observatories
Private Bag, Woden P.O., A.C.T. 2606
Australia


The evolution of a 17.5 $M_\odot$ star, chosen to be similar to the precursor of SN1987A, has been studied using the input physics described in Wood and Faulkner (1987). The calculations: use opacities from the *Astrophysical Opacity Library* of Huebner *et al* (1977) with H, He, C, N, O and other metals in LMC ratios; treat semi-convection in the manner of Lamb, Iben and Howard (1976); and assume the mass loss rate to be the minimum of (a) $\alpha$ times the rate give in Waldron (1985) (this rate applied in the blue part of the HR diagram), and (b) $L/(cv)$ (this rate applied in the red), where $v$ is the stellar wind expansion velocity which was taken to be 12 km s$^{-1}$.

Typical evolutionary tracks resulting from these calculations are shown in Figure 1. The track shown as a continuous line is the one shown in Wood and Faulkner (1987) and corresponds to $\alpha = 2.5$, while the track shown as a dotted line results from a slightly different treatment of semi-convection on the main-sequence and to $\alpha = 1$.

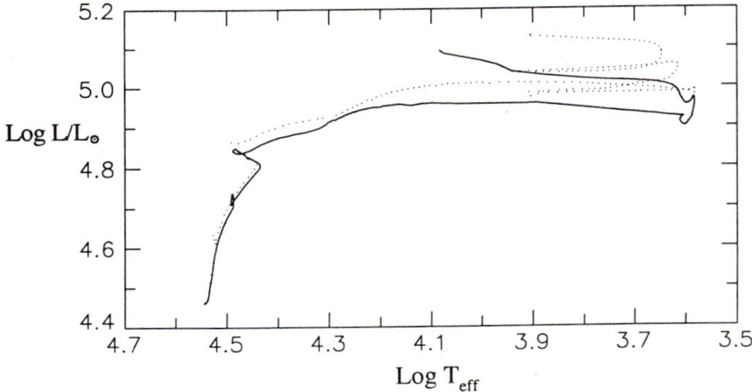

Figure 1. Evolution of a 17.5$M_\odot$ star for two different treatments of the input physics.

From these tracks, and a number of others which we have computed, we find the following.
(1) During the main-sequence phase, the star has a shrinking fully convective core surrounded by a semi-convective zone; once only, a fully convective region in the semi-convective zone makes contact with the central convective zone, causing an increase in the central hydrogen abundance and giving rise to the brief loop seen during the main-sequence phase. At the end of the main-sequence phase, a fully convective zone develops outside the core and mixes the former semi-convection zone. The resultant helium abundance profile is shown in Figure 2; this profile has a strong effect on subsequent evolution.

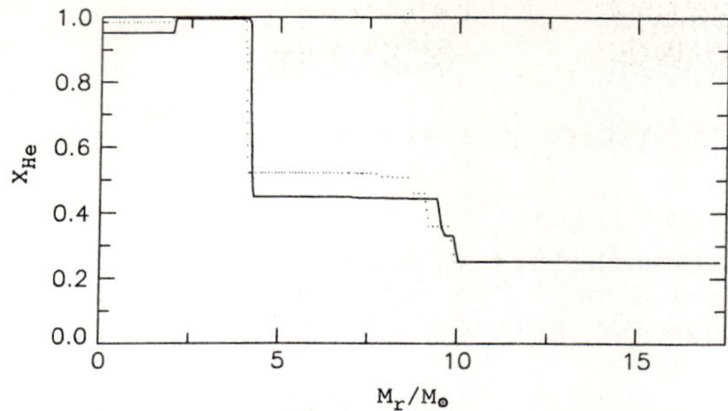

Figure 2. The helium abundance profile left at the end of the main-sequence phase for the two sequences in Fig. 1.

(2) The star always begins helium core burning in the blue (log $T_{eff} \approx 4.3$) but *always* evolves to the red supergiant domain some time during core helium burning.

(3) The star *always* makes a final trip back to the blue and undergoes a supernova explosion when the hydrogen envelope is reduced to a few tenths of a solar mass (assuming sufficient mass loss has occurred). Final mass is ~$5.5 M_\odot$.

(4) The details of the intervening evolution are sensitive to the choice of convective mixing on the main-sequence and mass loss. Two blue loops may occur (a) when mass loss reveals the material of high helium content within ~10 $M_\odot$ of the stellar centre, and (b) when the hydrogen envelope is reduced to ~0.6 $M_\odot$ while the hydrogen shell is still active (when the H shell is extinguished by further mass loss the star takes a final trip to the red). Figure 3 shows time variation of some stellar properties.

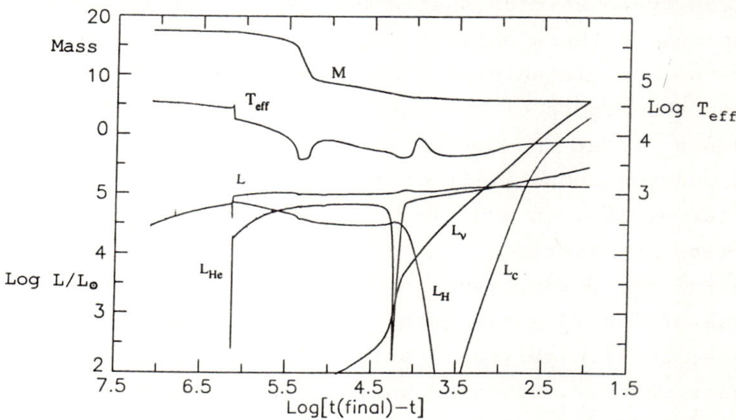

Figure 3. Time dependence along the model sequence shown as a dotted line in Figs. 1 and 2.

References


Huebner, W.F., Merts, A.L., Magee, N.H. and Argo, M.F. 1977, *Astrophysical Opacity Library*, Los Alamos Scientific Laboratory, La-6760-M.

Lamb, S.A., Iben, I., and Howard, W.M. 1976, *Ap.J.*, **207**, 209.

Waldron, W.L. 1985, *The Origin of Nonradiative Heating/Momentum in Hot Stars*, A.B. Underhill and A.G. Michelitsianos (eds.), NASA Conf. Publ. CP-2358, p.95.

Wood, P.R. and Faulkner, D.J. 1987, *Proc. Astr. Soc. of Australia*, in press.

Calculated Energy Distributions for SN II

P.H. Hauschildt, W. Spies, R. Wehrse
Institut für Theoretische Astrophysik, Universität Heidelberg
Im Neuenheimer Feld 561, D-6900 Heidelberg

G. Shaviv
Dept. of Physics, Technion, Israel Institute of Technology
IL-32000 Haifa

ABSTRACT

We have calculated a large grid of hydrogen-rich supernova photospheres, in which radii, effective temperatures, density profiles, and expansion velocities have been varied. Spherical geometry, radiative equilibrium and LTE level populations are assumed. In the quasi-exact radiative transfer, the dilution of the radiation field, and scattering as well as absorption (by all relevant continuous processes and up to 150 000 lines in some models) are accurately considered. Good agreement can be obtained with the UV and IR spectra of supernovae 1979C, 1980K, and 1987A as observed during the coasting phase. Potential methods of parameter determinations for SN II are briefly discussed.

Since supernovae of type II are believed to be explosions of massive stars with large hydrogen-rich shells (see e.g. Woosley and Weaver, 1986), the spectra of these objects contain important information on the abundances of the elements of the parent stars (and therefore implicitly on the abundances of the parent galaxy) during the first days up to a few weeks  after the outburst and at later times information on the amount and the composition of the processed matter returned to the interstellar medium. In addition, the atmospheric parameters can be used to determine the distance to the object with a minimum of assumptions by means of the Baade-Wesselink method. Since - due to the very high brightness of SNe II - the data can be obtained for a very large volume of space, it is highly desirable to model in detail the photospheres of these objects. However, until recently realistic models were inhibited by the required complexity of the radiative transfer (spherical geometry, differential expansion, dominance of scattering, line blanketing etc.).

In order to obtain an overview on the significance of these effects and on the influence of the parameters on the emergent fluxes we have calculated a large grid of spherical hydrogen-rich supernova photospheres with the following assumptions: (i) the radial density distribution is given by a power law (exponents n = 5...14), (ii) the expansion velocities are proportional to the radial distance (the values at absorption

optical depths of unity are varied between $5.10^3$ and $15.10^3$ km $s^{-1}$),
(iii) the level populations follow Boltzmann distributions of the
local temperature, (iv) the temperature stratifications are derived
from the condition of radiative equilibrium. In the non-grey quasi-
exact radiative transfer all relevant opacity sources for the continuum
(cf. Hauschildt et al., 1987) and in some models up to 150 000 individual lines are taken into account. We consider ranges of effective
temperature $6000 \leq T_{eff} \leq 15000$ K and radius $10^{13} \leq R \leq 10^{15}$ cm. The
composition is solar, except for two sets in which the metal content
is reduced to 1/3 and 1/10 solar.

The resulting energy distributions show that for $T_{eff} > 10^4$ K line
blanketing is of minor importance (even in the UV), but for lower
temperatures it is highly significant both for the spectrum and for
the temperature stratification. In most wavelength ranges the fluxes
are strongly influenced simultaneously by n, R, and $T_{eff}$, i.e. a
change in one parameter can be compensated by changes in the other ones.
However, rather accurate determinations of $T_{eff}$ seem to be possible
from the slope in the red and infrared range. The radii and density
parameters should be derived from a simultaneous fit of the Blamer
discontinuity and the flux gradient in the ultraviolet. It seems
difficult to obtain accurate estimates for the abundances of individual
metals from UV and optical spectra because usually many lines are
blended and most lines have absorption and emission components. But
from such features the expansion velocities can be determined rather
reliably.

Comparison of our calculated fluxes with the observed spectra of the
SNe 1979C and 1980K shows that good agreement can be obtained for the
UV and the IR during the coasting phase (see e.g. Hauschildt et al.,
1987). In the visible a good fit is possible as long as the Balmer
lines are weak indicating NLTE effects for hydrogen in the outer line
forming layers as expected (Höflich et al., 1986). For the heavily
blanketed SN 1987A the main features in the UV can be reproduced for
metallicities 1/3 to 1/1 solar; however, details indicate that the
relative abundances in the irongroup differ from the solar values.

ACKNOWLEDGMENT
This work was supported by the Deutsche Forschungsgemeinschaft
(SFB 328).

REFERENCES

Hauschildt, P., Spies, W., Wehrse, R., Shaviv, G.: 1987, Proc. ESO
  Workshop "SN 1987A", Garching, in press
Höflich, P., Wehrse, R., Shaviv, G.: 1986, Astron. Astrophys. 163,105
Spies, W.: 1987, Synthetische Spektren für Supernovae vom Typ II,
  Diplomarbeit Heidelberg
Spies, W., Hauschildt, P., Wehrse, R., Baschek, B., Shaviv, G.: in
  Nuclear Astrophysics, Lecture Notes in Physics 287,316, Springer,
  1987
Woosley, S.E., Weaver, T.A.: 1986, Ann. Rev. Astron. Astrophys. 24,
  205

L$\alpha$ RADIATION AND CaII IONIZATION IN THE TYPE II SUPERNOVAE
AT LATE TIMES


Nikolai N.Chugai
Astronomical Council of the Academy of Sciences of the USSR
Pjatnitskaja 48, Moscow 109071, USSR


Abstract: The ionization of CaII by the L$\alpha$ quanta in the envelope
of the type II supernova 1970g on 270th day is considered.  The ratio
CaII/CaIII is found to be very low ($< 0.1$). This results in  the low
theoretical intensities of the CaII emission lines; they are at least
an order of magnitude weaker than the observed ones. The implications
are discussed.


An interpretation of the strong[CaII] $\lambda$ 7300 and $\lambda$ 8600 emissions
in the late time spectra of SNII, e.g. SN1970g on 270th day after the
explosion (see: Kirshner et al., 1973; Kirshner and Kwan, 1975) is
related closely to the value of CaII fractional ionization. Yet the
situation with the ratio CaII/Ca is indefinite because the contribution
of L$\alpha$  quanta into the ionization of CaII has been ignored so far.
This process is studied here for the case of SN1970g on 270th day.

A following simple model for the envelope is adopted (see: Chugai,
1987):  a homogeneous sphere with the expansion law v $=r/t$, boundary
velocity $v_o$ = 4000 km s$^{-1}$, hydrogen mass $M_H$ = 2.5$M_\odot$ (total mass is
$\approx 4M_\odot$), electron concentration $n_e$ = $3.5 \cdot 10^7$cm$^{-3}$, electron temperature
$T_e$ = 7000K. The hydrogen population on the second level is $n_2 > 7$ cm$^{-3}$
(Kirshner and Kwan, 1975; Chugai, 1987).

L$\alpha$ quanta ionize CaII from 3d level ($\lambda_{3d}$ = 1218$\overset{\circ}{\text{A}}$). The density
of L$\alpha$ quanta with $\lambda < \lambda_{3d}$ at some point inside the envelope is
determined by their creation and scattering in the local region of a
"sound" radius $v_{th}t \ll v_o t$. The L$\alpha$  spectrum is specified by:  the
local optical depth $\tau$ = k$\lambda$t = $2.5 \cdot 10^9$, Voight parameter a = $5.6 \cdot 10^{-4}$,
collisional destruction probability $\varepsilon$ = C(2p,2s)/A(2p,1s) = $10^{-5}$,
continuous absorption parameter $\omega_c$ = $k_c \Delta \nu_D / k \lesssim 10^{-12}$. In the selec-
tive absorption of L$\alpha$ the most pronounced effect is produced by FeII
$\lambda$ 1217.85 line with the absorption parameter $\omega_s$ = $k_s/k \approx 2 \cdot 10^{-9}$. For
the parameters indicated above the rate of CaII photoionization with
the L$\alpha$  quanta is found to be $P_{3d} > 0.020(M_H/2.5M_\odot)$ s$^{-1}$.

The photoionization balance of CaII in the SN1970g on 270th day
is calculated in the "two level (4s and 3d) plus continuum" approximation. A lower limit for $P_{3d}$ and the solar ratio for Ca/H are adopted.
We found that: (1) the ionization of CaII with $L\alpha$ radiation in SN
1970g on 270th day is the very effective process which results in
CaII/CaIII < 0.1; (2) the theoretical intensity of [CaII] $\lambda$7300 emission is at least an order of magnitude lower than the observed one.
The latter refers to CaII $\lambda$8600 as well. Both results are stable
against variation of the hydrogen mass in the range 1.5 + 3.5$M_\odot$.

The discrepancy between predicted and observed intensities of the
CaII emissions in the case of SN1970g on 270th day requires some modification of the model applied above. Three possibilities are concei-
vable.

1) Ca is overabundant, i.e. $Z_{SN}(Ca) \approx 10Z_0(Ca)$.

2) The CaII fractional ionization is controlled with the charge
exchange $Ca^{++} + H + 1.7eV \rightleftarrows Ca^+ + H^+$ which results in CaII/CaIII $\approx$
1.5 provided the cross-section for the reaction from the right to the
left $\sigma \gtrsim 2\cdot10^{-17}$ cm$^2$. Unfortunately the value of $\sigma$ is not known.

3) The SN1970g envelope is chemically inhomogeneous and consist of
the two main components: (a) H-rich gas and (b) He-rich filaments
which are inserted into the H-rich background. CaII emission lines
presumably originate from He-rich component where Ca is ionized singly.
To account for the observed CaII emission an order of 1$M_\odot$ in the form
of He-rich filaments is required. Such inhomogeneous structure might
be produced by the Rayleigh-Taylor instability during the supernova
explosion when the H-rich red supergiant envelope is shocked with the
fast expanding He-rich matter.

The third possibility is favoured by the absence of the strong
absorption component of $H\alpha$ in the spectra of SN1970g at earlier phase
which is consistent with the idea of the inhomogeneous distribution
of the hydrogen in the supernova envelope (see: Chugai, 1982).


## References


Chugai N.N. 1982, Astron. Zh., 59, 1134.
Chugai N.N. 1987, Pis'ma Astron. Zh., 13, 671.
Kirshner R.P., Oke J.B., Penston M.V., Searle L. 1973. Ap. J.,
185, 303.
Kirshner R.P. and Kwan J. 1975, Ap. J., 197, 415.

## The Effect of General Relativity and Equation of State on the Adiabatic Collapse and Explosion of a Stellar core.


N. Sack and I. Lichtenstadt

Racah Institiute of Physics

The Hebrew University of Jerusalem, ISRAEL 91904


The collapse of the iron core of massive stars ( $M \geq 8\ M_o$ ) is initiated by photodissociation and electron capture. The collapse of the inner core proceeds homologously until it is stopped by the stiffness of the equation of state (hereafter EOS) at nuclear density and it stops or rebounds. A shock forms at the edge of homology. The initial strength of the shock increases with the velocity difference between the inner and outer cores, i.e. it increases with a larger rebound of the inner core. The uniterrupted propagation of this prompt shock through the remainder of the core to the stellar mantle, where it can deliver enough energy to blow off the loosely bound outer layers, has long been proposed as the mechanism of type II supernovae explosions. However most authors did not get an explosion as a result of the prompt mechanism. Recently Baron et al. (1985) reported that the combination of General Relativity (GR) with a relatively soft EOS at nuclear densities leads to a much greater blow off than they got with Newtonian hydrodynamics. In order to see where purely hydrodynamical effects are important, namely for what EOS the GR outburst is greater than the Newtonian, we did a set of pure hydrodynamical adiabatic calculations (complete neutrino trapping) with different EOS above nuclear densities, turning the GR terms on and off. Neutrino leakage, which we do not incorporate, usually leads to harmful energy losses.

We used a 1.35 $M_o$ iron core formed from a 12 $M_o$ star calculated by Woosley et al. (1984), and the compressible liquid drop model EOS of Lamb et al. (1978) up to nuclear densities. At higher densities we added to the lepton pressure a cold pressure as given by Baron et al. in the

following form:
$$P_{cold} = \frac{K_o \cdot \rho_o}{9 \cdot \gamma} \left[ \left[ \frac{\rho}{\rho_o} \right]^\gamma - 1 \right]$$

where $K_o = 9 \cdot (dP/d\rho)_{\rho=\rho_o}$ is the nuclear incompressiblity at saturation, $\rho_o$ is the nuclear density and $\gamma$ is the high density adiabatic index. Thus we have a two parameter ($K_o$ and $\gamma$) form for the cold nuclear EOS. We used three different values for $K_o$ : 125, 200 and 275 MeV, where $K_o$ =200 MeV is a conservative value (Blaizot 1980) and the other two represent a distinct change to the low and high incompressibilities. For $\gamma$ we used the values 2, 2.5 and 3. All calculations were done twice, turning GR terms off and on. The results are represented in the following table:

| $K_o$ (MeV) | $\gamma$ | $\rho_B^{NR}/10^{14}$ g·cm$^{-3}$ | $M_{ej}^{NR}/M_o$ | $E_{ej}^{NR}/10^{50}$ (erg) | $\rho_B^{GR}/\rho_B^{NR}$ | $M_{ej}^{GR}/M_{ej}^{NR}$ | $E_{ej}^{GR}/E_{ej}^{NR}$ |
|---|---|---|---|---|---|---|---|
| 125 | 2.0 | 7.7 | 0.093 | 2.2 | – | – | – |
| 125 | 2.5 | 6.6 | 0.093 | 3.8 | 2.6 | 1.1 | 0.66 |
| 125 | 3.0 | 5.9 | 0.093 | 3.6 | 1.8 | 0.01 | 0.036 |
| 200 | 2.0 | 5.7 | 0.093 | 2.6 | 3.1 | 1.4 | 0.77 |
| 200 | 2.5 | 5.2 | 0.093 | 2.5 | 1.9 | 0.04 | 0.03 |
| 200 | 3.0 | 4.9 | 0.083 | 2.5 | 1.6 | 0 | 0 |
| 275 | 2.0 | 4.9 | 0.083 | 4.7 | 2.2 | 0.03 | 0.13 |
| 275 | 2.5 | 4.6 | 0.083 | 4.7 | 1.7 | 0 | 0 |
| 275 | 3.0 | 4.4 | 0.083 | 2.9 | 1.5 | 0 | 0 |

Superscripts NR and GR denote the Newtonian and GR results respectively. $\rho_B$ is the maximum density at bounce, $M_{ej}$ is the ejected mass and $E_{ej}$ is the total energy associated with it.

In the Newtonian calculations, we see the dependency of the bounce density $\rho_B$ on $K_o$ and $\gamma$. Softer nuclear cold EOS results in a deeper penetration. The ejected mass is almost independent of $K_o$ and $\gamma$. It depends mostly on the size of the homologous core , which does not change here. The size of the homologous core depends on the effective adiabatic index $\Gamma_{eff}$ during the collapse and not on $K_o$ and $\gamma$ , and is about 0.95 $M_o$. The prompt shock always originates at the edge of the homologous core moving into the looslier bound mantle, and the mass bifurcation point does not change much with $K_o$ and $\gamma$ . On the other hand the energy of the ejecta, does depend on the details of penetration at bounce. General relativity changes the Newtonian picture : as expected, $\Gamma_{eff}$ decreases and therefore $\rho_B$ is increased and the size of the homologous core decreases to about 0.6 $M_o$ . For $K_o$ = 125 MeV and $\gamma=2$ the collapse procceeds to a black hole, so that we do not represent GR values. The high incompressibility $K_o$= 275 MeV ,in GR supresses the ejected mass for all $\gamma$ used. For the intermediate $K_o$= 200 MeV, only $\gamma=2$ benefits from GR and for the low $K_o$= 125 MeV we need $\gamma=2.5$ to get an ejecta. All the other cases resulted in negative or marginal explosions. Still even in the cases of the two GR successful explosions the ejected mass is greater for GR while the ejected energy is reduced by GR effects.

As Baron et al. pointed out there is a constraint on the EOS from the observations of neutron-star gravitational masses to be about 1.5 $M_o$. There is a maximum for the neutron-star mass for a given EOS and this value increases monotonically with the stifness of the EOS. They summarized their results in the following formula:

$$M_{max} = 1.07 \cdot (\gamma-1) \cdot (K_o/230)^{1/(2\gamma-2)} M_o$$

Using this formula we obtained the minimal $\gamma$ ($\gamma^{min}$) necessary to get a 1.5 $M_o$ neutron star for the different incompressibilities we used. The minimal $\gamma$ values are 2.7, 2.5 and 2.3 for $K_o$= 125, 200 and 275 MeV respectively. These minimal adiabatic indices are <u>larger</u> than the values we find necessary to produce energetic explosions, suggesting that the relativistic amplification of the rebound amplitude may not be the mechanism responsible for supernova explosions.

The differences between our results and those of Baron et al. might be due the different EOS, the usage of a leakage scheme and due to their dependency of $K_o$ on $Y_e$.

## CONCLUSIONS

1. GR effects are important in the collapse and explosion of an iron stellar core. Even when the net ejected mass+energy is similar, the inner hydrodynamics is different, mainly by forming a smaller homologous core. We found that the extra energy gained by the deeper bounce penetration is more than wasted on heating up the thicker mantle.

2. There is a narrow regime near the points: ($K_o$=125 MeV, $\gamma$=2.5) and ($K_o$=200 MeV, $\gamma$=2) that permits a successful explosion. However, as Baron et al. themselves realize, these values for the EOS parameters are ruled out by the masses observed for cold neutron stars, if the EOS of neutron-star matter is not stiffer than our EOS.

3. Allowing neutrino leakage would allow some small reduction in $K_o$, but would lead to neutrino energy losses, larger in the GR case than in the Newtonian, because of the smaller inner core and the larger hot overlaying mantle.


References:
Baron, E., Cooperstein, J., and Kahana, S. 1985, Phys. Rev. Letters, **55**, 126.
Blaizot, J.P. 1980, Phys. Rep., **64**, 171.
Lamb, D. Q., Lattimer, J.M., Pethick, C., and Ravenhall, D.G. 1978, Phys. Rev. Letters, **41**, 1623.
Woosley, S. E., and Weaver, T. A. 1984, Bull. Am. Astron. Soc., **16**, 971.

QUASI-STATIC AND STEADY-STATE PICTURES
FOR COLLAPSING CORE OF TYPE II SUPERNOVA


D. Sugimoto, A. Sasaki, and T. Ebisuzaki
College of Arts and Sciences, University of Tokyo


1.  _Introduction_: Explosion of type II supernova is, in principle, a difficult process: The presupernova star was in gravitationally bound state with negative energy but it has to be divided into two parts, the collapsed core of still lower (negative) energy and the ejected envelope of positive energy.  This process is against nature in the sense most of the phenomena in nature proceed towards equipartition of energies.  Thus, some finely tuned mechanism should be necessary for successful explosion.  Two different mechanisms have been proposed; one is the _prompt_ explosion where the gravitational energy release by the core collapse is transferred to the mantle by a shock wave, and the other is _delayed_ explosion where it is transferred slowly by neutrinos diffusing out of the core.  In what follows we shall concentrate in the case of the delayed explosion.

2.  _Slow Timescale and Quasi-Homologous Collapse_:  According to Wilson's calculation (Wilson 1982; Wilson et al. 1985; Bethe and Wilson 1985) the core collapse proceeds typically with the timescale of $t_{coll} \sim 0.05$ sec while the free-fall time in the central core is $t_{ff} \sim 0.01$ sec.  Since the ratio of the inertial term to the graviational accerelation is $(t_{ff}/t_{coll})^2$, it is as small as $\sim 1/25$.  This implies that the well known concepts and theories of quasi-static stellar structure are applicable. If we plot Wilson's (1982) numerical model of 10 $M_\odot$ star on a plane of homology invariants we see that the structure of the core stays very close to that of polytrope of index N=3 except for the very outer mantle.

It may be due to the equation of state for which the deviation of the adiabatic exponent from 4/3 is small, i.e., $\varepsilon = 4/3-\gamma < 0.05$.  However, we have to show that this proximity to N=3 should be practically be kept even during the collapse.  For the structure close to N=3 the linearized stability theory for stellar structure has a wider applicability, since the polytrope of N=3 with $\varepsilon = 0$ has the eigen value of $\omega_0^2 = 0$ and the eigen function of $\xi_0$=constant (to be normalized as unity).  For the deviation in the equation of state we expanded the linearlized stability equation still to the order of $\varepsilon$ by putting $\omega^2 = \omega_0^2 + \varepsilon\omega_1^2$ and $\xi(r_0) = \xi_0(r_0) + \varepsilon\xi_1(r_0)$.

We solved it and obtained $\omega_1^2 = -19.9(2GM/R^3)$, and, thus, $t_{coll} = 1/i\omega = 0.22\varepsilon^{-1/2} t_{ff}$ (at $M_r = M$). Note that the free-fall time calculated at the surface ($M_r = M$) is appreciably longer than that in the bulk of the core. Here we see that the factor $\varepsilon^{-1/2}$ brings about the relatively slow collapse. As for the eigen function the value of $-\xi_1(r_0)$ remains as small as 3 within the inner 90 percent of the mass of the core though it grows from 5 to 13 within the outer 2 percent. If we multiply it with $\varepsilon$, we see how small is the deviation from the homologous contraction for the inner core.

3. <u>Later Quasi-Steady Expansion</u>: After the core collapse a weak shock is generated. It heats the outer core and the structure of the region with $M_r > 0.6 M_\odot$ deviates greatly from $N=3$. The shock wave once stalls but after about 0.3 second it revives (Bethe and Wilson 1985). Then the structure of the outer core becomes close to $N=3$ again up to the neutrino sphere ($M_r \simeq 0.4 M_\odot$). This is due to the trapped neutrinos.

In the layers exterior to the neutrino sphere, the temperature gradient is less steep. It is characterized by the thermal balance of neutrino heating and cooling. The heating rate is proportional to $T_\nu^6/r^2$ where $T_\nu$ is the effective temperature of the neutrino sphere and $\underline{r}$ is the radial distance, and where the neutrino opacity proportional to the square of the neutrino energy is taken into account (Bethe and Wilson 1985). The cooling rate is proportional to $T_m^6$ where $T_m$ is the matter temperature. Balancing them, we obtain $T_m \sim r^{-1/3}$.

In order to realize this thermal balance the outer layers have to expand against the gravity by absorbing energy. It results in a small deficit in energies in still outer layers and thus somewhat stronger temperature gradient. Finally, the quasi-steady mass loss is realized which is described by Parker-type neutrino-driven stellar wind having temperature gradient of $T \sim r^{-2/3}$ (Duncan et al. 1986).

The rate of mass loss is proportional to $T_\nu^{10}$ (Duncan et al. 1986). Thus, an analogue of HR diagram with neutrino luminosity versus neutrino effective temperature is desirable to represent numerical results. If we assume the structure of polytrope with $N=3$ and take account of the definition of the neutrino sphere, we can calculate a neutrino Hayashi-line analogously to that for red giant stars. We have shown that numerical results could be understandable if analyzed with appropriate concepts.



Bethe, H.A., and Wilson, J.R. 1985, Ap. J., <u>295</u>, 14.
Duncan, R.C., Shapiro, S.L., and Wasserman, I. 1986, Ap. J., <u>309</u>, 141.
Wilson, J.R. 1982, private communications.
Wilson, J.R., Mayle, R., Woosley, S.E., and Weaver, T. 1985, Ann. N.Y.
  Acad. Sci., <u>470</u>, 267.

Analysis of Neutrinos from Supernova 1987A

H. Y. Chiu, K. L. Chan and Y. Kondo
Goddard Space Flight Center
Greenbelt, MD 20771, U. S. A.

We have developed a time-energy correlation method[1] to bring
forth the mass signature from Supernova 1987a neutrino
observations, if the neutrino has any mass at all. This method is
particularly effective in analyzing data sets with a small number
of events, such as the Kamiokande II[2] and the IMB[3] observations
of neutrino bursts from Supernova 1987A.  The time dispersion
$\Delta t_{12}$ between two simultaneously emitted neutrinos of energies $E_1$,
$E_2$ ($E_1 < E_2$) and a neutrino mass energy $m_\nu$ is given by:

$$\Delta t_{12} = (1/2)\ (L/c)\ (m_\nu/E_1)^2 \left[ 1 - (E_1/E_2)^2 \right], \qquad (1)$$

where $L$ is the distance of the source and $c$ is the velocity of
light. Conversely, Eq.(1) can also be used to establish time
relationships of detected neutrinos and the existence of a mass.
Applying Eq.(1) to all pairs for which real values of $m_\nu$ (called
the correlation mass) are obtained from the observed $\Delta t_{12}$, $E_1$,
$E_2$, the existence of a cluster of pairs with essentially the same
mass $m_\nu$ will indicate (a) many pairs of neutrinos were emitted
within a narrow time window, and (b) the existence of a mass at
$m_\nu$.  If a group of neutrinos were emitted within a narrow window,
these groups will show a strong time correlation.  Thus, this
method of analysis does *not* impose a condition for the emission
mechanism - rather, if the result of this analysis indicates the
existence of a mass, there must exist a time correlation among
the neutrinos.

A pre-requisite for the applicability of this method of
analysis is that the dead time for neutrino detection be small.
The lower limit for the applicability of this method is
determined by the energy range and the dead time of the detection
system, and can be worked out from Eq.(1).  No clustering on any
value of $m_\nu$ should be observed below this lower limit, including
the case that a nonzero mass exists.

The two nearly simultaneous neutrino detections, the
Kamiokande II and the IMB data are analyzed.  The energy range of
the IMB experiment is from 20 to 40 *MeV*, and the experimental
dead time is 0.1 *sec*.  The lower limit for detectability of
a mass energy is at least 5 *eV*. The energy range of the
Kamiokande II experiment is from 7.5 *MeV* to 35 *MeV*, and the dead
time is 50 nano*sec*, so that the lower limit of the detectable
mass is well below 1 *eV*.

A mass signature was found on the nominal energies of the
KII neutrinos at 3.6 *eV*, and none was found in the IMB data.
This is consistent with our previous discussion on the
detectability of a mass signature.  Out of the 12 KII neutrinos,
5 showed intimate time correlation which are, in the time
sequence of detection, 1,2,3,4 and 6.  Using a Monte Carlo
calculation which included experimental errors in the energies, a
strong mass signature is found, which peaked at 2.8 *eV* (see
Figure 1).  This Monte Carlo calculation results in the relative
probability for the masses; compared with that at the peak value
2.8 *eV*, the relative probability drops to 0.5 at 4.8 *eV* and 1.4
*eV* respectively.  The relative probability for a mass, say, below
0.5 *eV*, is less than 0.01.  Neutrino 6 has the lowest energy, 6.3
*MeV* and there is a good chance that this event might be a
background.  Removing this event from our analysis slightly
reduces the probability peak in our Monte Carlo calculation, but
essentially the same conclusion can be drawn.  On the basis of
this strong correlation, we suggest that Neutrino 6, too, is a
true event, and not a background.

Because of the paucity of events, it is necessary to
evaluate the probability for chance coincidence.  Several
statistical tests were made with random numbers, and the
probability for chance coincidences ranges between 1 % and 3 %.

We thus conclude that, (a) the mass of the neutrino is
either 0 (with a probability of < 3 %), or (b) the mass energy of
the neutrino is around 2 - 4 *eV* (with a probability of > 97 %).

Details of this work will be published elsewhere[4].

References:


1. H. Y. Chiu, K. L. Chan, and Y. Kondo, BAAS, *19*, 2, 740(1987).
2. K. Hirata *et al*, Phys. Rev. Ltrs. *58*, 1490 (1987).
3. R. M. Bionta *et al*, Phys. Rev. Ltrs. *58*, 1494 (1987), also Y.
   Totsuka, this proceeding.
4. H. Y. Chiu, K. L. Chan, and Y. Kondo, Astrophys. J. (in press)
   (1987).


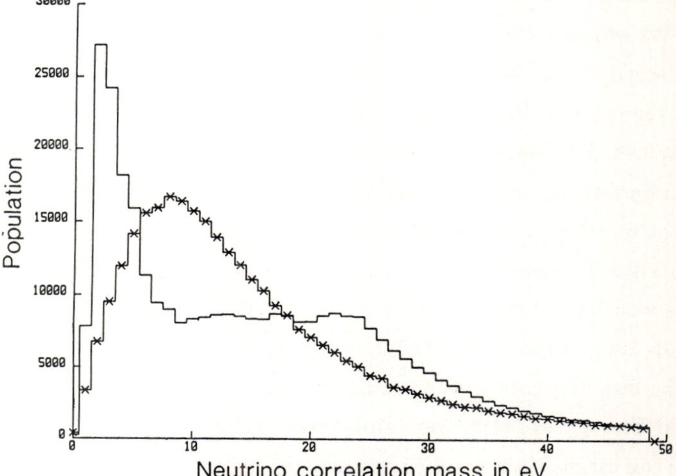


Figure 1. Cummulative distribution of KII data with energies
distributed according to a Gaussian.  Curve with x is obtained by
randomizing the time of the neutrinos.

# Analysis of Neutrino Burst from the Supernova in the Large Magellanic Cloud


Hideyuki Suzuki and Katsuhiko Sato

Department of Physics, Faculty of Science, University of Tokyo

Bunkyo-ku, Tokyo 113, Japan


A massive star has been believed to end his life with the collapsed driven supernova explosion and the formation of the compact object such as a neutron star or a black hole. When the compact object is formed, a large amount of energy corresponding to the binding energy of the object must be released. It has been considered that most of the energy is emitted by neutrinos because of their adequate coupling with the matter. The observation of the neutrino burst from SN1987A by Kamiokande[1] and IMB[2] offered us the first chance to test these scenarios of the collapse driven supernova explosion directly. We began to analyze the data just after their publication and got many important results which are presented below. In our analysis the distance of SN1987A is assumed to be 50kpc.

**The First 2 Events** : In all the detected 19 events( Kamiokande : 11, IMB : 8), only the first 2 events of the Kamiokande data may be scattering events, that is, $\nu_e + e^- \longrightarrow \nu_e + e^-$. One may consider that these 2 events due to the neutronization burst of $\nu_e$. Taking into account the small cross section of the scattering, however, it is doubtful to treat them as the scattering events. We estimate the energy of $\nu_e$ corresponding to the first 2 events and got the large value, $1.9 \cdot 10^{53}$ erg[3]. In all simulations ever published, the energy corresponding to the neutronization burst is the order of $10^{51}$ erg and the expected number of events in the Kamiokande detector is less than 0.04. Furthermore the duration time of the neutronization burst is about 10msec while the second event occurred 100msec after the first one. Hereafter we treat the all events as ones due to the reaction $\overline{\nu}_e + p \longrightarrow e^+ + n$.

**Neutrino Temperature** : Assuming the neutrino spectrum to be Fermi-Dirac distribution with the vanishing chemical potential, we got the $\overline{\nu}_e$ temperature, $T_\nu$[3,4]. For the Kamiokande data, $T_\nu$ is $2.6 \sim 3.1$MeV, for IMB $3.9 \sim 5.3$ MeV. These values are a little close to the values by those who insist the late time neutrino heating mechanism of the explosion.

**Total Energy of Neutrinos and Mass of Neutron Star** : Using above neutrino temperature, we can estimate the total energy emitted by neutrinos as six times $\overline{\nu}_e$ energy[4]. It becomes very reasonable value, $2.9^{+0.6}_{-0.4} \cdot 10^{53}$ erg for the Kamiokande data and $1.5^{+1.2}_{-0.6} \cdot 10^{53}$ erg for IMB. This energy can be interpreted as the binding energy of the remnant neutron star. Once we know the binding energy of the neutron star, we can determine the mass of the neutron star. Due to the ambiguity of the equation of state and the small statistics of the data, we cannot specify one value for the mass. But we may conclude that a neutron star whose mass is in the range of $0.6 \sim 1.9 M_\odot$ has been formed by SN1987A not a black hole (see Figure 1).

**Bunched Structure of the Data** : Kamiokande and IMB data show some bunched structure. We calculate the neutrino temperature and the the radius of the neutrinosphere for each bunch[5]. The first 8 events of Kamiokande and the first 6 events of IMB can be well understood in the scenario of the protoneutron star contraction. That is, the protoneutron star born with the radius of several times 10km contracts to the normal neutron star whose radius is about 10km in a few seconds and in this stage neutrino temperature once increases with the reduction of the volume.

**The Last 3 Events** : As described above, whole feature of the neutrino burst is well explained by the standard model. But the last 3 events of Kamiokande which occurred after the 7sec gap is mysterious. For example, the corresponding energy is large and the radius of the neutrinosphere of several times 10km is necessary for this energy[5].

The concept that neutrinos diffuse out of the supernova core carrying out its thermal energy is confirmed by the observation of the neutrino burst from SN1987A for the first time. But we must keep in mind that there are some puzzles such as the last 3 events of Kamiokande. There may exist some unknown mechanism of neutrino emission at the late time.

### References


[1] K. Hirata *et al.*, Phys. Rev. Lett. **58** (1987) 1490.
[2] R. Bionta *et al.*, Phys. Rev. Lett. **58** (1987) 1494.
[3] K. Sato and H. Suzuki, Phys. Rev. Lett. **58** (1987) 2722.
[4] K. Sato and H. Suzuki, Phys. Lett. in press.
[5] H. Suzuki and K. Sato, Publ. Astron. Soc. Japan **39** (1987) 521.


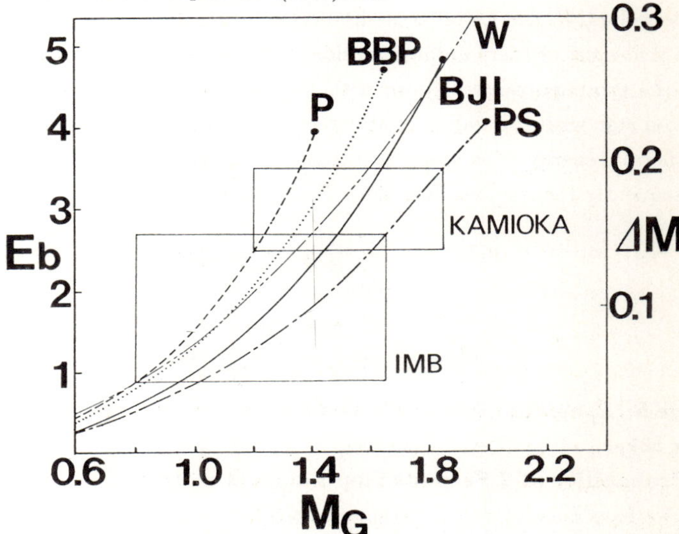



Figure 1:
The mass of neutron star vs. the binding energy for several EOS.

$M_G$ in units of $M_\odot$
$E_b$ in units of $10^{53}$erg

# Time Profile of the Neutrino Burst from the Supernova 1987A

Hideyuki Suzuki and Katsuhiko Sato
Department of Physics, Faculty of Science, University of Tokyo
Bunkyo-ku, Tokyo 113, Japan

SN1987A gave us the first opportunity to study the supernova core directly by providing us the neutrino signal from the core. The observational data of the neutrino flux detected by Kamiokande[1] and IMB[2] show surprisingly good agreements with the theoretical predictions as a whole[3,4]. The fundamental concept of the collapse driven supernova explosion is confirmed for the first time. On the other hand, there are some puzzles. The most peculiar feature of the data is the 7 seconds gap of the Kamiokande data. The first 8 events of Kamiokande were detected in 2 seconds, following the 7 seconds gap and the last 3 events in 4 seconds. Of course just only 7 seconds gap is not unnatural if small neutrino flux come. But there were detected 3 events after the gap. These 3 events may not be produced by the weak flux. We can estimate the time integrated luminosity of $\bar{\nu}_e$ corresponding to the last 3 events and get the large value such as $7 \cdot 10^{52}$erg [5]. Can we get out of this inconsistency, 3 events after the 7 seconds gap? If not, we may need to consider some nonstandard mechanism of the neutrino emission at the late time. In order to investigate the probability of the case in which there is a 7 seconds gap before 3 events, we have performed Monte Carlo simulations for the simple model of neutrino flux.

**Simple Model of Neutrino Flux** : Many authors study the data of Kamiokande and IMB and insist that their simple models of the neutrino flux can well explain the data. For example, Spergel *et al.*[6] investigate the cooling model of the protoneutron star and got the best fit parameters such as the radius, $R$, the initial neutrino temperature, $T_i$, and the decay time of the temperature, $\tau_T$. Note that most of them did not consider the contraction of the protoneutron star. It is considered that the protoneutron star born with the radius of several times 10 km contracts to the normal neutron star with the radius of about 10 km in a few seconds with emitting neutrinos and losing the thermal energy. Therefore we introduce one more parameter, the decay time of the radius, $\tau_R$. Consequently the number flux of $\bar{\nu}_e$, $F(E, t)$ is given by

$$
\begin{aligned}
F(E, t) &= \frac{c}{4} \frac{4\pi}{(hc)^3} \frac{E^2}{exp[E/T(t)] + 1} \frac{R(t)^2}{d^2} \qquad (\ t < t_m\ ) \\
&= 0 \qquad\qquad\qquad\qquad\qquad\qquad\qquad (\ t > t_m\ ) \quad (1) \\
T(t) &= T_i \cdot exp(-t/\tau_T) \qquad\qquad\qquad\qquad\qquad\qquad\quad (2) \\
R(t) &= R_f + (R_i - R_f) \cdot exp(-t/\tau_R) \qquad\qquad\qquad\quad (3)
\end{aligned}
$$

where $R_i$, $R_f$ are the initial and final radius respectively and the distance of SN1987A, $d$ is assumed to be 50kpc.

**Probability of 7 Seconds Gap before 3 Events** : Using the above models of the neutrino flux, we have done Monte Carlo simulations for the Kamiokande detector. Then we investigate the

### Table I : Results of Monte Carlo Simulations.

| $T_i$ (MeV) | $\tau_T$ (sec) | $R_i$ (km) | $R_f$ (km) | $\tau_R$ (sec) | $t_m$ (sec) | $N$ | $E_{\bar\nu}$ ($10^{52}$erg) | Probability |
|---|---|---|---|---|---|---|---|---|
| 4.2 | 18 | 27 | 27 | $\infty$ | 16 | 17.5 | 5.6 | 1/2813 |
| 4.2 | 18 | 20 | 10 | 10 | 16 | 7.5 | 2.3 | 1/2891 |
| 4.2 | 18 | 20 | 20 | $\infty$ | 13 | 9.5 | 3.0 | 0/7083 |
| 4.2 | 18 | 37 | 10 | 1 | 16 | 7.9 | 2.1 | 2/3505 |
| 4.2 | 18 | 40 | 20 | 1 | 16 | 15.5 | 4.5 | 1/5633 |
| 4.6 | 18 | 20 | 10 | 1 | 13 | 6.3 | 1.6 | 0/1611 |
| 4.6 | 18 | 20 | 10 | 10 | 16 | 12.4 | 3.2 | 2/8409 |
| 4.6 | 18 | 20 | 20 | $\infty$ | 13 | 15.7 | 4.3 | 1/5350 |
| 4.6 | 18 | 37 | 10 | 1 | 16 | 12.9 | 3.1 | 3/9862 |
| 5.0 | 18 | 20 | 10 | 1 | 16 | 9.8 | 2.3 | 6/6261 |
| 5.0 | 18 | 20 | 10 | 10 | 16 | 19.3 | 4.5 | 1/1413 |
| 5.0 | 18 | 37 | 10 | 1 | 16 | 19.9 | 4.3 | 0/1054 |

$N$ is the expected number of events in the Kamiokande detector and $E_{\bar\nu}$ is the total energy emitted as $\bar\nu_e$.

probability of the case in which there is a long gap longer than 7 seconds before the last 3 events. The results are summarized in Table I. Spergel *et al.*'s best fit flux is also included. In 2813 trials with their model, there is only one trial in which 3 events occurred after 7 seconds gap. In all other cases the probability is also very small, less than 0.1 percent. It seems that the probability is not very sensitive to the decay time of the radius. It is reasonable when we consider the dependence of the event rate to the temperature and the radius. The event rate is proportional to the neutrino number flux and the cross section of the reaction, $\bar\nu_e + \mathrm{p} \longrightarrow \mathrm{e}^+ + \mathrm{n}$, and the detection efficiency. The radius affects only the number flux and if the temperature remains constant the event rate is not reduced by the factor less than $(R_f/R_i)^2$. On the other hand the temperature strongly affects the event rate. The number flux, the cross section and the efficiency vary in the manner proportional to $T^3$, $T^2$, $T^\alpha$ respectively where $\alpha$ is some positive value. Consequently the event rate decays rapidly in the time scale of $\tau_T^{5+\alpha}$ and it becomes very difficult for the flux to produce more than 3 events after the long gap. Of course more systematic analysis is required, but it is cleared by this analysis that the rapidly decreasing $\bar\nu_e$ flux is inconsistent with the data of Kamiokande. If the protoneutron star actually cools down rapidly, the Kamiokande data indicate some unknown mechanism of neutrino emission 10 seconds after the collapse. We must be allowed to consider some nonstandard mechanism of the neutrino emission.

**Note added** : After this colloquium, we analyze the the probability of the long gap systematically. We find that the almost constant neutrino flux whose decay time of the event rate is longer than 10 seconds can caused the 7 seconds gap before 3 events with the probability of $1 \sim 2$ percent. But it seems not very natural to consider the constant neutrino flux.

### References


[1] K. Hirata *et al.*, Phys. Rev. Lett. **58** (1987) 1490.

[2] R. Bionta *et al.*, Phys. Rev. Lett. **58** (1987) 1494.

[3] K. Sato and H. Suzuki, Phys. Rev. Lett. **58** (1987) 2722.

[4] K. Sato and H. Suzuki, Phys. Lett. in press.

[5] H. Suzuki and K. Sato, Publ. Astron. Soc. Japan **39** (1987) 521.

[6] D. N. Spergel *et al. preprint* (1987).

# SN1987A AND CONSTRAINT ON THE MASS AND LIFETIME OF TAU NEUTRINOS


Mariko TAKAHARA[a] and Katsuhiko SATO[b]

[a]Department of Physics, Tokyo Institute of Technology, Tokyo, Japan
[b] Departments of Physics, University of Tokyo, Tokyo, Japan


Recently it has been suggested that neutrinos have nonvanishing mass from both experimental and theoretical investigations. Accelerator experiments and cosmology, however, impose the constraints on mass $m_\nu$ and lifetime $\tau_\nu$ of $\nu_\tau$ (see Fig. 1). As we showed in the previous paper [1], supernova explosion also imposes the stringent constraint on them. The purpose of this paper is to investigate the constraint imposed on $m_\nu$ and $\tau_\nu$ of $\nu_\tau$ from SN1987A.

Tau neutrinos are emitted from the central part of the iron core together with the electron neutrinos because they are also in thermal equilibrium with matter there. When tau neutrinos decay on the way of streaming out through the decay modes of $\nu_\tau \to \nu_e + \gamma$ and $\nu_\tau \to e^+ + e^- + \nu_e$, $\gamma$-rays are produced in both cases (in the latter case, the positron annihilates in the ambient matter and produces $\gamma$-rays).

If tau neutrinos emitted from SN1987A decayed outside the star, the $\gamma$-ray burst following the $\nu$-burst could be expected. Therefore observation of the $\gamma$-ray burst from SN1987A imposes a strict constraint on the mass and lifetime of tau neutrinos.

We estimated the $\gamma$-ray flux produced by the decay mode of $\nu_\tau \to \nu_e + \gamma$ because it is difficult to estimate the $\gamma$-ray flux produced by the decay mode of $\nu_\tau \to \nu_e + e^+ + e^-$ due to the ambiguity of the amount of the matter which surrounded the progenitor of SN1987A.

We assumed that tau neutrinos are emitted from their neutrinosphere with the spectrum of Fermi-Dirac distribution with zero chemical potential determined by the temperature at the neutrinosphere. This temperature is estimated to be 5 MeV from the spectrum calculated by Wilson and his collaborators.

The radius of the neutrinosphere and duration time of neutrino emission are adjusted so that the total energy lost by neutrinos should be the binding energy of neutron stars $3 \times 10^{53}$ erg.

If $\tau_\nu$ is much longer than the time scale of streaming out $t_0$, it is assumed that tau neutrinos decay outside the star at the typical distance from the center of the star $R = v\tau_\nu/(1-v^2/c^2)^{1/2}$ (v and c denote the velocity of electron neutrinos and light, respectively) and that the energy of $\nu_\tau$ is shared equally between $\gamma$-ray and $\nu_e$. The distance to SN1987A is assumed to be 48kpc. It is to be noticed that the arrival time of $\gamma$-rays spreads to the order of 2R/c.

On the other hand, if $\tau_\nu \ll t_0$, almost all neutrinos decay inside the star. Therefore in this case $\gamma$-rays are produced at the surface of the star and $\gamma$-ray burst spreads to the time scale of $2t_0$. As for the details of the calculation of the $\gamma$-ray flux, see Ref. [2].

At present the $\gamma$-ray detectors in operation are those aboard the Solar Maximum Mission (SMM) and Pioneer Venus Orbiter (PVO). We estimate the typical detection limit of the counter of SMM to be 0.1 counts/s/cm$^2$ and that of PVO 10 counts/s/cm$^2$, respectively. Since the evidence of $\gamma$-ray burst from SN1987A has not been reported yet, we investigated the constraint on $m_\nu$ and $\tau_\nu$ of $\nu_\tau$ on the assumption that the $\gamma$-ray burst was not observed.

In Fig. 1 we showed the prohibited region of the $m_\nu$ and $\tau_\nu$ of $\nu_\tau$ by the horizontally hatched regions (SMM) and dashed lines (PVO) for the various radii of the progenitor of SN1987A, $R_0$.

The region prohibited by the experiment, cosmology and $\gamma$-ray background radiation from the supernovae are also shown in Fig. 1 by the dotted region, the vertically hatched region, and obliquely hatched region, respectively.

Comparing these constraints with that imposed by the present paper, it is obviously seen that the region allowed by the previous constraints ($m_\nu \sim 50$ MeV, $\tau_\nu \sim 10^3$ sec) is completely prohibited by the present constraint one if $R_0 < 10^{13}$ cm, which is supported by the observation of the light curve of SN1987A.

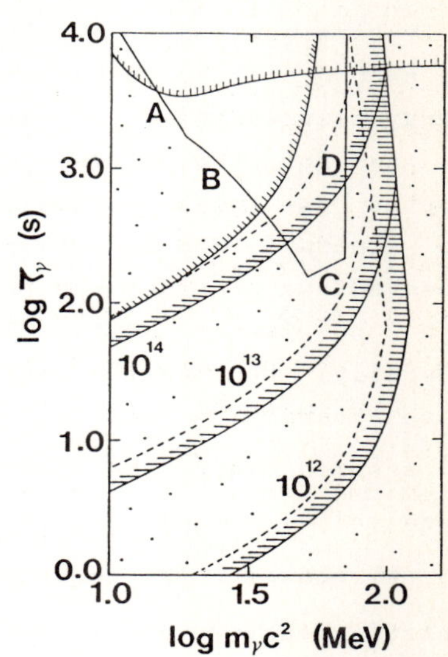



Fig. 1 Constraints on mass and lifetime of tau neutrinos. For details, see in the text.


[1] M. Takahara and K. Sato, Phys. Lett. 174B (1986) 373.
[2] M. Takahara and K. Sato, Modern Phys. Lett., A2 (1987) 293.

# COLLAPSE OF THE NEUTRON STAR INDUCED BY PHASE TRANSITIONS AND NEUTRINO EMISSION FROM SN1987A


Mariko TAKAHARA[a] and Katsuhiko SATO[b]

[a]Department of Physics, Tokyo Institute of Technology, Tokyo, Japan
[b]Department of Physics, University of Tokyo, Tokyo, Japan


The neutrino burst from SN1987A were detected by KAMIOKANDE and also IMB in February this year. The neutrino events detected by KAMIOKANDE were clustered into three bunches; event number 1-5, 7-9, 10-12. The first two bunches of the neutrino events can be understood by the standard scenario of supernova explosions.

However, it is very difficult to explain the last bunch of the neutrino events by the standard scenario, especially the total energy of the neutrinos and the radius of the neutrino sphere; it is necessary at least several times of $10^{52}$ ergs to explain the total energy of the last three events. In order to explain this energy, we must consider the non-standard mechanism of the neutrino emission.

As a possible mechanism we investigated the energy release produced by the phase transitions of the high density matter (transitions to the pion condensed state or to the quark matter); the newly born neutron star collapses due to the phase transitions between the second and the third bunch of the neutrino events and releases the energy necessary for the last three events.

As for the normal equation of state (EOS) of neutron stars we adopted the ones suggested by Bethe and Johnson of Model I (BJ) for the standard EOS and Cohen et al. (CLRC) for the very stiff EOS. To these EOS we consider the phase transitions of the high density matter with no latent heat. To represent the phase transition very generally, we treated the critical densities (the density at which phase transition starts, $\rho_{c1}$, or terminates, $\rho_{c2}$) as the parameters. We assumed that the pressure is constant at $\rho_{c1} < \rho < \rho_{c2}$ and that the sound velocity equals to the light velocity at $\rho > \rho_{c2}$.

We consider the neutron star of $1.4 M_\odot$ with normal EOS and set the central density equal to $\rho_{c1}$. Then $\rho_{c1}$ depends on the normal EOS and, as the stiffness of the normal EOS increases, $\rho_{c1}$ decreases; $\rho_{c1} = 9.8 \times 10^{14} \mathrm{gcm}^{-3}$ for BJ and $5.6 \times 10^{14} \mathrm{gcm}^{-3}$ for CLRC. Among the various choice of $\rho_{c2}$, we search the EOS which satisfy the following conditions; 1) the critical mass of the neutron star calculated in use of the EOS, $M_{cr}$, is larger than the observed mass of the neutron star

1.4 $M_\odot$, 2) the released energy by the collapse of the neutron stars induced by phase transitions, $\Delta E$, is larger than several times of $10^{52}$ ergs necessary for last three events detected by KAMIOKANDE, 3) the EOS must be stiff enough so that the collapse of neutron star can be halted if a neutron star is left after this supernova explosion. The last condition is checked by the hydrodynamical calculation of the collapse of the neutron star.

In Fig. 1 we plotted $M_{cr}$ and $\Delta E$ versus $\rho_{c2}$. As phase transition becomes stronger, i.e., $\rho_{c2}$ becomes larger, $M_{cr}$ becomes smaller and $\Delta E$ becomes larger. However, if phase transitions are too strong, the collapse of the neutron star cannot be halted. For example, for the normal EOS of BJ, the region of $\rho_{c2} > 1.8 \times 10^{15} \text{gcm}^{-3}$ is excluded by this condition. Therefore, very strong phase transitions are avoided by the conditions 1) and 3), while very weak transitions are avoided by the condition 2). Consequently moderately strong phase transitions are adequate for our purpose.

However, as is seen in Fig. 1, the absolute value of $\Delta E$ depends sensitively on the normal EOS; the stiffer EOS can release larger energy. If we adopt BJ, the releases energy is too small to explain the last three events. However, the enough energy is released for the very stiff EOS of CLRC. Therefore, we conclude that, if the normal EOS is sufficiently stiff, $\Delta E$ is large enough to explain the last three events detected by KAMIOKANDE by this mechanism.

Fig. 1. The critical mass of neutron stars $M_{cr}$ (dashed lines) and the released energy by the collapse of the neutron star induced by phase transition $\Delta E$ (solid lines) versus the strength of phase transition ($\rho_{c2}$). As for the normal EOS, two normal EOS is considered; EOS of Bethe and Johnson (BJ) for standard EOS and Cohen et al. (CLRC) for very stiff EOS.

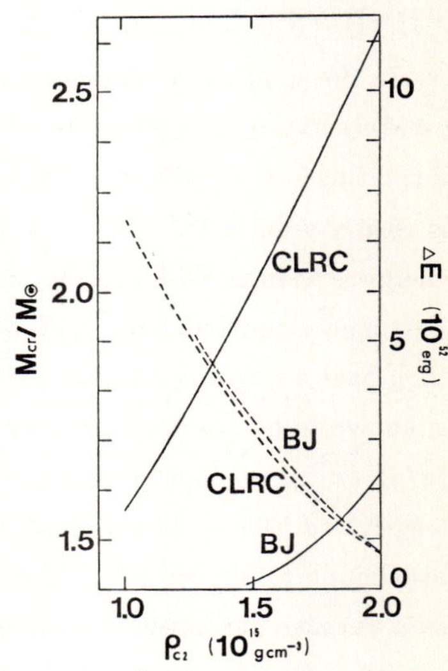

# A Rotating Stellar Collapse Model for Supernova 1987a

TAKASHI NAKAMURA

*Department of Physics, Kyoto University, Kyoto 606, Japan*

MASATAKA FUKUGITA

*Research Institute for Fundamental Physics*
*Kyoto University, Kyoto 606, Japan*

**It is shown that the bunch structure of the Kamiokande neutrino events associated with SN1987a can be naturally understood, if one assumes that the core of the progenitor star was rotating moderately with $q(\equiv Jc/GM^2) \approx 3$ with J the total angular momentum and M the gravitational mass of the core.**

We assume that the presence of the observed gap , at least that between the second and the third bunch, is real and consider its implications in the dynamics of the core collapse. Let us define a nondimensional angular momentum $q(\equiv Jc/GM^2)$ with J the total angular momentum and M the gravitational mass of the core. We assume that the value of q for the core of SN1987a was about 3. Then we expect that the effect of the angular momentum plays an important role when the size of the core $r$ shrinks to $< 10^7 cm$. In the spherically symmetric collapse model, the size of the unshocked homologous core is of the order of $10^7 cm$. Hence we expect that the core at the bounce is essentially governed by the spherically symmetric dynamics. To verify this point we may refer to the simulation by Symbalisty[1] Our model almost corresponds to Model ROT2 in which one obtains a neutrinosphere with a roughly spherical radius of approximately 50km. We then expect that the infalling nonhomologous matter onto the core will liberate a gravitational energy of the order of $10^{52} ergs$.

In the spherically symmetric model, the core enters the Kelvin-Helmholtz contraction phase , and its radius is expected to be $10^6 cm$ within 1 sec or so after the bounce [2] The evolution of a rotating star is different. The contraction takes place mainly along the rotational axis. This is clearly seen in the case with Model ROT2. In this Kelvin phase the released gravitational energy is at most $\frac{(\pi/2-1)\cdot 3\cdot GM^2}{5r} \approx 2\cdot 10^{52} ergs$. Using the cooling rate calculated by Burrows and Lattimer[2] , one finds that the core becomes a disk with thickness d $\approx 2\cdot 10^6 cm$ within 0.5sec. We then have a rotating disk-like proto neutron star with an aspect ratio $r/d \approx 5$. Such a thin disk is known to be gravitationally unstable irrespective of the equation of state[3] It fragments into $(r/d)^2/\pi^2$ pieces in a free fall time scale $\approx 0.01 sec$, because the most unstable mode has a wave length $2\pi d$. Three dimensional numerical simulations of collapse of rotating isothermal clouds also support this result[4] In our case we will have two or three fragments. Since the specific spin angular momentum is reduced by a factor of 3 to 5,[4] the effect of angular momentum is not important for the Kelvin contraction of each fragment. Then the binding energy of neutron stars $\approx 10^{53} ergs$ will be liberated after all of the order of one second according to the cooling rate in Ref.2. We consider that the second neutrino bunch corresponds to this phase.

Let us assume for simplicity that a binary-like system with masses $m_1$ and $m_2$ with a separation distance r is formed as a result of the fragmentation. Then the distance r decreases due to the emission of gravitational radiation at the rate of $\dot{r} = -7.7 \cdot 10^5 cm/sec \cdot \frac{f \cdot (m_1/0.7M_\odot) \cdot (m_2/0.7M_\odot) \cdot ((m_1+m_2)/1.4M_\odot)}{(r/10^7 cm)^3}$, with f the efficiency factor. Equation (2) for f=1 is derived by assuming that two bodies are point particles. As a value of f we adopt f= 0.1 $\approx$ 0.5 to take into account non-point like effects because an extended object is a weaker emitter of the gravitational radiation due to phase cancellation effects [5,6] Now the characteristic time of the decrease of the separation r due to the emission of the gravitational radiation is given by

$$t = 9sec \cdot (f/0.3)^{-1} \cdot \frac{(r/10^7 cm)^4}{(m_1/0.7M_\odot) \cdot (m_2/0.7M_\odot) \cdot ((m_1+m_2)/1.4M_\odot)}.$$

If the two fragments have almost equal mass $0.7 M_\odot$, then two proto neutron stars will collide after 9sec with a subsonic speed $2.2 \times 10^8 cm/sec$ . Eventually a rapidly rotating neutron star of mass $1.4 M_\odot$ will be formed and the difference of the binding energy up to $10^{53} ergs$ between proto neutron stars and a final single neutron star should be liberated as neutrinos, gravitational radiation and electromagnetic waves. Alternatively, if the two fragments have different masses, say, $0.9 M_\odot$ and $0.5 M_\odot$ then the binary system enters a stable mass stripping phase under the plausible parameters of mass loss and angular momentum loss [7] In a numerical calculation for the case of two neutron stars of mass $1.3 M_\odot$ and $0.8 M_\odot$ the integrated energy of the neutrino emission by the mass accretion is estimated to be $2 \cdot 10^{53} ergs$ with time duration about 2sec[7] We can estimate the average energy of neutrinos in the accretion phase to be $\approx$ 10MeV[7] In any case we expect that a neutrino flux with total energy $\approx 10^{53} ergs$ is liberated almost 10sec after from the first burst.

REFERENCES


1. E.D. Symbalisty, Asrtrophys. J. **285**(1984),729.
2. A. Burrows and J.M. Lattimer, Asrtrophys. J. **307**(1986) 178.
3. P. Goldreich and D. Lynden-Bell, Mon. Not. R. A. Soc. **130**(1965) 97.
4. S.M. Miyama, S. Narita and C. Hayashi Asrtrophys. J. **279**(1984) 621.
5. T. Nakamura and M. Sasaki, Phys. Letters **106b**(1981) 69.
6. T. Nakamura and K. Oohara, Phys. Letters **98A**(1983) 403.
7. J.P.A. Clark and D.M. Eardley, Astrophys. J. **215**(1977) 311.

NEUTRINO EMISSION PROCESSES IN THE WEINBERG-SALAM THEORY

NAOKI ITOH

Department of Physics, Sophia University,
Tokyo, Japan

The neutrino emission processes play essential roles in stellar evolution as expemplified by the observations of the neutrinos from SN 1987a by the KAMIOKANDE-II and IMB experiments. Recently a very extensive study of the various neutrino emission processes based on the Weinberg-Salam theory has been completed by the present author and his collaborators. The neutrino emission processes calculated by the author's group include pair, photo-, plasma, and bremsstrahlung neutrino processes. The neutrino energy loss rates due to pair, photo-, and plasma processes in the framework of the Weinberg-Salam theory are found to be substantially lower than the result obtained by Beaudet, Petrosian, and Salpeter. The reduction factor $\alpha$ is in the range $0.35 < \alpha < 0.88$ depending on the neutrino masses, density, and temperature. The ionic correlation effects play important roles in the bremsstrahlung neutrino process. The present author and his collaborators recently calculated the bremsstrahlung neutrino energy loss rate taking into account the ionic correlation effects in the crystalline lattice state as well as in the liquid metal state. They found that the ionic correlation effects suppress the bremsstrhlung neutrino energy loss typically by a factor 2-20. The present findings will bear great importance in neutrino astronomy.

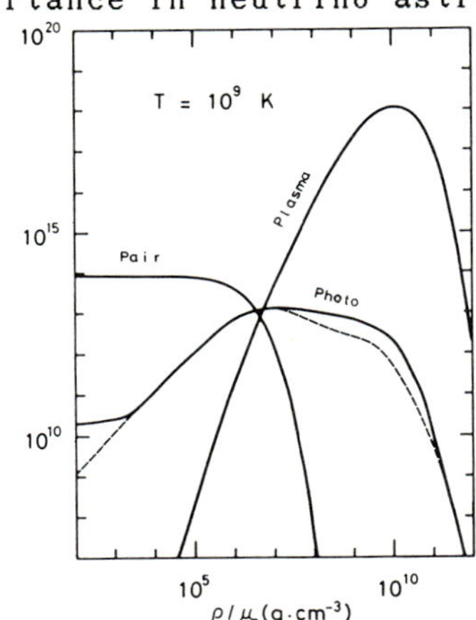


FIG. 1. - Pair, photo, and plasma neutrino energy loss rates. ( ergs s$^{-1}$ cm$^{-3}$) for T=10$^9$K.

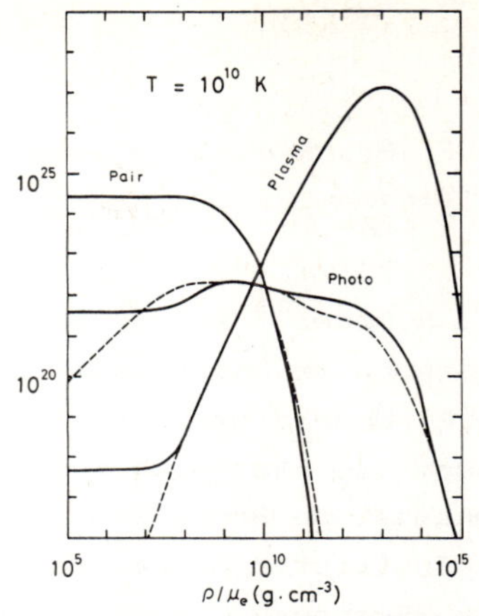

FIG. 2. - Pair, photo, and plasma neutrino energy loss rates.
( ergs s$^{-1}$ cm$^{-3}$) for T=10$^{10}$K.

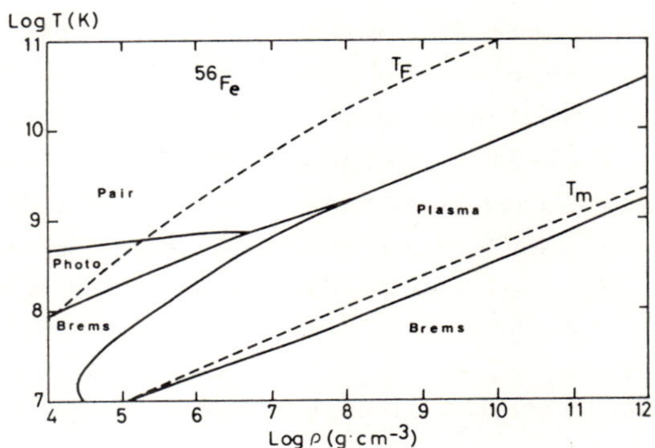

FIG. 3. - Most dominant neurtino process for a given density and
temperature for the case of  n=1 and $^{56}$Fe matter.


REFERENCES


[1]  Itoh,N., and Kohyama,Y. 1983, Ap.J., 275, 858.
[2]  Itoh,N., Kohyama,Y., Matsumoto,N., and Seki, M. 1984a, AP.J., 280, 787.
[3]  Itoh,N., Kohyama,Y., Matsumoto.N., and Seki,M. 1984b, Ap.J., 285, 304.
[4]  Itoh,N., Matsumoto,N., Seki, M., and Kohyama.Y. 1984c, Ap.J., 279, 413.
[5]  Kohyama,Y., Itoh,N., and Munakata,H. 1986, Ap.J.,310, 815.
[6]  Munakata,H., Kohyama,Y., and Itoh,N. 1985, Ap.J., 296, 197.
[7]  Munakata,H., Kohyama,Y., and Itoh,N. 1986, Ap.J., 304, 580.
[8]  Munakata,H., Kohyama,Y., and Itoh,N. 1987, Ap.J., 316, 708.

# A SIMPLE TREATMENT OF THE PROBLEM OF RADIATIVE TRANSFER IN SUPERNOVA-LIKE ENVELOPES


J.Isern[1,2], R.López[2,3] and E.Simonneau[4]

1) Instituto de Astrofísica de Andalucía(C.S.I.C.) Apdo.2144.
18080 Granada (Spain).2) Grup d'Astrofísica (Institut d'Estudis
Catalans). 3) Departament de Física de l'Atmosfera. Astronomía
i Astrofísica. (Universitat de Barcelona) Diagonal 647. 08028
Barcelona (Spain). 4) Institut d'Astrophysique de Paris 98bis
Bd. Aragó. 75014 Paris (France).


Since the only direct information on the physics of supernova explosions comes from spectrophotometry of their light curves, it is wortwhile to put considerable effort on understanding the properties of supernova atmospheres. Two problems appear, however, in doing so. First, the mean free path of a photon in the region where the spectrum forms (the atmosphere) is an important fraction of the supernova size and the plane parallel approximation breaks down. Second, the roles of scattering and absorption in the formation of the continuum are not clearly determined.

The complications introduced by spherical geometry arise from the angular derivative terms in the transfer equation and from the high degree of anisotropy introduced by the geometry. The complications introduced by scattering arise from the presence of the mean intensity in the source term of the transfer equation. As an outcome,its numerical solution is time consuming , and classical simplifications are not possible due to the strong anisotropy of the radiation field. In hydrodynamical calculations it is better to calculate the moments of the radiation field, but the success of this method depends on the possibility of replacing the infinite set of differential equations by the equations of low order moments plus a closure relationship that keeps all the properties of the higher moments.

The equations for the moments (frequency integrated) are, in the comoving frame (Castor 1972):

$$\frac{1}{c}\frac{\partial J}{\partial t} + 4\pi r^2 \rho \frac{\partial H}{\partial M_r} + \frac{2H}{r} - (3K-J)\frac{v}{cr} - (J+K)\frac{1}{c}\frac{\partial \ln\rho}{\partial t} = -\int_0^\infty (k_\nu J_\nu - i_\nu)\,d\nu \qquad (1)$$

$$\frac{1}{c}\frac{\partial H}{\partial t} + 4\pi r^2 \rho \frac{\partial K}{\partial M_r} + \frac{3K-J}{r} - \frac{2v}{cr}H - \frac{2}{c}\frac{\partial \ln\rho}{\partial t}H = -\int_0^\infty k_\nu H_\nu\,d\nu \qquad (2)$$

which take into account the geometrical dilution of the energy and momentum fluxes. In order to solve both equations it is necessary to have a linear relationship between the moments appearing in both equations.

As the intensity strongly peaks inside the cone $\mu \leq \mu_c$ subtended by the opaque interior, the Eddington relationship is no longer valid since it lies on the hypothesis that the intensity is quasiisotropic in the regions $(0 < \mu < 1)$ and $(-1 < \mu < 0)$. However, this relationship can be easily generalized to the spherical case by defining a two zone model $(\mu \leq \mu_c ; \mu > \mu_c)$ (Simonneau 1980; Isern, López, Simonneau 1987; López, Simonneau, Isern 1987; Simonneau, Isern, López 1987). This two region model gives the relationship:

$$3K-J = 2\mu_c H \qquad (3)$$

$$\frac{d\mu_c}{dr} = \frac{1-\mu_c^2}{r\mu_c} e^{-\tau/\mu_c} \qquad (4)$$

with the initial condition $\mu_c = 0$ for $r = 0$. It is easy to see that (1) and (2) with the boundary conditions: $r = 0$, $H = 0$; $r = R$, $H = \frac{1}{2}J(1+\mu_c)$ and the linear relationship (3) give the correct behaviour at the center and at the outer boundary of the envelope (López, Simonneau, Isern 1987).

Equation (4) only describes the anisotropy due to the geometry and the opacity distribution, but not that introduced by the scale height of the source function. This means that the accuracy of the solutions obtained by the method proposed here improves when temperature gradients are smouth or/and scattering is dominant. The accuracy, in the worst case, for a supernovalike envelope is better than 10% in J for $\tau < 1$, and better than 5% in $H(\tau=0)$.

This work has been supported by the "Subdirección General de Cooperación Científica y Técnica", the CAICYT grant 400-84 and the CSIC program Nuclear Astrophysics.

REFERENCES

Castor, J.I., 1972, Astrophys. J.,178, 779.
Isern,J., López,R., and Simonneau,E., 1987, Rev.Mex.Astron. Astrof., 14 in press.
López,R., Simonneau,E., and Isern,J., 1987, Astr. Astroph., 184, in press.
Simonneau,E. ,Isern,J., and López,R., 1987, preprint.
Simonneau,E., 1980, Bull. Soc. Roy. des Sci. de Liège.n9-10, p.281.

# THE LIGHT CURVE OF SN 1987 A

R. Schaeffer[1], M. Cassé[2], R. Mochkovitch[3], S. Cahen[2]

1 — Service de Physique Théorique, C.E.N. Saclay, France
2 — Service d'Astrophysique, C.E.N. Saclay, France
3 — Institut d'Astrophysique du CNRS, Paris, France

**Abstract.** We use a semi–analytical model of supernova light curves that extends Arnett's scheme [1,2] to include the effect of recombination of hydrogen, helium and heavy elements in the expanding ejecta. Introducing in the model the physical parameters of Sanduleak –69 202, the salient characteristics of the light curve of SN 1987 A are reasonably reproduced over 100 days.

## 1. Relevant physical parameters.

The light curve is sensitive to several critical parameters. Those are mainly the mass M and radius R of the progenitor star and the energy E of the explosion. Other important parameters are the composition of the exploding star, on which depend the opacity and the details of recombination in the ejecta, the mass of $^{56}$Ni ejected and the energy which can be injected by a newly born pulsar.

We have adopted M = $15M_\odot$, R = $3\ 10^{12}$ cm and the energy E, expected to be of the order of $10^{51}$ erg, has been ajusted to fit the supernova luminosity in its plateau phase. The composition has been deduced from the supernova models of Woosley [3], assuming for Sk–69 202 a ZAMS mass of 20 $M_\odot$, which means that we suppose that about 5 $M_\odot$ of hydrogen–rich material have been lost during the stellar lifetime. We have also assumed a radioactive energy output corresponding to 0.01 $M_\odot$ of $^{56}$Ni (the preferred value is however now 0.07 $M_\odot$ [4,5]) and no pulsar energy injection.

## 2. Results.

The results given by our semi–analytical model are illustrated in figures 1,2, 3 and 4. The explosion energy is E=1.2 $10^{51}$ erg and the 0.01 $M_\odot$ of $^{56}$Ni are located in the very inner region of the exploding star. The early light curve exhibits a first abrupt rise in the V band followed by a 10 day rather flat part. The bolometric luminosity is constant during that period. The gentle increase after ten days occurs because the photospheric temperature is then fixed at the recombination temperature, whereas the photospheric radius is still increasing. The second plateau at 50–100 days, is due to the progress of recombination in the expanding envelope that stabilizes the photosphere at a given radius and temperature. The prevailing contribution to the bolometric luminosity is, over most of the evolution, the initial energy imparted by the explosion. The recombination energy and the $^{56}$Ni radioactivity however begin to play a significant role just before the drop of luminosity after about 100 days.

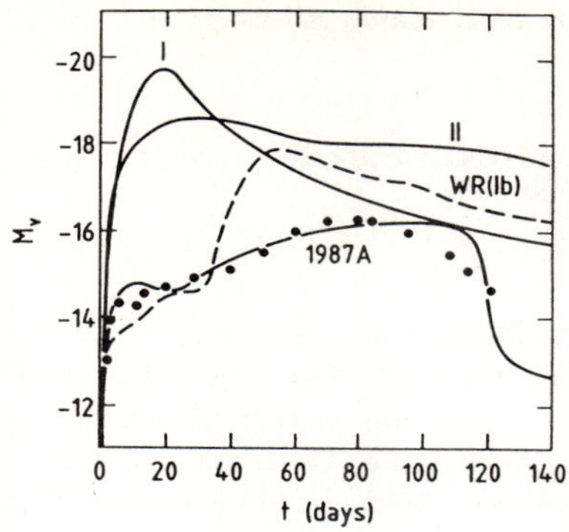

Figure 1: Visual light curve of SN 1987 A compared to typical SN I and SN II. The dashed line corresponds to the Wolf-Rayet supernova considered in [6] and the lower full line to the light curve computed with the Sk −69 202 parameters.

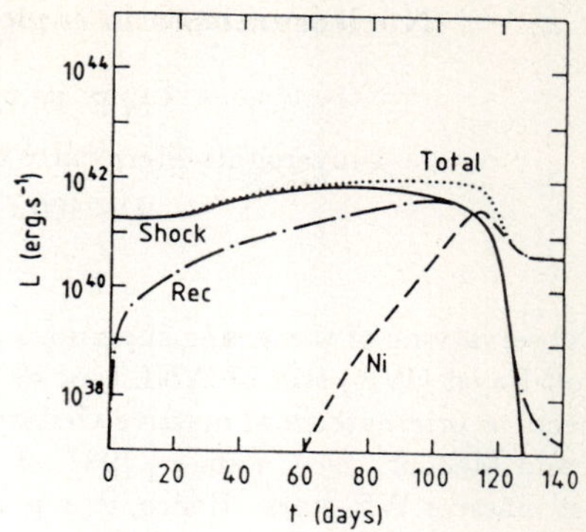

Figure 2: Bolometric light curve of SN 1987 A for the assumed parameters of Sk −69 202. The dots indicate the total luminosity, the full line the luminosity due to the shock energy, the dashed line the luminosity coming from the decay of $^{56}$Ni and the dash-dotted line the recombination energy.

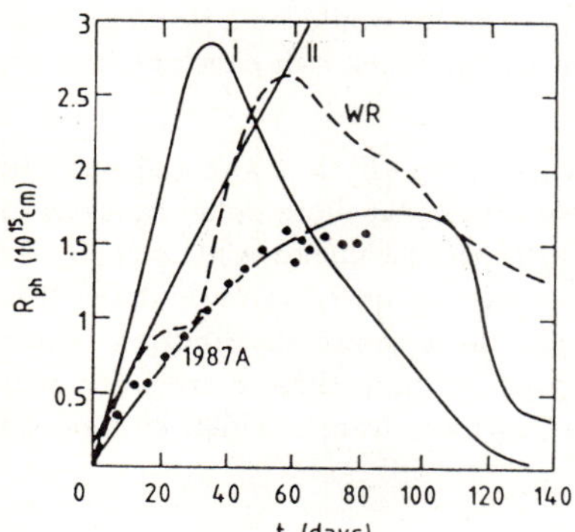

Figure 3: Photospheric radius for the different models considered in Fig.1, as compared to the observational data for SN 1987 A obtained by Dopita et al. [7].

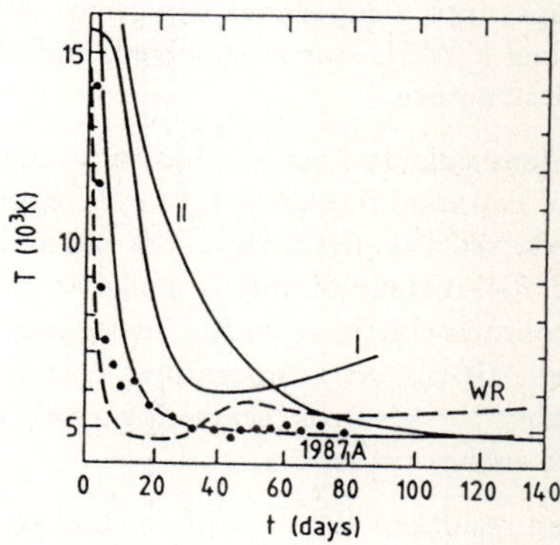

Figure 4: Effective temperature for the different models considered in Fig.1, as compared to the data for SN 1987 A of Dopita et al. [7].

References


[1] Arnett, W.D. : 1980, Astrophys. J. 237, 541

[2] Arnett, W.D. : 1982, Astrophys. J. 253, 785

[3] Woosley, S.E. : 1987, in $16^{th}$ Advanced Course of Saas–Fee, ed. B. Hauck, A. Maeder, G. Meynet, p. 127

[4] Nomoto, K. : 1987, this conference

[5] Woosley, S.E. : 1987, this conference

[6] Schaeffer, R., Cassé, M., Cahen, S. : 1987, Astrophys. J. 316, L31

[7] Dopita, M.A., Achilleos, N., Dawe, J.A., Flynn, C., Meatheringham, S. J. : 1987, to be published in Proc. Astron. Soc. Austr.

# Nucleosynthesis in exploding massive Wolf-Rayet stars

M.F. El Eid and N. Langer

Universitäts-Sternwarte Göttingen, Geismarlandstraße 11

D–3400 Göttingen, F.R.G.

Recent observations of the young supernova remnant Cas A (Fesen et al., 1987) suggest an exploding Wolf-Rayet (WR) star of WNL type as a progenitor of this object. The majority of the WR stars seems to originate from massive O-stars of $M > 40\,M_\odot$. According to current investigations (Schild and Maeder, 1984; Langer, 1987; cf. also: Langer, this volume) WNL stars rank among the most massive WR stars. Hence, it is possible to assume that the stellar progenitor of Cas A was indeed a very massive star.

As shown by Langer and El Eid (1986), (see also Woosley, 1986) a population I star of initially $100\,M_\odot$ may loose enough mass during its evolution up to core He exhaustion to become a WN star of $\sim 45\,M_\odot$, which then mainly consists of oxygen (more than 80%) synthesized during He burning.

In a previous work (El Eid and Langer, 1986) we have shown that such a star encounters the $e^\pm$-pair creation instability at central oxygen ignition. During the ensuing collapse oxygen burning proceeds explosively on a typical time scale of $\sim 40\,s$, and may release enough energy to reverse the collapse into a supernova explosion. We note that such explosion would be a peculiar type II event when a WNL star is envolved, since it retains part of its hydrogen rich envelope but has a compact structure.

Our present calculations are basically similar to the previous ones (El Eid and Langer, 1986) except of two modifications: (i) we have followed the nucleosynthesis in more detail by extending the nuclear reaction network (cf. El Eid and Prantzos, 1987), which now includes 39 nuclei between $^4$He and $^{56}$Ni instead of only $\alpha$-nuclei as previously. (ii) The opacities have been calculated taking into account inelastic compton scattering and degeneracy in the electron distribution (Buchler and Yueh, 1976). At temperatures encountered during oxygen burning these corrections reduce appreciably the electron scattering cross section over the Thomson value, which enhances the radiative energy transport.

The main results of the present calculations are sketched as follows:

During the collapse phase of the $45\,M_\odot$ WN star higher peak values of temperature ($3.37 \cdot 10^9\,K$) and density ($2.05 \cdot 10^6\,g\,cm^{-3}$) are achieved mainly due to the reduced opacities. Consequently, the amount of consumed oxygen is higher ($\sim 3\,M_\odot$). Hence, the energy release from explosive oxygen burning is enough to disrupt the star. The resulting kinetic energy was $\sim 3 \cdot 10^{51}\,erg$, and the surface velocity $\sim 6900\,km\,s^{-1}$.

The resulting nucleosynthetic yield from such explosion is compared with the abundances inferred for Cas A from the optical observations of specific fast moving knots (FMKs). In Fig. 1, the abundances of several FMKs are superimposed on the calculated profiles. Several features are compatible with observations: (a) the observed ratios for S, Ar, and Ca of various knots correspond to layers in the model which have undergone incomplete oxygen burning. (b) Knot KB115 lies between that of [OIII] filament and knot KB33. (c) The layers accommodating the observed ratios for S, Ar, and Ca show substantial amounts of both, O and Si. (d) The upper limits of C and Ne are found in the outer layers around $M_r = 20\,M_\odot$. This means that the layers which are needed to explain all observed abundances of the FMKs comprise enough mass to account for the estimated lower limit ($\sim 15\,M_\odot$) of the remnant mass inferred from the X-ray observations (Fabian et al., 1980; Markert et al., 1983).

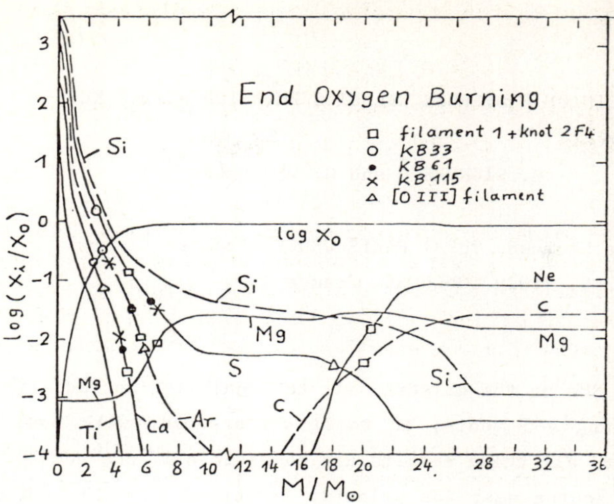


**Fig. 1:** Comparison of the observed abundance ratios relative to oxygen in the fast moving knots of Cas A with calculated values for the exploding $45\,M_\odot$ WN star (only $36\,M_\odot$ are shown). The observed values taken from Chevalier and Kirshner (1979) are plotted on the calculated profiles. The oxygen profile ($\log X_O$) is also shown.

In summary, though we cannot claim to have found the evolutionary scenario for the complicated object Cas A, several features in the present model of an exploding WN star are compatible with the available observations of this object. It appears, that the explosion of WNL stars as a pair creation supernova can be envisaged as a possible explanation for some extragalactic supernova remnants of Cas A type, like the one in NGC 4449 (Kirshner and Blair, 1980), or for supernovae showing spectra resembling the WN classification. It seems worthwhile to extend the present calculations by including detailed radiation transport in order to calculate the light curve of such explosions.

## References


Buchler, J.R., and Yueh, W.R.: 1976, Astrophys. J. **210**, 440

Chevalier, R.A., and Kirshner, R.P.: 1979, Astrophys. J. **233**, 154

El Eid, M.F., and Langer, N.: 1986, Astron. Astrophys. **167**, 274

El Eid, M.F., Prantzos, N.: 1987, in: *The Origin and Distribution of the Elements*, New Orleans, ed. G.J. Mathews, World Sci. Publ., in press

Fabian, A.C., Willingale, R., Pye, J.P., Murray, S.S., and Fabbiano, G.: 1980, M.N.R.A.S. **193**, 175

Fesen, R.A., Becker, R.H., and Blair, W.P.: 1987, Astrophys. J. **313**, 378

Kirshner, R.P., Blair, W.P.: 1980, Astrophys. J. **236**, 135

Langer, N.: 1987, Astron. Astrophys. (Letters) **171**, L1

Langer, N., El Eid, M.F.: 1986, Astron. Astrophys. **167**, 265

Markert, T,H., Canizares, C.R., Clark, G.W., Winkler, P.F.: 1983, Astrophys. J. **268**, 134

Schild, H., Maeder, A.: 1984, Astron. Astrophys. **136**, 237

Woosley, S.E.: 1986, in: *Nucleosynthesis and Chemical Evolution*, Proc. $16^{th}$ Saas-Fee Course, B. Hauck et al., eds.

ORIGIN OF A DIFFUSE GALACTIC EMISSION AT 511 KEV

M. SIGNORE [*] and G. VEDRENNE [**]

[*] E.N.S.  24, rue Lhomond, 75231 PARIS CEDEX, France
[**] C.E.S.R.  B.P. 4346, 31029 TOULOUSE, France

Recent results of SMM on the galactic 511 keV annihilation radiation (1) exhibit a constant flux during 4-5 years, in total disagreement with post 1980 balloon observations. However, all these results are compatible if one admits the existence of a variable point source near the Galactic Center of $F_p \sim 10^{-3}$ ph cm$^{-2}$s$^{-1}$ and a diffuse interstellar source of positrons: $F_D \sim 1.5\text{-}1.8\ 10^{-3}$ ph cm$^{-2}$s$^{-1}$rad$^{-1}$. Evaluations of $^{26}$Al and $^{56}$Co decays contributions have been given last year (2) and that of $^{44}$Ti decays more recently (3). We have studied the nature of the sources of the main $e^+$ – emitting radioisotopes – SN and novae – and their galactic distributions using the observational data at 511 keV and at 1809 keV, the galactic angular distributions considered by Leising and Clayton (4) and the nucleosynthesis of the models of novae and SN of Woosley and his collaborators (5). Because there are two sets of data, the $e^+$-flux, for the central radian, produced via $^{26}$Al decays can be separated from the flux produced via non $^{26}$Al decays :

$$(1)\quad F_{Al} = 1.33\ \sigma^{Al}_{SN} + 2.6\ \sigma^{Al}_{n} = 4\ 10^{-4}e^+\ \mathrm{cm}^{-2}\mathrm{s}^{-1}\mathrm{rad}^{-1}$$

$$(2)\quad F_{non\,Al} = 1.33\ \sigma^{non\,Al}_{SN} + 2.6\ \sigma^{non\,Al}_{n} = 19\ 10^{-4}e^+\ \mathrm{cm}^{-2}\mathrm{s}^{-1}\mathrm{rad}^{-1}$$

The right hand side of Eq.(1): $4\ 10^{-4}e^+$ cm$^{-2}$s$^{-1}$rad$^{-1}$ is deduced from the observed $^{26}$Al line: $4.8\ 10^{-4}\gamma$ cm$^{-2}$s$^{-1}$rad$^{-1}$ while the right hand side of Eq.(2) corresponds to the observed annihilation line with a "Ps-fraction" f = 0.9 (the annihilation is supposed to occur in molecular hydrogen). It becomes 11 or $3.5\ 10^{-4}e^+$ cm$^{-2}$s$^{-1}$rad$^{-1}$ if respectively $f \sim 0.65$ or $f \sim 0$ (annihilation in dust grains). The value of f deduced from UNH's observations is $f \sim 0.7$;

- $\sigma^{Al}_{SN}$ is the local rate of $e^+$ ejected by SN, per unit area, and idem for others – see Table 1;

Table 1

| | | $^{26}$Al | $^{56}$Ni | $^{44}$Ti | $^{22}$Na |
|---|---|---|---|---|---|
| $\frac{\sigma}{\odot}$ $10^{-4}$ $e^+$ cm$^{-2}$ s$^{-1}$ | novae | 1.4 $a_1 t_1$ | | | 1.8 $b_1 t_1$ |
| | SN I | | 7.5 $\varepsilon_1 y_1$ | $\begin{cases} 0.75\ z_1 \\ 0.75\ y_1 \end{cases}$ | |
| | SN II | 1.4 $x_1$ | | 0.75 $x_1$ | |

- $\varepsilon = 10^{-2}\varepsilon_1$ is the escape fraction of $e^+$ from SN I via $^{56}$Co decays;
- novae eject ($a_1\ 10^{-7}\ M_\odot$) of $^{26}$Al and ($b_1\ 10^{-7}\ M_\odot$) of $^{22}$Na at a local rate of $t_1\ 10^{-8}$ pc$^{-2}$ yr$^{-1}$ – outbursts on a O–Ne–Mg WD –;
- there are $x_1\ 10^{-11}$ pc$^{-2}$ yr$^{-1}$ SN II explosions of over 15 $M_\odot$ stars, $y_1\ 10^{-11}$ pc$^{-2}$ yr$^{-1}$ standard SN I (SN Ia) and $z_1\ 10^{-13}$ pc$^{-2}$ yr$^{-1}$, Peculiar SN I–Helium dwarf detonating SN I that experience explosive helium burning (5);
- we have considered many working hypothesis (6) for the relative frequencies and the angular distributions of the various types of SN and novae.

In any case, it is possible to explain the diffuse 511 keV and 1809 keV lines with the present models of nucleosynthesis for SN and novae. The contributions to the $e^+$ background from Al, Co, Ti, Na decays depend of the values of the diffuse flux, the values of the Ps fraction f, the nature of the SN distributions, the rate of occurence of the Peculiar SN I. In most cases, $^{56}$Co decays are the main contributor to the diffuse $e^+$-flux. But $^{44}$Ti decays become the main contributor if a part of SN Ia and the Peculiar SN I have a nova type distribution and if the rate of these Peculiar SN I leads to the solar abundance of $^{44}$Ca produced during the last $10^6$ years. Moreover, if one assumes that SN Ib – which seem to be correlated with regions of recent star formation – have a CO distribution, we can reconcile the whole observational results on the 511 keV flux. Future observations such as SIGMA and GRO can have strong implications on the nucleosynthesis models of novae and SN and on galactic distributions and relative frequencies of their progenitors : in particular Wolf–Rayet but also all sorts of helium star cataclysmics.

REFERENCES


(1) Share G. H. et al., 1986, Advances in Space Res., 6, no 4, 145.

(2) Ramaty R., Lingenfelter R. E., 1986, in "The Galactic Center", D. C. Backer Ed., AIP Publ., 155, N. Y., p. 155.

(3) Woosley S. E., 1987, preprint.

(4) Leising M. D., Clayton D. D., 1985, Astrophys. J., 294, 591.

(5) Woosley S. E., 1986, "16th Advanced Course of the Swiss Academy of Astron. and Astrophys", B. Hauck, A. Maeder Ed., Genova and References.

(6) Signore M., Vedrenne G., 1987, sumbitted to Astron. and Astrophys..

CHEMICAL COMPOSITION OF HIGH-ENERGY COSMIC-RAY NUCLEI
AND ITS POSSIBLE ORIGIN IN TYPE-II SUPERNOVA


Yoshiyuki Takahashi

Department of Physics, University of Alabama in Huntsville
Huntsville, Alabama 35899, USA


High energy cosmic ray spectrum has been known to have an interesting bump in the energy range $10^{14} - 10^{16}$ eV. Various models to explain the spectral break in this energy range have been so far proposed; which incorporate either a large-scale termination of galactic wind [1], shocks with greater age and spatial extent associated with hypothetical super-bubbles powered by multiple supernovae [2], intersection of two quantum-gravitational components [3], extra-galactic component [4], red-shift of big-bang remnant [5], or a proton component from pulsars [6]. More recently, their possible origin in type-II supernovae with magnetic acceleration mechanism is proposed [7] by considering direct observational results of chemical composition near the bump regime of cosmic ray spectrum.

To discriminate these models detailed observation of the chemical composition at the bump regime is imperative. The latest data include those obtained by the Spacelab-2 (CRNE-experiment of the University of Chicago) [8], and the 1986 balloon-flight experiment by the Japanese-American-Cooperative-Emulsion-Experiment (JACEE) Collaboration [9]. The latter still indicates high A/Fe ratios for A being medium heavy to calcium nuclei, while the former indicates a less steep power spectrum for iron than the known spectrum for protons (as measured by the JACEE experiments). Plans to advance the observation are recently encouraged by a promotion of an exposure facility JEM S-003 in the Japanese Space Station Program [10], which is promising in revealing elemental and isotopic details of high energy nuclei at around the bump.

Among many remaining theoretical problems in defining supernova-pulsar origin of high energy cosmic rays, we illustrate here an importance of x-ray photo-disintegration in the vicinity of a supernova remnant, which substantially modifies elemental and isotopic composition in the SN-II pulsar component. X-rays originating from both comptonization of line gamma-rays of $Co^{56}$ decays and pulsar's electro-magnetic radiation are sufficiently abundant to induce copious photo-disintegration of ultra-relativistic nuclei in flight. Due to steep power-law nature in the energy spectrum of a pulsar component over the life of a pulsar, different Q-values in $\gamma - \alpha$ nuclear reaction will significantly alter the abundance ratio in the highest energy regime of pulsar-oriented nuclei. Fig. 1 illustrates relative loss of low-threshold component such as $Zn^{66}$, which, otherwise, should be prominent in SN-II component [7,11]. It is interesting to note that several nuclei with high Q-values, such as neutron-rich $Ca^{48}$, can survive at the highest energy range of the bump. The x-ray luminocity of $3 \times 10^{36}$ erg/s of the crab pulsar is used in this calculation. Much higher luminosity of $10^{37} - 10^{38}$ erg/s in SN1987a would certainly reduce high energy components further. This "milking" of nuclei in the vicinity of a neutron star must be taken into account for consideration of elemental composition from SN-II with a pulsar remnant, which consequently can serve as a discriminator of SN-II (pulsar) origin of cosmic rays in the bump regime, and may provide some information on elemental abundances near the surface of a neutron star.

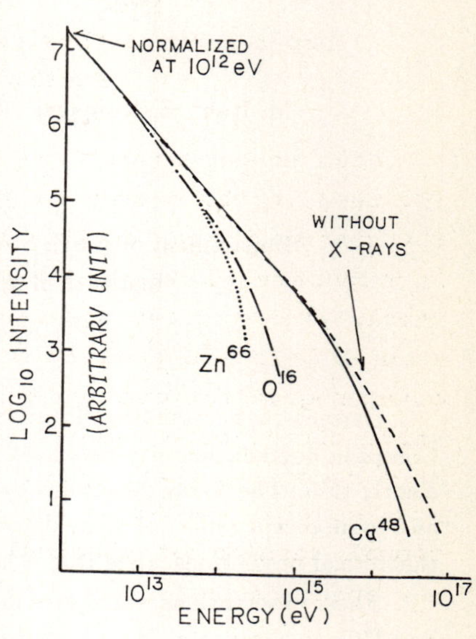


Fig. 1

References:

[1]. J. R. Jokipii and G. E. Morrfill, Proc. 19th Int. Cosmic Ray Conf., La Jolla, 3, 132 (1985), (NASA, Washington, DC, 1985).

[2]. R. E. Streitmatter et al., Proc. 18th Int. Cosmic Ray Conf., Bangalore, 2, 183 (1983).

[3]. Y. Tomozawa, Proc. 19th Int. Cosmic Ray Conf., La Jolla, 2, 354 (1985).

[4]. Stecker and Wolfendale, Proc. 19th Int. Cosmic Ray Conf., La Jolla, 2, 354 (1985).

[5]. A. M. Hillas, Can. J. Phys. 46, 5623 (1968).

[6]. S. Karakura et al., J. Phys. A7, 437, (1974); Gunn and Ostriker, Phys. Rev. Lett. 22, 728 (1969).

[7]. Y. Takahashi et al., Nature 321, 839 (1986).

[8]. D. Muller et al., Proc. of 20th Int. Cosmic Ray Conf., Moscow, (1987).

[9]. The JACEE Collaboration, T. H. Burnett, et. al, Proc. 20th Int. Cosmic Ray Conf., Moscow, (1987).

[10]. K. Higuchi (NASDA), private communication, 1987.

[11]. D. Hartman et al., Ap. J. 297, 837 (1985).

# HARD X-RAYS AND GAMMA-RAYS FROM SN 1987A

M. Itoh[2], S. Kumagai[1], T. Shigeyama[1], K. Nomoto[1], and J. Nishimura[2]

[1]Department of Earth Science and Astronomy, University of Tokyo, Tokyo
[2]Institute of Space and Astronautical Science, Tokyo

Gamma-rays originating from radioactive decays of $^{56}$Ni and $^{56}$Co and hard X-rays due to Compton degradation of $\gamma$-rays have been predicted to emerge when the supernova becomes sufficiently thin. The X-ray detections by Ginga (Dotani et al. 1988) and Kvant (Sunyaev et al. 1988) and more recent report of $\gamma$-ray detections by SMM (Matz et al. 1988) were much earlier than the theoretical predictions. (See Itoh et al. 1987 and references therein.)

These observations would give important constraints on the distribution of the heavy elements and $^{56}$Co in the ejecta. We adopted the hydrodynamical model 11E1Y6 (Nomoto et al. 1988) and carried out Monte Carlo simulation for photon transfer. A step-like distribution of $^{56}$Co was assumed where the mass fraction of $^{56}$Co in the layers at $M_r \leq 4.6\ M_\odot$, 4.6 - 6 $M_\odot$, 6 - 8 $M_\odot$, and 8 - 10 $M_\odot$ are $X_{\mathrm{Co}} = 0.0128, 0.0035, 0.0021$, and $0.0011$, respectively. Other heavy elements were distributed with mass fractions in proportion to $^{56}$Co.

Our calculation can consistently account for the early emergence and the subsequent light curves of $\gamma$- and hard X-rays and the spectral evolution (see Nomoto et al. 1988 for the light curves).

Figures 1a, b show the hard X-ray and $\gamma$-ray spectra for the above model at $t = 200$ and 250 d as compared with the Ginga and Kvant observations at $t = 200$ d and the balloon borne observations (Wilson et al. 1988) at $t = 250$ d. The calculations clearly show that the spectrum becomes harder as the ejecta expands and the number of Compton scattering decreases. The power law spectra, $E^{-\alpha}$, at 30 - 200 keV with the index $\alpha \sim 1.3$ (200 d) and 1.1 (250 d) are in reasonable agreement with observations.

The most important improvement compared with the earlier models (M. Itoh et al. 1987 and references therein) is that the flux at 16 - 30 keV is consistent with the observations around $t = 200$ d. This is because more X-rays below 30 keV are absorbed by the enhanced amount of heavy elements which are mixed into the envelope.

As a prediction for future observations, X-ray and $\gamma$-ray spectra at $t = 400$ d and 600 d are shown in Figures 1c, d. The spectrum will become harder as the column depth of the ejecta decreases. Comparison with the observations will inform us more detailed abundance distribution and the clumpiness of the ejecta.

**References**

Dotani, T. et al.: 1987, *Nature*, **330**, 230.

Itoh, M., Kumagai, S., Shigeyama, T., Nomoto, K., Nishimura, J.: 1987, *Nature*, **330**, 233.

Matz, S.M., Share, G.H., Leising, M.D., Chupp, E.L., Vestrand, W.T., Purcell, W.R., Strickman, M.S., Reppin, C.: 1988, *Nature*, **331**, 416.

Nomoto, K., Shigeyama, T., Hashimoto, M.: 1988, in *IAU Colloquium 108, Atmospheric Diagnostics of Stellar Evolution*, ed. K. Nomoto, Springer- Verlag, p. 319.

Sunyaev, R. et al.: 1987, *Nature*, **330**, 227.

Wilson, R.B., Fishman, G., Meegan, C., Paciesas, W., Sandie, W., Chase, L., Nakano, G.: 1988, private communication.


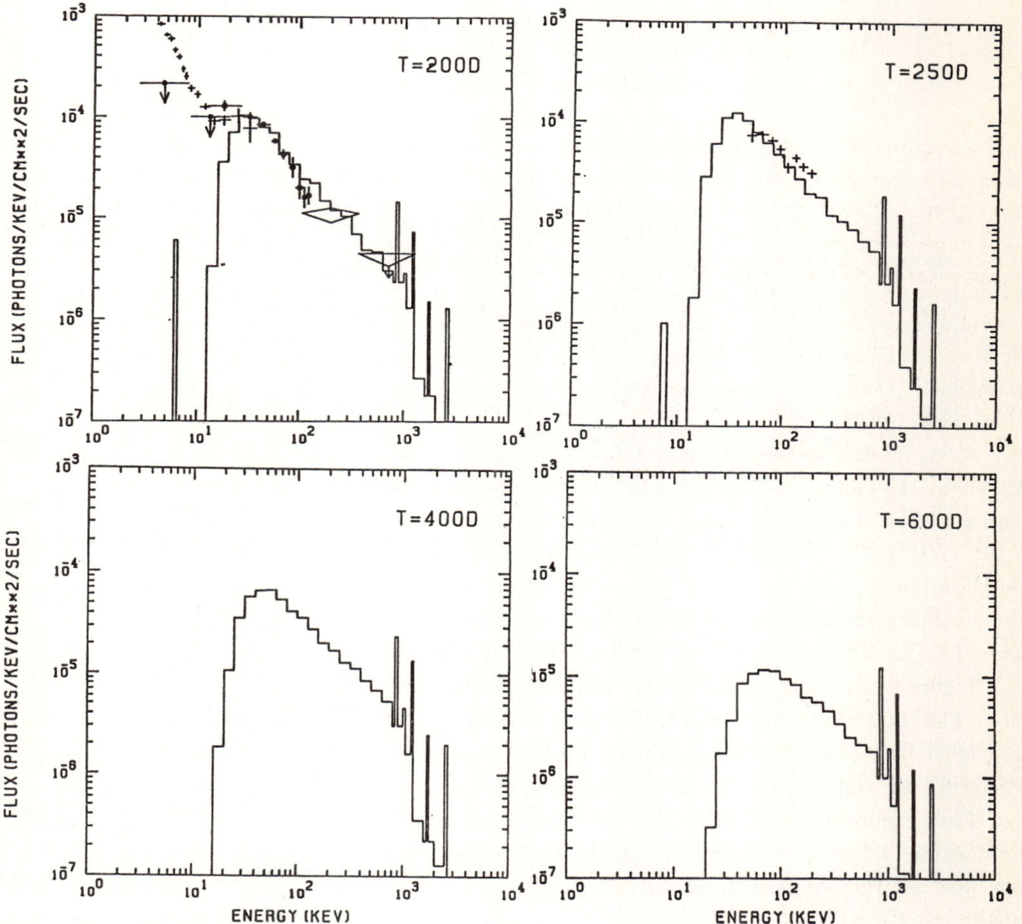



**Figure 1:** The calculated hard X-ray and $\gamma$-ray spectra for the the hydrodynamical model with $E = 1 \times 10^{51}$ erg and $M_{\mathrm{env}} = 6.7\ M_{\odot}$ at a) $t = 200$ d, b) $t = 250$ d, c) $t = 400$ d, and d) $t = 600$ d. At $t = 200$ d, the crosses indicate the spectrum observed by Ginga and the open circles and the diamonds are obtained by Kvant. At $t = 250$ d, balloon-borne observation (Wilson et al. 1988) is shown.

# THERMAL RADIATION FROM A NEUTRON STAR IN SN 1987A

Ken'ichi Nomoto[1] and Sachiko Tsuruta[2]

[1]Department of Earth Science and Astronomy, University of Tokyo
[2]Department of Physics, Montana State University

The supernova 1987A in the Large Magellanic Cloud has provided a new opportunity to study the evolution of a young neutron star right after its birth. A proto-neutron star first cools down by emitting neutrinos that diffuse out of the interior within a minutes. After the neutron star becomes transparent to neutrinos, the neutron star core with $> 10^{14}$ g cm$^{-3}$ cools predominantly by Urca neutrino emission. However, the surface layers remain hot because it takes at least 100 years before the cooling waves from the central core reach the surface layers (Nomoto and Tsuruta 1981, 1986, 1987).

From the hot surface, thermal X-rays are emitted. The detection limit for X- rays from SN 1987A by the Ginga satellite is $3 \times 10^{36}$ erg s$^{-1}$ (Makino 1987; Tanaka 1987). If the thermal X-rays are to be observed by Ginga, the surface temperature should continue to be as high as $T_s > 8 \times 10^6$ $(R/10\mathrm{km})^{-1/2}$ K until the ejecta becomes transparent. The exact value of the initial surface temperature depends on various factors during the violent stages of explosion, cooling stages of the proto-neutron star through diffusive neutrinos, and possible re-infalling of the ejected material. Therefore, until the surface layers become thermally relaxed $T_s$ may satisfy the above condition.

Figure 1 shows the cooling behavior of neutron stars during the first one year after the explosion. The total luminosity of the surface photon radiation $L_{\mathrm{ph}}$ (left) and the surface temperature $T_s$ (right), both to be observed at infinity, are plotted as a function of time, for three nuclear models PS (stiff), FP (intermediate), and BPS (soft). The temperature scale refers to the FP model.

From Figure 1 it is clear that the surface radiation falls significantly below the Ginga detection limit within a few to $\sim 20$ days for all the models considered. Such a decrease is caused by the plasmon neutrino emission from the outer layers of $\rho = 10^9$ - $10^{10}$ g cm$^{-3}$. The surface layers at the lower densities ($\rho < 10^{10}$ g cm$^{-3}$) are so thin that the time scale of thermal conduction is shorter than the time scale of the plasmon neutrino cooling. Consequently, the surface cooling quickly follows the plasmon neutrino cooling in the layers just beneath the surface. This mechanism of surface cooling by the plasmon neutrino process clearly wipes out the initial conditions. We note that during these early stages the surface temperature is independent of the complicated thermal behavior in the central core.

Consequently, it is unlikely that Ginga would detect the thermal X-ray emission directly from the surface of a neutron star in SN 1987A even if the ejecta should become transparent right now.

Looking beyond Ginga, it should be important to search by other future X-ray satellites for a thermal soft X-ray point source in SN 1987A. This is because due to the finite time scale of thermal conduction, it will take at least another 100 years before the efficient cooling of the central core by the Urca process will be registered at the surface. Until then the observed surface temperature will remain at the level of at least two to four million degrees (see Figure 9 in Nomoto and Tsuruta 1987). These temperatures correspond to the observed luminosity of $> 10^{35}$ erg s$^{-1}$. Note that the expected detection limit for ROSAT is $10^{34}$ erg s$^{-1}$ (Trumper 1981), while for AXAF it is less than $10^{32}$ erg s$^{-1}$ (NASA 1980).

We conclude that future satellite observations of SN 1987A with high sensitivity soft X-ray detectors should be critical in order to test the evolution theories of young neutron stars. This goal may very well be realized within the next 10 years, by the programs such as ROSAT, SXO, GRANAT, SPECTRA, and AXAF.

## References


NASA: 1980, TM-78285 (NASA, MSFC)

Nomoto, K., Tsuruta, S.: 1981, *Astrophys. J.*, **250**, L19.

Nomoto, K., Tsuruta, S.: 1986, *Astrophys. J. Letters*, **305**, L19.

Nomoto, K., Tsuruta, S.: 1987, *Astrophys. J.*, **312**, 711.

Makino, F., the Ginga team: 1986, Astrophys. Lett. Commun., **25**, 223.

Tanaka, Y.: private communication.

Trumper, J.: 1981, in Proc. Uhuru Memorial Symposium (NASA, GSFC).


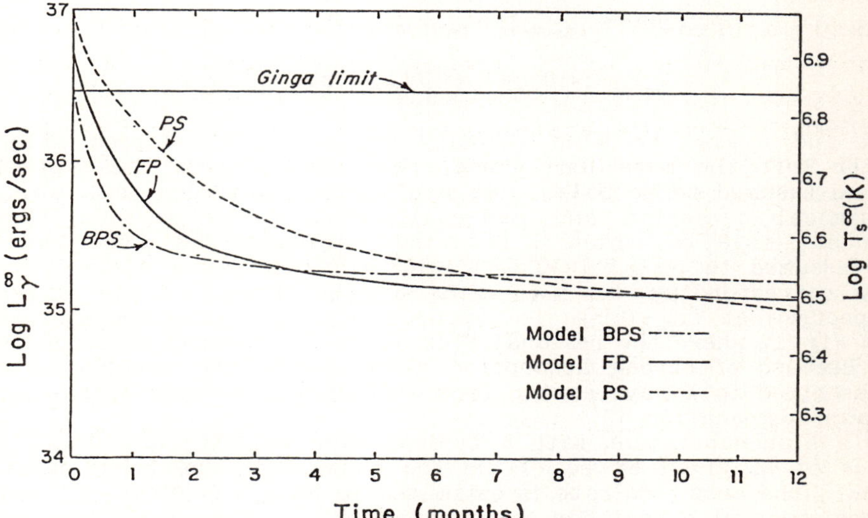


Figure 1: Observed photon luminosity vs. time during a year after the supernova explosion predicted from standard cooling theory of neutron stars. Shown are three representative nuclear models PS (dashed), FP (solid), and BPS (dot-dashed). The detection limit from Ginga is shown as a horizontal line.

# THERMAL X-RAYS DUE TO EJECTA/CSM INTERACTION IN SN 1987A


K. Masai[1], S. Hayakawa[2], H. Itoh[3], K. Nomoto[4] and T. Shigeyama[4]

[1]Institute of Plasma Physics, Nagoya University, Nagoya 464, Japan
[2]Department of Astrophysics, Nagoya University, Nagoya 464, Japan
[3]Department of Astronomy, University of Kyoto, Kyoto 606, Japan
[4]Department of Earth Science and Astronomy, University of Tokyo,
 Tokyo 153, Japan


The X-ray spectrum observed by Ginga[1] is characterized by a component below $10\,keV$ which decreases with increasing photon energy, and a component above $10\,keV$ which is nearly flat. This unusual X-ray spectrum may be understood as follows; X-rays below $10\,keV$ is likely to be due to thermal emission coming from the shock-heated ejecta, and X-rays above $10\,keV$ to be due to $\gamma$-ray degradation inside the ejecta. If thermal emission due to the collision of the ejecta with circumstellar matter (CSM)[2)-4)] is responsible for X-rays below $10\,keV$, the epoch of the collision can be estimated to be $\sim 0.2\,yr$ after the explosion if $\sim 0.5\,yr$ is the time when the X-ray flux at $\sim 10\,keV$ reaches its maximum. The X-ray light curve then requires the inner radius of CSM to be $\sim 1\times10^{16}\,cm$ for an expansion velocity, $V_{ex} \simeq 2\times10^9\,cm\,s^{-1}$.

X-rays emitted from the reverse-shocked ejecta dominate those from the blast-shocked CSM because of much higher density in the ejecta[3]. The electron temperature of the shocked ejecta is raised to $> 10\,keV$ and then decreases with time to $\sim 10\,keV$ at $\sim t_{max}$, at which the flux at $\sim 10\,keV$ reaches its maximum. Free-free emission dominates the thermal spectrum except for a contribution of iron K-line emission around $7\,keV$. If CSM concerned is the remnant of a stellar wind, the mass loss rate $\dot{M}_w$ can be estimated from the observed thermal flux at $10\,keV$ as[4],

$$\dot{M}_w \sim 3\times10^{-6}\,(v_w/10^6\,cm\,s^{-1})^{3/4}(I_{10keV}/10^{-4}\,photons\,cm^{-2}\,keV^{-1}\,s^{-1})^{1/2}$$

$$\times(V_{ex}/2\times10^9\,cm\,s^{-1})^{3/4}(t_{max}/0.5\,yr)^{3/4}\,M_\odot\,yr^{-1}, \tag{1}$$

where $v_w$ and $\tau$ are the wind velocity and the time elapsed after the progenitor left the mass loss phase, respectively, and the distance to SN 1987A is assumed to be $55\,kpc$. We performed a numerical calculation of the dynamical evolution and non-equilibrium X-ray emission for the collision at $1.1\times10^{16}\,cm$, which is the inner radius of CSM. The density of CSM was assumed to be $\simeq 5.1\times10^4\,(r/1.1\times10^{16}\,cm)^{-2}\,H\,cm^{-3}$ at a distance $r$ from SN 1987A, corresponding to $\dot{M}_w/v_w \simeq 2.8\times10^{-6}\,M_\odot\,yr^{-1}/10\,km\,s^{-1}$. The calculated X-ray spectrum at $t_{max} \sim 0.5\,yr$ is compared with the spectrum observed by Ginga in Fig. 1, where the residual flux above the thermal spectrum is also shown. Because of strong absorption below $20\,keV$, the residual spectrum may be ascribed to X-rays coming from the optically thick ejecta through the Compton degradation of $\gamma$-rays[5].

For a blue supergiant with a typical wind velocity $v_w \simeq 100\text{-}1000\,km\,s^{-1}$, the value of $\dot{M}_w$ given by eq. (1) is too large. On the other hand, for $v_w \simeq 30\,km\,s^{-1}$, the mass loss rate is estimated to be $\dot{M}_w \sim 7\times10^{-6}\,M_\odot\,yr^{-1}$, which is not inconsistent with that for a red supergiant. This result implies that the CSM may be attributed to the stellar wind in the latest phase of a red supergiant or in the transition phase to a blue supergiant. Then, the time which has passed since the stellar wind ceased is estimated as, $\tau \sim 1\times10^2\,(v_w/3\times10^6\,cm\,s^{-1})^{-1}\,yr$. In any case, the lifetime in the blue supergiant phase would be of the order of $10^2\,yr$ if X-rays below $10\,keV$ arise from the interaction between the ejecta and the wind remnant.

The material related with the secondary light source $\sim 4\times10^{16}\,cm$ away from the primary (IAU Circ. No. 4382) can be one of the candidates responsible for thermal X-ray emission. If the outermost envelope of the ejecta expands with the velocity $(3\text{-}4)\times10^{9}\,cm\,s^{-1}$, it has reached this material $0.3\text{-}0.4\,yr$ after the explosion. Also molecular clouds in the vicinity of SN 1987A can be candidates to interact with the ejecta. Fig. 2 shows 6-10 $keV$ light curve expected from the wind remnant model adopted here. The light curve of thermal X-rays depends on the density structure of the colliding material. Therefore, continual watching can provide information on CSM. Recent near-infrared speckle data (IAU Circ. No. 4481) may suggest the existence of shell-like dusty CSM at $(1\text{-}2)\times10^{17}\,cm$ from SN 1987A. This CSM may be a remnant of the stellar wind in the red supergiant phase. The expanding ejecta will collide with this material and emit thermal X-rays 3-6 years hence. These X-rays will be an important diagnostic tool for the study of environments of Sk $-69\,202$ and of chemical composition and the stellar evolution of similar blue supergiants in LMC. It is desired to make a future plan of X-ray observatory succeeding to active instruments.

References

1) Dotani, T., and the Ginga team, Nature 330, 230-231 (1987).
2) Chevalier, R.A., and Fransson, C., Nature 328, 44-45 (1987)
3) Itoh, H., Hayakawa, S., Masai, K., and Nomoto, K., Publ. Astron. Soc. Japan 39, 529-537 (1987).
4) Masai, K., Hayakawa, S., Itoh, H., and Nomoto, K., Nature 330, 235-236 (1987).
5) Itoh, M., Kumagai, S., Shigeyama, T., Nomoto, K., and Nishimura, J., Nature 330, 233-235 (1987), and references therein.


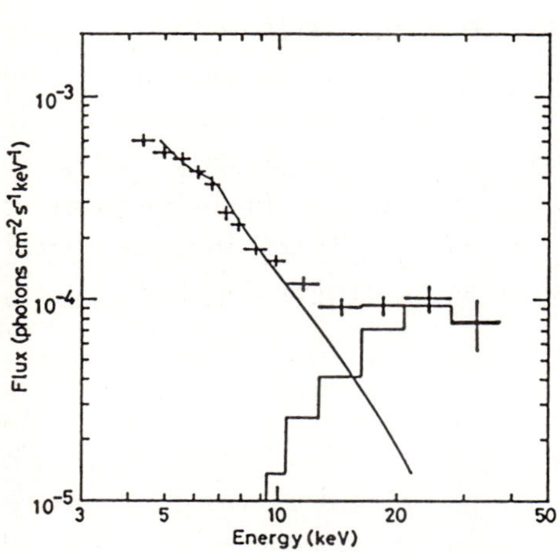

Fig. 1 Thermal spectrum (solid line) calculated with an energy resolution of $0.49(E/keV)^{1/2}\,keV$ (FWHM) is overlaid on the observed spectrum with Ginga[1] (crosses). The residual flux over the thermal emission is represented by the histogram.

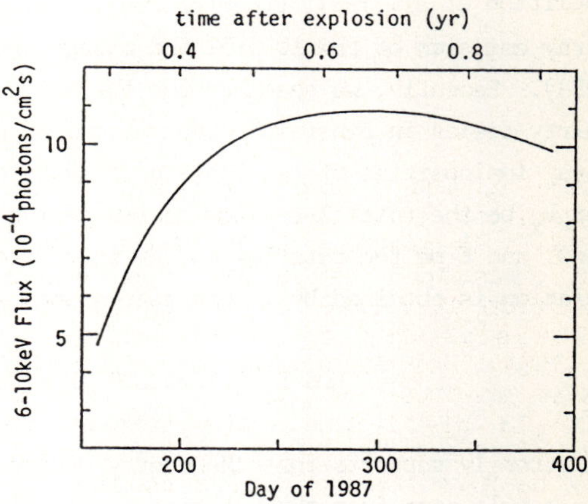

Fig. 2 6-10 $keV$ light curve of thermal X-rays expected from a wind remnant model (see text). $Z = 1/3\,Z_{\odot}$, the hydrogen column density of $4\times10^{21}\,cm^{-2}$ and the distance of $55\,kpc$ to SN 1987A are assumed in Figs. 1 and 2.

AN X-RAY INVESTIGATION OF CRAB-LIKE SUPERNOVA REMNANTS

Z. R. WANG

Center of Astron. & Astrophys., CCAST(World Lab.),Nanjing
Department of Astronomy, Nanjing University,China
and
F. D. Seward
Harvard-Smithsonian Center for Astrophysics

The Crab Nebula is the most noticeable object in our Galaxy, and the remnant of
the famous Chinese guest star appeared in 1054AD, the best association between supernova
remnants(SNR) and ancient guest stars. Before seventies, the Crab Nebula was considered
as a special SNR with different morphology and physical features from that of most SNRs,
Now more and more Crab-like SNRs have been detected(Weiler 1985). It is necessary to
make a systematical investigation for the Crab-like SNRs, especially for those with central pulsars because they offer us more physical messages than others.

The two associations SNR and PSR, the Crab Nebula with PSR 0531+21 and the Vela
SNR with PSR 0833-45, were confirmed at the end of sixties and still were the only two
associations before eighties. Recent years, PSR 1509-58 in MSH 15-52 and PSR 0540-69 in
a SNR of LMC(Seward & Harnden 1984, Seward etal. 1984) as well as PSR 1951+32 in CTB 80
( Clifton et al. 1987, Fruchter etal. 1987 ) were discovered. On the other hand  , the
X-ray emission of PSR 1055-52 and others were detected (Cheng & Helfand 1983, Helfand
1983).  Recently, we searched for the X-ray emission of isolated pulsars over more than
thirty fields in Einstein  observation. By use of the same instrument calibrstions, the
X-ray luminosities of ten isolated pulsars and their surrounding nebulae are obtained.
Let $L_x$ be the total X-ray luminosity of a pulsar and its nebula in the 0.2-4.0 kev
band, and $\dot{E}$ be the rotating energy loss rate of  the pulsar, the following statistical
relation is obtained by linear regression  and shown as the line in Fig. 1.

$$\text{LOG } L_x = 1.30 \text{ LOG } \dot{E} - 13.2 \tag{1}$$

It strongly supports that the energy source of X-rays both from an isolated pulsar and
its nebula come from the rotating energy loss of the pulsar, and the energy transformation may be simply and directly. Eq.(1) offers a good empirical relation to further investigate their X-ray radiation mechanism.

Besides the above Crab-like SNRs with PSRs, ther are some Crab-like SNRs with
central compact sources but no pulsating radiation being observed, such as 3C 58 ,G 21.5
-0.9, Kes 73 and Kes 75 et al.. If we assume that such compact sources are also pulsars
in nature but only hidden and Eq(1) is  approximately satisfied for them, their periods

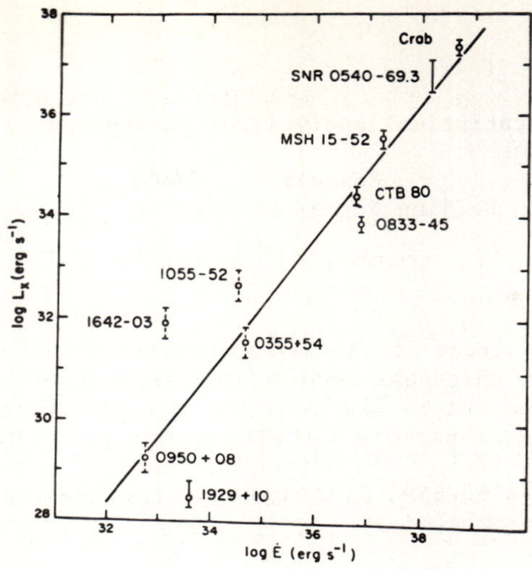


Figure 1.

can be deduced from their X-ray luminosities and their ages which are estimated from their identification with guest stars (Wang et al.1986, Wang 1987). Some of the pulsed signals will be expected by the future observations with higher resolution and sensitivity.

References


Cheng, A.F. and Helfand, D.J., 1983, Ap.J., 271, 271
Clifton, T.R. etal., 1987, IAU Circ. No. 4422
Fruchter, A.S. et al., 1987,, IAU Circ. No.4426
Helfand, D.J., 1983, in Supernova Remnants and Their X-Ray Emission, Danziger, J. and Gorenstein, P. Eds. (Dordrecht Reidel) , 471
Seward, F.D. and Harnden, Jr. F.R., 1984, Ap.J., 281, 650
Seward, F.D. Harnden, Jr. F.R. and Helfand, D.J., 1984, Ap.J., 287, L19
Strom, R., 1987, Ap. J. in press
Wang, Z.R., Liu, J.Y., Gorenstein, P. and Zombeck, M.V., 1986, in Highlights of Astronomy, 7, Swings, J.P. Ed., 583
Wang, Z.R., 1987, in The Origin and Evolution of Neutron Stars, Helfand, D.J. and Huang, J. Eds., (Dordrecht Reidel), 305
Weiler, K.W., 1985, in The Crab Nebula and Related Supernova Remnants, Kafatos, M.and Henry, R. Ed. (Cambridge Univ. Press, London)

# The statistical analysis of Supernovae

Zongwei Li
Beijing Normal University

## 1. Introduction

Due to the growing interest in the SN(supernovae) phenomenon in recent years, and to large amount of data which has been accumulating lately,there is now a need for a new publication in which to find all the essentail data on SNe and their parentgalaxies.A revised 568 supernova catalogue has published by Barbon et al (1984).

The paper supplements 65 SNe, discovered since 1984 up to update,and some statistical studies were completed.

## 2.The catalogue of SNe

The first two comprehensive lists of SNe were published by Zwicky(1958,1965) and included the 54 and 111 objects respectively reported up to 1956 and 1962. A more complete list was issued by Zwicky(1964),154 SNe is inclused.

This list was kept updated by the Palomer SN search and published (Kowal et al 1971,Sargent et al 1974),the total number of reported SNe reached 378.The Palomar SNe Master List has been kept updated by Kowal.

A <<Preliminary catalogue of SNe>> was published by Karpowicz and Rudnicki(1968), and an improved and updated version,including later on (Flin et al 1979).Besides a large body of data on the 454 SNe known up to that time and a list of 69 suspected and false SNe, a complete bibliography for each object was given.

A revised SNe catalogue was published by Barbon et al (1984).We summarize a new catalogue.

## 3. Statistical Analysis

To quickly show some of information contained in the catalogue ,and to present the kind of statistics which can be made with the data,some figures and tables have been prepared and discussed in the following.

The distrbution of SNe according to type (table 2).Only 34% (219/633) of thediscovered SNe have been classified. SNI (type I SN) outnumber the remaining objects in ratio 3:2.SNI/SNII in ratio 2:1.The observational result is biased by lower intrinsic luminosity and higher absorption present in SNII,besides the fact that these latter objects avoid elliptical and early spiral galaxies.All these effects must be taken into account in order to derive meaningful frequencies.

**Table1 Distribution of SNe according to type**

| type I | 120 | type II | 61 | type III | 2 |
|---|---|---|---|---|---|
| I: | 8 | II: | 9 | IV | 1 |
| Ipec | 8 | IIpec | 2 | V | 3 |
| | | | | pec | 5 |
| all SN I | 136 | all SN II | 72 | other | 11 |

The distribution of SNe with types,according to the class of their parent galaxies (table2).

**table 2 distribution of SNe with type of their parent galaxies**

| | E | So | Sa | Sab | Sb | Sbc | Sc | Scd | Sd | s | Io | Im | I | not | total |
|---|---|---|---|---|---|---|---|---|---|---|---|---|---|---|---|
| SNI | 16 | 12 | 4 | 2 | 23 | 14 | 36 | 2 | 2 | 7 | 6 | 2 | - | 10 | 136 |
| SnII | - | - | 1 | 1 | 11 | 6 | 42 | 2 | - | 2 | 2 | 1 | - | 4 | 72 |
| other | - | - | 1 | 1 | 1 | - | 6 | - | - | - | - | - | 1 | 2 | 11 |

RC2 contain 4,364 galaxies, 423 galaxies from RC2 with Vo<1200 km/s,(Ho adopted 55 km/s Mpc ) with 89 SNe  of which 33 are classified as SNI,and 37 as SNII.From the sample of galaxies and the sample of SNe, the relative frequency of the type of SNe vs. the type of galaxies is prersented.

**table 3 SN type vs. galaxies type**

| type | $N_{galaxies}$ | % | SNI | SNII | other | $N_{SN}$ | % | $N_{SN}/N_{gal}$ |
|---|---|---|---|---|---|---|---|---|
| E | 40 | 10.0 | 5 | 0 | 1 | 6 | 6.7 | 0.15 |
| So | 67 | 17.0 | 2 | 0 | 2 | 4 | 4.5 | 0.06 |
| So/a,Sa | 18 | 4.5 | 1 | 1 | 2 | 4 | 3.4 | 0.22 |
| Sab,Sb | 50 | 12.7 | 9 | 9 | 3 | 21 | 24.7 | 0.42 |
| Sbc,Sc,Scd | 112 | 28.5 | 11 | 27 | 11 | 49 | 55.1 | 0.44 |
| Im | 98 | 25.0 | 2 | 0 | 0 | 0 | 2.2 | 0.02 |
| Io | 8 | 2.0 | 3 | 0 | 0 | 3 | 3.4 | 0.38 |
| all | 393 | 100 | 33 | 37 | 19 | 89 | 100 | 0.23 |

The main conclusions from table 3 are foloowing:
1) the overall SN rate increases from E galaxy to spiral galaxies.
2) as is well known (Tammann 1982) SNI occur in all types of galaxies. They do belong to the old stellar population.
3) SNII occur only in Sab and later galaxies,if 1984e is abolished.

We also resarch the galaxies of mulifrequency supernovae,it be noted that the most of these galaxies are the spiral galaxies.The location of explosion is near spiral.

The active galaxies of explosion supernovae are explored,it is about 14 active galaxies that there are more 20 supernovae which were detected.

It must be noted that the discovery of the 1983 SNe in NGC 4753 is of interest for the solution of the problem of the nature of SNI.NGC 4753 became the second type Io galaxies,after NGC 5253 in which two SNI have been detected.This sopports the conclusion that there is an increased frequency of SNI outbursts in type Io galaxies. The true Type Ib and Type Ipec tractions are likely to be significantly higher than the observed fractions because both Ib and Ipec are fainter and therefore less likely to be discovered than Ia. The possibility that type Ib and Type Ipec each may be as numerous as Type Ia,at least in spiral galaxies,can not excluded.

Refrences
Borbon,R.,Capaccioli,M.,West,R.M.,Astron. Astrophys.Suppl. 49,73(1984)
Flin,et al, Acta Cosmologica. part. 8 (1979)
Zwicky,F., A list of SNe discovered since 1885, (1964)

# THE X-RAY SPECTRUM OF THE CYGNUS LOOP WITH GSPC

Hiroshi Tsunemi, Makoto Manabe and Koujun Yamashita
Department of Physics, Faculty of Science, Osaka University, 1-1,
Machikaneyama-cho, Toyonaka, Osaka 560 Japan

We observed the Cygnus Loop with Gas Scintillation Proportional Counter (GSPC) on board Tenma satellite. GSPC has an energy resolution two times better than that of a proportional counter (PC). Fig. 1 shows the spectrum with the crosses being the pulse height data with $\pm 1\sigma$ statistics. Superposed upon the data point is the best fit model spectra folded through the detector response. We found that two emission line features at 1.9 keV and 2.5keV, respectively corresponding to Si-$K\alpha$ and S-$K\alpha$ line blends, are needed to obtain an acceptable fit. The parameters for the emission lines are summarized in table 1. The abundances of these elements are consistent with those of cosmic values. The continuum spectrum in the energy range $1\sim3$ keV can be represented with thermal bremsstrahlung spectrum with an electron temperature $T_e$ of $7\times10^6$K.

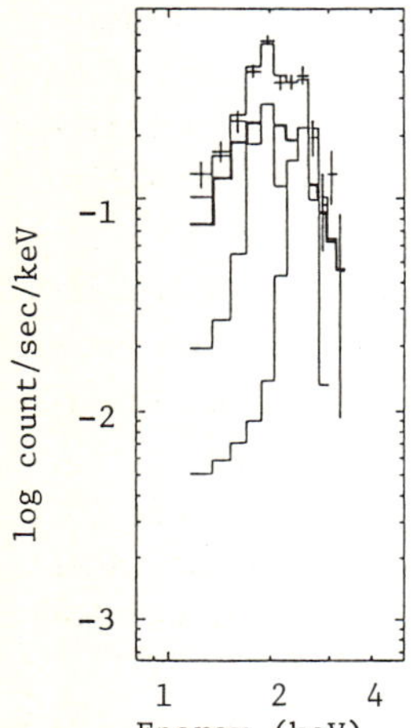

Fig. 1. X-ray spectrum observed with Tenma are shown with crosses.

Table 1.  Emission line features in the Cygnus Loop

| Element | Intensity (photonssec$^{-1}$cm$^{-2}$) | Line energy (keV) |
|---|---|---|
| Si $K\alpha$ | $8.4\pm2.3\times10^{-3}$ | $1.92\pm0.04$ |
| S $K\alpha$ | $2.7\pm1.0\times10^{-3}$ | $2.45\pm0.06$ |

Errors are 90% confidence level.

We performed a sounding rocket experiment in 1977 with GSPC (Inoue et al. 1979) and obtained the Loop spectrum in the energy range of $0.1\sim1.5$keV. Combined the results with the sounding rocket flight shown in fig. 2 gave us a wide band of X-ray spectrum for the whole Cygnus Loop with the best energy resolution reported so far.

We fitted the combined data with model spectra based on the atomic data compiled by Raymond and Smith (1977). The model spectra employed here are both for collisional ionization equilibrium (CIE) and non-equilibrium ionization (NEI) models with cosmic abundances (Allen, 1977). Single $T_e$ spectrum for both models can not fit the data. Two components of different $T_e$ models can reproduce the data well for both models. The physical parameters obtained with CIE models are self inconsistent because the ionization parameter $\tau$ ( the electron density n $\times$ the elapsed time t the after shock heating ) is about $10^{11}$cm$^{-3}$sec which is too short by an order of magnitude for the CIE condition to be reached.

Superposed upon the data point in fig. 2 is the best fit NEI model spectra.

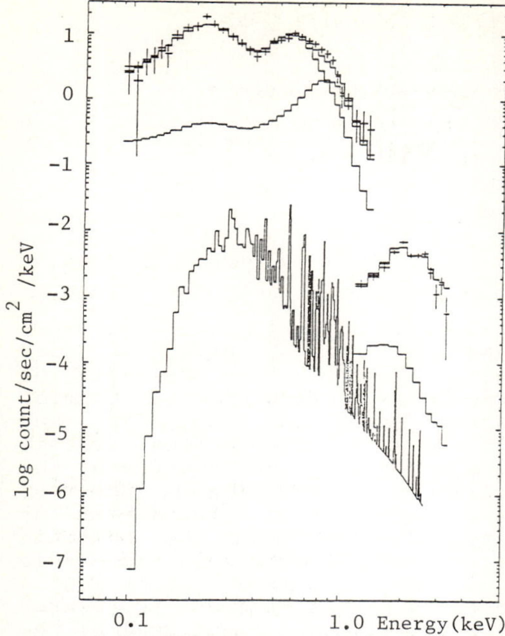


Fig. 2. Wide band X-ray spectrum for the whole Cygnus Loop with GSPC. Superposed the best fit NEI model spectra with two T$_e$ components.

The model spectra contain two component of thin thermal spectra with different T$_e$ and $\tau$. The 90% confidence level contour in log$\tau$-T$_e$ plane are shown in fig. 3.

Previous observations so far with employing PC reported that the Cygnus Loop could be represented with a single T$_e$ component of $2\sim4\times10^6$K (Gorenstein et al. 1971). The high spatial obser-vation of the Loop with the Einstein Observatory (Charles et al.1985) found T$_e$ in the limb to be lower than that of the interior. Vedder et al. (1986) ob-served a limited portion of the Loop with FPCS on the Einstein Observatory and found that the CIE condition has not been reached. Their results are shown in fig. 3 in dashed line.

From this context, we conclude that the low T$_e$ component is from the shell region while the high T$_e$ component from inside the Loop, since both of them are thermal in origin. If the width of the shell region is assumed to be R/12 from the strong shock theory, where R is the radius of the Loop, we found that n for low and high T$_e$ plasma are $0.25\sim0.6$ cm$^{-3}$ and $0.06\sim0.07$ cm$^{-3}$, respectively. The obtained range of $\tau$ restricts t as t $\geqq$ $2.5\times10^4$years.

## References


Allen, C. W. 1973, Astrophysical Quantities.

Charles, P. A., Kahn, S. M., and McKee, C.F. 1985, Ap. J., <u>295</u> 456.

Gorenstein,P., et al. 1971,Science, <u>172</u> 369.

Inoue, H., et al. 1979, X-ray Astronomy, (Oxford, Pergamon Press.) 309.

Raymond, J. C., and Smith, B. W., 1977 Ap. J., suppl., <u>35</u> 419.

Vedder, P. W., et al. 1986, Ap. J., <u>307</u> 269.


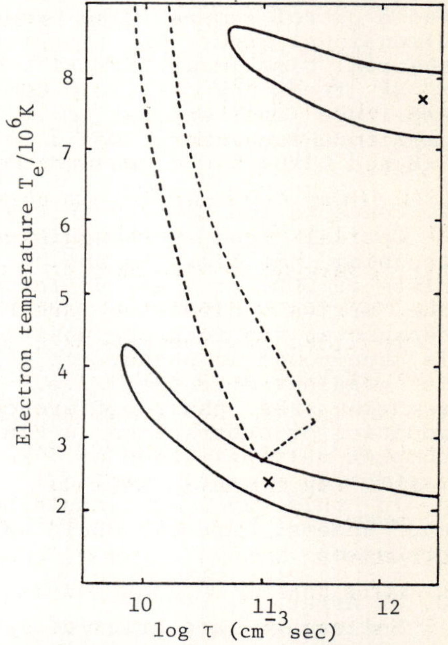


Fig. 3. Solid lines show 90% confi-dence level contour for the NEI model parameters. See text.

Radio Patrol Camera for Supernovae Search

T. Daishido, K. Asuma, S. Inoue, K. Nishibori,
H. Ohara, S. Komatsu
Waseda University, Shinjuku-ku, Tokyo
K. Nagane
Suginami-ku, Tokyo

## 1. Introduction

Zwicky started extragalactic supernovae patrol using 10 inch Schmidt camera about fifty years ago. After that the research of supernovae was accelerated, because the wide view of the Schmidt camera made it possible to watch large field of the sky. The key technology of the Schmidt camera was its sophisticated optical system.

Anticipated next supernova in our Galaxy may be undetectable by the optical instrument due to the Galactic extinction. However, supernovae are now known to be intense radio sources after a year or so of the explosion. Even if the positions are beyond the Galactic center, the radio supernova could be observed using middle size radio telescope.

We are planning to construct the radio patrol camera to search transient radio sources like supernovae or Cyg X-3. Final goal of the radio patrol camera is 2 dimensional 64x64 = 4096 elements filled aperture, and it will be able to map the whole sky once a week with 10 arcmin resolution and the sensitivity of 30 mJy.

Using a pilot system at Waseda University, we have been able to develop the technology and the concept of digital optics for the radio patrol camera. The present pilot system is an eight elements 1 dimensional array. So, the picture points are eight. The wide view and the real time image formation have been established in this system.

RF frequency is 10.6 GHz and bandwidths are 20 MHz. Phase and Amplitude are automatically controlled by the digital complex amplitude equalizers. Design of two dimensional systems in progress of 8x8 and 64x64 are discussed.

## 2. Concepts and Devices in Digital Optics

Optical lens distinguishes the arrival directions of light by focusing the light to the corresponding positions on the photographic plate or CCD. Phase differences against position on the lens due to the arrival direction are removed through the focusing process, because at the focusing position only the power of a certain direction is superposed in phase. In other words, the arrival direction of wave is gradient of phase; i.e. k = grad(phase). The above process of removing the phase difference is Fourier transformation. The phase addition or subtraction in Fourier transformation could be replaced by complex multiplication or 4 real multiplication and additions, since exp(ip+iq) = exp(ip)exp(iq).

In this way, lens could be constructed by the digital multipliers and adders. And one could say "a digital lens" for this imaging FFT processor.

## 3. 2D Array

Followings are brief design of 8 x 8 = 64 elements patrol camera which is now planning before the full system of 4096 elements. Collective aperture is 15m x 15m, which is same as the full system. So, the same sensitivity of 30 mJy and the same resolution is

expected. Picture points are 8 x 8 = 64, and mapping speed is 64 times
fast   as that of a single    dish.  Room temperature receivers will be
used, and their system temperatures will be about 100K.


Table 1.     A Design of 8 x 8 = 64   2D Radio Patrol Camera
------------------------------------------------------------------------
Analog part

Number of elements         8 x 8 = 64
RF          frequency      10.6          GHz
1st local frequency        9.6           GHz
1st IF      frequency      1.0 - 1.1   GHz
2nd local frequency        1.05          GHz
Baseband  frequency        -50 -- +50   MHz

Digital part

             A/D     converters    Complex   amplitude      Digital     lens
                                      equalizer           (FFT processor)
Number       64 x 2                    64                2D complex(8x8)
Clock      20 -- 100   MHz          20      MHz           20     MHz
Dynamic       8        bit           8      bit            8     bit
range

Image integrators                          64
2nd integrator and switching       Digital Signal Processor
------------------------------------------------------------------------


                          References


Daishido, T., Asuma, K., Ohara, H., Komatsu, S., and Nagane, K., 1986,
IEEE/ICASP86, Tokyo, 53.4.1, 2855.

Daishido, T., Asuma, K., Nishibori, K., and Inoue, S.,
1987,  Prc.IAU Symp.  No.129,  "The Impact of VLBI on Astrophysics and
Geophysics",(Reidel,in press)

# IV. Concluding Remarks

CONCLUDING REMARKS OF THE I.A.U. COLLOQUIUM NO.108

Keiichi Kodaira
Tokyo Astronomical Observatory
Mitaka, Tokyo 181

The discussions during the colloquium clearly pointed to the recent rapid progress in applying the atmospheric models to the detailed diagnostics of stellar evolution.

The high-quality spectroscopic data now available enable us to evaluate the abundances of scarce elements and isotopes in the classical stellar atmospheres. Many of them are found to be useful to sound the mixing and the gravity settling phenomena in the stellar interior. Extensive contributions were presented about the former process in the AGB stars, and about the latter process in the Li-defficient F stars.

The evolutionary interactions between the stellar interior and the atmosphere become distinct when the mass-loss processes set in. It was pointed out that high-rate mass loss can influence the internal evolution and mixing on one side, and that on the other side the mixing (dredging-up) can change the abundances of heavy elements in the atmosphere and further modulate the mass-loss rate. Many speakers presented ample of observational data from UV to mm-wave range indicating occurences of substantial mass loss from various kinds of stars. Self consistent models with spherical geometry were proposed for stellar winds which are driven by the radiation pressure acting on ions or dust grains. This involves comprehensive calculation related to the molecular chemistry and the grain physics. So far as the flow is stationary and spherically symmetric, the escape velocity and the mass-loss rate determine the density structure of the wind in radiative equilibrium.

The colloquium was enlivened by the timely appearance of SN1987A whose topics dominated in the session of exploding stars. In order to obtain the direct information of the chemical structure of the progenitor, several contributors modelled the expanding envelope using a simple power law or a similarity solution for density structure. The spherical symmetry and the radiative equilibrium were mostly assumed in fitting the models to the observed spectra and energy distributions. Except for few cases the investigators found satisfactory solutions of parameter sets and chemical abundances for the expanding atmosphere of SN1987A at various epochs. They were most elaborated model calculations ever done for dynamical atmospheres to provide the diagnostics of SN explosion.

These models of expanding atmospheres, however, involve more parameters than the classical static atmospheres. The influences of each physical parameters on the observational quantities must be carefully investigated. The assumption of spherical symmetry and the radiative equilibrium may not be valid over all phases of the development of SN1987A whose speckle image was reported to be elongated.

This colloquium was one of the rare chances where the researchers specialized for stellar atmospheres and those for internal structure could be engaged in direct discussions over a wide range of concrete subjects. It is certainly not only the present reviewer but also all of the participants who found this conference for fruitful and pleasant, with the most effective poster- and the "beer" sessions.

In conclusion, on behalf of all the attendants I would like to congratulate SOC chairmen, Drs. Nomoto and Kudritzki, and LOC chairman, Dr. Tsuji, on their success in organizing this colloquium.

CONCLUDING REMARKS - II

D. Sugimoto

College of Arts and Sciences, University of Tokyo


In the Opening Remarks Dr. Kudritzki asked what tricks were used in
calculating stellar structure.  Though many fine models and discussions
have been presented, answer has not been given yet to this question.
The theory of          stellar structure consists of two very much different kinds of building blocks.  One is the local physics such as equation
of state, opacity, nuclear reaction rate, and so on.  They are determined
only locally when the values of temperature and density at a point are
given, i.e., they are irrelevant to the conditions whether they are considered in the stellar interior or not.  Another is the physics of self-
gravitating systems in which the global spatial structures are discussed.
Characteristics of astrophysics are often said to lie in the fact that it
covers much wider parameter ranges than in laboratories.  This, however,
grasps only one side of the problem, i.e., only local physics.  Those
which are not encountered in laboratory physics lie mainly in the global
physics.  It shows characteristics out of common sense; examples are
negative specific heat and associated gravothermal catastrophe which are
related with the gravitational contraction of the stars, formation of core-
halo structures which is observed not only in red giant stars but also
many celestial objects, and tendencies dividing a system into two sub-
systems one with high energy and/or high entropy and the other with low
energy in the deep potential well and/or low entropy as seen in supernova
explosion of type II and formation of jet in various objects.  They behave
contrary to the general trend, i.e., against equipartition and thermal
equilibration.

For these phenomena numerical models are constructed.  Usually,
however, one discusses in detail the local physics as the fundamental
assumptions, and then jumps into presenting numerical model.  The logical
connection between them are not clear at all.  To make it worse people are
apt to be fond of more detailed models.  Though it is easy for computers
to take less essential effects into account, they often obscure important
physical processes.  In the traditional physics, on the contrary, ideali-
zation was intentionally done to extract essentials.   In astronomy, of
course, standard models in which everything is taken into computation are

necessary to compare with observations.  However, this does not deny the
necessity of idealized models.  Anyway, more efforts seem necessary in the
side of global physics characteristic to self-gravitating systems.

Things which make it more difficult are some natures of non-linear
systems.  Let me give an example.  In relation with the supernova 1987A
the size of the progenitor star and its evolutionary implications were
discussed in this colloquium.  Some said that the star spended presupernova stages as a blue supergiant for the low metalicity as in the
Large Magellanic Cloud.  Others paid attention to the existence of red
supergiants in LMC.  Theoretically, the blue star can not swell to a red
giant in a short time as in the presupernova stages because of too long
time scale of heat transport in its envelope.  Dr. Woosley gave a talk
concerning the possible envelope mass.  It lies in the range which Dr.
Wheeler excluded in his talk.  This contradiction might arise because the
former took account of thermal disequilibrium of the envelope during rapid
evolution, while the latter assumed thermal equilibrium envelope.  One
might think the apparent contradiction could thus be understood.

However, the situation is not so simple when the heat transport is
coupled with stellar structure.  Here, I would give another example.  It
is a problem of X-ray bursting neutron star.  The X-ray burst proceeds in
tens of second and it might be much shorter as compared with the time
scale of heat transport in the envelope of the neutron star.  During the
burst the X-ray luminosity of the neutron star becomes very close to the
Eddington luminosity and the outer layers of the envelope are pushed up by
the radiation coming from the interior.  Then the neutron star is puffed
up and the time scale of heat transport becomes shorter and shorter.
Finally, the envelope solution with steady mass flow in thermal equilibrium becomes a good approximation and such situation is also observationally confirmed.  Before this has become understood, a specialist tried to
calculate such expansion of the envelope all the way as an initial value
problem by means of stellar evolution code, but it was found impracticable.

Here, we see limitations of straight-forward approach of modeling as
well as peculiar behaviors of non-linear systems which sometimes show very
strong or even practically unstable response even to the slightest change
in the initial conditions.  Such a hyperbolic instabilities are recently
discussed for non-linear systems in relation with their chaotic behaviors.
Since the self-gravitating systems are one of the typical non-linear
systems in the sense that the whole system interacts coherently even
between the most distant points because of long-range nature of the gravitational interaction.  Thus, another approach seems also necessary which
aims at some general understanding on the global behaviors in celestial
objects.

IAU Colloquium No.108

Participants List

| Name | | Institute | Country |
|---|---|---|---|
| T. | Aikawa | Tohoku-Gakuin University | Japan |
| K. | Akiyama | University of Tokyo | Japan |
| H. | Ando | University of Tokyo | Japan |
| H. | Andrillat | Univ. des Sci. et Tech. du Languedoc | France |
| Y. | Andrillat | Univ. des Sci. et Tech. du Languedoc | France |
| J. | Arafune | University of Tokyo | Japan |
| K. | Arai | Kumamoto University | Japan |
| T.S. | Axelrod | Lawrence Livermore Natinal Lab. | USA |
| G.B. | Baratta | Osservatorio Astronomico di Roma | Italy |
| D. | Branch | University of Oklahoma | USA |
| R.M. | Catchpole | South African Astron. Obs. | South Africa |
| K.L. | Chan | Applied Research Corporation | USA |
| H.Y. | Chiu | Goddard Space Flight Center | USA |
| N. | Chugaj | Astr. Council Acad. Sci. USSR | USSR |
| T. | Daishido | Waseda University | Japan |
| M. | de Groot | Armagh Observatory | U.K. |
| C. | de Jager | Lab. for Space Research | Netherlands |
| C.B. | de Loore | Vrije Universiteit Brussel | Belgium |
| D. | Dearborn | Lawrence Livermore National Lab. | USA |
| S. | Deguchi | University of Tokyo | Japan |
| C. | Doom | Vrije Universiteit Brussel | Belgium |
| M.A. | Dopita | Mt. Stromlo and Siding Spring Obs. | Australia |
| Y. | Eriguchi | University of Tokyo | Japan |
| Y. | Fadeyev | Astron. Council of USSR Acad. Sci. | USSR |
| C. | Fransson | Stockholm Observatory | Sweden |
| M. | Fujimoto | Niigata University | Japan |
| Y. | Fujita | University of Tokyo | Japan |
| I. | Fukuda | Kanazawa Institute of Technology | Japan |
| M. | Fukugita | University of Kyoto | Japan |
| H. | Hanami | Hokkaido University | Japan |
| R. | Harkness | University of Texas | USA |
| S. | Hayakawa | University of Nagoya | Japan |
| H. | Hill | University of Arizona | USA |
| R. | Hirata | University of Kyoto | Japan |
| P. | Hoeflich | MPI fur Astrophysik | FRG |
| D. | Hollowell | University of Illinois | USA |
| D. | Husfeld | Universitatssternwarte Munchen | FRG |
| J. | Isern | Inst. de Astrofisica de Andalucia | Spain |
| K. | Ishida | University of Tokyo | Japan |
| H. | Itoh | University of Kyoto | Japan |
| M. | Itoh | Inst. Space and Astronaut. Science | Japan |
| N. | Itoh | Sophia University | Japan |
| J. | Jugaku | University of Tokyo | Japan |
| S.H. | Kahana | Brookhaven National Laboratory | USA |
| E. | Kanbe | University of Tokyo | Japan |
| K. | Kasahara | University of Tokyo | Japan |

| K. | Kawabata | Science University of Tokyo | Japan |
| M. | Kato | Keio University | Japan |
| L. | Khyanni | Tartu Astrophysical Observatory | USSR |
| T. | Kifune | University of Tokyo | Japan |
| M. | Kiguchi | Kinki University | Japan |
| R. | Kirshner | Center for Astrophysics | USA |
| M. | Kitamura | University of Tokyo | Japan |
| R.H. | Koch | University of Pennsylvania | USA |
| K. | Kodaira | University of Tokyo | Japan |
| T. | Kogure | University of Kyoto | Japan |
| M. | Kondo | Senshu University | Japan |
| Y. | Kondo | NASA/Goddard space Flight Center | USA |
| K. | Koyama | Kinki University | Japan |
| R.P. | Kudritzki | Universitatssternwarte Munchen | FRG |
| S. | Kumagai | University of Tokyo | Japan |
| G.S. | Kutter | Goddard Space Flight Center/NASA | USA |
| J.D. | Landstreet | University of Western Ontario | Canada |
| N. | Langer | Universitatssternwarte Goettingen | FRG |
| U. | Lee | University of Tokyo | Japan |
| Z. | Li | Beijing Normal University | China |
| I. | Lichtenstadt | Hebrew University | Israel |
| A.P. | Linnell | Michigan State University | USA |
| M. | Livio | University of Illinois | USA |
| R. | Lopez | Universidad de Barcelona | Spain |
| A. | Maeder | Geneva Observatory | Switzerland |
| H. | Maehara | University of Tokyo | Japan |
| K. | Makishima | University of Tokyo | Japan |
| M. | Makita | University of Tokyo | Japan |
| K. | Masai | Nagoya University | Japan |
| T. | Matsuda | University of Kyoto | Japan |
| G. | Michaud | Universite de Montreal | Canada |
| M. | Minakata | Tokyo Metropolitan University | Japan |
| R. | Mitalas | University of Western Ontario | Canada |
| S. | Miyaji | Chiba University | Japan |
| R. | Mochkovitch | Inst. Astrophys. Paris | France |
| T. | Nakamura | University of Kyoto | Japan |
| Y. | Nakamura | Komaba Senior High School | Japan |
| N. | Nakata | Chiba-Keizai Junior College | Japan |
| K. | Nariai | University of Tokyo | Japan |
| K. | Nomoto | University of Tokyo | Japan |
| A. | Okazaki | Tsuda College | Japan |
| T. | Onaka | University of Tokyo | Japan |
| Y. | Osaki | University of Tokyo | Japan |
| C. | Pilachowski | National Optical Astron. Obs. | USA |
| A.V. | Raveendran | Indian Institute of Astrophysics | India |
| J. | Sahade | FCAG UNLP and IAR | Argentina |
| H. | Saio | University of Tokyo | Japan |
| W. | Schmutz | Universitat Kiel | FRG |
| M. | Scholz | Institute f. Theor. Astrophys. | FRG |
| E. | Sedlmayr | Technische Universitat Berlin | FRG |
| T. | Sekii | University of Tokyo | Japan |
| H. | Shibahashi | University of Tokyo | Japan |
| T. | Shigeyama | University of Tokyo | Japan |
| B. | Shustov | Astr. Council of USSR Acad. Sci. | USSR |
| M.G. | Signore | Radioastr. Ecole Norm. Superieure | France |

W.     Sparks              Los Alamos National Laboratory        USA
K.     Suda                Tohoku University                     Japan
D.     Sugimoto            University of Tokyo                   Japan
P.G.   Sutherland          McMaster University                   USA
H.     Suzuki              University of Tokyo                   Japan
M.     Takahara            University of Tokyo                   Japan
Y.     Takahashi           Univ. of Alabama in Huntsville        USA
M.     Tamura              University of Kyoto                   Japan
S.     Tamura              Tohoku University                     Japan
Y.     Tanaka              Ibaraki University                    Japan
Y.     Tanaka              Inst. Space and Astronaut. Science    Japan
Y.     Tohdo               University of Tokyo                   Japan
T.     Tsuji               University of Tokyo                   Japan
H.     Tsunemi             Osaka University                      Japan
N.     Ukita               University of Tokyo                   Japan
W.     Unno                Kinki University                      Japan
K.     Utsumi              Hiroshima University                  Japan
K.A.   Van der Hucht       SRON Space Research Laboratory        Netherlands
S.     Vauclair            Observatoire de Toulouse              France
R.     Wagoner             Stanford University                   USA
N.R.   Walborn             Space Telescope Science Institute     USA
Z.     Wang                Nanjing University                    China
A.     Wehlau              University of Western Ontario         Canada
W.     Wehlau              University of Western Ontario         Canada
R.     Wehrse              University of Heidelberg              FRG
J.C.   Wheeler             University of Texas                   USA
R.     Williams            Cerro Tololo Inter-American Obs.      Chile
J.R.   Wilson              Lawrence Livermore National Lab.      USA
B.     Wolf                Landessternwarte Koenigstuhl          FRG
P.R.   Wood                Mt. Stromlo and Siding Spring Obs.    Australia
S.E.   Woosley             University of California              USA
Y.     Yamada              University of Kyoto                   Japan
Y.     Yamashita           University of Tokyo                   Japan
H.     Yokoo               Kitasato University                   Japan
M.     Yuasa               Kinki University                      Japan

# Astronomy and Astrophysics Library

**K. Rohlfs**

## Tools of Radio Astronomy

1986. 127 figures. XII, 319 pages.
ISBN 3-540-16188-0

**H. Scheffler, H. Elsässer**

## Physics of the Galaxy and Interstellar Matter

Translated from the German by A. H. Armstrong

1987. 207 figures. XI, 492 pages.
ISBN 3-540-17314-5

**P. Lena**

## Observational Astrophysics

Translated from the French by A. R. King

1988. 200 figures. Approx. 350 pages.
ISBN 3-540-18433-3

**M. Harwit**

## Astrophysical Concepts

2nd edition. 1988. 175 figures. XV, 625 pages.
ISBN 3-540-96683-8

**G. L. Verschuur, K. I. Kellermann** (Eds.)

## Galactic and Extragalactic Radio Astronomy

2nd edition. 1988. 207 figures. XXII, 698 pages.
ISBN 3-540-96575-0

**S. Laustsen, C. Madsen, R. M. West**

## Exploring the Southern Sky

**A Pictorial Atlas from the European Southern Observatory (ESO)**

1987. 240 photographs, partly in colour, 31 diagrams and a Fould-out-Plate. VI, 274 pages. ISBN 3-540-17735-5

**Contents:** Foreword. Introduction. Acknowledgements. A Guide to This Book.
**Part 1: The Universe and Its Galaxies**
A Look into Fornax. Galaxy Types in an Ordered Sequence. The Local Group. The Sculptor Group. Multiple Galaxies. Clusters of Galaxies. Peculiar Galaxies.
**Part 2: The Milky Way Galaxy**
Panorama of the Milky Way. The Milky Way from Orion to Puppis. The Milky Way from Vela to Carina. The Milky Way from Crux to Norma. The Milky Way from Scorpius to Scutum. Milky Way Objects at High Galactic Latitude.
**Part 3: Minor Bodies in the Solar System**
Meteorites and Minor Planets. Comets.
**Part 4: The Southern Sky and ESO**
A European Organization for Astronomy. The La Silla Observatory. The Headquarters in Garching. The Next Generation Telescopes. Glossary. Plate Data. Index of Objects.

**Exploring the Southern Sky** is not only a superb pictorial atlas and guide to the astronomical opportunities of the Southern Hemisphere but also a treasured addition to every stargazer's library. It is the timely new way to expand your astronomical horizons and add to your picture of the Universe.

German edition available by Birkhäuser-Verlag, Basle.

Springer-Verlag
Berlin Heidelberg New York
London Paris Tokyo